# ELECTRONICS
# TECHNOLOGY
# HANDBOOK

# ELECTRONICS TECHNOLOGY HANDBOOK

**NEIL SCLATER**

**McGraw-Hill**

New York   San Francisco   Washington, D.C.   Auckland   Bogotá
Caracas   Lisbon   London   Madrid   Mexico City   Milan
Montreal   New Delhi   San Juan   Singapore
Sydney   Tokyo   Toronto

**Library of Congress Cataloging-in-Publication Data**

Sclater, Neil.
   Electronics technology handbook / Neil Sclater.
      p.     cm.
   Includes index.
   ISBN 0-07-058048-0
   1. Electronics—Handbooks, manuals, etc.   I. Title.
  TK7825.S33   1999
  621.381—dc21                99-13541
                                CIP

## McGraw-Hill

*A Division of The* **McGraw·Hill** *Companies*

  2 3 4 5 6 7 8 9 0  DOC/DOC    0 4 3 2 1 0

ISBN 0-07-058048-0

*The sponsoring editor for this book was Steve Chapman, the editing supervisor was Frank Kotowski, Jr., and the production supervisor was Tina Cameron.*

*It was set in Times New Roman at North Market Street Graphics.*

*Printed and bound by R. R. Donnelley & Sons.*

McGraw-Hill books are available at special quantity discounts to use as premiums and sales promotions, or for use in corporate training programs. For more information, please write to the Director of Special Sales, McGraw-Hill, Professional Publishing, Two Penn Plaza, New York, NY 10121-2298. Or contact your local bookstore.

 This book is printed on recycled, acid-free paper containing a minimum of 50% recycled de-inked fiber.

# CONTENTS

# PREFACE

Electronics technology has become an important part of the fabric of our daily lives. Newspapers and general-interest magazines that never previously reported on electronics, computers, and communications now regularly run columns and sections on them, and TV channels and radio stations include these subjects in their science programming. The mainstream media has covered such issues as the future of the Internet; the possible merger of computers, TV receivers, and telephones; and the problems posed by the introduction of digital TV.

In considering the flood of announcements and advertising for electronic products from computers to cellular telephones, it is not surprising that there is talk of "technological overload." Some people seem overwhelmed by these rapid advancements in technology, and are dismayed by the forced obsolescence of so much electronic hardware. However, many others really want to learn more about these leading-edge developments, how they work, and how they will affect their lives, but they are finding that it can be difficult.

Unfortunately, the mass media is not the best source of educational information on high technology. Some publications include tutorial paragraphs in their articles, but all too often technology reporting is oversimplified and presented as futuristic science-fiction hype. Thus, persons seeking a better understanding of these subjects must find sources that they can understand if they want to be brought up to speed. They will need to become familiar with certain key concepts and be able to interpret the language or jargon used in describing them. This can be a challenge for those who have not had recent formal instruction or training in electronics and computers. Some who have attempted self-education by consulting engineering textbooks and other professional references have been put off by the unfamiliar technical theory, language, and math that they found there.

This handbook is a user-friendly reference book for persons seeking a better understanding of modern electronics technology without requiring that they wade through a jungle of incomprehensible text. Its 30 sections reflect the existing organization of the electronics industry, and each contains an overview of that sector and many short articles describing major topics within that sector. The book contains more than 400 illustrations and tables based on actual devices, circuits, and systems. Each section has been written so that it can be read or browsed in any order. The key technical terms presented in each section are highlighted in italics and defined in that section. Cross-references make it easy for the reader to find additional information on topics covered in more than one section.

While no formal educational prerequisites are needed to learn from this handbook, readers possessing at least a high-school-level knowledge of the physical sciences and some understanding of the basics of electronics will be more comfortable with its contents. It has been written at a reading comprehension level comparable to that found in most popular magazines on science, electronics, and computing, and the math content is limited to a few simple formulas that are helpful in explaining important concepts.

The articles cover a wide range of subjects from electronic components and circuits to advanced electronic products and systems and communication techniques. It also includes the manufacture of semiconductor devices and circuits. There are, for example, articles on computers, microprocessors, GPS receivers, lasers, night-vision scopes, ATM machines, bar-code readers, radar, cellular telephones, and communication satellites.

This handbook will be valuable for anyone who wants to learn more about electronics technology for personal reasons or occupational advancement. It will appeal to students, hobbyists, teachers, technicians, reference librarians, and persons working in such fields as advertising, marketing, sales, training, and technical publications, in or out of the electronics industry. However, the handbook's practical up-to-date information makes it a valuable desktop reference for engineers and managers in all disciplines. For some the handbook will be an introduction to the broad spectrum of electronics technology, and for others it will be a handy source of detailed information in areas of special interest.

Tutorial articles on elementary physics, electricity, and electronics were not included because they are beyond the scope of this reference book. Moreover, obsolete topics such as receiving tubes and circuits were omitted because they are now primarily of historical interest. However, articles on subjects such as computer programming, operating systems, and applications software were restricted to those necessary for understanding how they interact with computer circuitry.

The first two sections discuss basic components including resistors, capacitors, diodes, and transistors. A separate section describes electromechanical components such as switches, relays, and motors that are either part of electronic circuits or are dependent upon them for their operation. A section is devoted to amplifiers and oscillators, and another covers important "building-block" circuits. Antennas and radio-frequency propagation are the subjects of separate sections, as are components and circuits for applications in the UHF and microwave bands.

Linear and analog ICs and logic and digital ICs, including computer memory chips, are covered in separate sections, as are the energy sources of batteries and power supplies, electronic test instruments, optoelectronics, and electronic displays. Computer technology is covered in three sections: microprocessors and microcontrollers, computer systems, and computer peripherals, including disk drives, monitors, modems, printers, and scanners.

Sensors and transducers, radio and television broadcasting and receiving, and telecommunications are the subjects of individual sections. The next five sections discuss consumer electronics products and services, industrial electronics, military and aerospace systems, and important electronic instrumentation in medicine and science. The manufacture of semiconductor devices and their packaging are the subjects of individual sections, as are the manufacture and population of circuit boards. A section covers wire, cable, and connectors, and another discusses the products and techniques for protecting sensitive semiconductor devices and circuits from damage or destruction by excessive voltages and currents.

A bibliography identifies the principal reference sources used in the preparation of this handbook. This listing includes books on elementary electronics as well as those devoted to the more specialized fields. It will be a valuable resource for those who wish to delve deeper into the subjects selected for inclusion in this reference book.

Neil Sclater

# PASSIVE ELECTRONIC
# COMPONENTS

# Overview

A *passive electronic component* is a circuit part that functions without an external power requirement. The most common passive components are resistors, capacitors, and inductors. Most of them have two leads. An axial-leaded component, as shown in Fig. 1-1, has leads projecting from each end of the component body aligned with the long axis of the part, while a radial-leaded component, as shown in Fig. 1-2, has parallel leads projecting at right angles from its body. Axial leads must first be bent 90° to insert them into the holes of circuit boards, while radial leads can typically be inserted directly into those holes without bending. However, both axial- and radial-leaded parts can be inserted by automatic machines.

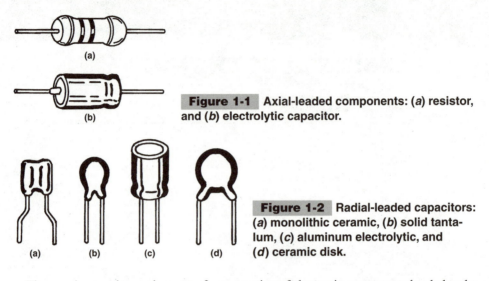

**Figure 1-1** Axial-leaded components: (*a*) resistor, and (*b*) electrolytic capacitor.

**Figure 1-2** Radial-leaded capacitors: (*a*) monolithic ceramic, (*b*) solid tantalum, (*c*) aluminum electrolytic, and (*d*) ceramic disk.

The ongoing trend toward more surface mounting of electronic components has led to the introduction of more active and passive "leadless" components that can be soldered directly to tinned or plated pads on "hole-less" or surface-mount technology (SMT) circuit boards. Passive SMT components such as capacitors and resistors are leadless rectangular chips or cylinders with metallized end surfaces that are reflow soldered to the circuit boards, but many active components, such as transistors and integrated circuits, are in cases with bent stub or "gull wing" leads that can also be reflow soldered directly to circuit board pads.

# Fixed Resistors

A *resistor* is a circuit component that provides a fixed value of resistance in ohms to oppose the flow of electrical current. Resistors can limit the amount of current flowing in a circuit, provide a voltage drop in accordance with Ohm's laws, or dissipate energy as heat.

Fixed resistors are discrete units typically made in cylindrical or planar form. The most common cylindrical style is the axial-leaded resistor, as shown in Fig. 1-1. The resistive element is wound or deposited on a cylindrical core, and a cap with a lead wire is positioned on each end. The resistive elements include *resistive wire* (*wirewound*), *metal film, carbon film, cermet,* and *metal oxide.* Resistor networks and chip resistors are examples of planar resistors. All fixed resistors are rated for a nominal resistance value in ohms over the range of fractions of an ohm to thousands of ohms (kilohms), or millions of ohms (megohms). Other electrical ratings include:

- *Resistive tolerance* as a percentage of nominal value in ohms
- *Power dissipation* in watts (W)
- *Temperature coefficient* (*tempco*) in parts per million per degree Celsius of temperature change (ppm/°C)
- *Maximum working voltage* in volts (V)

Some resistors also have additional ratings for electrical noise, parasitic inductance, and parasitic capacitance. Resistors exhibit unwanted parasitics of inductance and capacitance because of their construction. These effects must be considered by the designer when selecting resistors for unusual or specialized applications such as their use in instrumentation.

A resistor's ability to dissipate power is directly related to its size. With the exception of those specified for power supplies, most resistors for electronic circuits are rated under 5 W, usually less than 1 W. A 5-W cylindrical resistor is about 1 in (25.4 mm) long with a diameter of ¼ in (6.4 mm). The ½-, ¼-, and ⅛-W resistors are correspondingly smaller.

## CARBON-COMPOSITION RESISTORS

A *carbon-composition resistor,* as shown in Fig. 1-3, is made by mixing powdered carbon with a phenolic binder to form a viscous bulk resistive material, which is placed in a mold with embedded lead ends and fired in a furnace. Because their resistive elements are a bulk material, they can both withstand wider temperature excursions and absorb higher electrical transients than either carbon- or metal-film resistors. These qualities are offset by their typically wider resistive tolerances of ±10 to 20 percent and tendency to absorb moisture in humid environments, causing their values to change. However, the benefits of carbon-composition resistors are less important in low-voltage transistorized circuits, so demand for them has declined. These resistors have ratings of 1 ohm to 100 megohms, but values in the 10- to 100-ohm range were most popular. Power ratings are ⅛ to 2 W.

## CARBON-FILM RESISTORS

A *carbon-film resistor,* as shown in Fig. 1-4, is made by screening carbon-based resistive ink on long ceramic rods or mandrels and then firing them in a furnace. The rod is then sliced to form individual resistors. After leaded end caps are attached, the resistance values are set precisely in a laser trimming machine that trims away excess resistive film under closed-loop control. The trimmed resistors are then coated with an insulating plastic jacket. Resistive tolerances of carbon-film resistors are typically ±10 percent. Standard resistors have power ratings of ½, ¼, and ⅛ W.

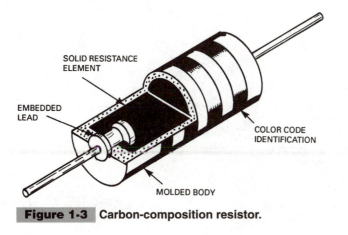

**Figure 1-3**  Carbon-composition resistor.

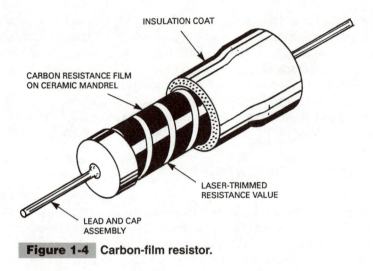

**Figure 1-4**  **Carbon-film resistor.**

## WIREWOUND RESISTORS

A *wirewound resistor,* as shown in Fig. 1-5, is made by winding fine resistive wire on a plastic or ceramic mandrel. The most commonly used resistance wire is nickel-chromium (nichrome). The axial leads and end caps are attached to the ends of the wire winding and welded to complete the electrical circuit. There are both general-purpose and power wirewound resistors. General-purpose units have resistive values of 10 ohms to 1 megohm, resistance tolerances of $\pm 2$ percent, and temperature coefficients of $\pm 100$ ppm/°C. Power units rated for more than 5 W have tolerances that can exceed $\pm 10$ percent.

Wirewound resistors are generally limited to low-frequency applications because each is a solenoid that exhibits inductive reactance in an AC circuit, which adds to its DC resistive value. The inductive reactance can be reduced or eliminated at low or medium frequencies by bifilar winding. This is done by folding the entire length of resistive wire back on itself, hairpin fashion, before winding it on the mandrel. As a result, opposing inductive fields cancel each other, lowering or eliminating inductive reactance.

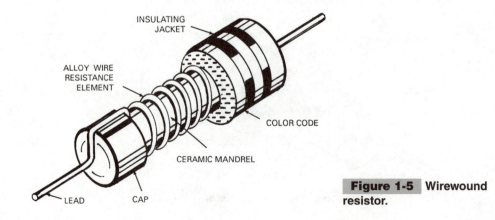

**Figure 1-5**  **Wirewound resistor.**

Wirewound resistors are made with both axial and radial leads. Epoxy or silicone insulation is applied to some low-power wirewound resistors, but high-power units are encased in ceramic or placed in heat-dissipating aluminum cases. This reduces the danger of the hot resistor igniting nearby flammable materials or burning fingertips if accidentally touched.

## METAL-FILM RESISTORS

A *metal-film resistor,* as shown in Fig. 1-6, is made by the same general method as a carbon-film resistor. A thin metal film is sputtered or vacuum deposited on an alumina (aluminum-oxide) mandrel in a vacuum chamber, or a thick metal film is applied in air. Tin oxide or nickel-chromium are widely used thin films, and a thick film made from powdered precious metal and glass (frit) in a volatile binder is a common cermet resistive ink. These resistors are laser trimmed to precise values under closed-loop control after firing. Metal-film resistors are offered in two grades: (1) those with resistive tolerances of $\pm 1$ percent and temperature coefficients of 25 to 100 ppm/°C, and (2) those with resistive tolerances of $\pm 5$ percent and temperature coefficients of 200 ppm/°C. Demand is highest for ¼- and ⅛-W units, but ¹⁄₂₀-W units are available. Resistive values up to 100 megohms are available as catalog items, but they are generally rated for less than 10 kilohms.

## RESISTOR NETWORKS

A *resistor network,* as shown in Fig. 1-7, consists of two or more resistive elements on the same insulating substrate. These networks are specified where 6 to 15 low-value resistors are required in a restricted space. Most commercial networks contain thick-film resistors, and they are packaged in dual-in-line packages (DIPs) or single-in-line packages (SIPs). Standard DIPs have 14 or 16 pins, and standard SIPs have 6, 8, or 10 pins. Resistor networks are used for "pull-up" and "pull-down" transitions between logic circuits operating at different voltage values, for sense amplifier termination, and for light-emitting diode (LED) display current limiting.

Alumina ceramic is the most widely used network substrate. Conductive traces are formed by screening an ink made from a powdered silver-palladium mix in a volatile binder

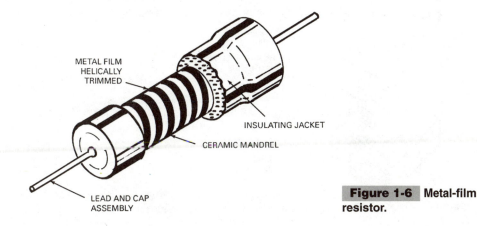

METAL FILM
HELICALLY
TRIMMED

INSULATING JACKET

CERAMIC MANDREL

LEAD AND CAP
ASSEMBLY

**Figure 1-6** Metal-film resistor.

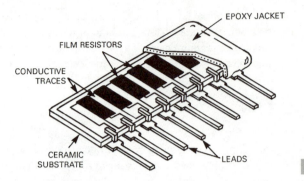

**Figure 1-7**  **Resistor network.**

on the bare ceramic substrate. After firing, the ink bonds with the ceramic to form hard, low-resistance paths. Resistive inks made from a powdered ruthenium-cermet mix with a powdered glass frit and a volatile binder are then screened over the ends of the conductors to form the resistive elements. This ink is also fired, and when it bonds with the ceramic it forms a hard, resistive element. Network resistors are laser trimmed under closed-loop control to precise resistance values. Standard network resistance values are from 10 ohms to 10 megohms with tolerances of $\pm 2$ percent. Most networks can safely dissipate less than ½ W.

Where more precise resistance values are required, thin-film networks are specified. They are made formed from compositions that include nickel-chromium, chrome-cobalt, and tantalum nitride, deposited or sputtered on alumina ceramic substrates. Unpackaged thin-film resistor networks are also sold as hybrid-circuit substrates. Thin-film resistive-capacitive (*RC*) networks are also packaged in metal and ceramic flatpacks.

## CERAMIC-CHIP RESISTORS

A *ceramic-chip resistor,* as shown in Fig. 1-8, is made by screening and firing cermet resistive inks or sputtering tantalum nitride or nickel-chromium on an alumina substrate. The deposited resistive surface is then coated with glass for protection. The substrate is then diced into individual chips, and a silver-based ink is applied to the end surfaces and fired as the first step in forming leadless terminals. A barrier layer of nickel plating is then applied to prevent the migration of silver from the inner electrode. Finally, the terminations are coated with lead-tin solder for improved adhesion during reflow soldering.

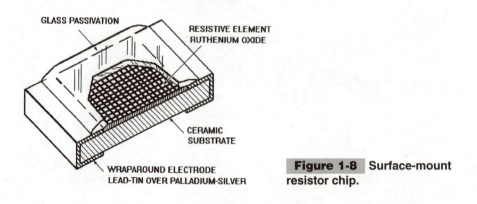

**Figure 1-8**  **Surface-mount resistor chip.**

Chip resistors were originally made for hybrid circuits, but surface-mount technology (SMT) has increased demand for them. Surface-mount chip resistor dimensions have been standardized to $1.6 \times 3.2$ mm for handling by automatic pick-and-place machines. (This is the same size as the 1206 chip capacitor that measures $0.063 \times 0.125$ in.) Chip resistors are typically rated for $\frac{1}{8}$ W or less. An alternative form of SMT resistor is the leadless cylinder with solder-coated bands around each end for reflow solder bonding.

# Variable Resistors

## POTENTIOMETERS

A *potentiometer* is a variable resistor whose resistance value can be changed by moving a sliding contact or wiper along its resistive element to pick off the desired value. A potentiometer has terminals at each end of its fixed resistive element, and the third terminal is connected to a moveable wiper. If the wiper is moved back to the beginning of the resistive element, the potentiometer's resistance value is minimal, but if it is moved across the full length of the element, the value reaches its maximum. There are three different mechanisms for moving the wiper along the resistance element:

**1.** Sliding the wiper by finger pressure
**2.** Turning a leadscrew on the case to drive the wiper back and forth
**3.** Rotating a screw or knob attached to the wiper to sweep it around a curved element

Potentiometers for electronic circuits are classified as follows:

- Precision
- Panel or volume-control
- Trimmer

The common abbreviation for potentiometer is *pot,* so there is a *precision pot* and a *panel* or *volume-control pot.* However, a trimmer potentiometer is usually called a *trimmer* (to be distinguished from a trimmer capacitor). These variable resistors share the same schematic symbol and are made from many of the same kinds of materials.

## PANEL OR VOLUME-CONTROL POTENTIOMETERS

A *panel* or *volume-control potentiometer,* as shown in Fig. 1-9*a,* is made to have a long rotational life performing such functions as tuning radio frequencies, controlling audio volume, and adjusting brightness, intensity, or contrast in video circuits. Panel pots are used in many different electronics products, including radios, stereos, TV receivers, tape recorders, computer monitors, oscilloscopes, and other electronic test equipment.

Panel pots permit the user to make personal adjustments of a physical variable, so no attempt is made to relate shaft position and output. These pots are in cylindrical cases with axial shafts and are similar in size and appearance to precision potentiometers. Panel pots

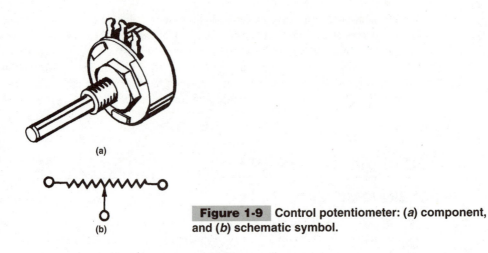

**Figure 1-9**  Control potentiometer: (*a*) component, and (*b*) schematic symbol.

are typically mounted behind the front panel of a case or enclosure with a threaded bushing projecting through a cutout in the panel, and they are fastened with a ring nut and lockwasher. But some control pots have threadless bushings for mounting on a circuit board behind the front panel. The pot is mounted on a circuit board that is fastened behind the panel so that the bushing and control shaft can project through a hole in the panel. Some panel pots include on-off switches to reduce part counts, as on small portable radios.

The resistive elements for panel pots can be hot-molded carbon, cermet, or conductive plastic. Each has a different resistive range, tolerance, and power rating. Tolerances are typically ±10 to 20 percent, and both carbon and conductive-plastic elements can have resistive tapers. Cermet elements permit the highest power dissipation. Panel pots are made both as standard and custom components, and they are made to conform to either commercial or military standards. Some are assembled from modular, interchangeable parts, permitting a wide selection of resistive elements. Modular assemblies can be ganged with two or more resistive module elements controlled by the same coaxial shaft to save front-panel space. The schematic symbol for all potentiometers is shown in Fig. 1-9*b*.

## TRIMMER POTENTIOMETERS

A *trimmer pot* is a small "set-and-forget" variable resistor for making infrequent, post-manufacture adjustments, usually in linear circuits. Adjustments are normally done during the final testing of entertainment products and instruments. But they might be reset during calibration procedures to compensate for changes in resistive and capacitive values that occur as circuitry ages. Trimmers are used in radios, TV sets, audio equipment, computer monitors, and many different kinds of test and communications equipment. There would be little need for them if all components were precisely made and not subject to value changes due to exposure to elevated temperatures, high humidity, or degradation with age. Trimmers are usually mounted inside a product's case where they are inaccessible to users. There are many variations in trimmer designs, styles, sizes and resistive elements, and they are made to conform to either military or commercial standards. Two common types are *rotary* and *linear* or *rectangular*.

## ROTARY TRIMMER POTENTIOMETERS

A *single-turn rotary trimmer,* as shown in Fig. 1-10, includes a semicircular resistive element and a wiper that can be swept over its length with a single turn of a shaft or knob. The styles suitable for circuit-board mounting are in round open cases with typical diameters of ¼ and ⅜ in (6 and 10 mm), and the resistive elements are exposed. Larger ½-in (13-mm) diameter units are available. A *multiturn rotary trimmer* also has a semicircular resistive element, and its resistive value is set by turning a slotted leadscrew mounted either on the top, side, or end of the case for accessibility in restricted spaces. Rotating mechanisms permit the wiper to be swept around the element to cover the complete resistive range in up to 20 turns. The popular sizes are the square ¼- and ⅜-in cases with pins spaced for PC board mounting. Surface-mount versions of both of these trimmer styles are available.

## RECTANGULAR TRIMMER POTENTIOMETERS

A *rectangular* or *linear trimmer* has a linear resistive element whose resistive value is set by turning an internal leadscrew. The wiper can traverse the entire element in up to 20 turns. Popular units are in rectangular packages ¾ in (19.1 mm) long. PC-board mounting pins project from the case. Other versions have wipers that can be pushed back and forth along the resistive element by finger pressure. The resistive elements of these trimmers can be carbon film, bulk carbon, resistive wire, cermet, conductive plastic, or bulk metal. Most

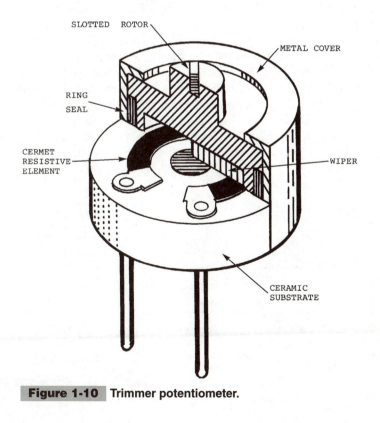

**Figure 1-10** Trimmer potentiometer.

rectangular trimmers can dissipate ½ W, but some large 1¼-in (32-mm) multiturn units can dissipate 1 W. Power rating is determined by trimmer size and the choice of resistive element. Both leaded and surface-mount versions are made to conform to military and commercial standards.

## PRECISION POTENTIOMETERS

A *precision pot,* as shown in Fig. 1-11, is an instrument-grade variable resistor. It can provide repeatable resistive accuracy of at least 1 percent. These pots were widely used in analog computers, instruments, and military and aerospace systems, but they now function primarily as sensors. They can provide precise and resettable voltages corresponding to each setting of the control shaft. Vernier dials make it possible to return its shaft to a specific position to obtain a repeatable output voltage within close tolerances.

Most precision pots have cylindrical cases and an axial rotating shaft. The resistive material in a single-turn precision pot is cut in a C shape and fastened inside the case. However, the resistive element of a multiturn precision pot is formed as a helix or spiral which is also attached to the inside of the case, as shown in Fig. 1-11. A sliding leadscrew assembly mounted on the control shaft advances and retracts the wiper assembly with shaft motion. This causes the wiper to track around the inside of the helix.

Precision pots are identified by their resistive elements. Most are wound resistive wire (wirewound) or resistive plastic. The wirewound element is formed by winding fine resistance wire on a heavier wire form or mandrel. These elements have low temperature coefficients, but they exhibit finite resolution. As the wiper slides along the resistive element, it spans resistance increments equal to the resistive value of an individual turn of fine wire wound around the mandrel. While accuracy improves with helix length, the element always has a tolerance of ±1 wire turn. But infinite resolution can be obtained with a hybrid helix, a wirewound element coated with resistive plastic. The coating compensates for the resistive increments.

Because bulk resistive plastic resistors made from sheets can have infinite resolution, elements can easily be cut from it to form nonlinear elements. They can be contoured or tapered to produce an output voltage that varies with respect to shaft setting. For example, tapers can be designed to produce output voltages that express sine, cosine, square law, or logarithmic functions.

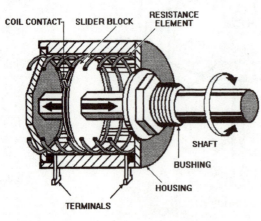

**Figure 1-11**  Precision potentiometer.

Ceramic-metal (cermet) elements, also capable of infinite resolution, are specified when the precision pot will be operated in a high-temperature environment. Unfortunately, these elements are abrasive and can wear down the wiper, thus limiting the pot's useful life.

Precision pots are also classified as single-turn or multiturn. Because of the diversity in resistive materials and the conventions accepted for their manufacture, wirewound and hybrid pots can either be single-turn or multiturn, but all precision pots with conductive plastic or cermet resistive elements are single-turn.

The principal specifications for precision potentiometers are:

- Starting or running torque
- Resistance range
- Power rating
- Ambient temperature range
- Rotational life

These factors determine the choice of number of turns and resistive element. If a single-turn pot has a resistive element that is too short to give the desired accuracy, a multiturn element is selected. The effective rotation of a single-turn pot is about 320°. The most common multiturn potentiometers are the 3-turn (1080°) and 10-turn (3600°), but 5-, 15-, 25-, and 40-turn units are available.

Both single- and multiturn pots with linearities of 0.025 percent or better are standard items. The low-resistance range for single-turn precision pots is about 10 to 150 ohms, and their high-resistance range is about 200 kilohms to 1 megohm. Similarly, the low-resistance range of multiturn precision pots is about 3 ohms to 1 kilohm, and their high-resistance range is about 200 kilohms to more than 5 megohms.

Precision pots are made as panel- or servo-mounted units. Panel-mounted units, like control pots, are positioned behind the panel with their shafts and threaded bushing projecting through a formed hole, and they are fastened with ring nuts and lockwashers. Servo-mounted units are positioned facedown on metal baseplates and clamped with screw-type lugs secured in the clamping groove that runs around the circumference of the precision pot's case. Precision pots are made as either standard or custom products.

# Capacitors

A *capacitor,* as shown in Fig. 1-12, is an electronic component capable of storing electrical energy. The simplest form of capacitor is two metal plates insulated from each other by some dielectric. Capacitors are the second most widely purchased passive components next to resistors. There are both fixed and variable capacitors for electronics, and their capacitance values vary from a few picofarads (pF) to thousands of microfarads (μF). The schematic symbol for a fixed capacitor is shown in Fig. 1-12b and that for a variable capacitor is shown in Fig. 1-12c.

Capacitors are classified as either *electrostatic* or *electrolytic.* Electrostatic capacitors have dielectrics that are either air or some solid insulating material such as plastic film, ceramic, glass, or mica. (Paper dielectric capacitors are no longer specified in electronics.)

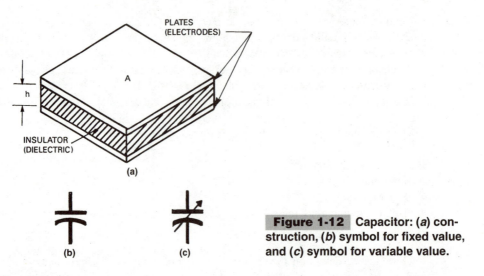

**Figure 1-12** Capacitor: (*a*) construction, (*b*) symbol for fixed value, and (*c*) symbol for variable value.

Electrolytic capacitors are further classified as aluminum or tantalum because those metals form thin oxide film dielectrics by electrochemical processing. They can have wet-foil, wet-slug, or dry-slug anodes.

The capacitance value of fixed capacitors remains essentially unchanged except for small variations caused by temperature changes. By contrast, the capacitance value of variable capacitors can be set to any value within a preset range of values. Variable capacitors are usually used in RF circuits.

## ELECTROSTATIC CAPACITORS

An *electrostatic capacitor* has a dielectric made from plastic film, mica, or glass, and its plates or electrodes are made from metal foil or metal deposited on the dielectric. Ceramic capacitors have plates formed from precious-metal inks that have been screened on the raw ceramic prior to furnace firing.

## PLASTIC-FILM CAPACITORS

A *plastic-film capacitor,* as shown in Fig. 1-13, is typically made by rolling a thin film of plastic dielectric with metal foil or a metallized dielectric film into a cylindrical form and attaching leads. The dielectrics include polyester, polypropylene, polystyrene, and polycarbonate. Film thickness can range from 0.06 mil (1.5 μm) to over 0.8 mil (20 μm). The most popular film capacitors have capacitance values of 0.001 to 10 μF, although values from 50 pF to 500 μF are available as standard products. Working voltages range from 50 to 1600 VDC, and capacitance tolerance is from ±1 to ±20 percent.

In film-and-foil construction, tin or aluminum foil about 0.00025 in (0.00635 mm) thick is wound with the dielectric film, but in metallized-film construction, aluminum or zinc is vacuum deposited to thicknesses of 200 to 500 Å (20 to 50 nm) on the film. Film capacitors can also be made by cutting and stacking metallized foil with attached leads. A capacitor with metallized film is smaller and weighs less than a comparably rated film-and-foil unit. Moreover, metallized-film capacitors are *self-healing;* that is, if the capacitor dielec-

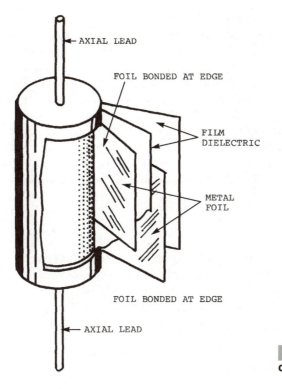

AXIAL LEAD

FOIL BONDED AT EDGE

FILM
DIELECTRIC

METAL
FOIL

FOIL BONDED AT EDGE

AXIAL LEAD

**Figure 1-13**  Plastic-film capacitor.

tric is pierced by a transient overvoltage, the metal film around the hole will evaporate, effectively lining the hole with molten plastic dielectric. This prevents short-circuits between adjacent metal layers and preserves the capacitor.

After rolling or stacking is complete, the capacitor is dipped in or conformally coated with an insulating plastic jacket. Some units are also hermetically sealed in tubular or rectangular metal cases for added environmental protection. Both film-and-foil and metallized-film capacitors are available with axial or radial leads in a wide variety of case styles.

## FILM DIELECTRICS

*Polyester film* (tradenamed Mylar) is the most popular general-purpose dielectric in film-type capacitors. It permits smaller capacitors than comparably rated units made from other films, and these capacitors exhibit low leakage, moderate temperature coefficients over the −55 to 85°C range, and moderate dissipation factors. Capacitance tolerance is typically ±10 percent. The film-and-foil versions are widely used in consumer electronics products while the metallized units perform general blocking, coupling, decoupling, bypass, and filtering functions.

*Polypropylene film* provides capacitor characteristics that are superior to those of polyester. Polypropylene capacitors have both high- and low-frequency applications. The plastic has properties that are similar to those of polystyrene, but capacitors made from it have higher AC current ratings. Polypropylene capacitors can operate at 105°C, and their volumetric efficiency is better than those made of polyester. Foil and polypropylene capacitors are used in CRT deflection, pulse-forming, and RF circuits. The capacitance tolerance for polypropylene capacitors is ±5 percent, and their temperature coefficients are linear.

*Polystyrene film* has characteristics that are similar to those of polypropylene. Capacitors made from the film exhibit a low dissipation factor, small capacitance change with temperature, and very good stability. But they are larger than comparably rated polypropylene units. Used in timing, integrating, and tuning circuits, their maximum operating temperature is 85°C.

*Polycarbonate film* capacitors offer dissipation factors and capacitance stability which approaches those of polystyrene capacitors. They also offer high insulation resistance stability. Operating temperatures are −55 to 125°C with capacitance tolerances of ±5 percent. These capacitors are widely used in military applications.

## MICA CAPACITORS

A *mica capacitor* has dielectrics of thin rectangular sheets of mica, a natural mineral. Mica has a dielectric constant from 6 to 8. The electrodes are either thin sheets of metal foil interleaved between mica sheets, or thin films of silver that have been screened and fired on the mica. Silvered mica capacitors have greater mechanical stability and offer more uniform properties than foil and mica capacitors. Both are used primarily in RF applications. Mica capacitors perform satisfactorily over temperature ranges as wide as −55 to 150°C, and they have high insulation resistance. Their capacitance values range from about 1 pF to 0.1 µF. However, they have a low ratio of capacitance to volume or mass.

## CERAMIC CAPACITORS

*Ceramic dielectric capacitors* are classified by dielectric constant $k$, as Classes I, II, and III. Class I dielectrics exhibit low $k$ values, but they have excellent temperature stability; Class II dielectrics have generally high $k$ values and volumetric efficiency but lower temperature stability; and Class III dielectrics are prepared for the lower-cost disk and tube capacitors.

Class I dielectrics include negative positive zero (NPO) ceramics, which are designated COG and BY. These ceramics are made by combining magnesium titanate (with a positive coefficient) and calcium titanate (with a negative coefficient) to form a dielectric with excellent temperature stability. Their properties are essentially independent of frequency, and they have ultrastable temperature coefficients of $0 ± 30$ ppm°C over the range of −55 to 125°C. These dielectrics show a flat response to both AC and DC voltage changes. Low-$k$ multilayer ceramic capacitors (MLCs) are used in resonant circuits and filters.

Class II dielectrics are high-$k$ ceramics called *ferroelectrics* made from barium titanate. The addition of barium stannate, barium zirconate, or magnesium titanate lowers the dielectric constant from values as high as 8000. These compounds stabilize the capacitor over a wider temperature range. Class II dielectrics include the general-purpose X7R (BX) and Z5U (BZ). X7R is stable but its capacitance can vary ±15 percent over the temperature range of −55 to 125°C. Its capacitance value decreases with DC voltage but increases with AC voltage. Z5U compositions exhibit maximum temperature-capacity changes of +22 and −56 percent over the range of 10 to 85°C.

Class III dielectrics, developed for ceramic-disk capacitors, give high volumetric efficiency but with the tradeoff of high leakage resistance and dissipation factor. Capacitors made with Class III dielectrics have low working voltages.

Ceramic dielectric capacitors are constructed in three styles: (1) *single-layer disk,* (2) *tubular,* and (3) *monolithic multilayer.*

## MONOLITHIC MULTILAYER CERAMIC (MLC) CAPACITORS

A *monolithic multilayer ceramic (MLC) capacitor,* as shown in cutaway view Fig. 1-14, is a multilayer ceramic chip capacitor that offers high volumetric efficiency because a large capacitor area is compressed into a small block. Preformed metallized layers are stacked and fired to form MLCs in a wide range of sizes and values with different properties. Originally developed for hybrid circuits, MLCs are widely used in surface mounting because they can substitute for larger capacitors with comparable capacitance values. They offer low residual inductance values and low resistance, a wide range of capacitance values in a given size, and a wide selection of temperature coefficients. They also exhibit lower inductance and resistance values than tantalum capacitors with comparable ratings. MLCs are used for timing and frequency selection.

MLCs are made as sandwiches of "green" (unfired) barium-titanate ceramic strips 0.8 mils (20 μm) thick that have been imprinted with silver-palladium ink to form plates. Up to 40 layers of the soft doughlike strips are stacked, compressed, diced, and furnace fired to form the monolithic chips.

End terminals for solder bonding MLCs to a circuit board or attaching leads are made by plating successive layers of silver-palladium, nickel, and tin or lead-tin on the ends of the chips. The process used depends on whether the chip is to be leaded and coated with insulation or is to remain bare for bonding directly to a circuit board.

Bare MLCs are used on hybrid microcircuits and in surface-mount assembly. They will withstand the 232°C reflow-soldering temperatures and the 282°C wave-soldering temperatures. Bare MLC chip sizes are standardized. Examples include 0.08 × 0.05 in (2.0 × 1.3

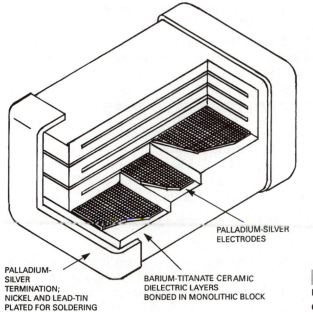

PALLADIUM-SILVER
ELECTRODES

PALLADIUM-
SILVER
TERMINATION;
NICKEL AND LEAD-TIN
PLATED FOR SOLDERING

BARIUM-TITANATE CERAMIC
DIELECTRIC LAYERS
BONDED IN MONOLITHIC BLOCK

**Figure 1-14** Monolithic multilayer ceramic (MLC) capacitor.

mm), designated 0805; 0.125 × 0.063 in (3.2 × 1.6 mm), designated 1206; and 0.225 × 0.05 in (5.7 × 1.3 mm), designated 2225. Standard MLCs have capacitance values of 10 pF to 3.5 μF, capacitance tolerances of ± 1 to 20 percent, and maximum voltages of 50 V.

## CERAMIC-DISK CAPACITORS

A *ceramic-disk capacitor* is a radial-leaded capacitor made as a metallized ceramic disk. Silver-based ink is screened on both sides of the ceramic disk to form the plates and sites for attaching the radial leads. After firing and lead bonding, the capacitors are dipped or conformally coated with a protective jacket of phenolic resin or epoxy. These capacitors are used in tuning circuits.

## CERAMIC TUBULAR CAPACITORS

A *ceramic tubular capacitor* is a length of ceramic tube whose inner and outer surfaces are painted with silver ink to form its plates. They have replaced ceramic disk capacitors in surface-mounted circuits to save board space and permit automatic placement. They are protected with a coat of protective resin.

## ELECTROLYTIC CAPACITORS

*Electrolytic capacitors* are specified where high values of capacitance are required in the least amount of space (high volumetric efficiency). This property is called *high CV ratio*. They are formed by electrochemical processes in which oxide dielectrics are grown in and on porous aluminum and tantalum foil and pellets. The metal foils are acid etched to make them porous, increasing their effective exposed areas from 6 to 20 times. High CV ratios are made possible by the thin oxide layers formed on the plates of the capacitors. The pellets are also made so that they are porous or spongelike and have large exposed surfaces.

However, electrolytic capacitors have higher leakage current than electrostatic capacitors because of the impurities embedded in the foil and the electrolyte. This current increases with temperature while voltage breakdown decreases with temperature. Electrolytic capacitors also have higher power factors than electrostatic capacitors, causing losses called *equivalent series resistance* (ESR).

## ALUMINUM ELECTROLYTIC CAPACITORS

An *aluminum electrolytic capacitor* is made by sandwiching a paper separator soaked in electrolyte between two strips of etched aluminum foil, as shown in Fig. 1-15. The paper spacer prevents a short circuit between the cathode and anode foils. The layers of materials are wound in jelly-roll fashion and inserted in an aluminum case. External connections are made from the electrodes to the outside terminals of the case. Direct current is passed through the terminals of the capacitor, causing a thin dielectric layer of aluminum oxide to form on the anode. The electrolyte in contact with the metal foil is the cathode. A plus sign marks the positive terminal of an aluminum electrolytic capacitor.

These capacitors offer high CV ratios and are low in cost. but they exhibit high DC leakage and low insulation resistance. They also have limited shelf lives, and their capacitance values deteriorate with time. Standard units are available in radial- or axial-leaded cases in a

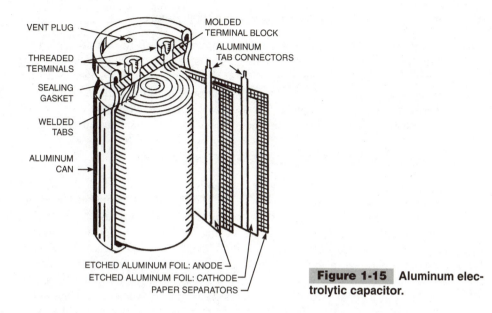

VENT PLUG

MOLDED
TERMINAL BLOCK

ALUMINUM
TAB CONNECTORS

THREADED
TERMINALS

SEALING
GASKET

WELDED
TABS

ALUMINUM
CAN

ETCHED ALUMINUM FOIL: ANODE
ETCHED ALUMINUM FOIL: CATHODE
PAPER SEPARATORS

**Figure 1-15** Aluminum electrolytic capacitor.

wide range of sizes and values. The most commonly specified values are between 4.7 and 2200 μF with working voltages up to 50 VDC. These capacitors are polarized, and this property must be observed when connecting the capacitor in a circuit or it will be destroyed.

*Nonpolarized aluminum electrolytic capacitors* are available for use in AC circuits for such applications as speaker crossovers and audio filtering. Two polarized capacitors are placed in series with their cathode terminals connected. The anode terminals form the external circuit connections, and the cathode terminals are isolated from the external circuit by an insulator. These capacitors are rated from 1 to 10 μF with maximum working voltages of 50 VDC.

## TANTALUM ELECTROLYTIC CAPACITORS

*Tantalum electrolytic capacitors* are made in three styles: (1) *wet foil,* (2) *wet anode,* and (3) *solid anode.* Tantalum capacitors typically have higher CV ratings than aluminum electrolytic capacitors with the same capacitance values. The dielectric formed, tantalum oxide ($Ta_2O_5$), has nearly twice the dielectric constant of aluminum oxide. All tantalum capacitors are inherently polarized. As a group, they offer long shelf life, stable operating characteristics, high operating temperature ranges, and higher CV ratios than aluminum electrolytic capacitors. However, they are more expensive than comparably rated aluminum capacitors and have lower voltage ratings.

## WET-FOIL TANTALUM CAPACITORS

A *wet-foil tantalum capacitor* is made by a process similar to that used in making an aluminum electrolytic capacitor. These capacitors can withstand voltages of up to 300 VDC. Packaged in tantalum cases, they are primarily specified for military/aerospace and high-reliability applications.

## WET-ANODE TANTALUM CAPACITORS

A *wet-anode tantalum capacitor,* as shown in Fig. 1-16, is made from a porous tantalum pellet that is formed by pressing finely ground tantalum powder and a binder in a mold and firing it in a vacuum furnace at about 2000°C. Heat welds or *sinters* the powder into a solid spongelike pellet with a large effective surface area. A thin film of tantalum oxide is grown electrochemically on the pellet and electrolyte is added. Packaged in silver or tantalum cases, their CV ratios are about 3 times those of wet-foil tantalum capacitors.

## SOLID-ANODE TANTALUM CAPACITORS

A *solid-anode tantalum capacitor,* as shown in Fig. 1-17, is also made from a porous pellet anode. A thin film of manganese dioxide that is chemically deposited on the tantalum oxide dielectric serves as a solid electrolyte and cathode. Then a layer of carbon and conductive paint is applied to complete the cathode connection. The most popular and lowest-cost tantalum capacitors, they are available with either radial or axial leads. They are dipped or molded in plastic resin to form protective jackets. Some are also enclosed in tantalum cases for further environmental protection. These capacitors have the longest lives and lowest leakage current of any tantalum capacitors. They can have capacitive values of 0.10 to 680 µF, capacitive tolerances of ± 10 to 20 percent, and maximum voltages of 50 V. The popular ratings are 1 to 10 µF.

## SOLID-ANODE CHIP TANTALUM CAPACITORS

A *solid-anode chip tantalum capacitor,* as shown in Fig. 1-18, is made by the same methods as the radial-leaded version, but it is packaged in a leadless molded epoxy case for

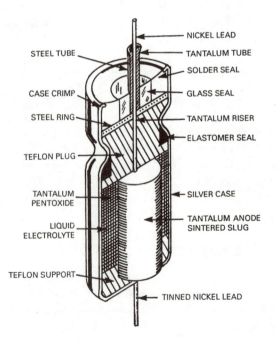

**Figure 1-16** Wet-slug tantalum electrolytic capacitor.

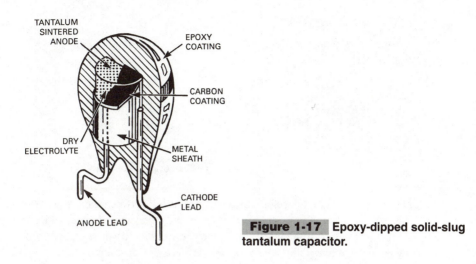

**Figure 1-17** Epoxy-dipped solid-slug tantalum capacitor.

bonding to surface-mount cards or hybrid circuits. They can have capacitive values of 100 pF to 100 μF, capacitive tolerances of ± 5 to 20 percent, and maximum voltages of 50 V.

## VARIABLE CAPACITORS

A *variable capacitor* is a capacitor whose capacitance value can be adjusted by turning a shaft or screw. Used almost exclusively in RF circuits, there are two classes: *tuning* and *trimmer*. Their dielectrics can be plastic, ceramic, glass, or air.

## TUNING CAPACITORS

A *tuning capacitor* is a variable air-dielectric capacitor with plates that move within other plates to change the overall capacitance value. A *single gang-tuning capacitor,* as shown in Fig. 1-19, has a set of aluminum plates called the *rotor* mounted on a shaft so that the plates interleave with a matching set called the *stator* mounted on a rigid spacer. When the rotor

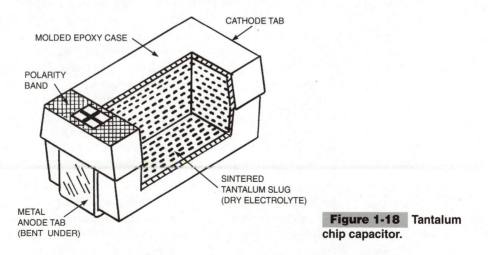

**Figure 1-18** Tantalum chip capacitor.

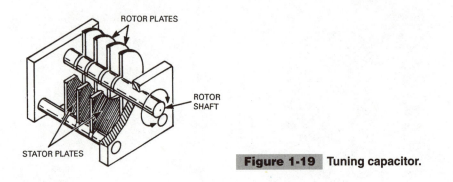

**Figure 1-19**  Tuning capacitor.

shaft is turned by a knob, the rotor plates move in or out between the stator plates without touching them. A change in knob position alters the capacitance value, which is directly proportional to the area of the interleaved plates. Capacitance values can be from 1 to 500 pF. They are used to tune radio receivers, transmitters, and oscillators.

## TRIMMER CAPACITORS

A *trimmer capacitor* is a small variable capacitor with air, ceramic, plastic, glass, or other dielectric that is used for fine-tuning RF circuits. They have capacitance values from 2 to about 100 pF. Made in many different styles, plate spacing is changed to alter the capacitance value by turning an adjustment screw.

# Inductors

An *inductor* provides a known amount of inductance in an AC circuit. It is made by winding a length of copper wire around a cylinder or other form to make a coil or toroid. The value of inductance can be increased by inserting a core of high magnetic permeability material such as iron or ferrite within the coil. Factory-made standard inductors have values that range from less than 1 µH to about 10 H. Small inductors are used in tuned RF circuits, and large inductors are widely used in tuned audio circuits. However, the inductors with the largest values are used as filter chokes in linear power supplies. A perfect inductor would have only pure inductive reactance, but real inductors have a finite resistance. The inductance value of a *variable inductor* can be adjusted over a finite range by changing the number of turns in the coil or moving a permeable core in or out of the coil. At high UHF and microwave frequencies, short lengths of copper or aluminum wire serve as inductors.

# Transformers

A *transformer* transfers electrical energy from one or more primary circuits to one or more secondary circuits by means of electromagnetic induction. It consists of at least one primary winding and one secondary winding of insulated wire on a common core. No electrical connection exists between any primary or input circuit and any secondary or output circuit, and no change in frequency occurs between the two circuits.

If an AC voltage is applied to the primary winding of a transformer, an electromagnetic field forms around the core and expands and contracts at the input frequency. This changing field cuts the wires in the secondary winding and induces a voltage in it. The voltage that appears across the secondary winding depends on the voltage at the primary winding and the ratio of turns in the primary and secondary windings. Schematic diagrams for three commonly specified transformer configurations are shown in Fig. 1-20.

A *step-up transformer,* as shown in Fig. 1-20*a,* has twice the number of turns in its secondary winding as it has in its primary winding, so the voltage across the secondary winding will be twice that of the voltage across the primary winding. Similarly, a *step-down* transformer, as shown in Fig. 1-20*b,* has half as many turns in its secondary as in its primary, so the secondary voltage will be half that of the primary voltage. A *multiple-winding* transformer, as shown in Fig. 1-20*c,* provides three separate output voltages that also depend on the ratios between primary and secondary windings.

All of these transformer configurations obey the law of conservation of energy. In transformers this can be interpreted as the equality of the products of voltage and current or power in both primary and secondary windings, except for losses. Thus, the power input at the primary winding is nearly equal to the power output at the secondary winding or the sum of the secondary windings if there are more than one.

If, for example, the voltage at the secondary terminals of the transformer is twice that of the primary terminals, the current at the secondary terminals must be about half that

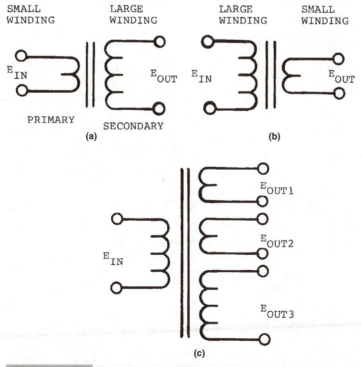

**Figure 1-20** Transformer schematic symbols: (*a*) step-up transformer, (*b*) step-down transformer, and (*c*) multiple-wound transformer.

at the primary terminals to keep the product of voltage and current, which is equal to power, constant. An ideal transformer would be 100 percent efficient because the power output would be equal to the power input. But, because losses reduce the efficiency of most transformers to about 90 percent, output power is about 10 percent less than input power. The total loss is the sum of *ohmic resistance loss, eddy-current induction loss,* and *hysteresis* (molecular friction) *loss,* all caused by the changing polarity of the applied current.

Most transformers transform voltage or current up or down, but an *isolation transformer* provides secondary voltage and current that are essentially the same as the primary voltage and current (except for resistive losses) because both windings have the same number of turns. These transformers prevent the transfer of unwanted electrical noise from the primary to the secondary windings, thus providing isolation.

The transformers closely associated with electronics are the power, audio, pulse, and RF transformers. They are rated according to the products of their secondary voltages and current in voltamperes (VA) or watts. The transformers specified for most electronic applications are rated for less than 100 VA or 100 W, but some switching power supplies have transformers rated to 1 kW.

Military Standard MIL-T-27 is the mandatory guide for workmanship on mil-spec transformers, but it is also widely used as a guide in the manufacture of commercial units. Commercial transformers that are connected to the AC power line are usually certified by a national organization for conformance to recognized safety guidelines because faults or failures in these transformers could cause electrocution or fires.

## POWER TRANSFORMERS

A *power transformer* can transform 50- to 60-Hz AC line power to voltages suitable for rectification to regulated DC. They are made in volume as standard products for the linear power supplies in such products as TV sets, VCRs, and stereos. Their laminated iron or steel cores are made from stacks of E- and I-shaped stampings assembled around toroidal bobbins. Power transformers intended for use in switching power supplies that switch at 400 Hz to 50 kHz are wound on ferrite cores because the reactance losses from laminated iron cores limit efficient operation to about 400 Hz.

## AUDIO OR VOICE TRANSFORMERS

An *audio* or *voice transformer* is similar to a power transformer, but it operates over a wider frequency range. These transformers can conduct DC in one or more windings, transform voltage and current levels, and act as impedance matching and coupling devices, or as filters. A limited range of voice frequencies within the 20 Hz to 20 kHz audio band can be passed by audio transformers.

## PULSE TRANSFORMERS

A *pulse transformer* is a miniature transformer that generates fast-rising output pulses for timing, counting, and triggering such electronic devices as thyristors (silicon controlled rectifiers) [SCRs] and triacs) and photographic flash lamps.

## CIRCUIT-BOARD TRANSFORMERS

A *circuit-board transformer* is made for circuit-board mounting. Classed in this group are miniature power, audio, and pulse transformers. Some have low profiles, as shown in Fig. 1-21, to permit circuit cards in card cages to be stacked closely together. Typically, these transformers are dipped in epoxy resin to seal them from dirt and moisture. Some windings have pin terminations for circuit-board insertion, and others have pads for surface mounting.

## RADIO-FREQUENCY TRANSFORMERS

A *radio-frequency transformer* is designed to function efficiently at radio frequencies. Unlike low-frequency transformers, they are wound on air-core bobbins because neither ferrite nor laminated iron cores are efficient at radio frequencies.

## TOROIDAL TRANSFORMERS

A *toroidal transformer* is wound on a ring-shaped core made by winding long thin continuous sheet metal strips around a cylindrical form. Both the primary and secondary windings are wound on the core by special machines designed to be able to pass wire through and around the open core. Toroidal transformers are more efficient and lighter than comparably rated laminated-core transformers, and they do not emit an audible chatter.

# Filters

A *filter* is a circuit that passes certain frequencies while suppressing others. This property is useful for eliminating unwanted frequencies and separating wide frequency bands into multiple channels. A *passive* filter does not require a power source, but because it dissipates input power it cannot provide either current or voltage gain. Moreover, it has a limited frequency range. Signal loss caused by filtering with a passive filter is called *insertion loss.*

By contrast, an *active filter* can perform the same functions as a passive filter, but it can perform those functions over a wider frequency range, and it can provide current or voltage gain. Although an active filter requires a power source, it does not need a bulky inductor.

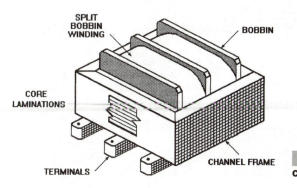

**Figure 1-21** Transformer for circuit-board mounting.

Thus, it can be smaller and lighter than a comparably rated passive filter. See "Active Filters" in Sec. 8, "Analog and Linear Integrated Circuits."

## BASIC FILTER TYPES

There are four basic types of filter:

**1.** A *low-pass filter* can pass all frequencies from zero to its cutoff frequency, and block all frequencies above the cutoff.
**2.** A *high-pass filter* can block all frequencies below its cutoff frequency, and pass all frequencies above the cutoff. Its response is the inverse of the low-pass filter.
**3.** A *bandpass filter* can pass all frequencies within a band defined by lower and upper cutoff frequencies, and block all frequencies above and below that band.
**4.** A *band-reject* or *notch filter* can block all frequencies between its lower and upper cutoff frequencies, and pass all frequencies above and below that band. Its response is the inverse of the bandpass filter.

## FILTER DESIGNATIONS

- The *constant-k filter* is so named because the product of its series and parallel impedances remains a constant designated *k* at all frequencies. These impedances can be inductive or capacitive reactances. A constant-*k* filter can be configured as any of the basic filter types.
- The m-*derived filter* is a modified form of a constant-*k* filter based on a constant called *m,* the ratio of the cutoff frequency to the infinite attenuation frequency. An *m*-derived filter exhibits a sharper attenuation or roll-off curve than a constant-*k* filter because it has more poles. It can also be configured as any of the basic filter types.
- The *Butterworth filter* exhibits an essentially flat ripple response in the passband and a sharp attenuation or roll-off curve at its cutoff frequency. It has a wide operating frequency range that extends from DC into RF. These filters can be configured as low-pass, high-pass, and bandpass. Their transient responses are much better than those of Chebyshev filters.

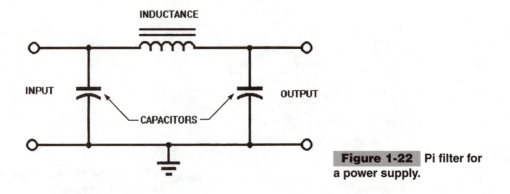

**Figure 1-22** Pi filter for a power supply.

Filters can be identified by one or more of the following classifications:

- The *Chebyshev filter* has characteristics that are similar to those of the Butterworth filter, but it trades off higher amplitude ripple response to obtain an even sharper frequency roll-off curve at its cutoff frequency. Because these are constant-*k* filters, they can be configured as low-pass, high-pass, and band-reject.
- The *Bessel filter* is named for the mathematical functions used to design it. Its frequency cutoff characteristics are not as sharp as those of the Butterworth filter.
- The *elliptical filter* is similar to a Chebyshev filter, but its passband contains even higher amplitude ripple response.
- A filter can be further characterized by its *number of poles,* as determined by the number of reactive components (inductors or capacitors) within the filter. (Resistors do not count as poles because they are not reactive.) The steepness of the attenuation curve or roll-off is determined by the number of poles. For example, a six-pole filter has a steeper attenuation curve than a two-pole filter.

# Passive Filters

A *passive filter* is a network of resistors, capacitors, and inductors configured to pass specific frequency bands while suppressing others. The upper and lower limits of the band are called *cutoff frequencies.* Filters are designed so that their input and output impedances match their source and load impedances. Roll-off or attenuation at the cutoff frequency is measured in decibels. A filter with high attenuation has a steep roll-off curve that is nearly a vertical slope.

Filters are configured by connecting capacitors and inductors in networks, and their schematics suggest letters or other familiar symbols. The four most common configurations are the L, T, pi, and ladder. The positions of the elements are determined by the desired function of the filter (e.g., low pass or high pass). The *L filter* schematic is shaped like an inverted letter L, and the T filter is shaped like the letter T. The *pi filter* schematic looks like the Greek letter $\pi$, as shown in Fig. 1-22, and the *ladder filter* looks like a ladder.

All capacitors can pass AC, and high frequencies pass with less opposition than low frequencies. (Capacitive reactance is inversely proportional to frequency.) But because a capacitor has conductive plates separated by an insulating dielectric, DC is completely blocked. By contrast, inductors, basically coils of wire, easily pass DC and very low frequency AC, but their ability to oppose AC is directly proportional to frequency because inductive reactance is proportional to frequency. Thus, passive filters exploit the frequency-response characteristics of capacitors and inductors.

## CHARACTERISTIC FILTER CURVES

The *characteristic curves* of the four basic types of filters are shown in Fig. 1-23. The frequency values on the horizontal axes are typical operating frequencies for the filters shown, and the positions on the curves labeled $f_C$ are the cutoff frequencies.

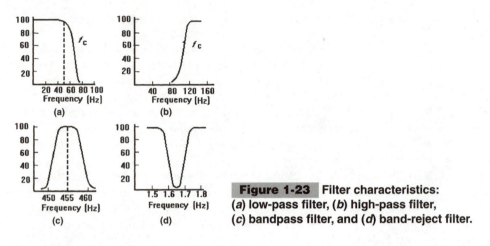

**Figure 1-23** Filter characteristics: (*a*) low-pass filter, (*b*) high-pass filter, (*c*) bandpass filter, and (*d*) band-reject filter.

# Power Supply Filters

A *power supply filter* is a passive filter for linear or switching power supplies to smooth ripples or pulsations in the raw DC output. A *line filter,* as shown in Fig. 1-24, suppresses RF interference (RFI) induced into or transmitted on the AC power line or induced into or conducted from within the host product. These filters are required in products powered by switching power supplies, such as personal computers, that must comply with Federal Communications Commission (FCC) regulations limiting EMI/RFI above 10 kHz.

# Surface Acoustic Wave (SAW) Filters

A *surface acoustic wave (SAW) filter* is a solid-state filter that can replace a conventional passive inductive-capacitive *LC* filter. It offers excellent amplitude and phase response over wide bandwidths and frequency ranges. SAW filters are made from piezoelectric materials such as lithium niobate ($LiNbO_3$) and quartz. A filter made from quartz offers excellent temperature stability over wide temperature ranges, and a lithium-niobate filter simplifies electromagnetic-to-acoustic coupling. These filters have relatively high insertion losses, so

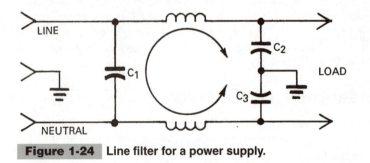

**Figure 1-24** Line filter for a power supply.

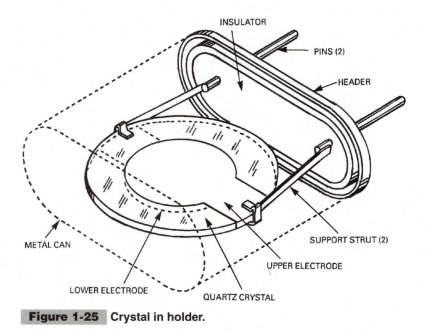

**Figure 1-25** Crystal in holder.

they typically require an amplifier in series with the SAW to recover lost signal strength. See also "Surface Acoustic Wave (SAW) Devices" in Sec. 17, "Electronic Sensors and Transducers."

# Crystal Frequency Standards

*Crystals* used as frequency standards are made from piezoelectric materials that resonate at high frequencies when subjected to an alternating current. Selectively cut quartz crystals generate more stable frequencies than coil-and-capacitor tank circuits. Crystals for generating frequencies for timing or other purposes are packaged in radial-leaded metal cases, as shown in Fig. 1-25.

Quartz wafers are ground to precise thicknesses, and metal-film electrodes are deposited on both sides. The electrodes are connected to the leads that extend through the base. When powered by AC, the quartz wafer vibrates at a frequency determined by its thickness. Thin crystals resonate at higher frequencies than thick crystals. The highest fundamental frequency of a quartz crystal wafer is 15 to 20 MHz. Harmonics or multiples of this frequency provide higher radio frequencies. Quartz crystals in holders serve as oscillator tank circuits. Crystals can also serve as selective filters because of their high $Q$ factors.

# ACTIVE DISCRETE
# COMPONENTS

## Overview

An *active electronic component* is a circuit component that requires external power to perform its function. The discussion of active components in this section is limited to discrete diodes, transistors, and thyristors. Integrated circuits (ICs), also active components, are covered in separate sections of this handbook. Analog and linear ICs are discussed in Sec. 8, digital ICs and semiconductor memories are covered in Sec. 9, and microprocessors and microcontrollers are covered in Sec. 14.

# Small-Signal Diodes

A *small-signal diode* is a two-terminal silicon PN junction that can rectify and clip signals. Rated to handle up to 1 W, these diodes are made by growing an N-type region on a P-type wafer so that there is a direct interface or junction between the two different materials. The wafer is then diced and packaged with terminals attached to both sides of the die. The P-type material is the *anode* and the N-type material is the *cathode,* as shown in the section view Fig. 2-1*a*. The P-type anode contains a surplus of "holes," or vacant sites that can be filled by electrons to conduct current, and the N-type cathode contains a surplus of electrons. The schematic symbol for a diode is shown in Fig. 2-1*b*. The arrowhead indicates the direction of conventional current flow, but this is opposite to electron flow, indicated by the arrow pointed in the opposite direction.

If a positive voltage is applied to the anode and a negative voltage is applied to the cathode, or it is connected to ground, the diode is *forward biased.* Electrons flow from the cathode across the PN junction to the anode, but conventional current is considered to flow in the opposite direction. However, if a negative voltage is applied to the anode and a positive voltage is applied to the cathode, or it is connected to ground, the diode is *reverse* or *back biased,* as shown in Fig. 2-2. Under these conditions there will be little or no electron flow across the PN junction. A reverse-biased diode effectively becomes an insulator with resistance measurable in megohms because of the expansion of the highly resistive *depletion region* that forms around the PN junction.

The characteristic curve for a conventional PN diode is shown in Fig. 2-3*a*. The effect of forward bias is shown by the essentially vertical curve moving toward the right, while the effect of reverse bias is shown by the essentially horizontal curve moving to the left.

Small-signal diodes are typically packaged in glass or plastic cases. A diode rated for more than 1 W is usually called a *rectifier* diode. Microwave diodes intended for much higher frequency operation are discussed in Sec. 7, "Microwave and UHF Technology," and photodiodes are discussed in Sec. 12, "Optoelectronics Sensing and Communication."

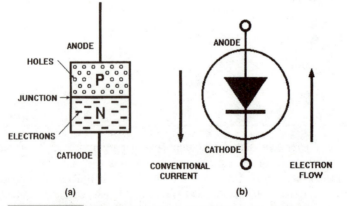

**Figure 2-1**  PN diode: (*a*) functional diagram, and (*b*) schematic symbol.

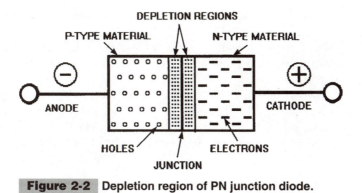

**Figure 2-2**  Depletion region of PN junction diode.

## ZENER DIODES

A *zener* or *reference diode* is a silicon PN junction made to operate only under reverse bias or voltage conditions. At a known reverse voltage an *avalanche breakdown* occurs, indicated by the *knee* in the curve shown on the left side of Fig. 2-3*a*. Beyond that point the reverse voltage remains constant enough to serve as a useful reference voltage. Zener diodes exhibit sharp reverse knees at less than about 6 V. Large quantities of electrons within the depletion region break the bonds with their atoms, causing a large reverse current to flow, as indicated by the vertical dropoff of the curve.

Zener diodes are stable voltage references because the voltage across the diode remains essentially constant for wide variations of current. These diodes are used as general-purpose voltage regulators and for clipping or bypassing voltages that exceed a specified level. Variations of the zener diode called *transient voltage suppressors* (TVSs) serve as circuit-protective devices because of their ability to bypass unwanted high-input voltage

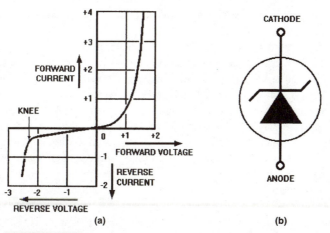

**Figure 2-3**  Characteristic curves for a PN diode: (*a*) forward bias (right) and reverse bias (left), and (*b*) symbol for zener diode.

transients. (See "Transient Voltage Suppressors" in Sec. 30, "Component and Circuit Protection."

The schematic symbol for a zener diode is shown in Fig. 2-3b. It differs from the conventional diode schematic symbol because of its S-shaped anode representation. Zener diodes have nominal reference voltage values from 1.8 to 200 V and power ratings from 250 mW to as high as 50 W. They are packaged in a variety of glass, metal, and plastic cases, some for surface mounting. TVS diodes have ratings from 5 to 300 V, and can handle up to 5 W steady-state or 1500 W peak power. Although both of these diodes can operate in the small-signal region, they are considered to be regulator and suppressor diodes rather than small-signal diodes.

## SCHOTTKY BARRIER DIODES

A *Schottky barrier diode* is a semiconductor diode formed by a semiconductor layer and a metal contact that provides a nonlinear rectification characteristic. Hot carriers (electrons for N-type materials or holes for P-type materials) are emitted from the Schottky barrier of the semiconductor and move to the metal coating that is the diode base. Majority carriers predominate, but there is essentially no injection or storage of minority carriers to limit switching speeds. These diodes are also called *hot-carrier* or *Schottky diodes*.

*Schottky-clamped transistors* used in some transistor-transistor logic (TTL) IC families include Schottky barrier diodes to prevent transistor saturation, thereby speeding up transistor switching. Also, the gates of gallium-arsenide MESFET transistors are actually Schottky barrier diodes.

## VARACTOR DIODES

A *varactor diode,* also known as a *voltage-variable capacitor diode* or *varicap,* is a reverse-biased PN junction whose operation depends on the variation of junction capacitance with reverse bias. Special dopant profiles are grown in the depletion layer to enhance this capacitance variation and minimize series resistance losses.

The varactor is made from a semiconductor material whose dopant concentration is graded throughout the device, with the heaviest concentration in the regions adjacent to the junction. The junction region is small to take advantage of the variation of junction capacitance with reverse voltage. Varactor diodes have very low internal resistance so that the PN junction, when reverse biased, acts as a pure capacitor. Because the junction is abrupt, junction capacitance varies inversely as the square root of the reverse voltage.

Most varactor diodes are made from silicon, but gallium-arsenide varactors offer higher-frequency response. Low-power varactors serve as voltage-variable capacitors in electronic tuners, and do phase shifting and switching in the VHF and microwave circuits. They also function as very low frequency multipliers in solid-state transmitters and do limiting and pulse shaping.

Standard varactors can provide 12 W at 1 GHz, 7 W at 2 GHz, 1 W at 5 GHz, and 50 mW at 20 GHz. Efficiencies of 70 to 80 percent have been obtained at 1 and 2 GHz. The dimensions of a varactor's package depend on its operating frequency and power dissipation.

# Rectifier Diodes

A *rectifier diode* is a diode capable of converting AC into DC. It can conduct 1 A or more or dissipate 1 W or more of power. Most rectifier diodes are now made from silicon. The dies have large PN junctions to eliminate or minimize damage from heat produced by power dissipation. Typically packaged as discrete devices, the rectifiers can be paralleled to increase their power-handling ability. Rectifiers rated for less than 6 A are usually packaged in axial-leaded glass or plastic cases. However, those with 8- to 20-A ratings are usually packaged in flat plastic cases with copper tabs that can act as heat sinks or metal-to-metal interfaces with larger heat-dissipating busbars. Rectifiers rated from about 12 to 75 A are usually packaged in metal cases. Some have threaded base studs for fastening the case directly to a larger heat-dissipating surface.

The most important electrical ratings for rectifier diodes are:

- Peak repetitive reverse voltage $V_{RRM}$
- Average rectified forward current $I_O$
- Peak repetitive forward surge current $I_{FSM}$

Standard PN junction rectifiers are specified for linear power supplies operating at input frequencies up to 300 Hz, but they are inefficient in switching power supplies that switch at frequencies of 10 kHz or higher because of their slow *recovery time*. This is the finite amount of time required for the minority and majority carriers—electrons and holes—to recombine after a polarity change of the input signal. The minority carriers must be removed before full blocking voltage is obtained.

Despite their slow recovery time, standard PN junction rectifiers have lower reverse currents, can operate at higher junction temperatures, and can withstand higher inverse voltages than faster rectifiers designed to overcome this speed limitation.

Three types of fast silicon rectifiers perform more efficiently at the higher-frequency switching rates:

**1.** Fast-recovery rectifiers.
**2.** Ultrafast- or superfast-recovery rectifiers.
**3.** Schottky rectifiers.

## FAST-RECOVERY RECTIFIERS

A *fast-recovery rectifier* is a PN junction rectifier made by diffusing gold atoms into a silicon substrate. The gold atoms accelerate the recombination of minority carriers to reduce reverse recovery time. These rectifiers can be switched in 200 to 750 ns. They have current ratings of 1 to 50 A and voltage ratings to 1200 V. Forward voltage drop is typically 1.4 V, higher than the 1.1 to 1.3 V of the standard PN junction. The maximum allowable junction temperature is about 25°C. This value is lower than that for a standard PN junction. The maximum reverse voltage for a fast-recovery rectifier is about 600 V.

## ULTRAFAST- OR SUPERFAST-RECOVERY RECTIFIERS

An *ultrafast-* or *superfast-recovery diode* is a PN junction rectifier whose reverse recovery time is between 25 and 100 ns. Gold or platinum is also diffused into the silicon wafers from which the rectifier is made to speed up minority carrier recombination. These rectifiers are specified for power supplies with output voltages of 12, 24, and 48 V.

## SCHOTTKY RECTIFIERS

A *Schottky rectifier* has a metal-to-semiconductor junction rather than a PN junction, so it does not have minority charge carriers. The die is in direct contact with one metal electrode, so recovery time, although not specified, is typically less than 10 ns. Recovery current is principally caused by junction capacitance. Schottky rectifiers provide lower forward voltages ($V_F$) than the PN rectifiers (0.4 to 0.8 V vs. 1.1 to 1.3 V). Hence power dissipation is lower and efficiency is higher. One drawback of the Schottky rectifier is its low blocking voltage, typically 35 to 50 V. However, Schottky rectifiers with maximum blocking voltages of 200 V are available. These rectifiers require transient protection, and they have inherently higher leakage current ($I_{RRM}$) than PN junction rectifiers. This makes them more susceptible to destruction by overheating (*thermal runaway*). Schottky rectifiers can be paralleled in the output stages of switching power supplies, where they are usually used with output terminals rated for 5 V or less.

# Signal-Level Transistors

A *transistor* is a three-terminal semiconductor device capable of amplification and switching. It is essentially the solid-state analogy of the triode vacuum tube. There are two principal classes of transistors: *bipolar junction transistors* (BJTs) and *field-effect transistors* (FETs). These transistors are made as discrete small-signal and power devices. Variations of them are integrated into digital and analog or linear ICs. Small-signal discrete BJTs remain popular in low-frequency circuits, while small-signal discrete FETs meet the requirements for high-input impedance transistors. Discrete power BJTs are still popular in low-frequency and linear circuits, but discrete metal-oxide semiconductor (MOSFET) transistors are preferred for high-frequency switching.

# Bipolar Junction Transistors (BJTs)

The term *transistor* implies a silicon *bipolar junction transistor* (BJT) unless modified by an adjective such as JFET or MOSFET. BJTs can be can be made in two different configurations: NPN and PNP. Figure 2-4 shows a section view of an NPN BJT transistor. Here the letter *N* indicates silicon doped with an N-type material, which, by convention, means that it contains an excess of negatively charged electrons. The letter *P* indicates

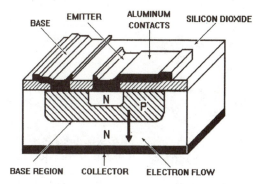

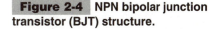

**Figure 2-4** NPN bipolar junction transistor (BJT) structure.

silicon doped with a P-type material, which means it has an excess of positively charged holes.

A voltage applied to the P-type *base* in the NPN transistor causes electrons to flow from the N-type emitter through the base to the N-type collector. (Conventional current is considered to flow in the opposite direction). This BJT has vertical topology, so its metal base contact is deposited on the P-type base next to the metal emitter contact on the N-type emitter, while the collector contact is a metal layer on the bottom of the N-type collector.

Electrons in an NPN transistor cannot flow from the emitter to the collector through the P-type base unless a positive bias is placed on the base contact and a positive voltage is applied to the collector contact. Then holes, repelled by the positive bias, enter the emitter region while electrons flow from the emitter region to the base region. Most of the injected electrons complete the transit through the base region into the N-type collector region and are collected at its contact.

Figure 2-5a shows a simplified section view of the NPN BJT, and Fig. 2-5b shows its schematic symbol. The direction of the arrow represents conventional current flow directed from its P-type base to its N-type emitter.

Figure 2-6a shows a simplified section view of a PNP BJT, and Fig. 2-6b shows its schematic symbol. It can be seen that the polarities and doping of NPN and PNP transistors are reversed. The PNP BJT schematic symbol has its arrow directed from its P-type emitter to its N-type base.

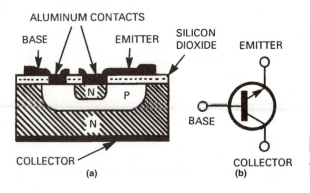

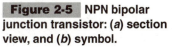

**Figure 2-5** NPN bipolar junction transistor: (*a*) section view, and (*b*) symbol.

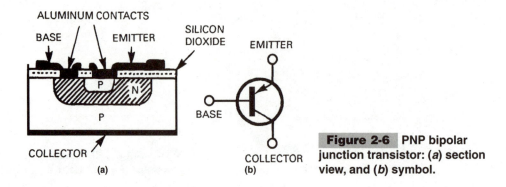

ALUMINUM CONTACTS

BASE          EMITTER          SILICON DIOXIDE

P          N

P

COLLECTOR

(a)

EMITTER

BASE

COLLECTOR

(b)

**Figure 2-6** PNP bipolar junction transistor: (*a*) section view, and (*b*) symbol.

# Darlington Transistor Pairs

A *Darlington pair,* as shown in the schematic Fig. 2-7, is a pair of BJTs in which the emitter of the first transistor is connected to the base of the second transistor. This configuration provides far higher current gain than a single transistor through direct coupling. The pair can be made on a single die and it is packaged in a three-terminal transistor case. The pairs are often used in linear ICs, such as operational amplifiers, and in power amplifier output stages. Its most common application is that of an *emitter follower.* The output is taken across a resistor from the emitter of the second transistor to ground. The input resistance at the base of the first transistor is raised to a higher value than that of a single-transistor emitter-follower circuit.

# Field-Effect Transistors

A *field-effect transistor* (FET) is a voltage-operated transistor. Unlike a BJT, a FET requires very little input current, and it exhibits extremely high input resistance. There are two major classes of field-effect transistors: *junction FETs* (JFETs) and *metal-oxide semiconductor FETs* (MOSFETs), also known as *insulated-gate FETs* (IGFETs). FETs are further subdivided into P- and N-type devices. FETS are *unipolar transistors* because, unlike the BJT, the drain current consists of only one kind of charge carrier: electrons in N-channel FETs and holes in P-channel FETs.

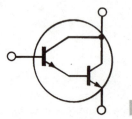

**Figure 2-7** Darlington transistor pair symbol.

FETs and MOSFETs are both made as discrete transistors, but MOSFET technology has been adopted for manufacturing power FETs (see "Power Transistors" later in this section) and ICs. There are both NMOS and PMOS ICs. When both P- and N-channel MOSFETs are integrated into the same gate circuit, it is a *complementary MOS* (CMOS). See Sec. 9, "Digital Logic and Integrated Circuits."

## JUNCTION FETs (JFETs)

The N-channel *junction FET* (JFET), shown in section view Fig. 2-8*a,* has an N channel diffused into a P-type substrate and a P-type region diffused or implanted into the N channel to form the P-type *gate.* Metal deposited directly on the *gate, source,* and *drain* regions forms their contacts. Because a JFET has a symmetrical structure, the drain and source are interchangeable. Thus, depending on the location of the ground and the +*V* power source, the JFET will work in either direction.

If a positive voltage is applied at the drain contact and a negative voltage is applied at the source contact with the gate contact open, a drain current flows. If the gate is then biased positive, channel resistance decreases and drain current increases. However, if the gate is biased negative with respect to the source, the PN junction is reverse biased and a *depletion region* depleted of charge carriers is formed. Because the N-type channel is more lightly doped than the P-type silicon, the depletion region penetrates into the *channel,* effectively narrowing it and increasing its resistance. If the gate bias voltage is made even more negative, drain current is cut off completely. A gate bias voltage value that will cut off the drain current is called the *pinch-off* or *gate cutoff voltage.* The schematic symbol for an N-channel JFET is shown in Fig. 2-8*b.* The arrow points from the P-type gate to the N-type channel.

The P-channel JFET, shown in Fig. 2-8*c,* has characteristics similar to those of the N-channel JFET except that the polarities of the voltage and current are reversed. A P-type

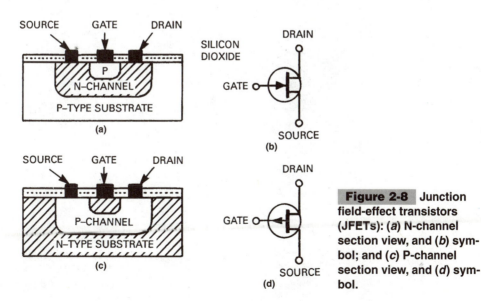

**Figure 2-8** Junction field-effect transistors (JFETs): (*a*) N-channel section view, and (*b*) symbol; and (*c*) P-channel section view, and (*d*) symbol.

channel is diffused into an N-type substrate and then an N-type gate region is diffused or implanted into the P-channel to form the N-type gate. If a negative voltage is applied to the drain and a positive voltage is applied to the source, current flows between source and drain. But if the gate is made more negative more current will flow, while if it is made positive with respect to the source, current will be cut off.

The schematic symbol for a P-channel JFET is shown in Fig. 2-8*d*. The arrow points from the P channel to the N gate.

## METAL-OXIDE SEMICONDUCTOR FETs (MOSFETs)

The *metal-oxide semiconductor FET* (MOSFET) offers a higher input impedance than a JFET. A section view of an N-channel MOSFET is shown in Fig. 2-9*a*. An insulating layer of silicon dioxide is grown on top of the region between the N-type source and the N-type drain. The gate is electrically isolated from the source and gate contacts and the source-to-drain channel beneath it. The schematic symbol for an N-channel MOSFET is shown in Fig. 2-9*b*. The two kinds of MOSFETs are *enhancement mode* and *depletion mode*. The depletion-mode MOSFET has a lightly doped source-to-drain channel, whereas the enhancement-mode version does not.

## ENHANCEMENT-MODE MOSFETs

An *enhancement-mode MOSFET* is normally off because it requires a gate bias signal to cause current flow because of the high impedance of its substrate source-to-drain channel.

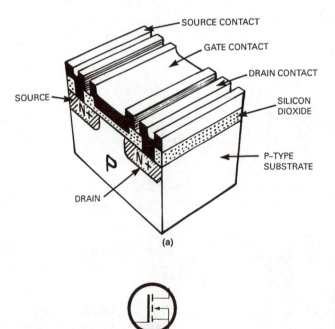

(a)

(b)

**Figure 2-9** Metal-oxide semiconductor FET (MOSFET): (*a*) section view, and (*b*) symbol.

In the N-channel enhancement-mode MOSFET shown in Fig. 2-10a, the substrate is P-type silicon and both the source and drain regions are heavily doped N-type silicon. The metal gate, the insulation layer, and the channel act like a capacitor, so if a bias is placed on the gate, a charge of opposite polarity will appear in the channel below it. For example, if the drain voltage is positive with respect to the source voltage, and the bias on the gate is zero, no current will flow.

But, if the gate is then made positive, negative charge carriers (electrons) are induced in the channel between the source and drain regions. Further increases in positive bias induce more electrons into the channel, where they accumulate to form an N-type channel between source to the drain. The value of drain current depends on channel resistance, so gate voltage controls drain current. Because channel conductivity is *enhanced* by a positive gate bias, the transistor is called an *enhancement-mode* MOSFET.

Figure 2-10b shows the schematic symbol for an N-type enhancement-mode MOSFET. The vertical line connected to the gate pin represents the gate, and the broken lines connected to the drain and source pins indicate that a channel does not exist until a gate voltage is applied. The arrowhead representing conventional current points from the P-type substrate to the induced N-type channel

A *P-channel enhancement-mode MOSFET* has the same geometry as the N-channel enhancement-mode MOSFET except that both the material dopants and the applied voltage polarities are reversed. Its schematic symbol is identical except that the direction of the arrowhead is reversed.

## DEPLETION-MODE MOSFETS

A *depletion-mode MOSFET* is normally on and does not require a gate bias to conduct because of the doping in the source-to-drain channel. In the N-channel depletion-type

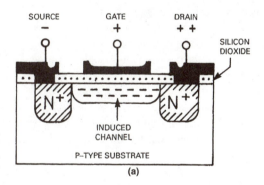

(a)

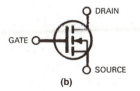

(b)

**Figure 2-10** Enhancement-mode
N-channel MOSFET: (*a*) section view,
and (*b*) symbol.

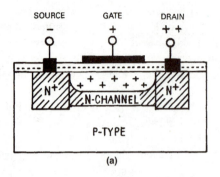

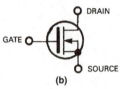

**Figure 2-11** Depletion-mode N-channel MOSFET: (*a*) section view, and (*b*) symbol.

MOSFET, as shown in Fig. 2-11*a*, the substrate is P-type silicon, both the source and drain regions are heavily doped N-type silicon, and the drain-to-source channel is lightly doped N-type silicon. It is called a *depletion-mode* MOSFET because if the drain is made positive with respect to the source, a drain current will flow even with zero gate bias voltage. The metal gate, insulation layer, and N-type channel act like a capacitor, so any gate bias will induce a charge of opposite polarity in the channel.

For example, if both gate bias and drain voltage are positive, the MOSFET acts as an *enhancement-mode* FET. But a negative gate bias must be applied to the gate to shut or pinch off the current. A negative gate bias induces positive charge carriers (holes) in the N-type channel that will combine with the electrons to increase channel resistance. When negative bias reaches the pinch-off value, all drain current ceases.

Figure 2-11*b* shows the schematic symbol for the N-channel depletion-mode MOSFET. It is similar to the symbol for an N-channel enhancement-mode MOSFET except that the line representing the channel is solid.

A *P-channel depletion-mode MOSFET* has the same geometry as the N-channel depletion-mode MOSFET except that both the material dopants and all applied voltage polarities are reversed. Its schematic symbol is identical except that the direction of the arrowhead is reversed.

# Gallium-Arsenide Transistors

The geometries of silicon BJTs and MOSFETs have been implemented in gallium arsenide (GaAs) to take advantage of the higher speed and operating frequencies made possible by the substitution of GaAs silicon. Because GaAs is a compound semiconductor material, it does not form natural oxides as silicon does, so this made it necessary to alter the silicon device geometries to devise different manufacturing methods.

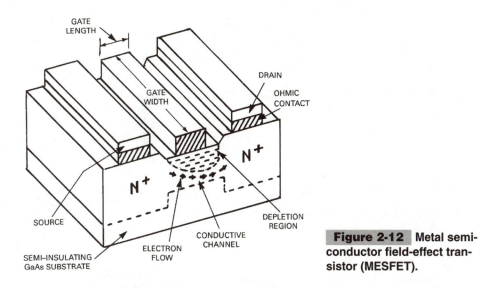

**Figure 2-12** Metal semi-conductor field-effect transistor (MESFET).

Three different gallium arsenide transistor designs have been developed: (1) metal semi-conductor field-effect transistor (MESFET), high-electron-mobility transistor (HEMT), (3) heterojunction bipolar junction transistor (HBT).

## METAL SEMICONDUCTOR FIELD-EFFECT TRANSISTORS (MESFETs)

The *metal semiconductor field-effect transistor* (MESFET) is a widely used discrete and integrated-circuit GaAs transistor geometry. Its structure is similar to that of a MOSFET, but its metal gate is deposited directly on the doped GaAs substrate, as shown in Fig. 2-12, to form a Schottky barrier diode. However, silicon oxides are deposited on the substrate for isolation and insulation. The length of the metallized gate (positioned between the source and drain) is critical in both discrete GaAs transistors and ICs.

Typically 0.5 to 1.0 μm in most discrete transistors, it could be as small as 0.2 μm in ICs. But the gate structure is usually much wider with respect to its length—typically 900 to 1200 μm. MESFETS can have interdigitated structures with multiple gates formed as comblike structures. Ion implantation is favored for doping active regions of MESFETs. A 0.1- to 0.2-μm-thick N-doped region is made for the most common depletion-mode MESFETs (D-MESFETs). The enhancement-mode MESFET (E-MESFET) and the enhancement-mode JFET (E-JFET) are other GaAs transistors that have been developed. Both E-MESFETs and D-MESFETs can be combined in ICs to form enhancement/depletion-mode (E/D) logic.

## HIGH-ELECTRON-MOBILITY TRANSISTORS (HEMTs)

A *high-electron-mobility transistor* (HEMT) is a GaAs transistor designed for IC integration. As shown in Fig. 2-13, it is fabricated on a layer of aluminum gallium arsenide (AlGaAs) grown on a GaAs substrate. This *heterojunction* design improves transistor performance and permits even higher levels of integration than are possible with the MESFET.

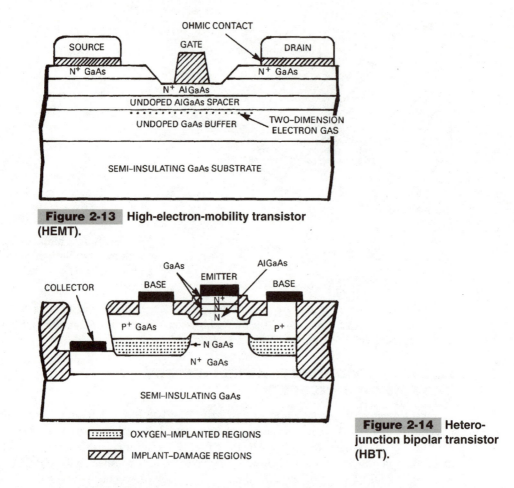

**Figure 2-13** High-electron-mobility transistor (HEMT).

**Figure 2-14** Hetero-junction bipolar transistor (HBT).

## HETEROJUNCTION BIPOLAR TRANSISTORS (HBTs)

A *heterojunction bipolar transistor* (HBT) is a bipolar GaAs transistor grown on a hetero-junction. Heterojunction E/D technology was developed to achieve cost-effective GaAs digital large-scale integrated (LSI) and very large scale integrated (VLSI) devices. The structure, shown in Fig. 2-14, permits high levels of integration. Both HEMTs and HBTs require special processing to achieve precise, sharp heterojunctions.

# Power Transistors

A *power transistor* is one that is capable of handling 1 W or more of power or drawing 1 A or more of current during normal operation without being damaged. Power transistors are used in such applications as amplification, oscillation, switching, and frequency conversion. The three small-signal transistor geometries adapted as power transistors are: (1) the *BJT,* (2) the *Darlington pair,* and (3) the *MOSFET.*

## POWER BJTs

A *power BJT* has a geometry similar to that of a discrete small-signal transistor. Most power BJTs have vertical geometries, with the bases and emitters grown on top of the substrate, which functions as the collector. There are both NPN and PNP power transistors. In the absence of a universal power BJT design that will perform all power functions equally well, many different variations have been developed to provide a range of electrical and thermal characteristics for different applications. Each design has advantages and disadvantages or tradeoffs. These structural variations can be classed by the number of diffused layers, the use of an epitaxial base, or combinations of these. BJTs can be made with mesa or planar structures. Some common power bipolar structures are:

- Single-diffused (hometaxial)
- Double-diffused (mesa, planar, epitaxial mesa, planar mesa, and multiple epitaxial mesa)
- Triple-diffused (mesa and planar)
- Epitaxial base (mesa)
- Multiple epitaxial base (mesa)

A *mesa* is a raised section of the die, with the emitter and base geometry in relief above the level of the silicon collector substrate. The mesa is formed by selectively chemically etching away all but the corners of a completed double-diffused die. A *planar* transistor is made in basically the same way as the mesa version, but the collector-base junction terminates under a protective oxide layer at the surface. Power transistors made with these topologies have different voltage ratings, switching speeds, saturation resistances, and leakage currents. The most advanced switching BJTs have multiple epitaxial, double-diffused structures.

Power bipolar transistors are specified by determining the required values for the following parameters:

- Voltage rating, collector to emitter
- Current rating of the collector
- Power rating
- Switching speed
- DC current gain
- Gain-bandwidth product
- Rise and fall times
- Safe operating area (SOA)
- Thermal properties

The popularity of the switching-regulated or *switch mode* power supply created a demand for power bipolar transistors capable of switching at frequencies in excess of 10 kHz. To qualify for this application, the power transistor must be able to withstand voltage that is typically twice its input voltage. It must also have collector current ratings and safe operating areas that are high enough for the intended application.

A bipolar transistor operated at high power densities is subject to *second breakdown* failure which occurs when a thermal hot-spot forms within the transistor chip and the emitter-to-collector voltage drops 10 to 25 V. Unless power is quickly removed, current concentrates in the small region and temperatures rise until the transistor is damaged or destroyed.

*Safe operating area* (SOA) is defined by a graph that indicates the ability of a power transistor to sustain simultaneous high currents and high voltages. It is the plot of collector current versus collector-to-emitter voltage. The curve defines, for both steady-state and pulsed operation, the voltage-current boundaries that result from the combined limitations imposed by voltage and current ratings, the maximum allowable dissipation, and the second breakdown limitations of the transistor.

## POWER DARLINGTON PAIRS

A *power Darlington pair* consists of two power bipolar transistors that are DC coupled internally as emitter-followers on the same die. This device, considered to be discrete, is packaged in a single case with three external leads. A power Darlington pair provides higher input resistance and more current gain than a single power bipolar transistor.

## POWER MOSFETs

A *power MOSFET* is a high-input-impedance, voltage-controlled transistor with an electrically isolated *gate*. Its structure is similar to that of a small-signal MOSFET, but it has multiple sources and gates and a single drain, as shown in Fig. 2-15a. As a majority-carrier device that stores no charge, it can switch faster than a bipolar transistor. The device shown as a section view in the figure is an N-channel, enhancement-mode power MOSFET. Most power MOSFETs, unlike the small-signal MOSFETs, are fabricated in a vertical geometry with the substrate as the drain and sources and gates formed on top of the device.

With no voltage applied between the gate and source terminals, the impedance between them is very high. But when voltage is applied between the gates and sources, electric fields are set up within the MOSFET, lowering the drain-to-source resistance. This permits conventional current to flow from the drain when a voltage is applied to it. (Electron flow is opposite, from the source to the drain.)

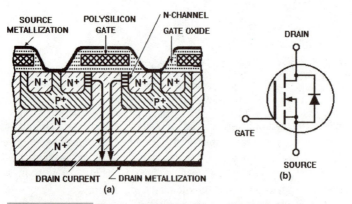

**Figure 2-15**  Enhancement-mode N-channel DMOS power MOSFET: (*a*) section view, and (*b*) symbol.

The schematic symbol for an N-channel enhancement-mode power MOSFET made by the DMOS process is shown in Fig. 2-15b. It includes a symbol for a diode in parallel with the MOSFET in recognition of the PN junction inherent in this transistor geometry.

A *P-channel enhancement-mode power MOSFET* is similar in construction but the polarities and doped regions are reversed. Conventional current flows from the source to drain in these devices, and electrons flow in the opposite direction.

The *vertical double-diffused* (*DMOS*) *process* used to make both of these power transistor types has replaced the *V-groove* (*VMOS*) *process,* popular in the 1970s. Multiple channels are formed by double diffusion at the periphery of each source cell. As shown in Fig. 2-15, an insulating gate oxide layer covers all of the channels and encloses the polysilicon gates, which are positioned over the channels. All of the source cells are than connected in parallel by an overall aluminum film, which forms the source terminal.

The multiple silicon gates are formed in basket-weave or hexagonal patterns on the top surface of the DMOS transistor die. The source cells consist of closed rectangular or hexagonal channels which separate a source region from the substrate drain body. They are formed by an integration process, and their density can exceed a half-million cells per square inch. The vertical DMOS process permits a saving of as much as 60 percent of the silicon substrate over the requirements of the earlier planar MOSFET fabrication processes.

Power MOSFETs are widely specified for high-frequency switching power supplies, chopper and inverter systems for DC and AC motor speed control, high-frequency generators for induction heating, ultrasonic generators, audio amplifiers, and AM transmitters. Power MOSFETs have the following advantages over bipolar transistors:

■ Faster switching speeds and low switching losses
■ Absence of second breakdown
■ Wider SOA
■ Higher input impedance
■ High, if not higher, gain
■ Faster rise and fall times
■ Simple drive circuitry

# Insulated-Gate Bipolar Transistors (IGBTs)

An *insulated-gate bipolar transistor* (IGBT), as shown in the section view Fig. 2-16, is a four-layer discrete power transistor that combines the characteristics of a power MOSFET and a thyristor. A PNP transistor drives an N-channel MOSFET in a pseudo–Darlington pair. An internal JFET conducts most of the voltage, permitting the internal MOSFET to have a lower voltage rating and a lower drain-to-source "on" resistance than similarly rated MOSFETs.

IGBTs can be used in circuits rated for more than 300 V. The IGBT cross section is similar to that of a power MOSFET. However, its P+ substrate allows the IGBT to function in a

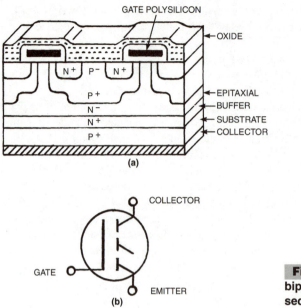

**Figure 2-16** Insulated-gate bipolar transistor (IGBT): (*a*) section view, and (*b*) symbol.

way that is more like the operation of a bipolar transistor than that of a power MOSFET. This power device is also called a COMFET, a GEMFET, and an IGT. In a switching power supply, it can convert DC from a battery to AC to perform such tasks as driving the motor of an electric vehicle.

# Unijunction Transistors (UJTs)

A *unijunction transistor* (UJT), as shown in the section view of Fig. 2-17*a,* is a three-terminal transistor with a single PN junction, a high-resistivity N-type substrate, an emitter, and two base terminals. The UJT's inherent negative resistance characteristic makes it useful in timing and oscillator circuits. Its operation differs from that of either the BJT or the MOSFET.

When a positive voltage is applied across the base terminals $B_1$ and $B_2$, the current that flows between them is determined by the high resistance value of the N-type silicon substrate. The P-type emitter forms a PN junction with the N-type substrate. If a positive bias is applied between the emitter terminal and $B_1$ so that current flows between $B_1$ and $B_2$, the PN junction becomes forward biased. Holes are injected into the N-type region and flow toward $B_1$. The resistance of this region decreases rapidly because of the presence of additional carriers. As a result, the $B_1$-$B_2$ voltage drop across the $E$-$B_1$ region will decrease, although the current through the region increases. This creates a negative resistance region that can be controlled by the $B_1$-$B_2$ voltage.

The emitter is formed by diffusing boron into the high-resistivity N-type silicon wafer to form a P-type region. A more negative base 1 region is formed independently by diffusing

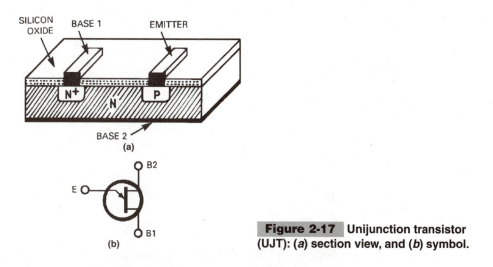

**Figure 2-17** Unijunction transistor (UJT): (*a*) section view, and (*b*) symbol.

phosphorous into the wafer. An N-type annular ring (not shown) is formed around it to protect the junction. Aluminum is evaporated onto the surface of the water to form the base 1 and emitter contacts, and gold is evaporated onto the bottom of the wafer to form the base 2 terminal. The wafer is then diced into many individual UJTs.

The schematic symbol for the UJT is shown in Fig. 2-17*b*. Conventional current, as indicated by the arrowhead, flows from the P-type emitter to the N and N+ bases.

A simple relaxation oscillator can be assembled from a UJT, a capacitor, and a resistor. Additional resistors, capacitors, and a potentiometer can make a variable-frequency UJT relaxation oscillator. UJTs can also control silicon controlled rectifiers (SCRs) and triacs.

The UJT is also called a *double-base diode.*

# Thyristors

The term *thyristor* applies to a class of four-layer semiconductor switching devices whose conduction can be controlled by applying a voltage to a *gate* electrode. The most important members of this family are the *silicon controlled rectifier* (SCR), the *triac,* and the *silicon controlled switch* (SCS).

## SILICON CONTROLLED RECTIFIERS (SCRs)

A *silicon controlled rectifier* (SCR) is a four-layer PNPN unidirectional rectifier diode suitable for bistable power switching. As shown in the diagram in Fig. 2-18*a,* it has three junctions and three terminals: *anode, cathode,* and *gate.* Current flowing to the gate determines the anode-to-cathode voltage for SCR conduction. A gate bias can hold the SCR off or cause conduction between the anode and cathode terminals to begin at any desired point in the forward half-cycle of the AC waveform.

The voltage on the anode of the SCR must be positive for conventional forward-biased operation. The SCR can then be turned on by connecting a positive voltage to the gate.

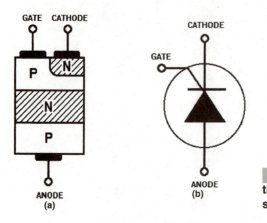

**Figure 2-18** Silicon controlled rectifier (SCR): (*a*) section view, and (*b*) symbol.

Once turned on, the SCR remains on even if the gate voltage is made negative or turned off. To shut off the SCR, the anode-to-cathode voltage must be reduced to its threshold level or the forward current must be reversed. Shutoff usually occurs when the AC across the anode and cathode terminals crosses the zero level after gate bias is switched.

Because the SCR can switch current in only one direction, it can control DC or half-wave AC. SCRs can also function as controlled rectifiers in power rectifier bridges. Some can handle hundreds of amperes or peak voltages of 1500 V with a triggering current of less than a few milliamperes. The most popular SCRs are rated for 40 A or less. Figure 2-18*b* shows the schematic symbol for an SCR.

## TRIACS

A *triac* (triode AC), as shown in the diagram Fig. 2-19*a,* is a bidirectional gate-controlled thyristor that provides full-wave control of AC power. With phase control of the gate signal, load current can be varied from about 5 to 95 percent of full power. The triac is the electrical equivalent of two back-to-back SCRs, and Fig. 2-19*b* shows its schematic symbol.

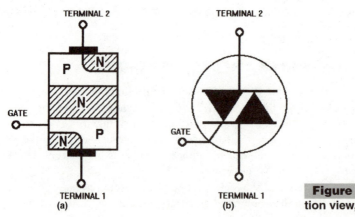

**Figure 2-19** Triac: (*a*) section view, and (*b*) symbol.

## SILICON CONTROLLED SWITCHES (SCSs)

A *silicon controlled switch* (SCS) is a low-current SCR with two gate terminals, an anode and a cathode. A negative pulse on the anode gate turns the SCS on, and a positive pulse on that gate turns it off. However, a positive pulse on the cathode gate can also switch the SCS on, but a negative pulse on that gate is required to turn it off.

# ELECTROMECHANICAL COMPONENTS

## Overview

Electromechanical components for electronics are those that are compatible with electronic circuitry in size, weight and ratings for mounting on circuit boards nearby in cases, cabinets or enclosures. Examples include relays, solenoids, switches, and fractional horsepower DC motors.

## Electromechanical Relays

A relay is the simplest form of *remote controller*. The input circuits of all relays are electrically isolated from their output circuits, and their contacts can be closed remotely by passing current sufficient to actuate the relay coils through wires controlled by a manual switch or another relay. Relays are classed as electromechanical (EM) and solid-state relays (SSRs). Solid-state relays are discussed in Sec. 5, "Fundamental Electronic Circuits."

All electromechanical relays are identified by:

- Output current rating
- Number of poles
- Composition of output contacts
- Packaging style and form factor
- Application

The general classifications for electromechanical relays are:

- General-purpose relays
- Power relays
- Sensitive relays
- Telephone-type relays
- Reed-switch relays

## GENERAL-PURPOSE ELECTROMECHANICAL RELAYS

A *general-purpose electromechanical relay* used in electronics applications is a modern version of the traditional relays invented more than a century ago and manufactured primarily for use in electrical power and lighting, telegraph, and telephone systems. A conventional coil-and-contact relay, as shown in Fig. 3-1, consists of an electromagnetic coil, movable spring-loaded armature, and electrically isolated input and output circuits. When the coil is energized, the electromagnet attracts the hinged and spring-loaded armature against the tension of the return spring, causing the upper normally closed (NC)

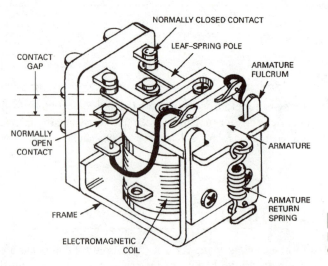

**Figure 3-1** General-purpose electromechanical relay.

contacts to *break* and the lower normally open (NO) contacts to *make* in a break-before-make sequence. In the closed position, the contacts apply power to the load. When the electromagnet is deenergized, the return spring pulls back on the armature opening and thereby closes the upper NC contacts. This action removes power from the load. The relay shown is a two-pole relay, meaning that it can control two separate circuits simultaneously.

Most general-purpose relays are single-, double-, or multipole, single- or double-throw units (SPST, DPDT, 3PDT, etc.). *Double throw* means that contacts are closed in one of two possible positions. The terms *normally closed* (NC) and *normally open* (NO) refer to contact positions when the relay coil is deenergized.

General-purpose relay contacts are made from low-resistance metal such as silver alloy bonded to the ends of the poles and base terminals. Fine silver contacts are rated to 5 A, and nickel-cadmium-silver boosts that rating to 10 A. The fine silver contacts of low-level or *dry* circuit relays are lightly gold plated or *flashed* to prevent oxidation that could form during periods of inactivity or storage and increase contact resistance.

General-purpose relays are rated to handle up to 10 A, although ratings of 2 to 3 A are more typical of those intended for PC-board mounting. There are many different styles, sizes, and ratings for commercial general-purpose relays, making possible the selection of:

- Number of poles (up to eight)
- Terminal form (solder, plug-in, quick-connect, or printed circuit)
- Packaging (open or covered)

Popular voltage ratings are 6, 12, and 24 V, AC or DC; 48 and 120 VDC; and 120 VAC. Mechanical life is typically 20 million operations and electrical life at full rated load is 100,000 closures or better. Most commercial relays rated for more than 50 V are Underwriters Laboratories (UL) recognized and Canadian Standards Association (CSA) certified.

## POWER RELAYS

A *power relay* is a general-purpose relay with high power-handling capability, typically rated 10 to 50 A at 28 VDC or 120/240 VAC. These relays are specified for large power-consuming systems such as computers, radars, and radio transmitters. Heavy-duty EM relays called *contactors* can switch higher power than power relays, but they are electrical power generation and transmission components.

## TELEPHONE-TYPE RELAYS

A *telephone-type relay* is a multipole, high-density relay for switching telephone voice or digital signals. The first models had AC or DC coils with up to eight poles. Contact forms were intermixed and made as flexible metal blades with bifurcated self-wiping tips. The split tips of the blades lightly scraped the surface of the mating contacts during both make and break actions to prevent oxidation buildup. Today miniature telecommunications relays perform the same functions as the earlier, bulky telephone-type relays. With typical ratings of 1 A or less, they are packaged as PC-board mounting flatpacks or sugar-cube-sized rectangular cases.

## SENSITIVE RELAYS

A *sensitive relay* is a general-purpose or telephone-type EM relay designed to switch with input signals from solid-state logic. It is any EM relay that can be switched with an input of less than 10 mW.

## TIME-DELAY RELAYS

A *time-delay relay* is a general-purpose AC and DC EM relay that includes circuits or mechanisms for delaying contact closure after actuation. Used primarily in industrial controls, these relays are available as *interval timers, flashers, one-shot,* and *slow-operate-slow-release* units. Time delays can be from 0.2 s to 120 min or longer. Timing intervals are obtained by setting the resistance value in an internal resistive-capacitive (RC) circuit or by counting down from the 50/60-Hz power line frequency.

## POLARIZED RELAYS

A *polarized relay* combines a movable magnet and an electromagnet to concentrate the magnetic field of an EM relay for improved volumetric and electrical efficiency. There are *pivoting core* and *magnetically biased* polarized relays. The magnetically biased relay has a samarium-cobalt permanent magnet that is about one-fourth the size of ferrite magnets used in other relays. It occupies about half the volume of a comparably rated miniature PC-board-mounted EM relay. Polarized relays are used in computers, copiers, industrial controls, home video systems, private branch telephone exchanges (PBXs), and other telephone equipment.

## REED-SWITCH RELAYS

A *reed-switch relay* or *reed relay* is an EM relay whose contacts are flexible ferrous metal strips or reeds sealed within a hermetically sealed glass capsule. The thin ferrous reeds are sealed axially in the ends of the capsule to protect them from oxidation and contaminants. Very high electrical isolation of $10^{12}$ ohms exists between the open reed contacts within the capsule, but when closed this value drops to about 0.75 ohm. The reeds can be opened or closed reliably for millions of switching cycles by magnetic fields that penetrate the capsule. The most popular contact arrangements in the capsules are Form A, single-pole, single-throw, normally open (SPST-NO) and Form C, single-pole, double-throw (SPDT).

The reed relay illustrated as a cutaway in Fig. 3-2 has a single capsule mounted coaxially within its solenoid coil. The magnetic field, induced when a DC current is applied to the coil terminals, makes or breaks the reed contacts within the capsule. The response depends on the form and arrangement of the contacts. The term *reed relay* implies the use of a dry reed-switch capsule, but some capsules contain small amounts of mercury for improved performance. See "Mercury-Wetted Reed Relays" in this section.

Reed relays are used in telecommunications equipment, medical instruments, and automated test equipment (ATE). A wide range of contact arrangements is available. Some applications require that the relay respond to very low level input currents such as those from a thermocouple. Contact arrangements for telephone switches and test equipment differ.

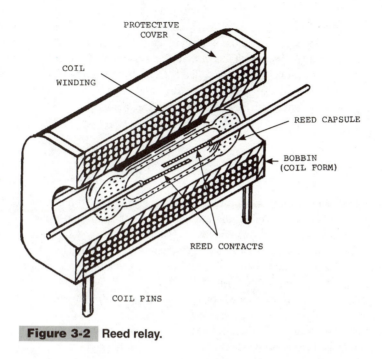

**Figure 3-2**  Reed relay.

Single-pole reed relays contain a single-reed capsule, but multipole relays can have two or more reed capsules mounted within the same coil. Control voltages are typically 5 to 24 VDC. Most reed relays today are packaged for PC-board mounting.

## OPEN REED RELAYS

An *open reed relay* is an unencapsulated assembly of a reed-switch capsule within an electromagnetic coil mounted on a base. These relays are specified where they will be protected from dust, moisture, and contaminants by a separate case or enclosure.

## ENCAPSULATED REED RELAYS

An *encapsulated reed relay* is an assembly of an electromagnetic coil and a reed-switch capsule that has been protected by potting or encapsulation against moisture and contaminants, making it suitable for applications in more demanding operating environments. The coil and capsule are mounted on a leadframe and both are typically molded in epoxy as a dual-in-line package (DIP) or a single-in-line package (SIP). Both styles can be socket mounted for ease of replacement. As an alternative to a molded package, open-style reed relays can be potted with an elastomeric compound in small plastic or metal tubs and inverted for PC-board mounting.

## MERCURY-WETTED REED RELAYS

A *mercury-wetted reed relay* is an assembly of an electromagnetic coil and a reed-switch capsule that contains a small amount of mercury to improve contact closure reliability. The mercury lowers the resistance path for contact closure and damps out contact *bounce* or *chatter* (the sequential opening and closing of contacts before final closure). Under low-level loads, mercury-wetted reed contacts provide consistent and predictable resistance over wide ranges of temperature and contact load current. But mercury slows contact closing to about 2 ms. These relays are position sensitive and must be mounted in an upright position to keep the reed contact surfaces immersed in the mercury.

# Electromechanical Switches

*Electromechanical switches* for electronics are small mechanisms with electrical contacts that permit manual opening or closing of circuits. The simplest and most basic of control devices, they are made in many forms and styles. They are packaged for panel or circuit-board mounting. A typical switch has movable metal contacts controlled by a lever-actuated spring mechanism. The contacts are insulated from the actuator by an insulating case. Spring action accelerates shutoff to minimize arcing.

Figure 3-3 shows a cutaway view of a typical miniature switch. Many of these miniature switches are scaled-down versions of electrical power switches. The most popular miniature and subminiature switches for electronics are: (1) pushbutton, (2) toggle, (3) rocker, (4) slide, (5) rotary, and (6) thumbwheel.

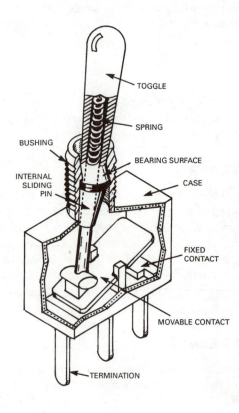

**Figure 3-3** Electromechanical switch.

## PUSHBUTTON, TOGGLE, AND ROCKER SWITCHES

*Pushbutton, toggle,* and *rocker switches* are generally made for both panel and circuit-board mounting. Some have threaded bushings for mounting through a hole in a metal or plastic panel with a lockwasher and ring nut. Others are mounted with snap-in bezels. The switch is inserted through a cutout and anchored with a snap-in clip. Many of these switches have been adapted for circuit-board mounting so that the actuator (pushbutton, handle, or rocker) projects through a hole in the panel or cover. These switches are mounted on the circuit board or card, which provides mechanical support. Pushbuttons, levers, and rockers are made in an almost infinite variety of colors, shapes, and sizes to meet specific customer requirements.

Pushbutton switches are combined with volume controls on dual-purpose switches. Others function as momentary key switches. Some circuit-board switches for "set-and-forget" encoding functions are mounted within the host product and are inaccessible to the user.

Switch ratings depend on their size, their contacts, and the dielectric properties of their cases. In addition, the spacing between conductors and contacts is considered. International safety standards apply to some switches rated for 50 VDC or more, but they are usually required for those rated 120 VAC or more. Panel switches for electronics are commonly rated for 5 A at 125 VAC or 3 A at 220 VAC.

Some pushbutton and rocker switches can be illuminated with incandescent lamps or light-emitting diodes (LEDs) but the term *illuminated pushbutton* (IPB) switch refers to families of industrial-grade switches with square or rectangular end caps that function as indicators or annunciators.

## SLIDE SWITCHES

A *slide switch* is a switch assembly that includes a sliding element which can make or break internal contacts. Lateral movement of the slider can turn an external circuit on or off. These switches are made in many different sizes, ratings, and styles. They are used in radios, electronic test instruments, and home appliances.

## ROTARY SWITCHES

A *rotary switch* is a switch assembly that includes an axial shaft on which a movable contact is mounted. Step rotation of the shaft causes the movable contact to make or break with internal fixed contacts, turning an external circuit on or off. Multideck versions contain multiple stacked wafers or decks, each with fixed contacts, and an equal number of shaft-mounted movable contacts. They permit the simultaneous switching of multiple electrically isolated circuits or functions. Typically panel mounted, rotary switches are used in radios, electronic test instruments, home appliances, and automobiles.

## THUMBWHEEL SWITCHES

*Thumbwheel switches* are encoders that convert numerical settings into binary digital code. They are typically made as stacked assemblies of two or more modules for panel mounting.

Each module contains a thumb-actuated wheel marked with the numbers 0 to 9 and a code-conversion circuit card. When powered and set to a desired number, the binary equivalent signal appears on its output pins. Thumbwheel and similar *lever switches* are used to enter commands into digital circuits for instrument, machine, and process control.

## SWITCH CONTACT ARRANGEMENTS

Commonly used switch contact arrangements are illustrated in Table 3-1. The choice of switch contact material depends on electrical rating and anticipated applications. Silver alloy is used for general-purpose applications and gold-flashed silver alloy is used for low voltage. The contacts in many switches are spring loaded to accelerate the switching action. (Slow switch closure causes contact burning, sticking, and arcing.) The spring also provides a tactile and sometimes an audible response to indicate that the contacts have closed.

Very low voltage circuits are called *dry circuits*. Microampere currents are blocked by resistive oxidation or contamination films on the contacts, so they are lightly gold plated or flashed to prevent this contamination buildup. It is particularly useful if the switch is inactive or in storage for long periods.

All switches are rated for contact resistance, working current and voltage, dielectric strength, insulation resistance, lifetime, and ambient temperature range. Switch contacts are derated for high ambient temperatures regardless of their electrical ratings. Contacts can be inadvertently welded together in DC circuits, but AC reversals minimize this effect. As a result, DC switch ratings are lower than AC switch ratings.

Switch cases are molded from various thermoplastic or thermosetting resins or ceramics. Flame-retardant thermoplastics cost less than the thermosetting plastics such as diallyl phthalate or phenolic. The metal parts of panel switches typically are formed or stamped

| Design | Sequence | Symbol | Form |
|--------|----------|--------|------|
| SPST-NO | Make [1] | | A |
| SPST-NC | Break [1] | | B |
| SPDT-BM | Break [1], Make [2] | | C |
| SPDT-MB | Make [1] before Break [2] | | D |
| SPDT-BMB | Break [1], Make [2] before Break [3] | | E |

**TABLE 3-1**  Commonly Used Switch Contact Arrangements

from sheet steel, stainless steel, or copper alloys such as phosphor-bronze or beryllium-copper.

# Special Switches

### REED SWITCHES

A *reed switch* is an assembly of a *reed-switch capsule* and a permanent magnet, as shown in Fig. 3-4. The magnet is mounted so that it moves with respect to the reed-switch capsule. When close to the reeds, the magnetic field penetrates the capsule, causing the reeds to open or close, thus switching the load on or off. The response depends on the contact arrangement of the reeds. Unless otherwise stated, the switch capsule is assumed to be a dry reed switch.

### MERCURY-WETTED REED SWITCHES

A *mercury-wetted reed switch* is a switch that includes a mercury-wetted reed-switch capsule containing a small amount of mercury to improve performance, as in mercury-wetted reed relays. For further information on the mercury-wetted switch capsule, see "Mercury-Wetted Reed Relays" in this section. These switches, like the relays, are position sensitive and must be mounted in an upright position to keep the reed contact surfaces immersed in the mercury.

### HALL-EFFECT SWITCHES

A *Hall-effect switch* is a magnetically actuated momentary keyswitch that contains a *Hall-effect transducer* (HET) and transistor amplifier with trigger circuits integrated on a silicon chip and a small permanent magnet. When the IC is powered by a DC current and the spring-

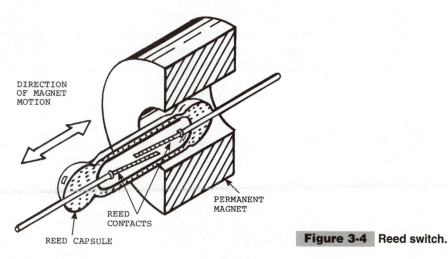

**Figure 3-4**  Reed switch.

loaded key is depressed, the IC moves into the magnetic field against spring pressure, and the IC switches the circuit. The keyswitch returns to its off position when the key is released.

## SHORT-TRAVEL KEYBOARDS

A *short-travel keyboard* is useful for performing data entry that does not require the tactile response and other features of a keyboard suitable for high-speed data entry. It might be organized in the QWERTY format or have customized legend or icon keys. Some short-travel, full-size keyboards substitute micromotion flexible-membrane switches for momentary action switches. Rugged, short-travel keyboards that are sealed against spilled liquids are suitable for use in retail store and fast-food restaurant point-of-sale terminals where coffee and other liquids can be spilled on them or in machine tools or industrial robots where oil can be spilled on them. The legends or characters on the keyboard are printed on the plastic membrane that is stretched over an array of micromotion keyswitches. The membrane need be depressed only about 0.005 in (0.13 mm) to actuate the keyswitch.

## KEYPADS

A *keypad* can be one of many different kinds of keyswitch assemblies primarily intended for numerical data entry into a digital system. The keys on standard, cordless, and cellular mobile telephones are arranged as keypads. Pocket calculators also have keypads, and the group of 17 keys on the right side of standard full-travel computer keyboards is organized as a numerical keypad.

General-purpose keypads are available as standard off-the-shelf items from commercial switch manufacturers. Some accept custom orders for variations of the standard catalog products and some will build them completely to customer specifications. Both standard and custom keypads can be assembled from arrays of full- or short-travel keyswitches on a circuit board or they can be made by covering arrays of short-travel switch contacts on a circuit board with individual caps or flexible membranes.

The simplest mechanical keypad consists of an array of spring-loaded, short-travel keycaps which, when depressed, short out metal contacts on a supporting circuit board. Keycap travel for actuation is typically 0.030 in (0.8 mm) or less. Membrane keypads permit the printing of alphanumeric characters or words on the membrane in different colors. The switch elements can be metallized rubber domes or plastic blisters that contact a matrix of conductors on the circuit board when depressed.

# Solenoids

A *solenoid* is an electromechanical actuator that converts electrical energy into mechanical motion. An iron core is mounted on a spring-loaded axial shaft which moves within the coil bobbin. When the coil is energized, the iron core pulls in against the spring. The solenoid shown in Fig. 3-5 has shaft extensions on both ends. On the right end of the core a clevis pulls in and on the left end a push rod pushes out or hammers when the solenoid is actuated. Most solenoids provide only one of these responses.

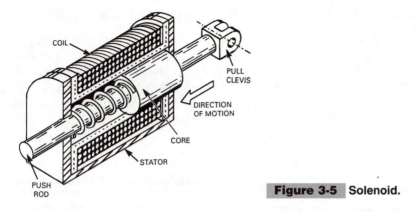

Figure 3-5 Solenoid.

Rotary solenoids operate on the same principle as linear solenoids, but shaft motion along the axis of the coil is translated by a mechanism into circular motion in a plane at right angles to the coil axis. Solenoids can be actuated either by DC or rectified AC.

# Stepper Motors

A *stepper* or *stepping motor* is an incremental open-loop actuator that converts digital pulse inputs into a sequence of step index motions. Classed as a DC motor, it is actually an AC motor operated by the pulse train. The motor rotor moves for a controlled fraction of a revolution each time an input step pulse is received. The movement of the shaft can be controlled and translated into precise rotational or linear motion. Input pulses are counted electronically to reach the programmed position without error correction by feedback from a closed-loop circuit. These motors can perform reliable, precise positioning at lower cost and with fewer components than a closed-loop servo system. The three types are:

**1.** Variable reluctance (VR)
**2.** Permanent magnet (PM)
**3.** Permanent-magnet hybrid

## VARIABLE-RELUCTANCE (VR) STEPPER MOTORS

A *variable-reluctance (VR) stepper motor* has a wound stator and a multipoled soft-iron rotor. Its step angles are determined by the number of its stator and rotor teeth, which can vary from 5 to 15°. Small, and inexpensive, these motors offer relatively low torque and inertial load capacity, but they are suitable as light-duty drives for instruments.

## PERMANENT-MAGNET (PM) STEPPER MOTORS

A *permanent-magnet (PM) stepper motor* has a step angle from 5 to 90°, and typically has a wound stator with a permanent magnet rotor that delivers low torque. Step rates are from 100 to 360 steps/s, and step accuracy is ±10 percent or better.

## PERMANENT-MAGNET HYBRID STEPPER MOTORS

This is the largest, fastest, and most powerful stepper motor, and it can be stepped with the highest accuracy. Combining the best features of both the PM and VR motors, these motors have toothed permanent-magnet rotors and toothed wound stators. Torque capacities are 50 to 2000 oz · in with step accuracies to ±3 percent. Their step angles range from 0.5 to 15°. Capable of being stepped at rates of 1000 steps/s or more, they are widely used in computer-controlled robots, machine tools, and process controls.

# Permanent-Magnet Electric Motors

## PERMANENT-MAGNET DC MOTORS

A *permanent-magnet DC motor,* as shown in the cutaway view Fig. 3-6, is used in a closed-loop control system because it can be precisely controlled with DC signals. It has a wound armature. These motors are used to drive precise machine tools and robots.

## BRUSHLESS DC MOTORS

A *brushless DC motor,* as shown in the cutaway view Fig. 3-7, is one whose commutation is performed electronically by switching transistors under the control of Hall-effect sensors. This design eliminates problems associated with brushes. These motors typically have

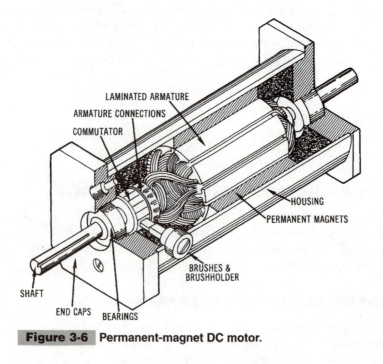

**Figure 3-6**   Permanent-magnet DC motor.

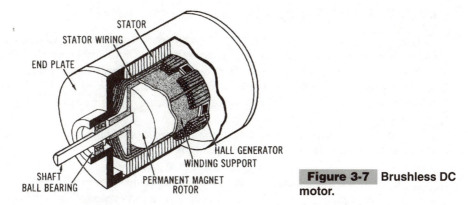

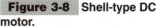
**Figure 3-7** Brushless DC motor.

permanent-magnet rotors and wound field stators. They can drive light loads and can substitute for conventional brush-type DC motors where flammable or explosive materials are present because they do not have brushes that could arc and ignite those materials.

## SHELL-TYPE DC MOTORS

A *shell-type DC motor,* as shown in the cutaway view Fig. 3-8, has a lightweight, low-inertia armature that can be rapidly accelerated to operating speed. It has a hollow ironless shell or cup-type armatures with a shell made from aluminum or copper coils that have been bonded by polymer resin and glass fiber. The armature rotates in an air gap with a very high flux density provided by permanent magnets mounted around the armature. It has conventional DC motor brushes for commutation. These motors are used in closed-loop servosystems where full rotational speed must be reached very rapidly.

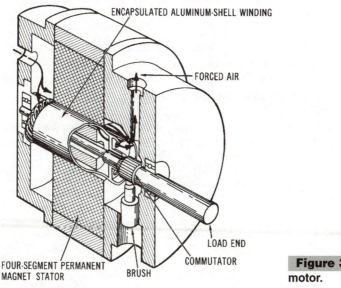

**Figure 3-8** Shell-type DC motor.

## DISK-TYPE DC MOTORS

*Disk-type DC motors* are so named because they have disk-type armatures and flat form factors. Their rotors are made from coil patterns punched from sheet copper and bonded with polymer resin to form rigid glass-fiber disks. These ironless armatures offer very low inertia and are rotated in an air gap by the high flux density provided by round permanent magnets mounted around the inner periphery of the case. These motors are commutated with brushes.

The first disk-type motors were called *printed-circuit* (armature) *motors* because their armatures were made like printed-circuit boards, but that method has been replaced by the punched coil process. The early motors were called *pancake motors* because their cases were round and flat like pancakes, with a small length-to-diameter ratio. This form factor, still used today, concentrates all of the motor's mass in a short space close to the front bell. This permits it to be flange-mounted to a flat surface or beam without supporting brackets, a desirable feature for motors that drive servo-controlled machine tools and robots where space is limited.

# Resolvers

A *resolver* is a rotary electromechanical transformer that senses position in a servo-controlled system. Resolvers, like synchros, contain a rotor and a stator. They accept AC reference excitation at a pair of rotor input terminals and produce voltages that are 90° (electrical) apart at two pairs of stator output terminals. The stator output signals are at carrier frequency, and their amplitudes are proportional to the sine and cosine, respectively, of the angular position of the rotor shaft.

Made like electric motors, the two stator windings are wound on slotted laminations. The common brushless resolvers are designed so that primary excitation voltage is coupled through a transformer.

Resolvers measure the phase angle difference between the AC reference excitation input and the output of the rotor coils. Rotor position is determined by comparing the time-phase-shifted output signal with the input signal. This position is then converted to a digital format by electronic circuitry that counts the number of pulses between the zero crossings of the two signals.

The three common techniques for converting resolver outputs into digital formats are:

**1.** Tracking
**2.** Successive approximation
**3.** Time-phase shift

# 4

# BASIC AMPLIFIER AND
# OSCILLATOR CIRCUITS

# Overview

Amplifiers and oscillators are important circuits that appear in many different configurations and perform many different functions in modern electronics. Most amplifier and oscillator circuits were originally designed during the vacuum-tube era, and they have been converted to transistors and ICs with modifications that relate primarily to the differences in characteristics of receiving tubes and transistors. This section on amplifiers and oscillators is confined to transistorized and integrated circuits. Amplifier and oscillator tubes that amplify and oscillate in the higher frequency UHF and microwave regions are discussed in Sec. 7, "Microwave and UHF Technology."

# Amplifier Circuits

An *amplifier* is any circuit or device capable of increasing the magnitude or power level of a time-variable signal without distorting its wave shape. Most low-power and low-frequency amplifiers today are electronic circuits that depend on transistors or ICs for their operation. However, there are also vacuum-tube, magnetic, electromechanical, and hydraulic amplifiers.

Electronic amplifier circuits are classified by:

- Application
- Circuit configuration
- Coupling (if more than one stage is used)
- Bandwidth and frequency of the signals being amplified
- Operating mode or bias

Amplification can be performed by many different kinds of discrete transistorized and integrated circuits. The active devices in these circuits include bipolar junction transistors (BJTs), junction field-effect transistors (JFETs), and metal-oxide semiconductor field-effect transistors (MOSFETs).

Amplifiers can be designed for either voltage or power amplification. *Voltage amplifiers* increase the voltage level of an applied signal because of the characteristics of the transistor. The output voltage of an amplifier is determined by the voltage drop across the output load, so the impedance of the load is made as large as practical. *Power amplifiers,* by contrast, are designed to deliver heavy current to the output load, so the load impedance must be low enough to allow a high-current output, but not so low that the signal will be distorted excessively. Power amplifiers are also called *current amplifiers.* Other amplifiers are classed as *buffer amplifiers, square-wave amplifiers,* and *frequency doublers.* Transistor amplifiers are also classified by the way their principal terminals or pins are returned to ground.

## NPN TRANSISTOR COMMON-EMITTER AMPLIFIERS

An *NPN transistor common-emitter amplifier,* shown in the schematic Fig. 4-1, has its emitter terminal in common with both its input and output ports. The input signal is applied

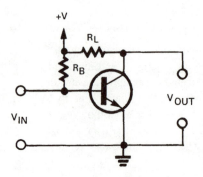

**Figure 4-1**  NPN transistor common-emitter amplifier.

between the base and ground, and the output signal is taken between the collector and ground. Also called a *grounded-emitter amplifier,* it offers medium input impedance, high output impedance, and both high voltage and current gain.

## NPN TRANSISTOR COMMON-BASE AMPLIFIERS

An *NPN transistor common-base amplifier,* shown in the schematic Fig. 4-2, has its base terminal in common with both its input and output ports. The input signal is applied between the emitter and ground, and the output signal is taken between the collector and ground. It is also called a *grounded-base amplifier.*

## NPN TRANSISTOR COMMON-COLLECTOR AMPLIFIERS

An *NPN transistor common-collector amplifier,* shown in the schematic Fig. 4-3, has its collector terminal in common with both its input and output ports. The input signal is applied between the base and ground, and the output signal is taken between the emitter and ground. The collector is normally connected to the power supply. The voltage gain from base to emitter is less than 1, but current gain is high. This amplifier offers high input impedance and low output impedance. It is also called an *emitter follower* or a *grounded-collector amplifier.* See also "Darlington Transistor Pairs" in Sec. 2.

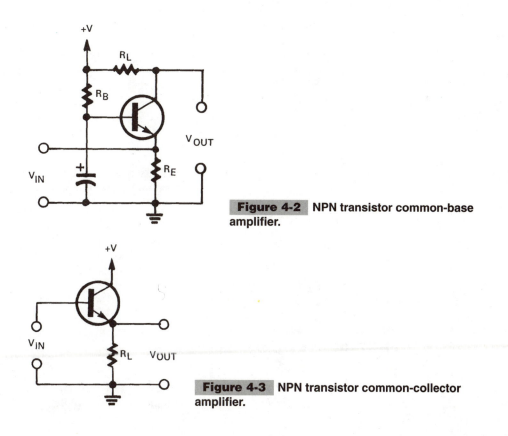

**Figure 4-2**  NPN transistor common-base amplifier.

**Figure 4-3**  NPN transistor common-collector amplifier.

## N-CHANNEL JFET COMMON-SOURCE AMPLIFIERS

An *N-channel JFET common-source amplifier,* shown in the schematic Fig. 4-4, has its source terminal in common with both its input and output ports. The input signal is applied between the gate and ground, and the output signal is taken between the source and ground. This amplifier offers high input impedance, medium to high output impedance, and voltage gain greater than unity.

## N-CHANNEL JFET COMMON-GATE AMPLIFIERS

An *N-channel JFET common-gate amplifier,* shown in the schematic Fig. 4-5, has its gate terminal in common with both its input and output ports. The input signal is applied between the source and ground, and the output signal is taken between the drain and ground. This amplifier can transform a low input impedance to a high output impedance and perform high-frequency amplification.

## N-CHANNEL JFET COMMON-DRAIN AMPLIFIERS

An *N-channel JFET common-drain amplifier,* shown in the schematic Fig. 4-6, has its drain terminal in common with both its input and output ports. The input signal is applied

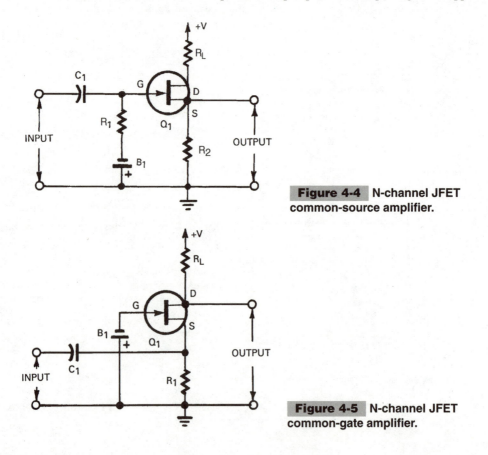

**Figure 4-4** N-channel JFET common-source amplifier.

**Figure 4-5** N-channel JFET common-gate amplifier.

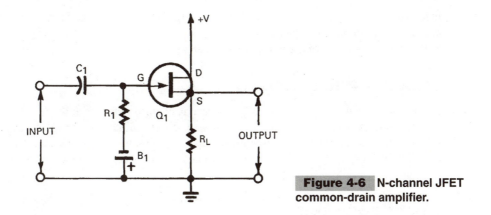

**Figure 4-6** N-channel JFET common-drain amplifier.

between the gate and ground, and the output signal is taken between the source and ground. Also called a *source-follower amplifier,* it is used where low input capacitance or the ability to handle large input signals is required.

# Amplifier Coupling

Amplifiers with more than one amplification stage are classified according to coupling method. These methods include *resistance capacitance* (RC) or *impedance coupling, transformer coupling,* and *direct coupling.*

## RESISTANCE-CAPACITANCE (*RC*) COUPLING

In *resistance-capacitance* (RC) *coupling,* the output load usually has a high resistance value. The *RC*-coupled amplifier, shown in the schematic Fig. 4-7, has good frequency response characteristics over a relatively wide frequency range. But its gain falls off above and below this range. The decrease in gain at the lower frequencies is caused by the

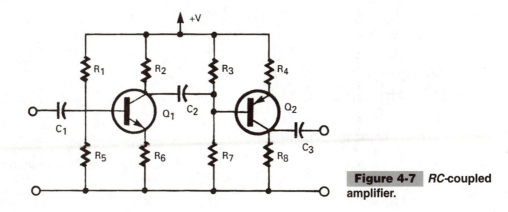

**Figure 4-7** *RC*-coupled amplifier.

increased reactance of the coupling capacitor. At higher frequencies, the decrease in gain is caused by the decreased reactance of the interelectrode capacitances of the stages as well as the stray capacitance of the wiring between stages.

## TRANSFORMER COUPLING

In *transformer coupling,* the output of one circuit is coupled to the input of the next circuit by a transformer. An example of a *transformer-coupled amplifier* is shown in the schematic Fig. 4-8. Additional amplification can be obtained if the transformer has a step-up turns ratio. Transformer coupling is widely used in tuned radio-frequency (RF) and intermediate-frequency (IF) amplifiers.

## DIRECT COUPLING

In *direct coupling* the output of the first stage is applied directly to the input of the second stage. The DC amplifier can amplify both DC and AC signals. Special circuitry in either the coupling network or the amplifier stage eliminates the need for a coupling capacitor. DC amplifiers are widely used for amplifying low-frequency signals.

# Amplifier Bandwidth

Amplifiers are classified by bandwidth as being either tuned or untuned. A *tuned amplifier* has a tuned (resonant) circuit in its input or output circuit (or both) that passes a relatively narrow band of frequencies. The center of this band is the resonant frequency of the tuned circuit. The width of the band depends on the $Q$ of the tuned circuit. Tuned amplifiers are important in RF and IF sections of radio and television receivers. The receiver is tuned to the carrier frequency of the desired signal, and the tuned amplifiers amplify that signal. The resonant frequency of some tuned amplifiers can be set by varying either the inductance or

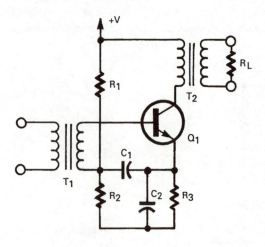

**Figure 4-8**  Transformer-coupled amplifier.

capacitance of the circuit. Gain is normally maximum at the resonant frequency, and it decreases for frequencies above or below the resonant frequency.

An *untuned amplifier* is not tuned to any specific band of frequencies. The range of frequencies that it can amplify is limited by the circuit components and stray capacitances. But an untuned amplifier can amplify a wider range of frequencies than a tuned amplifier.

# Amplifier Frequency

Amplifiers are also classified according to frequency as *direct current* (DC), *audio frequency* (AF), *intermediate frequency* (IF), *radio frequency* (RF), and *video frequency*. A DC amplifier can amplify DC signals, and an audio amplifier can amplify signals in the audio band from about 20 to 20,000 Hz. Video amplifiers amplify signals as high as 200 MHz. Broad frequency ranges are not given for IF or RF amplifiers because they are generally tuned and therefore amplify a relatively small band of frequencies.

# Amplifier Operating Methods

Amplifiers are also classified as *Class A, Class B, Class AB,* and *Class C* according to their operating method or biasing conditions. Operating class is determined by the quiescent point set by the bias of the sinusoidal input signal and by the magnitude of the signal voltage applied, as shown in Fig. 4-9.

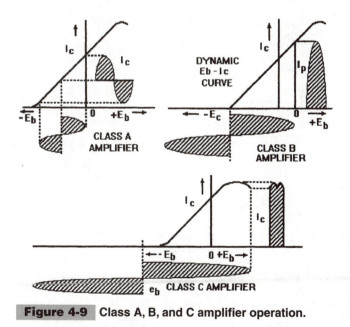

**Figure 4-9** Class A, B, and C amplifier operation.

## CLASS A AMPLIFIERS

A *Class A amplifier,* as shown in the schematic for the power amplifier in Fig. 4-10, is biased in the center of its operating curve so that its output current flows during the entire 360° cycle of the input signal, and no part of the signal is cut off. This class of operation causes the least amount of output signal distortion, but it is the least efficient operating mode because current flows continuously. Class A amplifiers are widely used in audio systems where low efficiency is accepted to obtain the least amount of signal distortion.

## CLASS B AMPLIFIERS

A *Class B amplifier,* as shown in the schematic for the push-pull amplifier in Fig. 4-11, is biased at cutoff so its output current flows for about one-half cycle (180°) of the input-signal voltage. When no input signal is present, no output current flows. Thus, a Class B amplifier cuts off half of its AC input-signal waveform. A single audio amplifier must be Class A to avoid output distortion, but two Class B audio amplifiers in a push-pull circuit can provide undistorted reception because each can supply opposite halves of the signal

## CLASS AB AMPLIFIERS

A *Class AB amplifier* is biased so that its output current flows for more than half of the input cycle but for less than the entire cycle. Its operation is Class A for small signals and Class B for large signals. In effect, Class AB amplifiers provide a compromise between the low distortion of a Class A amplifier and the high efficiency of a Class B amplifier.

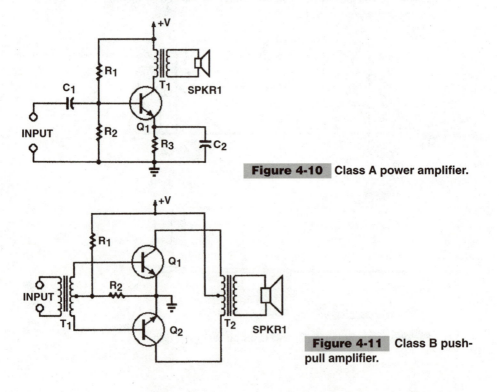

**Figure 4-10**  Class A power amplifier.

**Figure 4-11**  Class B push-pull amplifier.

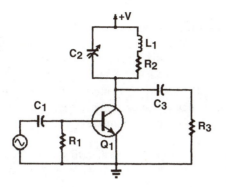

**Figure 4-12** Class C radio-frequency power amplifier.

## CLASS C AMPLIFIERS

A *Class C amplifier,* as shown in the schematic for the RF amplifier in Fig. 4-12, is biased beyond cutoff so that its output current flows only during the positive-going peak of the input cycle. This amplifier is capable of high power output, but it also introduces high distortion, preventing its use in audio applications. Class C operation is suitable for amplifying tuned radio frequencies because an inductance-capacitance (*LC*) circuit can reconstruct full sine waves at the output.

# Differential Amplifiers

A *differential amplifier,* as shown in the schematic Fig. 4-13, has outputs that are proportional to the difference between the voltages applied to its two inputs. It is also called a *difference amplifier.* Some operational amplifiers can operate in a differential mode.

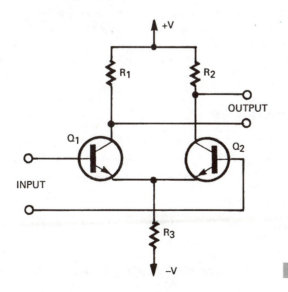

**Figure 4-13** Differential amplifier.

# Operational Amplifiers

An *operational amplifier* (op amp) is a voltage-feedback high-gain amplifier that requires feedback to become useful. Its gain and response characteristics are determined by external components such as resistors, capacitors, or diodes. Op amps are the basic circuits in many kinds of monolithic and hybrid linear circuits. An op amp can have a voltage gain as high as $10^9$. Moreover, op amps are front ends for many different sensors. Figure 4-14 shows the schematic symbol for an op amp.

The op amp has two inputs, *inverting* and *noninverting*. The inverting input, indicated by a minus (−) sign, provides a 180° phase shift between the input and output. A negative feedback loop, consisting of a resistor and/or capacitor between the inverting input and output, controls the gain of the amplifier. The noninverting input, indicated by a plus (+) sign, is in phase with the output. Some op amps require both negative and positive power sources.

The ideal op amp would have infinite input impedance, zero output impedance, infinite gain, and infinite bandwidth. In practice some op amps come close to realizing some of these properties, but no one circuit has realized all of them. An op amp can have very high input impedance, so it will draw almost no current, and it will have very low power consumption. Some op amps offer very low output impedance, extremely high gain (more than 100 dB), and wide bandwidth (up to several megahertz).

Op amps are used in many different analog circuits such as analog-to-digital converters (ADCs), differentiators, DC amplifiers, oscillators, and sweep generators. Monolithic ICs containing two or more amplifiers are available in dual-in-line packages (DIPs).

Three common circuits based on the operational amplifier are: (1) the *inverting amplifier,* (2) the *noninverting amplifier,* and (3) the *summing amplifier.*

## INVERTING AMPLIFIERS

An op amp *inverting amplifier,* as shown in the schematic Fig. 4-15, has its noninverting or plus terminal returned to ground, and its feedback resistor $R_2$ connected between the inverting and output terminals. Resistor $R_1$ is in series with signal source $V_{IN}$ and the inverting minus terminal. The output voltage change is inverted with respect to the input voltage change. Thus the output and input voltages are 180° out of phase. This circuit is also called an *inverter.*

## NONINVERTING AMPLIFIERS

A *noninverting amplifier,* as shown in the schematic Fig. 4-16, has its input signal fed directly to the noninverting (plus) terminal of the op amp. Resistors $R_1$ and $R_2$ form the

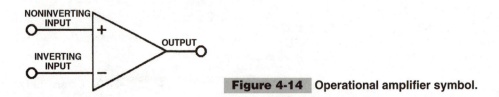

**Figure 4-14**  Operational amplifier symbol.

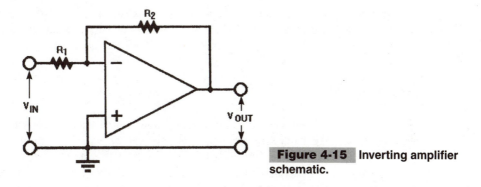

**Figure 4-15** Inverting amplifier schematic.

feedback network. Because of the virtual ground, the voltage at both the inverting (minus) and noninverting (plus) terminals is equal to $V_{IN}$. Both the output and input voltages are in phase in the noninverting amplifier.

## SUMMING AMPLIFIERS

A *summing amplifier,* as shown in the schematic Fig. 4-17, has an output that is proportional to the negative sum of its input voltages. In the schematic, this would be the negative sum of the three input voltages $V_1$, $V_2$, and $V_3$, if the values of all of the resistors $R_1$, $R_2$, $R_3$, and $R_4$ are equal. This will remain true if additional inputs are added and the resistor values

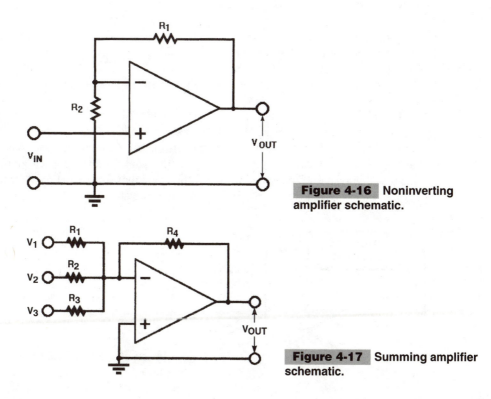

**Figure 4-16** Noninverting amplifier schematic.

**Figure 4-17** Summing amplifier schematic.

associated with those inputs are equal to the values of the existing resistors. However, if any of the resistor values are not equal, scale factors must be used to arrive at an accurate sum.

# Oscillators

An *oscillator* is a circuit that is capable of producing an AC output signal by converting DC power to AC power. Oscillators are either *sinusoidal* or *nonsinusoidal*. Audio- and radio-frequency oscillators produce sine waves and are sinusoidal, while multivibrators, flyback oscillators, and blocking oscillators produce square, sawtooth, or pulsed oscillations and are nonsinusoidal. Figure 4-18 shows a block diagram of a simple sinusoidal oscillator to illustrate the basic principles of oscillation. The output of the *amplifier* passes through a *phase-shift network* in a closed loop back to the input of the amplifier. This phase-shift network inverts the output signal so that the signal returned to the input is in phase with and augments the assumed input signal. The oscillator does not receive an external input signal because it generates an output from the amplifier's power source. Only a small part of the output signal will sustain oscillation when fed back to its input, but the amount of energy or power fed back must be sufficient to overcome the energy losses in the circuit. Thus round-trip gain, or *loop gain,* must be at least 1 or slightly more.

A second condition for oscillation is that the loop phase shift must be 0° (or 360°). Direct coupling is used in this circuit, and little or no phase shift is expected. These two conditions are known as the *Barkhausen conditions* for oscillation:

**1.** The loop gain must be slightly greater than 1.
**2.** The loop phase shift must be 0 or 360°.

To achieve these conditions the active device in the circuit must be an amplifier, and it should be biased for Class A operation.

Vacuum tubes were the original active elements in the oscillators described and illustrated schematically in this section. The circuits were redesigned for transistor amplifiers, including operational amplifiers. The output signals of both vacuum-tube and transistorized amplifiers are 180° out of phase with their input signals, so the feedback loop must provide the additional 180° phase shift to meet the conditions for oscillation. A practical oscillator's output frequency must not drift, and it must be stable.

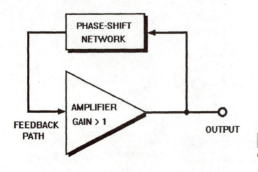

**Figure 4-18** Basic oscillator block diagram.

Some electron devices can fulfill the requirements for oscillation because, at high frequencies, the amplification and the feedback necessary for self-oscillation are provided by the inductive and capacitive effects of their internal construction. Power tubes such as the *magnetron* and *klystron* can serve both as amplifiers and as oscillators in the UHF and microwave bands. In addition, a semiconductor device, such as a *Gunn diode,* can oscillate because its geometry gives it the necessary inductive and capacitive properties that permit oscillation without external components. See Sec. 7, "Microwave and UHF Technology."

# Oscillator Examples

Descriptions and schematic diagrams of widely used and popular audio- and radio-frequency sinusoidal oscillators, as well as a nonsinusoidal oscillator, are presented here. The amplifiers shown in the schematics are transistors for ease in understanding the operation of the circuits, but most modern adaptations of these oscillators have operational amplifiers as active elements. Some even include integrated circuits. Because there are many variations in the design of these oscillators, there are no standard schematic diagrams.

## TUNED-EMITTER OSCILLATORS

A *tuned-emitter oscillator,* shown in the schematic Fig. 4-19, is an RF sine-wave oscillator with inductance-capacitance ($LC$) resonant-circuit feedback. The tuned resonant circuit consisting of $L_2$ and variable capacitor $C_3$ is coupled to the emitter, the base is bypassed to ground, and regenerative feedback is obtained by the close coupling of coil $L_1$ in the collector circuit to $L_2$ in the tuned circuit. The two coils are phased to provide the required feedback. The oscillator's frequency is determined by the tunable resonant circuit. It will oscillate when energy from the tank is coupled back to the base of transistor $Q_1$ through coil $L_1$. $Q_1$ must supply enough power to the emitter to maintain oscillations. Because the load is closely coupled to the frequency-determining tank circuit, both its frequency stability and waveforms are poor.

## TUNED-COLLECTOR OSCILLATORS

The *tuned-collector oscillator,* shown in the schematic Fig. 4-20, is an RF sine-wave oscillator with $LC$ resonant-circuit feedback. The resonant circuit formed by the secondary coil

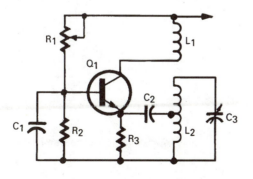

**Figure 4-19**  Tuned-emitter oscillator.

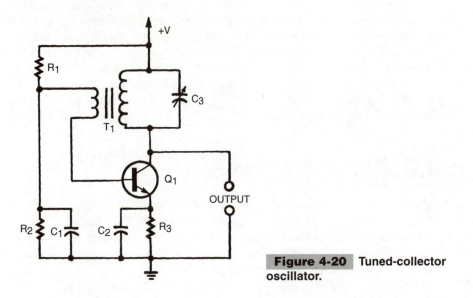

**Figure 4-20** Tuned-collector oscillator.

of transformer $T_1$ and variable capacitor $C_3$ is in the collector circuit, the base is bypassed to ground, and regenerative feedback is obtained from the primary coil of transformer $T_1$. The coils of $T_1$ are phased to provide the required feedback. The oscillator's frequency is determined by the tunable resonant tank. It will oscillate when energy from the tank is coupled back to the base of transistor $Q_1$. Because the load is closely coupled to the frequency-determining tank circuit, both the frequency stability and waveforms are poor.

## PIERCE OSCILLATORS

The *Pierce oscillator,* as shown in the schematic Fig. 4-21, is a crystal resonant-circuit sine-wave oscillator. It has a very high $Q$ piezoelectric quartz crystal that behaves as a series resonant circuit between the gate and drain of field-effect transistor $Q_1$. The quartz crystal holds the oscillator frequency to better than $\pm 0.01$ percent of its nominal value. The use of the FET as the active element has two advantages: its high-impedance gate does not adversely load the crystal and lower its $Q$, and its low drain voltage protects the crystal from damage by overexcitation or high-voltage breakdown. This oscillator requires a

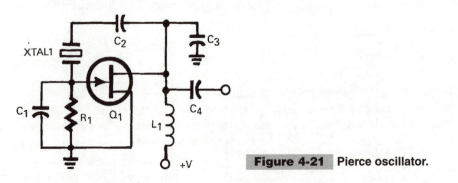

**Figure 4-21** Pierce oscillator.

capacitive reactance across both gate and drain, and the crystal provides the inductive reactance necessary to obtain the 180° phase shift. Coil $L_1$ is an RF choke.

## HARTLEY OSCILLATORS

The *Hartley oscillator,* shown in the schematic Fig. 4-22, is an RF sine-wave oscillator with *LC* resonant-circuit feedback. It has a tapped coil and a single tuning capacitor in its *LC* resonant tank circuit that couples the collector and base circuits of transistor $Q_1$. The emitter and tap on the coil are at ground potential. The collector and base circuits of $Q_1$ share a single resonant circuit, and the amount of the base drive is determined by the position of the tap on the coil. The collector and base of $Q_1$ are at opposite ends of the tuned circuit to provide the necessary 180° phase shift. Feedback occurs through mutual coupling between the two parts of $L_1$. It can generate sine-wave frequencies up to UHF.

## COLPITTS OSCILLATORS

The *Colpitts oscillator,* shown in the schematic Fig. 4-23, is an RF sine-wave oscillator with *LC* resonant-circuit feedback. It is similar to the Hartley oscillator except that the emitter of transistor amplifier $Q_1$ is tapped to a point on the capacitor side of the *LC* resonant tank circuit. Usually capacitor $C_2$ at the base end of the tuned circuit is fixed, and the oscillator frequency can be changed if $C_1$ is variable. The capacitive voltage dividers $C_1$ and $C_2$ provide the required phase shift. Some current flowing in the tank circuit is regeneratively fed back to the base of $Q_1$ through coupling capacitor $C_3$. It can generate sine-wave frequencies up to UHF.

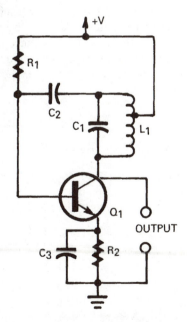

**Figure 4-22** Hartley oscillator.

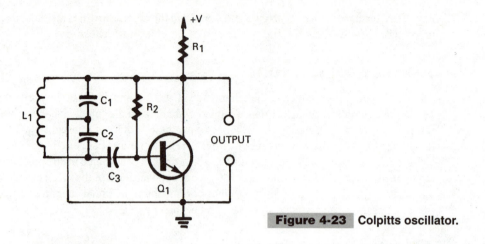

**Figure 4-23** Colpitts oscillator.

## CLAPP OSCILLATORS

The *Clapp oscillator,* shown in the schematic Fig. 4-24, is an RF sine-wave oscillator with *LC* resonant-circuit feedback. It is similar to the Colpitts oscillator except that the inductance is replaced by a series resonant circuit $L_1C_1$ in the resonant tank circuit. The values of the $C_1$ and $L_1$ determine oscillator frequency, and the output of transistor amplifier $Q_1$ is coupled to the base to provide the required 180° phase shift. It can generate sine-wave frequencies up to UHF.

## PHASE-SHIFT OSCILLATORS

The *phase-shift oscillator,* shown in the schematic Fig. 4-25, is an audio-frequency sine-wave oscillator with resistance-capacitance (*RC*) resonant-circuit feedback. It has a Class A transistor amplifier and an *RC* network consisting of three identical cascaded *RC* sections, $R_1C_2$, $R_2C_2$, and $R_3C_3$, each providing a 60° phase shift for the 180° required feedback. The output of $Q_1$ is connected to the network, which is also connected to the base of

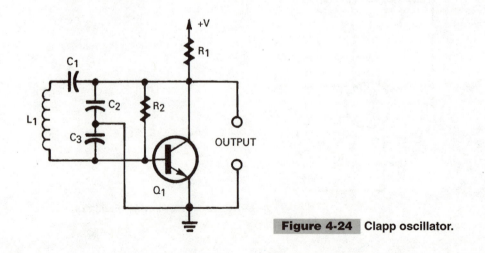

**Figure 4-24** Clapp oscillator.

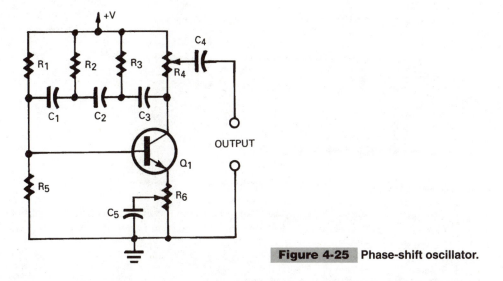

**Figure 4-25** Phase-shift oscillator.

$Q_1$. The phase-shift oscillator is simple and it provides stable frequencies and good wave-forms. It is typically used for fixed-frequency operation, and is one of the two most common audio-oscillator circuits.

## WIEN-BRIDGE OSCILLATORS

The *Wien-bridge oscillator,* shown in the schematic Fig. 4-26, is a two-transistor oscillator with *RC* feedback from a Wien bridge, its frequency-determining element. The Wien bridge is an AC bridge that can be balanced at a specific frequency. Losses in the feedback network make the two-stage transistor amplifier $Q_1$ and $Q_2$ necessary. The two-stage ampli-

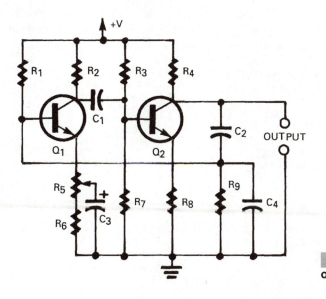

**Figure 4-26** Wien-bridge oscillator.

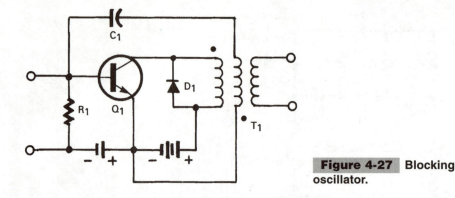

**Figure 4-27** Blocking oscillator.

fier provides an output signal that is in phase with the input. This oscillator offers a relatively pure sine-wave output and very stable frequency output. It is one of the two most common audio-oscillator circuits.

## BLOCKING OSCILLATORS

The *blocking oscillator,* as shown in the schematic Fig. 4-27, is a nonsinusoidal oscillator. It can generate a short high-energy pulse, followed by a long period in which the active element is cut off, or blocked. It can be used as a free-running or synchronized oscillator, as a source of steep wavefront pulses, and as a simple DC-to-AC inverter. Only a single transistor is required as an amplifier. When transistor $Q_1$ conducts, the closely coupled transformer $T_1$ drives the base so hard in the forward direction that capacitor $C_1$ is rapidly charged through the forward-biased diode. A negative base bias increases gradually during oscillation as capacitor $C_1$ is charged. Charging continues until a value is reached at which collector current is cut off and oscillations cease. A steep-sided output pulse is generated by the rapid turn-on of the collector current. Capacitor $C_1$ then discharges until the base is unblocked and oscillation resumes. The appropriate phase shift is fed back to the base of $Q_1$. The blocking oscillator produces sawtooth waveforms suitable for sweeping electron beams in cathode-ray tubes.

# FUNDAMENTAL
# ELECTRONIC CIRCUITS

## Overview

Certain basic electronic circuits have proven to be the sources from which many other practical circuit designs have evolved. These basic circuits perform such functions as wave shaping, pulse forming, and signal conversion. Most were originally developed in the receiving-tube era and have been transistorized with either bipolar junction transistors (BJTs) or field-effect transistors (FETs). Many of the circuits described are now available as standard monolithic integrated circuits, and many of these IC packages contain two or more circuits. Some of the circuits described in Sec. 4, "Basic Amplifier and Oscillator Circuits," are also fundamental electronic circuits.

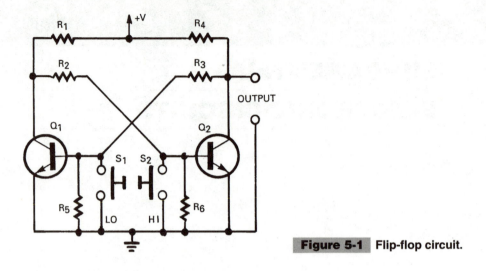

**Figure 5-1** Flip-flop circuit.

# Flip-Flop Circuits

A *flip-flop circuit,* as shown in the schematic Fig. 5-1, is a two-transistor multivibrator circuit that has two stable states. In one state, transistor $Q_1$ conducts and transistor $Q_2$ is cut off. In the other state, transistor $Q_1$ is cut off and transistor $Q_2$ conducts. A trigger signal switches the state of the circuit, and the next trigger signal switches it back to the first state. For counting and scaling, a flip-flop can deliver one output pulse for each pair of input pulses. The flip-flop is important because it is the simplest memory circuit. It acts like a toggle switch because when it is set in one state it will remain there until it receives another trigger signal. It is also known as a *bistable multivibrator, Eccles Jordan,* or *trigger circuit.* The four useful variations of the basic flip-flop, the *D flip-flop, J-K flip-flop, R-S flip-flop,* and *T flip-flop,* are building blocks in more complex circuits called *binary counters* and *shift registers.* See "Flip-Flop Circuit Variations" in Sec. 9, "Digital Logic and Integrated Circuits."

# Schmitt Trigger Circuits

A *Schmitt trigger,* as shown in the schematic Fig. 5-2, is a bistable trigger circuit that converts an AC input signal into a square-wave output by switching action. It is triggered at a predetermined point in each positive and negative swing of the input signal. The circuit remains off until a specified voltage threshold value is exceeded. When the circuit is triggered, its output voltage rises abruptly to a constant value and remains there until the input voltage falls below the fall threshold. The output voltage then drops back to zero almost instantly. The response recurs as long as the AC waveform is applied. It is applied when there is a requirement for square (or rectangular) waves with a constant amplitude. The circuit is widely used to convert sine waveforms into square or rectangular waveforms.

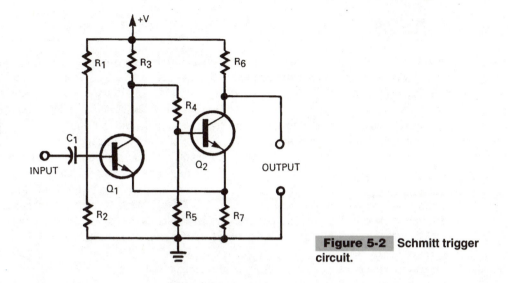

**Figure 5-2** Schmitt trigger circuit.

# Monostable Multivibrators

A *monostable multivibrator,* as shown in the schematic Fig. 5-3, is a multivibrator with one stable state and one unstable state. A trigger signal is required to drive the circuit into the unstable state, where it remains for a predetermined time before returning to its stable state. Switch $S_1$ initiates the response. The circuit is also called a *monomultivibrator* or a *one-shot multivibrator.* It is considered to be a nonsinusoidal oscillator.

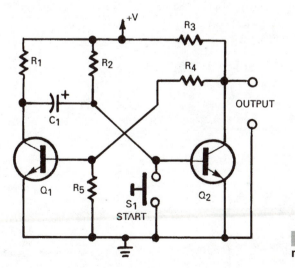

**Figure 5-3** Monostable multivibrator.

# Charge-Coupled Devices (CCDs)

A *charge-coupled device* (CCD) is a semiconductor shift register that can pass along a charge which has been created and stored within the device. The CCD is made as an interconnected string of metal-oxide semiconductor FET (MOSFET) stages sharing a common substrate. When a voltage is applied to the first MOSFET in the series, a depletion region capacitance causes a charge to form. The successive application of pulses to succeeding stages causes the charge to be moved down the line of stages, one stage at a time.

The simplified diagram Fig. 5-4 will help to explain CCD operation. The CCD consists of a P-type silicon substrate coated with an insulating silicon dioxide ($SiO_2$) layer. Closely spaced metal-gate electrodes are deposited in rows on the insulating layer and connected to a signal-voltage source. (Because the substrate is P-type, electrons are the minority carriers, but CCDs can also have N-type substrates and holes as the minority carriers.)

When the voltage applied to gate 1 is more negative than all of the voltages applied to the gates to the right side of gate 1, a depletion region or *well* forms in the substrate under the gate. This region, an electron charge, represents a packet of stored information. But if the voltage on gate 2 is then made more negative than the voltage on gate 1, the well and charge steps to the right, as shown in Fig. 5-4. By alternating the voltages applied to the gates in a controlled sequence, this process will continue, and the charge packets will be passed down the length of the device in a bucket-brigade sequence.

To overcome the poor charge-transfer efficiencies in the simplified CCD shown in Fig. 5-4 (based on early CCDs), two improvements were made. The first was the replacement of metal gates with a double-level polysilicon gate structure, and the second was the inclusion of a buried channel. An N-type silicon layer is formed by epitaxial growth, implantation, or diffusion on a P-type silicon substrate before the conductive gates and insulating $SiO_2$ are deposited. This process forms a buried N-type channel.

The stored charge packets in the CCD can represent either an analog or digital signal. When the charge represents a digital signal, the CCD is known as a *dynamic shift register* because it behaves like a shift register. It is considered to be a dynamic memory because it

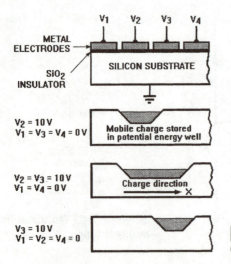

**Figure 5-4**  Charge-coupled device (CCD).

is volatile and requires periodic refreshing. A CCD digital memory is block-access oriented rather than random-bit accessed. A controlled time delay can be introduced into the signal path because the input appears as an output after a finite number of clock pulses. The delay can be controlled by adjusting the clock rate.

Stored charges represent analog signals in such circuits as delay lines, adaptive and fixed filters, matched filters, and optical imaging systems. Charge in the form of carriers can be generated optically by exposing the device to light at a frequency acceptable to the device.

A CCD array is the basic component of miniature solid-state video surveillance and TV cameras. It is also the imaging component in handheld camcorders and digital cameras. The CCD array can be scanned to convert the changing patterns of light intensities that are incident on the array into electrical signals to produce a TV image. The CCD cell packing density (the number of functions occupying the substrate) is hundreds of times greater than the packing density of other semiconductor devices. A CCD array consumes very little power, and it is capable of operating at high frequencies. See "Charge-Coupled Device Cameras" in Sec. 19, "Television Broadcasting and Receiving Technology."

# Rectifier Circuits

A *rectifier circuit* includes one or more rectifier diodes that rectify or convert AC (usually from the AC line) to DC. The three most widely used circuit configurations for single-phase rectification in power supplies are:

**1.** The *half-wave rectifier* circuit shown in the schematic Fig. 5-5a
**2.** The *full-wave center-tapped rectifier* circuit shown in Fig. 5-5b
**3.** The *full-wave bridge rectifier* circuit shown in Fig. 5-6

Two circuits are widely used for three-phase rectification of AC:

**1.** The *half-wave three-phase rectifier* circuit
**2.** The *full-wave three-phase rectifier* circuit

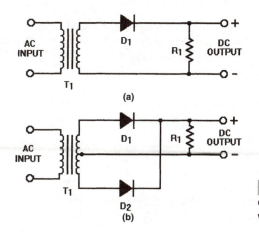

**Figure 5-5** Single-phase rectifier circuits: (*a*) half-wave rectifier, and (*b*) full-wave center-tapped rectifier.

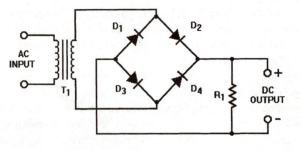

**Figure 5-6** Single-phase full-wave bridge rectifier.

# Rectifier Bridges

*Rectifier bridges* are available as commercial packaged components with four *rectifier diodes* connected for full-wave, single-phase rectification. Other versions with six rectifier diodes connected for full-wave, three-phase rectification of AC are also available. As factory-made products, they save the user's time and the cost of assembling discrete diodes, and they also conserve circuit-board space while improving heat dissipation from the diodes.

Low-power bridges are made by bonding glass-passivated or encapsulated rectifier diode dies to metal leadframes. These are then molded in epoxy to form flatpacks or dual-in-line packages (DIPs). Bridges are also made by connecting leaded glass-encapsulated rectifiers in bridge configurations and then potting them with an insulating compound in small copper cases which act as heat sinks. Potting keeps out moisture and contaminants.

Standard bridge rectifiers are rated from 1 to 40 A. Power bridges rated 20 A or higher have fast-connect, solder, or wire-wrap terminals for external connections. Half-bridges, composed of pairs of *Schottky* or *fast-recovery rectifiers* for switching power supplies, are packaged in standard metal and plastic power semiconductor cases. Rectifier bridges are factory tested to U.S. and international specifications and are recognized by the Underwriters Laboratories (UL) or certified by the Canadian Standards Association (CSA).

# Solid-State Relays

A *solid-state relay* (SSR) is an electronic circuit containing a signal-level trigger circuit coupled to a power semiconductor switch, either a transistor or a thyristor. These are factory-made packaged and tested products rather than circuits built from discrete components on a circuit board. Figure 5-7 shows a block diagram for a solid-state relay capable of switching AC. It has three functional sections:

**1.** An optocoupler consisting of a light-emitting diode and a phototransistor
**2.** A zero-voltage detector or trigger circuit
**3.** A solid-state load-switching device (transistor or thyristor)

In the block diagram, a triac serves as the load-switching device. An SSR differs from an electromechanical (EM) relay in both structure and operation, but both provide power gain.

Solid-state relays offer six advantages over EM relays:

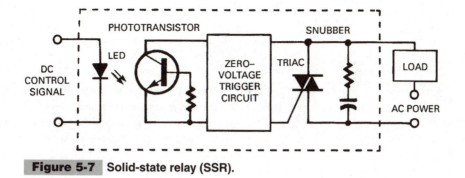

**Figure 5-7** Solid-state relay (SSR).

1. Longer life and higher reliability.
2. Better compatibility with logic-level circuits.
3. Higher switching speed.
4. High resistance to shock and vibration.
5. No contacts that can bounce or chatter to delay response time.
6. No arcing from contact opening that can generate electromagnetic interference (EMI) or pose fire or explosion hazards.

The drawbacks to SSRs are their higher cost than EM relays with equivalent ratings, and the requirement that power be supplied to them.

SSRs suitable for switching AC require either two inverse parallel (back-to-back) *silicon controlled rectifiers* (SCRs) or an electrically equivalent *triac,* as shown in Fig. 5-7. However, *power bipolar* or *MOSFET transistors* can switch DC.

The classification of an SSR is based on its input circuit or method for achieving input/output (I/O) isolation. True SSRs achieve electrical isolation between their input and output circuits with optocouplers, but hybrid SSRs use reed relays or transformers.

Solid-state relays can be classified in five general groups:

1. *AC power relays* capable of switching 24 to 530 VAC at 2 to 75 A with a DC input (typically 3 to 32 VDC) or AC input (typically 90 to 280 VAC) with triacs or two SCRs.
2. *DC power relays* capable of switching 100 to 500 VDC at 7 to 40 A under DC control with power transistors.
3. *AC low-power relays* for circuit-board mounting, capable of switching 60 to 240 VAC at 0.3 to 4 A with triac switches.
4. *DC low-power relays* for PC-board mounting capable of switching up to 60 VDC at 3 A with power transistors.
5. *Input/output modules,* miniature low-power AC and DC relays for circuit-board mounting that provide computer interfacing for industrial sensors and actuators.

## OPTOCOUPLED AC SSRs

An *optocoupled AC SSR* is a solid-state relay that contains an optocoupler. It can be controlled by either AC or DC input signals applied to its terminals. An infrared-emitting diode (IRED) conducts, transmitting an optical signal to a matched photodetector that provides

I/O isolation. The photodetector can be a *phototransistor, photodiode,* or *photocell.* The signal output from the photodetector triggers the output device to switch the load current.

### REED-RELAY COUPLED AC SSRs

A *reed-relay coupled AC SSR* is a hybrid SSR that offers most of the advantages of true SSRs except that input/output isolation is obtained from a reed-switch capsule. The input control signal is applied to the coil that encloses the reed switch, and the induced magnetic field closes the reed contacts. This actuates the circuitry that triggers the solid-state device to switch the load current.

### TRANSFORMER-COUPLED AC SSRs

A *transformer-coupled AC SSR* is a hybrid SSR that offers most of the advantages of true solid-state relays except that input/output isolation is obtained from a transformer. The input control signal is applied to the primary winding of a small transformer, and the output from the secondary winding actuates the circuit that triggers the solid-state device to switch the load current.

# Power Solid-State Relays

An AC *power solid-state relay* is an SSR capable of handling higher currents than typical SSRs. These SSRs have the same organization that is shown in Fig. 5-7. The load is typically switched by two back-to-back SCRs or a triac rated for 2 to 75 A.

The zero-voltage trigger circuit assures that the SCRs or triac will be triggered only when the AC voltage crosses the zero reference (in either the negative or positive direction). This minimizes the effect of surge currents when the load is switched. High currents can result from switching tungsten-filament incandescent lamp and capacitive loads. The cold resistance of a tungsten lamp is less than 10 percent of its illuminated resistance. If the SSR is turned on when the voltage is not at the zero crossing, the high instantaneous load current drawn by the lamp load could destroy the SSR.

The triac or SCRs, once triggered, will not stop conducting until the load current they are conducting falls to zero. A resistor and capacitor in series, called a *snubber,* bypasses voltage transients that occur with inductive loads when the current and voltage are out of phase. Most factory-made general-purpose AC SSRs contain triacs rated to 10 A at 120 to 240 V, but dual SCRs are capable of switching AC power loads in excess of 40 kW.

The principal technical considerations in the specification of AC SSRs are:

- Isolation voltage
- Operating temperature range
- Control-signal range
- Must-operate and must-release voltages
- Input current

Approved factory-made power SSRs carry the labels indicating UL recognition and CSA approval.

The four-terminal flatpack for AC SSRs rated 2 to 40 A has been accepted by industry. This rectangular panel-mounting package measures 2.25 × 1.75 × 0.90 in (57.2 × 44.5 × 22.9 mm). Four screw-type terminals permit the relay to be manually wired in the field.

## DIRECT-CURRENT SSRs

A *direct-current SSR* is an optically coupled unit that typically has a power MOSFET as an output switch. Some DC SSRs include optoisolators consisting of an IRED matched to a *photovoltaic cell* array. Voltage from the array turns on a MOSFET bidirectional output switch, permitting it to control AC and DC of either polarity. DC SSRs are available in a wide range of package styles, including dual-in-line packages (DIPs) and single-in-line packages (SIPs), but there is no package standardization. Some low-power DC SSRs have replaced reed relays for analog communications switching.

# 6

# ANTENNAS AND
# FEED HORNS

# Overview

An *antenna* is a component that can both radiate and receive radio waves. Because of its dual function, it is called a *reciprocal device*. An ideal antenna would radiate all of the power delivered to it by the transmitter (usually through a transmission line) in the desired direction or directions with the desired polarity. Practical antennas cannot achieve this ideal performance, so many different designs have evolved to approach this ideal at different frequencies for different applications.

Despite the antenna's reciprocity property, a convention has been established in the electronics industry that only the antenna's radiation function will be described. It is understood that the antenna can also receive the same kinds of signals that it transmits. That convention is followed in the descriptions of antennas in this section.

Antennas are classified by their application, operating frequency, or both. An antenna for low radio frequencies can be more than a mile (1.6 km) long, but one that has been optimized for use in the microwave band can have a length measured in inches or centimeters. In practice, however, antenna measurements are given in units of their principal transmission wavelengths rather than in standard measurement units. This is done because antennas that differ significantly in size can be described in similar terms. This convention is also followed in the descriptions of antennas in this section.

An antenna can be a single length of wire or an array of conductors that will both radiate and receive radio frequencies. The simplest antenna, or *aerial,* is a length of metal wire. When the output of a transmitter is coupled to an antenna, current flows back and forth along its length. But, because the antenna is not a closed circuit, current flow creates an unequal distribution of electrons. For example, in a simple center-fed, half-wavelength antenna, the current distribution curve is out of phase with the voltage or charge distribution curve. The charge at the ends of the antenna is maximum and the current is zero, while at the center of the antenna the charge is zero and the current is maximum.

Both current and charge buildup along the antenna vary sinusoidally with the input, and they produce fields in space around the antenna. Antenna current produces a magnetic field, while the charge produces an electric field. These fields are 90° out of phase with each other. If the frequency of the field is high enough, parts of both the magnetic and electric fields around the antenna detach themselves and move outward in space. A moving electric field $E$ then creates a magnetic field $H,$ and a moving magnetic field creates an electric field. These fields are in phase with and have a direction perpendicular to the fields that created them. The $E$ and $H$ fields add together vectorially in space to produce the single sinusoidally varying electromagnetic field called a *radio wave.*

The laws governing this radiation are described by *Maxwell's equations.* The field strength of a radio wave is maximum in the immediate vicinity of the antenna and it decreases inversely with distance from the antenna. The radiation pattern or *polar diagram* of an antenna shows how the field strength varies with distance and direction from the antenna. The radiation pattern obtained when an antenna is mounted away from the influence of nearby buildings, trees, hills, or earth is called a *free-space antenna pattern.*

A true *isotropic radiator* or *isotrope* is a theoretical antenna that radiates radio energy omnidirectionally in a perfect spherical pattern about a point source, as shown in Fig. 6-1a. Its radiation pattern is analogous to the inflation of a round balloon around a central point. The isotrope is theoretical because all real antennas have some radiation pattern irregularities, but the concept of perfect omnidirectionality is useful in making radiation measurements.

The *dipole* or *doublet* is the simplest wire radiator or antenna. If it is in free space it radiates a toroidal (doughnut-shaped) pattern, as shown in Fig. 6-1b. The height of the antenna above ground, the conductivity of the earth below it, and the shape and dimensions of the antenna all affect the radiated field pattern in space. In most applications, antenna radiation is directed between specified angles in both the horizontal and vertical planes.

Two important classical antenna designs are the Hertz antenna and the Marconi antenna. The *Hertz antenna,* shown in Fig. 6-2, is a simple dipole with a length equal to one-half of

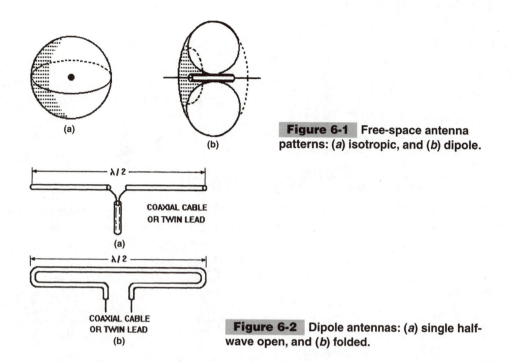

**Figure 6-1** Free-space antenna patterns: (*a*) isotropic, and (*b*) dipole.

**Figure 6-2** Dipole antennas: (*a*) single half-wave open, and (*b*) folded.

the wavelength of its operating frequency or any even or odd multiple of that value. This antenna can be installed above ground and mounted either vertically or horizontally. It does not need an electrical connection to ground. The *Marconi antenna,* shown in Fig. 6-3, is a grounded antenna one-quarter wavelength long that operates as a half-wavelength antenna. A transmitter can be connected between the bottom of the antenna and the ground. The ground, acting as a kind of mirror, provides a reflection of the current and voltage distribution established by the antenna. The wave emitted from the antenna-ground combination is the same as that emitted by a Hertz antenna operated at the same frequency.

# Antenna Power Transfer

To connect a transmission line to an antenna, a small gap is made in the antenna conductor, and the two wires of the line are connected to the terminals of the gap—the antenna input terminals. At this point of connection the antenna presents a load impedance to the transmission line. This impedance is also the *input impedance* of the antenna. If it is equal to the characteristic impedance of the line there will be no standing waves on the line, and the maximum amount of power can be transferred from the line to the antenna. Antenna input impedance $Z$ determines antenna current at the feed point for any given RF voltage.

In a half-wave antenna, the current $I$ is maximum at the center and zero at the ends. By contrast, the voltage $E$ is maximum at the ends and minimum at the center. By applying the equation $Z = E/I$, it can be seen that impedance varies along the length of the antenna and is, like the voltage, maximum at the ends and minimum at the center. Thus, if energy is fed

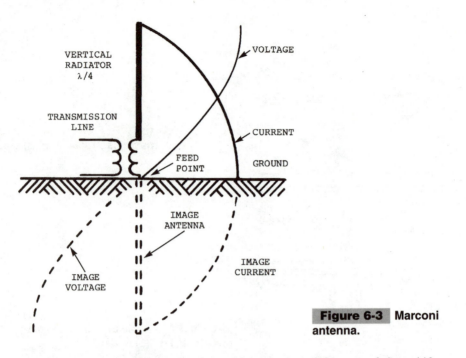

**Figure 6-3** Marconi antenna.

to a half-wave antenna at its center, it is said to be *center fed* (current fed), and if energy is fed at the ends it is said to be *end fed* (voltage fed). If a half-wave antenna is isolated in free space, the impedance is approximately 73 ohms at the center and 2500 ohms (allowing for losses) at the ends. The intermediate points have intermediate values of impedance.

The antenna at the end of the transmission line is equivalent to a resistance that absorbs some of the energy from the generator. Neglecting the losses that occur in the antenna, this is the energy that is radiated into space. The value of resistance that dissipates the same amount of power as the antenna is called *radiation resistance.*

# Antenna Polarization

The position of a simple antenna in space determines the *polarization* of the emitted wave. Polarization defines the orientation of the electric field component of the wave with respect to ground. Antennas can be polarized either horizontally or vertically. For low frequencies, the polarization is not disturbed and the radiation field has the same polarization at the distant receiving station that it had at the transmitting antenna. At high frequencies, however, polarization usually varies, sometimes quite rapidly, because the wave splits into several components which follow different paths. These paths will not be the same length, so on recombination the components generally will not be parallel. Such a radiated field is called either *circularly* or *elliptically polarized.*

The variation of signal strength around an antenna can be illustrated by a *polar diagram,* a circular plan-position graph divided into 360° coordinates, with the antenna assumed to be at the center. Computed or measured values of field strength can be plotted radially to show both magnitude and direction for a given distance from the antenna. Field strengths in the vertical plane are plotted on a semicircular polar chart called a *vertical polar diagram.*

# Beamwidth and Bandwidth

When the radiated power of an antenna is concentrated into a single major lobe, the angular width of this lobe is called the *beamwidth.* However, if the antenna has a pattern consisting of many lobes, this term does not apply. The *bandwidth* of an antenna is the range of frequency over which it will operate satisfactorily. Some antennas must operate only at one or two fixed frequencies with a signal that is narrow in bandwidth, but others that must be capable of scanning a wide frequency range have greater bandwidth. The factors that determine bandwidth are input impedance, radiation efficiency, power gain, beamwidth, beam direction, polarization, and side-lobe level. Antenna designs can be selected for wide inherent bandwidth.

# Electrically Small Antennas

Antennas whose dimensions are short compared with their operating wavelengths are called *electrically small antennas.* They typically exhibit low radiation resistance and high reactance, resulting in high $Q$ and narrow bandwidth. An example is an *end-fed monopole antenna* commonly used at low frequencies for long-range communication, commercial broadcasting, and mobile use. Where antenna height is a limiting factor at higher frequencies, the monopole antenna's height can be reduced by winding the conductor into a helix. Small *loop antennas* are widely used for direction finding and navigation.

Where there are restrictions on height, *slot antennas* can also be used. A slot in a conductive surface behaves like a conductive wire in space. The electric field radiated from the slot is the same as the magnetic field radiated from a wire with the same dimensions. The outside surface of a waveguide conducting radio energy is an example of a conductive surface, and the slot typically is narrow and a half wavelength long.

Both dipoles and monopoles are resonant antennas that exhibit approximate sinusoidal current distribution and pure resistance at their input terminals. However, where the ratio of diameter to length is small in these antennas, the input impedance varies widely and makes them unsuitable as broadband antennas. This limitation can be overcome by increasing the diameter of the antenna with fanned wires, triangular flat sheet-metal planes, and cones. Examples of these are the *biconical antenna* and the *bow-tie antenna.*

# Directional Antennas

Antennas can be made more directional by concentrating their transmitted energy. Directional antennas typically have many separate elements that function in unison to provide improved directivity. Multielement antennas are called *arrays*. Their characteristics are determined by the number and types of elements. Three widely used array elements are: (1) *dipoles,* (2) *reflectors,* and (3) *directors*. A dipole in an array is usually a half-wave antenna fed at its center. Reflectors and directors are *parasitic elements* that direct the dipole's radiation pattern. An antenna that includes them is called a *parasitic array,* and an example is the *Yagi-Uda antenna.*

*Driven arrays,* used for high-power applications, consist of two or more elements, usually half-wave dipoles. Each element is driven by the output of the transmitter. There are three basic types of driven arrays: (1) *broadside,* (2) *end-fire,* and (3) *collinear*. Other array antennas designed for maximum radiation or sensitivity in one direction include the *log-periodic antenna, multielement antenna, planar array,* and the *phased-array antenna.*

Dielectric materials such as plastics in solid rod or cylindrical form can be waveguides and antennas. The wavelength of a signal transmitted inside a dielectric rod is less than its free-space wavelength. If the rod diameter is large compared to the wavelength of the transmitted signal, most of the energy travels inside the dielectric. But if the diameter is reduced below a half wavelength in a gradual taper, the wave continues beyond the end of the rod and is propagated into free space with most of the energy moving in the direction of the rod. This is the principle behind the *dielectric-rod antenna.*

# Direct-Aperture Antennas

*Direct-aperture antennas* at high frequencies can be configured as horns, mirrors, or lenses. These antennas can make use of variously shaped conductive metal surfaces and solid dielectrics for radiation. This contrasts with low-frequency antennas that depend largely upon conductive wires or rods. Examples of direct-aperture antennas are the *conical* and *pyramidal horns* that produce high-gain beams over broad frequency bands.

The *Luneberg lens* is a dielectric sphere that can be an antenna because of its ability to focus radio energy for increased gain. Its index of refraction varies with the distance from the center of the sphere, so energy can be fed into the lens at the focal point for transmission and be removed from that point for receiving. The pencil beam formed by the lens can be steered by changing the position of the feed point.

*Three-dimensional reflectors* improve gain, modify patterns, and eliminate backward radiation at higher frequencies. Primary apertures such as *low-gain dipoles, slots,* or *horns* radiate toward larger reflectors called *secondary apertures*. The large reflector further shapes the radiated wave to produce the desired pattern. Examples include *plane sheet reflectors, corner reflectors,* and *parabolic reflectors.*

The *conical horn reflector antenna* is partly shaped as a paraboloid and was designed to eliminate the interference of the feed horn in the path of the reflected wave. These reflectors are used in point-to-point microwave systems and satellite communication ground stations because of their broadband and very low noise characteristics.

Directional RF beams can be formed in limited spaces with two-reflector systems. The RF energy emitted from a horn in the center of the main reflector is directed at a mirror that reflects it back to the main reflector, which forms it into a beam. The *Cassegrain antenna* is the most common two-reflector antenna, but the *Gregorian antenna* is similar in concept.

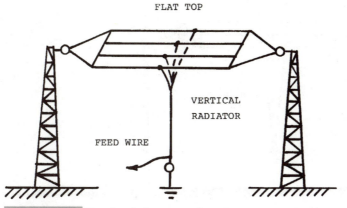

**Figure 6-4** Top-loaded monopole antenna.

# Basic Wire Antennas and Feed Methods

## TOP-LOADED SHORT-MONOPOLE ANTENNAS

A *top-loaded short-monopole antenna,* shown in Fig. 6-4, is a vertical antenna that is wider at the top to modify current distribution and improve the radiation pattern in the vertical plane.

## INVERTED L ANTENNAS

An *inverted L antenna,* shown in Fig. 6-5, is an antenna that consists of one or more long horizontal wires with vertical lead-in wires connected to one end.

## T ANTENNAS

A *T antenna,* shown in Fig. 6-6, has one or more horizontal wires, and its lead-in connections are made at the approximate center of each wire.

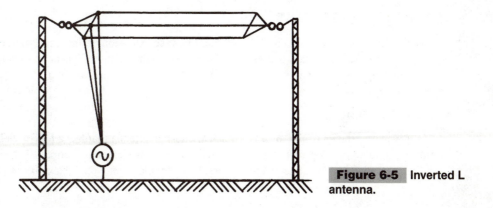

**Figure 6-5** Inverted L antenna.

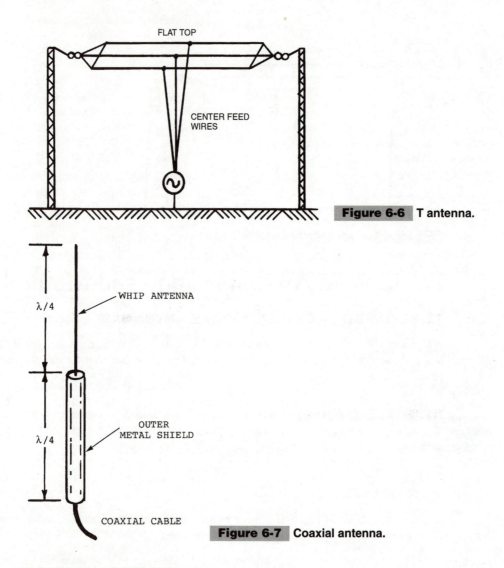

**Figure 6-6**  T antenna.

**Figure 6-7**  Coaxial antenna.

## COAXIAL ANTENNAS

A *coaxial antenna,* shown in Fig. 6-7, is a quarter-wave extension of the inner conductor of a quarter-wavelength section of coaxial cable whose outer conductor has been folded back the same length to form a radiating sleeve. It is also called a *sleeve antenna.*

## J ANTENNAS

The *J antenna,* shown in Fig. 6-8, is a vertical end-fed antenna that has a half-wavelength radiator mounted on top of one of two parallel quarter-wavelength sections. The second quarter-wave section is connected to the first by a common conductive base. It can radiate

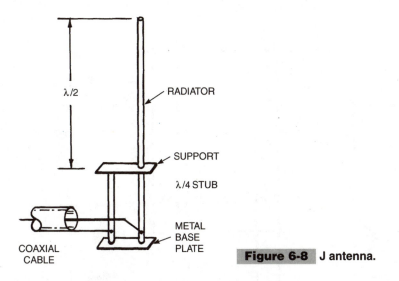

$\lambda/2$

RADIATOR

SUPPORT

$\lambda/4$ STUB

METAL
BASE
PLATE

COAXIAL
CABLE

**Figure 6-8**  J antenna.

a horizontal omnidirectional pattern, and it does not require a ground connection. This antenna is most efficient in the VHF and UHF frequencies above 7 MHz.

## LOOP ANTENNAS

The *loop antenna,* shown in Fig. 6-9, consists of one or more complete turns of a conductor in a loop. It is usually tuned to resonance by a variable capacitor connected to the terminals of the loop. Its radiation pattern is bidirectional, with maximum radiation or pickup in the plane of the loop and minimum radiation at right angles to the loop.

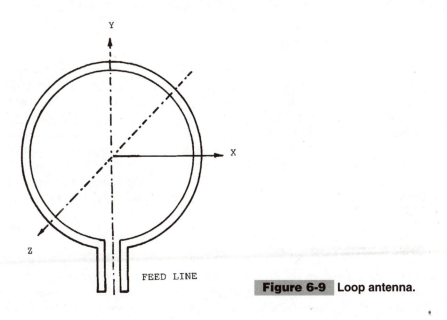

Y

X

Z

FEED LINE

**Figure 6-9**  Loop antenna.

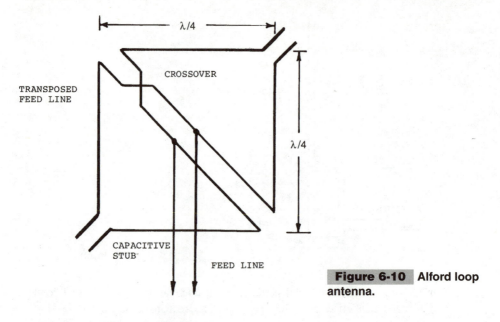

**Figure 6-10** Alford loop antenna.

## ALFORD LOOP ANTENNAS

An *Alford loop antenna,* shown in Fig. 6-10, is a square one-turn loop antenna with quarter-wavelength sides. It is fed with opposite phases at opposite corners, and the other two corners are capacitively connected. The out-of-phase feed is achieved by transposing one branch of the feed line. Open-end sections (*stubs*) of the transmission line act as capacitors. Its radiation pattern is uniform in the plane of the loop. It is much more efficient, especially at the higher frequencies, than the conventional loop antenna.

## HELICAL ANTENNAS

A *helical antenna,* shown in Fig. 6-11, is a radiator wound in the shape of a helix and connected at right angles but insulated from a flat metal sheet or grid reflecting disk. The helix is usually fed at one end by the center conductor of a coaxial transmission line whose outer conductor is connected to the ground plane. This antenna produces a circularly polarized wave that can rotate either clockwise or counterclockwise. Because these antennas are usually used at high frequencies, their dimensions are large with respect to wavelength. The disk diameter is sized to be equal to the lowest operating wavelength, and the helix is formed to a radius with dimensions about 15 percent of the center frequency wavelength with turn spacing about one-quarter wavelength long. Used in satellite communications, it is also called a *helix antenna.*

## HORN RADIATORS

A *Horn radiator,* shown in Fig. 6-12, is a microwave antenna made by flaring out the ends of a circular or rectangular waveguide to radiate radio waves directly into free space. One or both of the transverse dimensions of a rectangular horn antenna increase linearly from the small

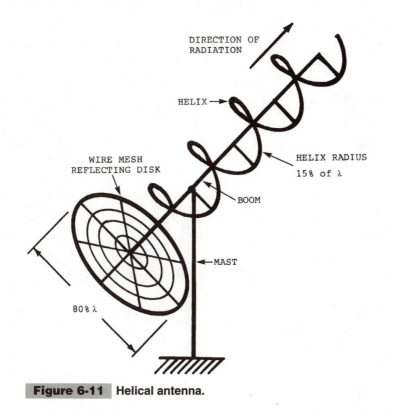

**Figure 6-11**  Helical antenna.

end or throat to the mouth. It is also called a *horn antenna.* The horns in Figs. 6-12*a* to *d* are an H-*plane horn,* a *pyramidal horn,* an E-*plane horn,* and a *conical horn,* respectively.

## BICONICAL HORNS

A *biconical horn,* shown in Fig. 6-13, is a radiator similar to the *biconical radiator,* but it is considered to be a *horn* if its vertex angle exceeds 90°.

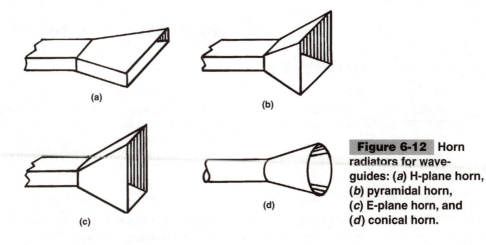

**Figure 6-12**  Horn radiators for wave-guides: (*a*) H-plane horn, (*b*) pyramidal horn, (*c*) E-plane horn, and (*d*) conical horn.

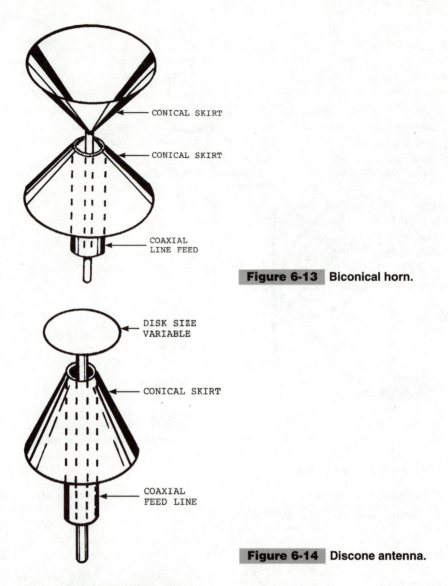

CONICAL SKIRT

CONICAL SKIRT

COAXIAL
LINE FEED

**Figure 6-13** Biconical horn.

DISK SIZE
VARIABLE

CONICAL SKIRT

COAXIAL
FEED LINE

**Figure 6-14** Discone antenna.

## DISCONE ANTENNAS

A *discone antenna,* shown in Fig. 6-14, is a biconical antenna in which one cone has been flattened to form a disk. The center conductor of a coaxial cable terminates at the center of the disk, and the cable shield terminates at the vertex of the cone. Both the input impedance and radiation pattern remain essentially constant over a wide frequency range. The disk is mounted horizontally so that its omnidirectional radiation pattern is in a horizontal plane.

## SLOT RADIATORS

A *slot radiator* is an antenna formed by cutting a narrow slot in a large flat sheet of metal connected to a source of RF power or a waveguide conducting RF power so that RF energy

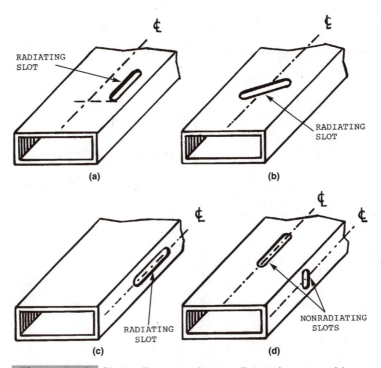

**Figure 6-15** Slot radiators and nonradiators in waveguides: (*a*) radiating slot parallel to the wide-face centerline, (*b*) radiating slot diagonal to the wide-face centerline, (*c*) radiating slot parallel to the narrow-face centerline, and (*d*) nonradiating slots.

will be radiated. The slot behaves like a dipole of the same length in space except that the electric field radiated from the slot is equivalent to a magnetic field from a wire dipole of similar dimensions and the magnetic field is equivalent to the electric field. If a two-wire transmission line is connected to the slotted sheet metal, both sides of the slot will radiate, but if radiation is wanted on only one side, the "back" side of the slot must be enclosed.

A unidirectionally radiating slot in a waveguide can be formed by cutting it in a specific position and orientation in the waveguide wall. The slot must be narrow and about a half wavelength long. Three slot radiators and two nonradiating slots in a $TE_{10}$-mode waveguide are shown in Fig. 6-15. Radiating slots must be positioned so they interrupt currents that would otherwise flow across the inner walls of the guide. The most effective slots will be resonant if they are about a half wavelength long. Only a fraction of the power flowing in the waveguide is radiated.

Figure 6-15*a* shows a radiating slot cut parallel to the centerline on the wide face of the waveguide, Fig. 6-15*b* shows a diagonal radiating slot cut at the centerline of the wide face, and Fig. 6-15*c* shows a radiating slot cut parallel to the centerline of the narrow side of the waveguide. However, the slots cut in the waveguide shown in Fig. 6-15*d* are nonradiating.

## DELTA-MATCHED ANTENNAS

A *delta-matched antenna,* shown in Fig. 6-16, is a single-wire antenna, usually one-half wavelength long, that is connected to an open-wire transmission line with leads formed in

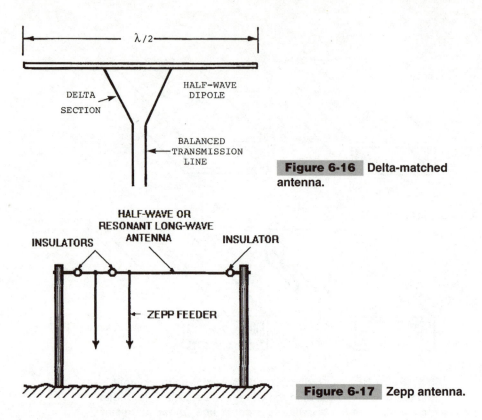

**Figure 6-16** Delta-matched antenna.

**Figure 6-17** Zepp antenna.

the shape of a Y. The flared parts of the Y match the transmission line to the antenna. Because the top of the Y is not cut, the matching section has the triangular shape of the Greek letter Δ (delta). It is also called a *Y antenna*.

## ZEPP ANTENNAS

A *Zepp antenna,* shown in Fig. 6-17, is a horizontal wire antenna whose length is some multiple of a half wavelength. It is fed at one end by one lead of a two-wire transmission line, which is also a multiple of a half wavelength long.

## CAGE ANTENNAS

A *cage antenna,* shown in Fig. 6-18, is a variation of the open *dipole antenna.* It has many active elements fastened horizontally and in parallel to circular conductive rings to form two cylindrical cagelike structures that function like single-wire dipoles. The antenna is fed at its center by a two-wire transmission line or coaxial cable.

# Antenna Arrays

*Parasitic arrays* are made up as *free* (nondriven) elements that absorb energy from the driven element and reradiate it so that gain is increased in one direction and greatly

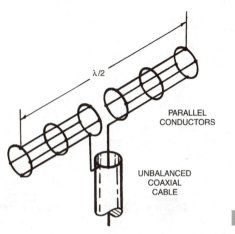

**Figure 6-18** Cage antenna.

reduced in the other direction. Short parasitic elements that are placed parallel to the driven element and in the direction of maximum radiation are called *directors.* Longer parasitic elements that are placed parallel to the driven antenna and in the direction of minimum radiation are called *reflectors. Driven arrays* have lower losses than parasitic arrays, and they can radiate higher power while retaining the narrow-beam characteristics of parasitic arrays.

If a number of driven half-wave antennas are positioned with respect to each other so that energy arriving at a distant point from the individual elements will add in certain directions and cancel in other directions, the antenna system is directional. By properly phasing the energy fed to the antenna and by properly locating the elements in space, the energy will add in the desired direction and be out of phase or opposed in the undesired direction.

Driven arrays are made from a number of half-wave dipoles positioned in space and phased so that a desired directional pattern is obtained. The three basic types are: (1) the *colinear array,* (2) the *broadside array,* and (3) the *end-fire array.*

## FRANKLIN ANTENNAS

A *Franklin antenna,* shown in Fig. 6-19, is a collinear dipole array of six half-wave elements. It is fed between the ends of the pair of dipoles, causing them to be in phase. Quarter-wave stubs are connected between the ends of the other adjacent pairs of dipoles to provide both a conductive feed connection and proper phasing.

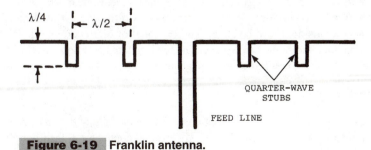

**Figure 6-19** Franklin antenna.

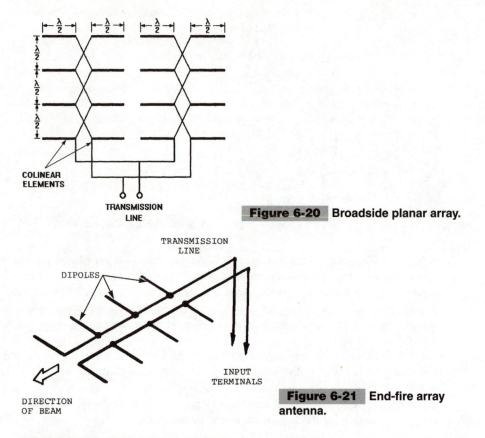

**Figure 6-20**  Broadside planar array.

**Figure 6-21**  End-fire array antenna.

## BROADSIDE PLANAR ARRAYS

A *broadside planar array,* as shown in Fig. 6-20, is an array of half-wave sections arranged to concentrate energy in a narrow beam. The maximum radiation direction is perpendicular to the plane of the array. The dipoles are usually stacked at half-wavelength distances. The array is also called a *billboard array* if it is backed up by a sheet-metal or screen reflector. Very high power gains can be obtained with broadside arrays, depending on the spacing, number, and tuning of its elements.

## END-FIRE ARRAY ANTENNAS

The *end-fire array antenna,* shown in Fig. 6-21, is a four-element linear array whose maximum direction of radiation is along the axis of the array—off the end—as indicated by the arrow, rather than broadside. It can be either unidirectional or bidirectional. The elements of the array are parallel and in the same plane. The feed system must be a matched impedance system to produce the required element phasing.

## HARP ANTENNAS

A *harp antenna,* shown in Fig. 6-22, a variation of the *ground-plane antenna,* consists of many vertical radiators which are connected to a horizontal conductor at their bottom ends. The longest radiator is one-quarter wavelength long and the others are incrementally

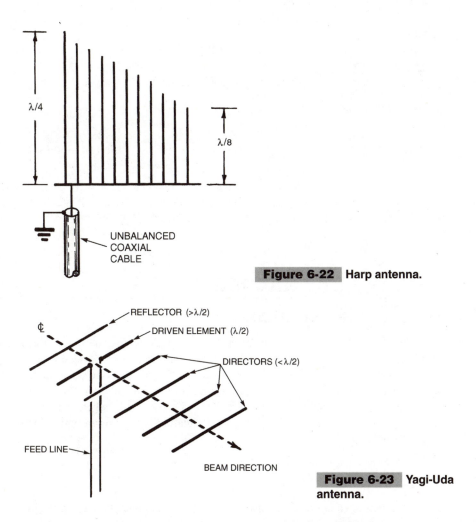

**Figure 6-22** Harp antenna.

**Figure 6-23** Yagi-Uda antenna.

smaller, down to one-eighth wavelength for the shortest radiator. This antenna's range is determined by the longest and shortest vertical radiators, and its pattern is omnidirectional.

## YAGI-UDA ANTENNAS

A *Yagi-Uda antenna,* shown in Fig. 6-23, is a *parasitic end-fire array* that usually has one driven half-wavelength dipole with a parasitic reflector on one side of the driven dipole and three or more parasitic directors on the other side of it. All of the elements are mounted in parallel in the same plane, about a quarter wavelength apart. The *reflector* is longer than a half wavelength, and the *directors* are shorter than a half wavelength. This antenna provides a unidirectional beam of moderate directivity with a simple feed. Useful up to 2.5 GHz, it is also called a *Yagi antenna,* a *Yagi array,* or just a *Yagi.*

## DIELECTRIC-ROD ANTENNAS

A *dielectric-rod antenna,* shown in Fig. 6-24, is an antenna whose end-fire radiation pattern is produced by the propagation of a wave from a cavity down the length of a tapered

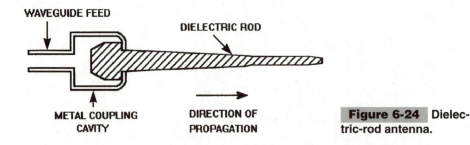

**Figure 6-24** Dielectric-rod antenna.

dielectric rod. If the rod diameter is less than a half wavelength, the wave continues beyond the end of the rod into free space. If the rod is made of polystyrene, it is called a *polyrod,* but if it is made of a ferrite material, it is called a *ferrod.*

# Long-Wire Antenna Arrays

### SINGLE-WIRE ANTENNAS

A *single-wire antenna* is usually one or more wavelengths long, and it is usually untuned or nonresonant. The radiation pattern of a long conductor in free space depends on its length in the wavelength units of its operating frequency. Single wires produce multilobed patterns with the number of lobes equal to the number of half wavelengths of the conductor. A long, single-wire antenna that is terminated with a resistor will radiate a major lobe in the direction of wave travel down the line, but a radiation pattern of an unterminated line is more symmetrical.

### V ANTENNAS

A *V antenna,* shown in Fig. 6-25, is a horizontal bidirectional antenna formed by positioning two long nonresonant antenna wires in a V formation on three masts and feeding them 180° out of phase at the apex. The angle between the V array is determined by the length of the conductors and the conditions necessary for in-phase major lobe addition. If each long wire is terminated with a matching resistor (nonresonant antenna), the antenna is unidirectional. But if the ends of the long wire are open (resonant antenna), the antenna can transmit and receive over the 7- to 300-MHz range.

### RHOMBIC ANTENNAS

A *rhombic antenna,* shown in Fig. 6-26, consists of four long-wire antennas strung between four masts in a diamond-shaped pattern with the signal feed introduced at one apex. If the apex opposite the feed is open (resonant antenna), it has a bidirectional response between the two apexes. But if the open end is terminated with a matching resistor (nonresonant antenna), it is unidirectional toward the terminated apex. Lobes of the four legs add in phase to form a major lobe. This antenna has a gain of 20 to 40 times that of a dipole.

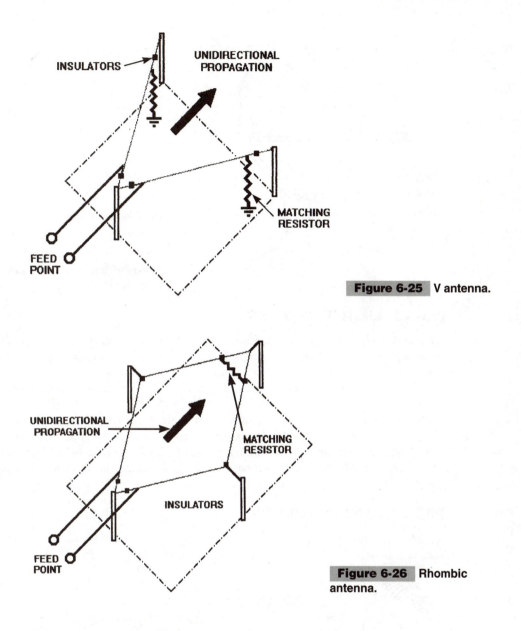

INSULATORS

UNIDIRECTIONAL
PROPAGATION

MATCHING
RESISTOR

FEED
POINT

**Figure 6-25**  V antenna.

UNIDIRECTIONAL
PROPAGATION

MATCHING
RESISTOR

INSULATORS

FEED
POINT

**Figure 6-26**  Rhombic antenna.

# Antenna Reflectors and Lenses

## PARABOLIC ANTENNAS

A *parabolic antenna,* shown in Fig. 6-27, includes a *parabolic reflector* to concentrate the transmitted or received RF energy into a parallel beam. This antenna is commonly used for terrestrial microwave communications, for satellite uplinks and downlinks, and for receiving TV broadcasts from satellites. It is also called a *dish antenna.*

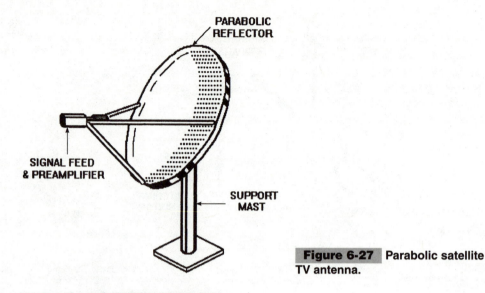

**Figure 6-27** Parabolic satellite TV antenna.

## PARABOLIC REFLECTORS

A *parabolic reflector* is a dish-shaped antenna reflector whose inner surface has a parabolic shape (a parabola rotated around its axis). When a microwave transmitting dipole, horn, or other antenna is placed at the reflector's focal point, its radiation is concentrated into a parallel beam. (This cannot be done with a spherical reflector.) Incoming signals are reflected to the reflector's focal point.

Three different parabolic reflectors are shown in Fig. 6-28. The *full parabolic reflector* (Fig. 6-28*a*) forms a round pencil beam, the *cut parabolic reflector* (Fig. 6-28*b*) forms a vertical fan-shaped beam, and the *parabolic cylindrical reflector* (Fig. 6-28*c*) forms a horizontal fan-shaped beam. These antennas can be formed from sheet or punched metal, or metal screening.

## PILLBOX ANTENNAS

A *pillbox antenna,* shown in Fig. 6-29, is a microwave antenna that produces a linear radiation pattern. It is a cylindrical parabolic reflector enclosed by two parallel plates perpen-

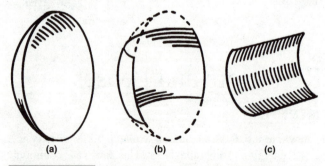

(a)              (b)              (c)

**Figure 6-28** Parabolic reflectors: (*a*) full paraboloid, (*b*) cut paraboloid, and (*c*) parabolic cylinder.

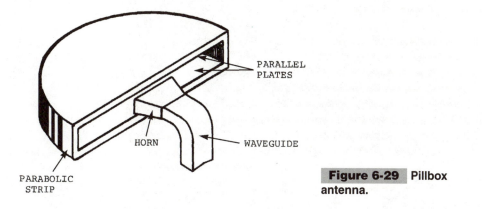

**Figure 6-29** Pillbox antenna.

dicular to the reflector spaced to permit the propagation of only one mode in the desired direction of polarization. RF energy is fed from a horn on its focal line.

## CASSEGRAIN ANTENNAS

A *Cassegrain antenna,* shown in Fig. 6-30, is a microwave antenna with a feed radiator mounted at or near the surface of its main paraboloidal reflector. The radiator is aimed at the convex hyperboloidal subreflector with its focal point colocated with the focal point of the paraboloid. Energy from the feed horn first illuminates the subreflector, then spreads outward to illuminate the main reflector. This design eliminates the need for mounting a heavy feed radiator out in front of the reflector on struts or a support beam.

## CORNER-REFLECTING ANTENNAS

A *corner-reflecting antenna,* shown in Fig. 6-31, has a reflector with two conducting planes that intersect at right angles (*square corner*). It is fed by a dipole or collinear dipole array located on the bisector of the angle between the reflecting planes. The planes can be made of sheet metal or wire mesh. Maximum pickup is obtained along the bisector.

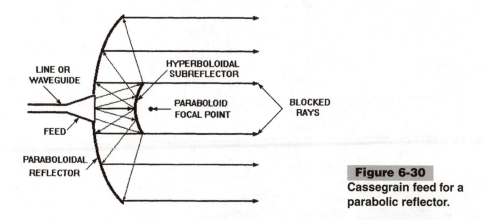

**Figure 6-30**
Cassegrain feed for a parabolic reflector.

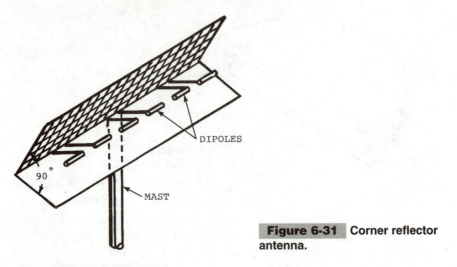

**Figure 6-31** Corner reflector antenna.

## LUNEBERG LENSES

A *Luneberg lens,* shown in Fig. 6-32, is a dielectric sphere that is classified as an antenna because it can focus high-frequency radio energy for increased gain. The spherical lens has a nonuniform internal distribution of *dielectric constant* and an *index of refraction* that varies with distance from the center of the sphere. Energy can either be fed into the lens at the focal point for transmission or be removed from it for receiving. The pencil beam formed by the lens can be steered by changing the position of the feed point.

# Broadband Antennas

## BICONICAL ANTENNAS

A *biconical antenna,* shown in Fig. 6-33, has two metal cones aligned on a common axis with their vertices coinciding and a coaxial cable or waveguide feed coupled to the vertices.

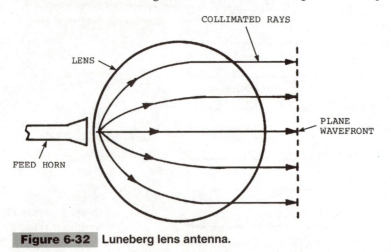

**Figure 6-32** Luneberg lens antenna.

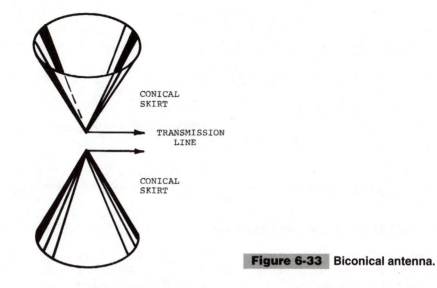

**Figure 6-33** Biconical antenna.

Its radiation pattern is circular in a plane perpendicular to the axis. This antenna is considered to be a dipole if its vertex angle is less than 90°. It is similar to the *biconical horn.*

## MONOPOLE ANTENNAS

A *monopole antenna,* shown in Fig. 6-34, is mounted on an imaging ground plane so that it produces a radiation pattern which approximates that of a dipole.

## FOLDED-DIPOLE ANTENNAS

A *folded-dipole antenna,* shown in Fig 6-35, is a pair of half-wave dipoles in shunt with their ends connected to increase its center impedance to 300 ohms so it will match a 300-ohm two-

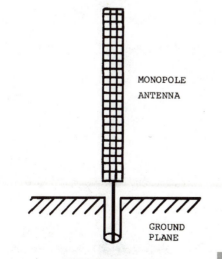

**Figure 6-34** Monopole antenna.

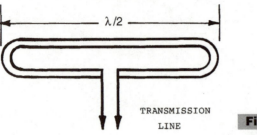

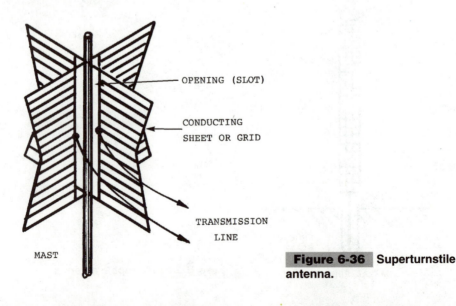

Figure 6-35    Folded-dipole antenna.

wire transmission line without a matching device called a *balun.* This antenna has a wider bandwidth than a single half-wave open dipole, which has an impedance of about 73 ohms.

## SUPERTURNSTILE ANTENNAS

A *superturnstile antenna,* shown in Fig. 6-36, is a modified *turnstile antenna* that has wing-shaped dipole elements in pairs mounted at right angles about a common vertical axis, typically a supporting mast. The dipole pairs are fed in quadrature to provide substantially omnidirectional radiation over a wide band for FM and television transmitters.

## BROADBAND DIPOLES

*Broadband dipoles,* as shown in Fig. 6-37, are made specifically for broadband transmission and reception. A standard dipole is shown in Fig. 6-37a. The *bow-tie dipole,* shown in Fig. 6-37b, is made either of stiff wire or flat sheet-metal triangles positioned vertically. The *fan dipole,* shown in Fig. 6-37c, can be stamped from flat sheet metal. Each vertex is the connecting point to the transmission line, twin leads, or coaxial cable. The bow-tie and fan dipoles are suitable for UHF television reception. Signal pickup is improved if a metal wire-mesh screen is positioned behind each antenna.

Figure 6-36    Superturnstile antenna.

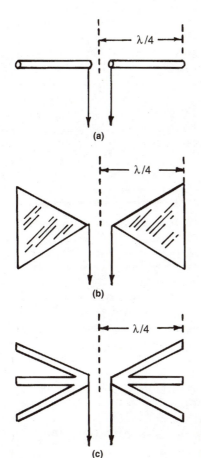

**Figure 6-37** Broadband dipole antenna styles: (*a*) conventional, (*b*) triangular or bowtie, and (*c*) fan.

# Special Antennas

## LOG-PERIODIC DIPOLE ARRAYS

A *log-periodic dipole array,* shown in Fig. 6-38, is a broadband antenna array whose dipole length and spacing increase with their distance from the transmission line source. The conductors are transposed between the adjacent dipole elements. Its radiation pattern is unidirectional in the backfire direction, toward the source. This antenna is one of many forms of the log-periodic array.

## TURNSTILE ANTENNAS

A *turnstile antenna,* shown in Fig. 6-39, consists of one or more layers of crossed horizontal dipoles on a mast. They are usually energized so that the currents in the two dipoles of a pair are equal and in quadrature. This antenna is capable of producing an essentially omnidirectional, horizontally polarized radiation pattern for TV, FM, and other UHF or VHF transmission. See also "Superturnstile Antennas," preceding.

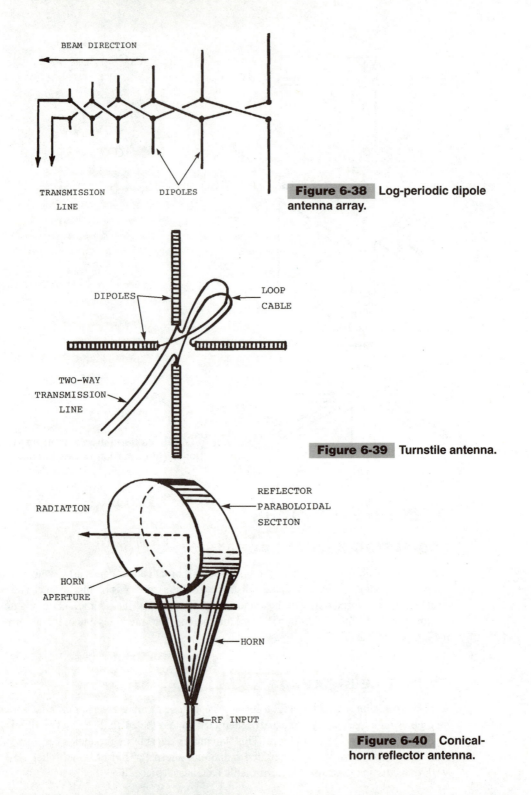

BEAM DIRECTION

TRANSMISSION
LINE

DIPOLES

**Figure 6-38** Log-periodic dipole antenna array.

DIPOLES

LOOP
CABLE

TWO-WAY
TRANSMISSION
LINE

**Figure 6-39** Turnstile antenna.

RADIATION

REFLECTOR
PARABOLOIDAL
SECTION

HORN
APERTURE

HORN

RF INPUT

**Figure 6-40** Conical-horn reflector antenna.

## CONICAL-HORN REFLECTOR ANTENNAS

A *conical-horn reflector antenna,* shown in Fig. 6-40, has a horn that is joined to a paraboloidal reflector with a circular aperture which can redirect radio energy 90°. The apex of the horn is at the focal point of the paraboloid. It can transmit and receive microwave signals at 4 to 6 GHz. The antenna's characteristics of low noise, wide bandwidth, high gain, and narrow beam formation make it useful in terrestrial microwave point-to-point relay links.

# MICROWAVE AND UHF TECHNOLOGY

## Overview

The ultra-high-frequency (UHF) band of the electromagnetic spectrum extends from 300 to 1000 MHz (1 GHz), and over that range wavelength decreases from 100 to 30 cm. The microwave band is considered to extend from the upper limit of the UHF band (1 GHz) to the lower limit of the millimeter region at 40 GHz. The wavelengths of all of the frequencies within this band can be measured in centimeters or millimeters. For example, the wavelength at 1 GHz is 30 cm and this decreases to 0.75 cm (7.5 mm) at 40 GHz. The millimeter band begins at 40 GHz and, for practical applications, ends at about 100 GHz. Table 7-1 relates frequency to wavelength and band designation.

The behavior of the radio frequencies at the high end of the UHF band and across the microwave and millimeter bands differs from that of the very high frequency (VHF) band and lower frequencies because different methods are required for generating RF power and transmitting it through conductors or free space. The first practical defensive radar systems developed during World War II operated in the UHF band, and long-range search radars still operate in this band.

Some experimental work was done in the microwave band before the turn of the twentieth century, but it was dropped in favor of lower frequencies because they offered more promise for commercial radio broadcasting and reception. Serious research into the characteristics of microwaves and the technology necessary to produce them with enough power for practical military radar applications began during World War II.

The early UHF radar systems were successful in detecting aircraft movements at distances of over 20 miles, but they provided very poor target definition and resolution. This made it difficult to discriminate between individual targets and a group of aircraft. It was apparent that higher-frequency microwaves would provide better target resolution and also be more useful in aiming antiaircraft guns.

Research in the early war years of the 1940s led to the development of a practical, reliable magnetron oscillator capable of producing enough microwave power to be useful in short-range search radars and fire-control radars for surface and antiaircraft artillery. Specialized vacuum tubes such as the lighthouse tube and klystron were put to use generating the intermediate frequencies necessary to receive and detect radar returns.

In the postwar years of 1945 through the 1950s, many new airborne, shipboard, and ground-based microwave radar systems were developed for applications ranging from target detection and tracking to fire control, navigation, and electronic countermeasures. During the Cold War years from the 1950s to the 1960s, many other types of microwave oscillators and amplifiers were developed. These included the forward- and backward-wave traveling-wave tubes (carcinotrons and TWTs), traveling-wave tube amplifiers

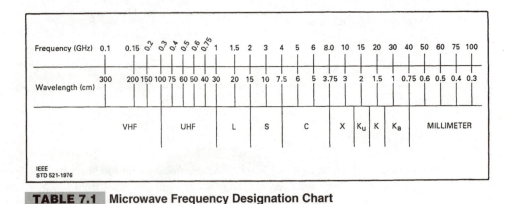

**TABLE 7.1**  Microwave Frequency Designation Chart

(TWTAs), and high-power multicavity klystrons. Later, solid-state microwave sources were introduced. TWTAs and solid-state input chains are now widely used in *phased-array radar* systems, and TWTs and power klystrons have been adapted for terrestrial and satellite telecommunications.

Microwave energy has also been put to work in industry for concentrated localized heating and accelerated drying of materials, while the microwave oven, powered by a continuous-wave magnetron, has become a standard appliance for cooking and reheating food in homes and restaurants.

# Coaxial Cable Transmission

Coaxial cable is widely used for the transmission of RF energy in the UHF and microwave bands. Fundamentally two-wire cable, it has an inner coaxial conductor wire enclosed by a cylindrical metallic sheath, typically made from woven copper wire braid. The conductors are separated by a dielectric sleeve, typically extruded from plastic resin.

The selection of coaxial cable for a specific signal transmission task depends on the electrical characteristics of the cable. These can be altered at the factory by making materials and dimensional changes. One of the most important coaxial cable characteristics is *characteristic impedance $Z_o$*, defined as the total opposition to the flow of current in the cable. It can also be defined in terms of the ratio of the sizes between inner and outer conductors and its inverse relation to the dielectric constant of the core material.

Maximum power can be transferred in a coaxial cable only when the characteristic impedances of transmitter, radio-frequency line, and receiver (or antenna) are equal. If the match is exact, losses are due only to resistance in the line (attenuation). If there is a mismatch, there will be reflection losses. The characteristic impedance of a coaxial cable is measured in ohms, but unlike conductor resistance, it does not vary with length. Coaxial cables generally are designed to match 50-, 75-, or 5-ohm impedances.

*Capacitance* in coaxial cable is the result of the mechanical construction of the cable and the properties of its conductors and dielectrics that permit electrical energy to be stored when the conductors are at different potentials. Capacitance, like impedance, is dependent upon the inner and outer conductor sizes and the dielectric constant of the core, but it is a reciprocal relationship. Thus capacitance increases as impedance decreases in a cable with the same dielectric constant. Capacitance in coaxial cable is expressed in picofarads per foot (pF/ft).

*Attenuation* in coaxial cable is defined as the loss of electrical power along a unit length of cable. Losses occur in the conductor and dielectric as well as from radiation, but an increase in conductor size will reduce attenuation because electrical loss is decreased. It is possible to increase conductor size while holding cable dimensions constant by using a dielectric material with a lower dielectric constant. Attenuation is measured in decibels per 100 feet (dB/100 ft).

*Velocity of propagation* is the speed of transmission of electrical energy in a coaxial cable as compared with its speed in air, considered to represent 100 percent. Velocity is inversely proportional to the dielectric constant, so a lower dielectric constant allows a velocity increase.

*Time delay* is the elapsed time between the initial transmission of a signal from one point to its appearance or detection at another point. The time delay calculation represents the maximum possible time delay, and it is measured in nanoseconds per foot (ns/ft). This measurement is useful in designing RF systems and selecting coaxial cable.

Because of its geometry, coaxial cable can exhibit *inductance,* and its value is useful in selecting the appropriate coaxial cable for a specific application. Inductance is the property of a circuit or circuit element, measured in microhenries ($\mu$H), that opposes a change in current flow, causing current change to lag behind voltage change.

See "Coaxial Cable" and "Coaxial Cable Manufacture" in Sec. 29, "Electronic Hardware: Wire, Cable, and Connectors," and Sec. 6, "Antennas and Feed Horns."

# Waveguide Transmission

A *waveguide* is a hollow metal pipe that transfers radio energy over a desired path in the form of electromagnetic fields rather than current flow. It is an alternative to coaxial cable and two-conductor transmission line for transmitting UHF and microwave signals. There are two principal types of waveguides, *rectangular* and *circular,* but parallel metal plates separated by air or a solid dielectric material, dielectric rods, and optical fiber are also considered to be waveguides.

The electric and magnetic fields are confined within the guides so that no power is lost through radiation. Some power is lost as heat in the conductive walls of the guide, but it is insignificant. The dielectric loss is negligible because the guide normally is filled with air or an inert gas. Radio energy can be launched into and removed from the guide by means of horns, loops, or probes.

Rectangular waveguides are more widely used than circular waveguides because the plane of polarization and the mode of operation are easier to control in rectangular waveguides. In addition, it is more difficult to join round pipes if their centerlines are not aligned because of the complexity of the joints required. Nevertheless, transitions to circular waveguides are made where rotary joints are required.

The minimum dimensions of a waveguide that will transmit a specified frequency is related to the free-space wavelength at that frequency. This relationship depends on the shape of the waveguide and the *mode,* or the way in which the electromagnetic fields are organized within the guide. For all waveguides there is a minimum or *cutoff frequency* that can be transmitted. It is determined by the inside width of the waveguide, as shown in Fig. 7-1. The cutoff-frequency wavelength is equal to twice the inside width of the guide.

If the frequency of the input signal is above the cutoff frequency, the electromagnetic energy can be transmitted through the guide without attenuation, but if it is below the cutoff frequency, it will be attenuated rapidly. Inner waveguide height is not critical, but it determines the waveguide's power-handling capability. If the power is excessive, the voltage will arc between the inside walls. Standard waveguide width is from 0.2 to 0.5 times the average or typical wavelength to be transmitted, and the height is about 0.7 times that value.

In the X band of 8 to 12 GHz, for example, the U.S. standard WR-90 rectangular waveguide has an inner width of 2.286 cm (0.9 in) and an inner height of 1.016 cm (0.4 in). This waveguide has a cutoff frequency of 6.6 GHz. Waveguides are rarely specified for fre-

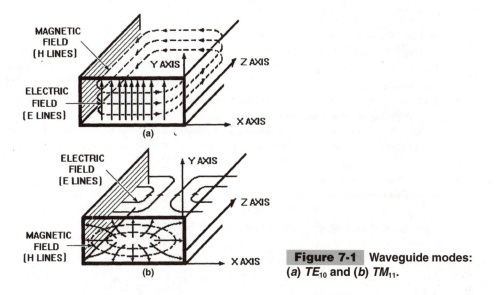

**Figure 7-1**  Waveguide modes: (*a*) $TE_{10}$ and (*b*) $TM_{11}$.

quencies below about 2 GHz (15 cm) because their dimensions become excessively large (approaching those of air-conditioning ducts).

The radius bends in the guide must be greater than two wavelengths to avoid excessive attenuation, and the cross section of the guide must be uniform around the bend. Dents or excess solder inside the guide increase transmission line attenuation, reduce the breakdown voltage, and can cause undesirable standing waves.

# Waveguide Transmission Modes

The electromagnetic wave in a waveguide consists of an electric vector and a magnetic vector, which are perpendicular to each other. There are many different patterns of these vectors, called *modes*. They depend on such factors as transmission frequency and the dimensions of the waveguide. There are two kinds of modes: (1) *transverse electric* (*TE*) and *transverse magnetic* (*TM*). In the *TE* mode the electric vector is *always* transverse or perpendicular to the direction of propagation; in the *TM* mode the magnetic vector is *always* transverse to the direction of propagation. In both modes the electric field must *always* be perpendicular to the waveguide wall at the surface, and the magnetic field must *always* be parallel to the waveguide wall.

The modes in rectangular waveguides are designated by two subscripts: the first indicates the number of half-wave variations in the electric field (in both *TE* and *TM* modes) across the wide dimension of the waveguide, and the second indicates the number of half-wave variations of the electric field across the narrow dimension. The $TE_{10}$ mode in a rectangular waveguide is shown in Fig. 7-1*a*. The electrical field $E$ is shown symbolically as vertical arrows ($E$ lines) parallel to the $y$ axis, concentrated near the center of the waveguide, and perpendicular to the $x$-$z$ axes or transverse plane. The magnetic field $H$ is shown as closed loops of dotted $H$ lines surrounding a group of $E$ lines of one polarity, and they lie parallel to the transverse plane. The electric field has a half-wave variation along its wide dimension or $x$ axis (1) and none (0) along its narrow dimension or $y$ axis.

The $TM_{11}$ mode has a different geometry, as shown in Fig. 7-1b. Here the magnetic field $H$ is represented as dotted closed ellipses that lie in the transverse $x$-$y$ plane enclosing back-to-back C-shaped patterns of $E$ lines shown in end view as radiating arrows parallel to the $x$-$z$ plane. The $TM_{11}$ mode has a magnetic field with a half-wave variation on both wide and narrow waveguide dimensions.

As previously stated in "Waveguide Transmission," the cutoff wavelength is the lower limit of frequency that can be transmitted through a waveguide of fixed size and shape. In other words, it is the longest wavelength corresponding to the lowest frequency that can be propagated. For the $TE_{10}$ mode in a rectangular waveguide, the cutoff wavelength is twice the wide dimension $a$. The $TE_{10}$ mode is the preferred mode for transmission in rectangular waveguides because it is easily excited, plane polarized, and easily matched to a horn or antenna. In addition, its cutoff frequency depends *only* on the waveguide's dimension $a$, making system design easier. Consequently, the $TE_{10}$ mode is called the *dominant mode,* and all other modes are the higher modes.

# Waveguide Coupling and Matching

The three principal ways of coupling RF energy into and out of a waveguide are with the following:

1. A *loop* of wire that cuts or couples to the $H$ lines of the magnetic field
2. A *probe* parallel to the $E$ lines of the electric field that acts as a monopole antenna
3. *Slots* or holes in waveguide walls that link internal and external fields

## WAVEGUIDE IRISES

A *waveguide iris* is a barrier formed by two thin metallic plates positioned within a waveguide to reduce its cross-sectional area. The plates are positioned perpendicular to the walls. An *inductive iris,* as shown in Fig. 7-2a, is formed when the opening is parallel to the narrow walls of the guide; it presents an inductive susceptance. A *capacitive iris,* as shown in Fig. 7-2b, is formed when the opening is parallel to the wide walls; it presents a capacitive susceptance.

## WAVEGUIDE POSTS

A *waveguide post,* as shown in Fig. 7-3a, is a pin inserted across the narrow dimension of a rectangular waveguide. It acts as an inductive shunt susceptance whose value depends on its diameter and position in the transverse plane of the waveguide.

## WAVEGUIDE TUNING SCREWS

A *waveguide tuning screw,* as shown in Fig. 7-3b, is a screw threaded through the wide wall of a waveguide so that it penetrates along the narrow dimension. It acts as a capacitive susceptance whose value can be set for precise tuning by adjusting the screw penetration.

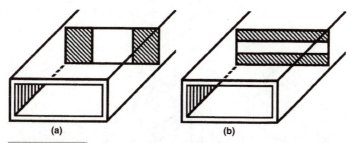

**Figure 7-2** Irises in rectangular waveguides: (*a*) inductive, and (*b*) capacitive.

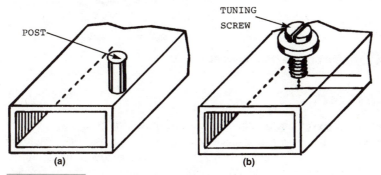

**Figure 7-3** Waveguide tuning adjustments: (*a*) post, and (*b*) screw.

# Microwave Striplines

A *microwave stripline,* as shown in Fig. 7-4, is an open-sided transmission line made as flat metal conductive strips or laminated foil separated by a dielectric layer. It is effective for short-range microwave transmission when space is too restricted for coaxial cable or waveguides. Also called *microstrip,* it is a two-conductor transmission line for microwave energy.

# Microwave Tubes

Most microwave power tubes are *crossed-field tubes* whose operation depends on the interaction of a DC electrical field with a perpendicular permanent magnetic field. The common examples are magnetrons, forward-wave crossed-field amplifiers (FWCFAs), backward-wave crossed-field amplifiers (BWCFAs or amplitrons), and backward-wave crossed-field oscillators (BWCFOs or carcinotrons). Other important microwave tubes are classed as linear-beam (O-type) tubes. These include the two-cavity and reflex klystron, the helix traveling-wave tube (TWT), the backward-wave amplifier (BWA), and the backward-wave oscillator (BWO).

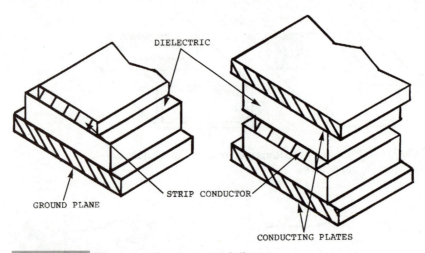

**Figure 7-4** Microwave-frequency stripline.

## MAGNETRONS

A *magnetron* is a power tube oscillator capable of converting electrical power directly into RF power in the microwave region. Technically a diode, it consists of a multicavity cylindrical anode and a coaxial cathode within a vacuum envelope. The strong permanent-magnet field is completed through the envelope with its lines parallel to the axial cathode, as shown in Fig. 7-5. This field can be produced by an external pair of magnets or a single C-shaped magnet.

When current flows in the filament the cathode is heated, driving off profuse amounts of electrons which form an electron cloud around the cathode. If a high negative voltage is applied to the cathode (making the anode positive with respect to the cathode) the electrons will be attracted to and will move off toward the positively charged anode. However, the electron paths will be modified by the strong perpendicular magnetic field.

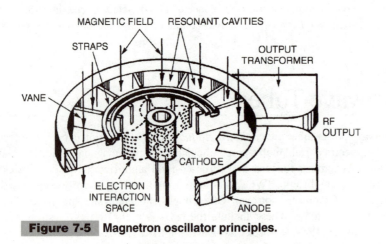

**Figure 7-5** Magnetron oscillator principles.

This crossed field forces the electrons to move in curved trajectories before reaching the ends of the reentrant cavity vanes. The cavities are formed by the rectangular vanes positioned radially from the cylindrical anode walls. The interaction between the electrons and the oscillating fields set up within each of the cavities causes the electrons to bunch into a pinwheel-shaped cloud that sweeps around the cathode at a speed determined by the resonant frequencies of the cavities.

As the spokes of the electron pinwheel turn, they repel free electrons in the walls of the resonant cavities, causing them to oscillate back and forth within the cavities. As a result, the inner surfaces of the cavities alternate in polarity at high frequency, creating and sustaining oscillating magnetic fields across the gaps. These fields, interacting in turn with the bunched electron cloud, sustain its rotation.

By properly adjusting the anode DC voltage, the average rotational velocity of the electrons will correspond to the phase velocity of the field in the slow-wave structure formed by the resonant cavities. Electron oscillations in the side walls of the cavities of the slow-wave structure create an oscillating radio-frequency field within each cavity. The electrons give up their energy to sustain these oscillations.

The frequency of electron oscillations within the cavities depends on the size and shape of the cavities. (Small cavities produce high frequencies and short wavelengths; large cavities produce low frequencies and long wavelengths.) The RF power generated by all of the cavities is coupled out of the slow-wave structure through one cavity containing either a loop or a *dielectric window.* The RF energy induced on the loop can be coupled directly to a coaxial cable (suitable for the S-band through L-band frequencies) or the energy can be coupled directly out through the window to a waveguide in X-band or higher-frequency magnetrons.

The anode shown in Fig. 7-5 has copper vanes that are alternatively connected by copper rings called *straps* to the copper anode to improve the magnetron's frequency stability. The vanes and straps are hard-soldered or *brazed* to the copper anode. Higher-frequency anode blocks are made by forcing fluted pins under high pressure through solid copper blocks in a *broaching* process.

Magnetrons were the first microwave tubes capable of producing enough power in the UHF and microwave bands to make high-definition radar possible. They can function as continuous-wave (CW) oscillators in radar altimeters and Doppler radars. CW magnetrons also power microwave ovens. Pulsed magnetrons can be pulsed at repetition rates as high as 1000 p/s to produce the bursts required for both short- and long-range search radars. Peak pulse power can be generated efficiently over the range of hundreds of watts to 3 MW. Magnetrons can operate over the frequency range of about 1 GHz to 95 GHz.

## CARCINOTRONS

The *carcinotron,* shown in the cutaway diagram Fig. 7-6, is a circular M-type backward-wave oscillator (M-BWO). Its principal parts are its electron gun (containing a filament, cathode, grid, and accelerator), a round interdigital delay line which slows the passage of RF energy moving around its circuitous pathway, and a round central electrode called the *sole.* Two halves of a split toroidal permanent magnet provide the perpendicular magnetic field.

The ribbon-shaped electron beam formed by the cathode is bent into a circular path by the magnetic field as it weaves back and forth through the DC field that exists between

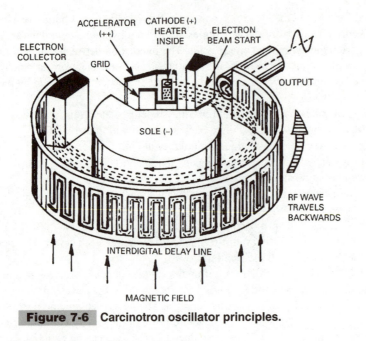

**Figure 7-6**  Carcinotron oscillator principles.

the negatively charged sole and the grounded delay line. The electron drift velocity of the beam is synchronized with a backward-wave space harmonic within the delay line so that the electrons give up their energy to the RF wave within the delay line. The carcinotron geometry is organized so that this backward-wave space harmonic is amplified as it flows in a direction opposing the direction of the electron beam. The backward RF wave generated is removed from the tube at an output terminal near the electron gun. Energy from the electron beam that is not given up to the backward wave is dissipated by a collector at the input end of the delay line.

A typical circular M-carcinotron with a diameter of about 8 in (20 cm) generates 200 W or more in the frequency range of 1.0 to 1.4 GHz. It has an efficiency from 30 to 60 percent. These tubes have been used as sources of high-frequency noise in military signal-jamming transmitters. Linear M-type BWOs operate according to the same principles as the circular M-BWO.

## TRAVELING-WAVE TUBES (TWTs)

A *traveling-wave tube* (TWT) is a microwave tube that can function as an oscillator or amplifier, depending on its physical structure. The tube, as shown in section view Fig. 7-7, contains an electron gun consisting of an anode, cathode, and heater; a slow-wave structure or delay line (a helix in the diagram); a focusing magnet; and a collector. The RF wave traveling in the circuitous slow wave is slowed by a factor of 10 to 20 percent so that its speed matches that of the electrons in the axial high-intensity beam formed by the electron gun. The synchronized passing electrons give up their energy to the RF wave, causing the tube to oscillate, and the collector dissipates any energy remaining in the electron beam.

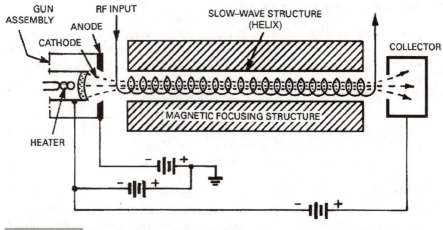

**Figure 7-7**  Traveling-wave tube (TWT).

The delay line can also be a serpentine interdigitated line. TWTs with that kind of slow-wave structure form electron beams that travel close to and parallel with the structure to give up their energy to it. Conventional TWTs are called *forward-wave traveling-wave tubes*. Their internal dimensions determine their operating frequencies.

## TRAVELING-WAVE TUBE AMPLIFIERS (TWTAs)

A *traveling-wave tube amplifier* (TWTA), as shown in the section view Fig. 7-8, is a TWT designed to operate as an amplifier. Microwave energy from a low-power source introduced into one end of the TWTA is amplified. These tubes can produce more than 200 W of pulsed output power over the range of 1 to 40 GHz.

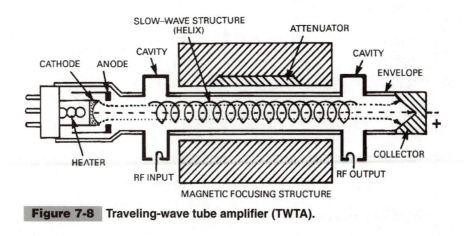

**Figure 7-8**  Traveling-wave tube amplifier (TWTA).

## TWO-CAVITY KLYSTRONS

A *two-cavity klystron,* as shown in the section view Fig. 7-9, is a microwave amplifier whose operation depends on velocity and current modulation. The tube includes an electron gun with a cathode and heater, an anode, two hollow disk-shaped *resonator cavities* (buncher and catcher), and a collector. Both resonant cavities have grids on each side that form gaps for the passage of an axial pencil electron beam.

Electrons emitted by the cathode travel at a uniform velocity until they reach the first grid of the buncher cavity. If they arrive at the grid when the signal voltage is zero, they will pass through the gap with no change in velocity. But if they arrive during a positive half-cycle, they will be accelerated, and if they arrive during a negative half-cycle they will be slowed.

These velocity changes imposed on the electrons cause them to bunch together in discrete groups as they transit the drift space between the buncher and catcher cavities. This means that electron density in the catcher cavity gap varies with respect to time. Maximum bunching occurs between the catcher cavity grids during its retarding phase when energy is transferred from the electrons to the RF field in the catcher cavity. The electrons leaving the catcher cavity are slowed and collide with the collector.

An RF signal introduced into the buncher cavity will appear as an amplified RF signal at the output of the catcher cavity. Two-cavity klystrons can produce 30 MW of pulsed power or 500 kW of continuous-wave power at 10 GHz. These amplifiers can have 40 percent efficiencies.

## REFLEX KLYSTRONS

A *reflex klystron,* as shown in the cutaway view Fig. 7-10, is a single-cavity oscillator whose operation depends on the velocity modulation of an electron beam. Its principal parts are an electron gun with cathode, heater, and beam-forming electrode; a drum-shaped resonator with three grids; and a reflector. Electrons from the cathode are first accelerated

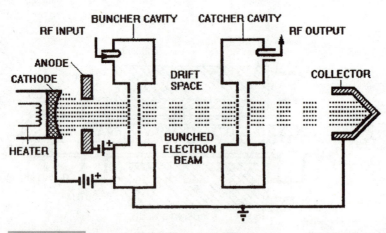

**Figure 7-9** Two-cavity klystron amplifier.

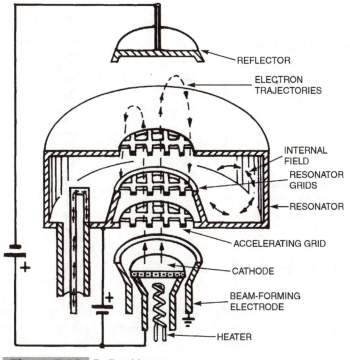

**Figure 7-10** Reflex klystron.

by the accelerating grid before passing through the first resonator grid. At that point they are velocity modulated by voltage across the resonator gap before being bunched and reversed in direction by the reflector voltage. As the electron bunches transit back through the cavity gap, they give up their energy to the electromagnetic field in the resonator to sustain oscillations.

Reflex klystrons are low-power microwave tubes that can generate 10 to 500 mW of RF power in the range of 1 to 25 GHz. They are used as microwave energy sources in laboratory bench test equipment and as local oscillators in radar systems.

# Magnetron Modulators

Specialized high-voltage circuits generate the pulses needed to trigger pulse-power magnetrons. There are two principal modulator types: *line type* and *magnetic*.

## LINE-TYPE MODULATORS

A *line-type modulator,* as shown in the simplified schematic Fig. 7-11, is the most popular magnetron pulsing circuit. The pulse-forming network (PFN) is charged between pulses,

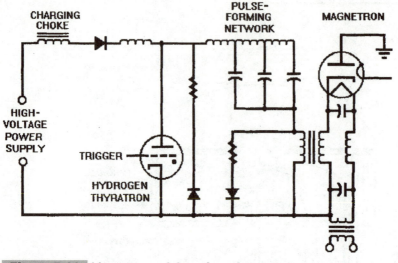

**Figure 7-11**  Line-type modulator for a magnetron.

and the modulator's thyratron is fired by a trigger signal, short-circuiting the input from the power supply. This causes the capacitor network to discharge and produce a rectangular pulse waveform at the primary winding of the transformer. The pulse from the secondary winding has sufficient voltage and current to cause the magnetron to oscillate when it is applied to the magnetron's cathode.

## MAGNETIC MODULATORS

A *magnetic modulator* is a cathode pulse modulator for radar magnetrons based on the saturation characteristics of inductors. It does not require a thyratron tube or other switch. The inductors transfer their energy resonantly through parallel capacitors in a pi network to the magnetron cathode.

# Microwave Diodes

Many different types of semiconductor diodes are made for use in UHF- and microwave-frequency circuits with characteristics that are similar to those of low-frequency diodes. They include the Gunn, IMPATT, PIN, tunnel, and TRAPATT diodes.

## GUNN DIODES

A *Gunn diode,* as shown in Fig. 7-12, uses the *Gunn effect* to oscillate at microwave frequencies or amplify an applied microwave signal. The diode is made from three layers of semiconductor material with different doping. When enclosed in an operating cavity, the diode operates in the transit-time mode, and its oscillation frequency depends on domain

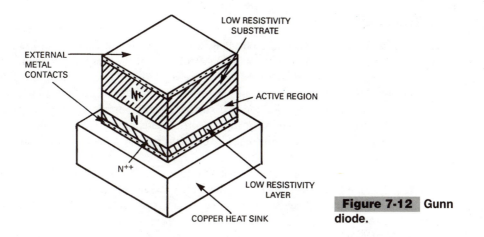

**Figure 7-12** Gunn diode.

transit time. A *transferred-electron diode,* its frequency can exceed 50 GHz. It is also called a *transferred-electron oscillator* (TEO).

## IMPATT DIODES

An *IMPATT* (impact avalanche transit time) *diode* exhibits negative resistance characteristics which result from a combination of impact avalanche breakdown and charge-carrier transit-time effects. It is made as a gallium-arsenide or silicon die. When mounted in a suitable tuned cavity or waveguide, the diode can function either as an oscillator or amplifier. An IMPATT amplifier can produce 100 W of pulsed or 20 W of continuous-wave power between 5 and 10 GHz.

## PIN DIODES

A *PIN diode,* as shown in the section view Fig. 7-13, is a junction diode whose heavily doped P and N regions are separated by a relatively thick layer of high-resistivity intrinsic (I) silicon. It can switch microwave transmission lines and function as a microwave limiter, replacing a TR tube in systems where peak powers are less than 100 kW. It can also act as a variable microwave attenuator and as an electronically controlled, rapid-acting phase shifter for microwave phased-array radar systems.

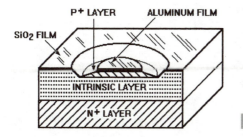

**Figure 7-13** PIN diode.

### TUNNEL DIODES

A *tunnel diode* is a PN junction diode that exhibits negative resistance, permitting it to function as an amplifier, oscillator, or switch. Because of its fast response to input signals, it is primarily used in the microwave band where it provides low-noise amplification. Its PN junction includes very heavily doped P and N regions which form an abrupt interface with a very thin barrier. As diode forward voltage increases from zero, majority carriers tunnel through the thin barrier. Very high electric fields are created with low voltages because the barrier is so thin. As the majority carriers cross the junction, the effective width of the junction expands.

With further increases in voltage, the tunneling current ceases, and normal forward current due to minority-carrier injection builds up. Between the tunneling current and the minority-carrier current, there is a negative-resistance region. Current flow decreases from a peak to a valley before rising again as voltage is increased, a characteristic that permits very fast switching.

Gallium-arsenide tunnel diodes are used in the first RF amplifier stages of receivers in microwave relay links and in Doppler navigation and weather radar. Very low power-handling capability limits their use as local oscillators in receivers, but they are unsuitable for power amplification. They are also called *Esaki diodes.*

### TRAPATT DIODES

A *TRAPATT* (trapped plasma avalanche transit time) *diode,* as shown in the section view Fig. 7-14, can oscillate at a frequency determined by the thickness of its active layer. It is a transit-time diode like the IMPATT diode, but it operates in a different mode: the avalanche zone moves through the drift region, creating a trapped space-charge plasma within the PN junction region.

## Microwave Transistors

Both silicon and gallium-arsenide (GaAs) RF transistors can function in the UHF and microwave frequency regions. Thus integrated circuits that include these RF transistors are

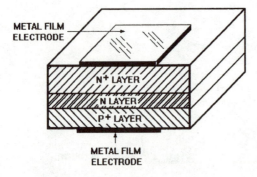

**METAL FILM ELECTRODE**

N⁺ LAYER

N LAYER

P⁺ LAYER

**METAL FILM ELECTRODE**

**Figure 7-14** TRAPATT diode.

also useful in those high-frequency regions. Silicon transistors function efficiently in the UHF and lower microwave frequencies, but GaAs transistors are more efficient at the higher microwave frequencies.

## METAL FIELD-EFFECT TRANSISTORS (MESFETs)

A *metal field-effect transistor* (MESFET), as shown in Fig. 7-15, is a common type of GaAs transistor for operation in the UHF and microwave bands. MESFET geometry is also used in GaAs ICs. Similar in structure to a MOSFET, its metal gate is deposited directly on the doped GaAs substrate, forming a Schottky barrier diode. Oxides of silicon are deposited on the substrate for isolation and insulation. See also "Metal Field-Effect Transistors (MESFETs)" in Sec. 2, "Active Discrete Components."

## HIGH-ELECTRON-MOBILITY TRANSISTORS (HEMTs)

A *high-electron-mobility transistor* (HEMT) is a GaAs transistor suitable for integration in microwave ICs. It is fabricated as a *heterojunction,* a layer of aluminum gallium arsenide (AlGaAs) grown on a GaAs substrate. This geometry improves transistor performance and permits higher levels of IC integration than MESFETs. See also "High-Electron-Mobility Transistors (HEMTs)" in Sec. 2, "Active Discrete Components."

# Microwave Monolithic Integrated Circuits (MMICs)

A *microwave monolithic integrated circuit* (MMIC) can be made from either silicon or GaAs. True integrated circuits, they should be distinguished from microwave *hybrid circuits.* Silicon MMICs can amplify or oscillate efficiently only to about 2 GHz, but GaAs MMICs can operate at higher frequencies. Consequently, they have replaced silicon monolithic and hybrid devices where their higher cost is justified.

A GaAs MMIC is illustrated in Fig. 7-16. These MMICs are cost-effective in the UHF and microwave bands from 500 MHz to 2 GHz, but essential at the higher microwave frequencies. GaAs MMICs are now included in military phased-array radars and electronic

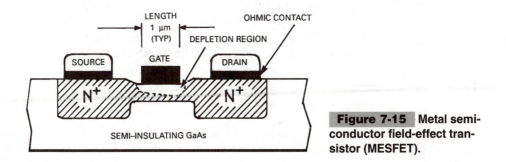

**Figure 7-15** Metal semiconductor field-effect transistor (MESFET).

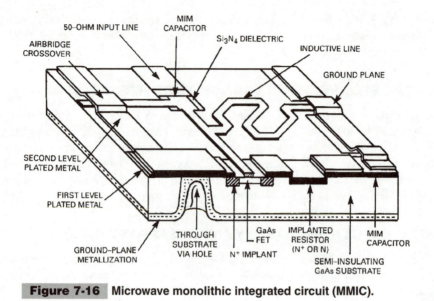

**Figure 7-16**  **Microwave monolithic integrated circuit (MMIC).**

warfare systems that operate in the C and X bands. GaAs MMIC amplifiers, oscillators, and mixers are now available.

# Microwave Couplers

Microwave couplers based on slotted sections of waveguide are able to extract small amounts of the energy transmitted through a waveguide for use in power and frequency measurements and for application where a small sample of the energy is required.

## DIRECTIONAL COUPLERS

A *directional coupler,* as shown in the cutaway Fig. 7-17, extracts a fixed small fraction of the energy flowing in one direction in the waveguide to determine the output power being transmitted. It consists of a short length of closed rectangular waveguide attached along the narrow side of the main waveguide and coupled to it by means of two small holes or *apertures* whose centers are a quarter wavelength apart. The short section contains a matched load in one end and a coaxial transition in the other end. The amount of coupling between the main waveguide and the short section is determined by the aperture diameters. Energy that is extracted can be fed to a *wattmeter* to determine RF power.

## BIDIRECTIONAL COUPLERS

A *bidirectional coupler* can measure both direct and reflected power. It is a straight section of waveguide with an enclosed section attached to each side along its narrow dimension.

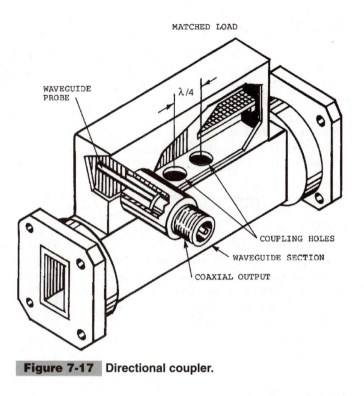

MATCHED LOAD

WAVEGUIDE
PROBE

λ/4

COUPLING HOLES

WAVEGUIDE SECTION

COAXIAL OUTPUT

**Figure 7-17** Directional coupler.

Each section contains an RF pickup probe at one end and an impedance termination at the other end. Microwave energy from the main waveguide passes through three apertures, each spaced one-quarter wavelength apart on centers. The RF probe farthest away from the transmitter can be used to measure direct power, and the one nearest the probe can be used to measure reflected power.

# Microwave Phase Shifters

Microwave phase shifters shift the phase or slow down the velocity of microwave signals passing through them. Some phase shifters are based on the physical properties of ferromagnetic materials, typically ferrites, and some are based on PIN diodes. Stacked arrays of phase shifters in phased-array radar antennas permit the beam to be scanned electronically.

The most common phase shifters are ferrite units that are placed in series with waveguide sections. Alternatives are microstrip versions that include ferrites and garnets or ceramic materials with magnetic properties. Phase shifters can be either *reciprocal* or *nonreciprocal*. Reggia-Spenser and Faraday rotator phase shifters are the most popular reciprocal units, and the toroidal phase shifter is the most common nonreciprocal type.

## REGGIA-SPENCER PHASE SHIFTERS

A *Reggia-Spencer phase shifter,* as shown in the cutaway view Fig. 7-18 is a waveguide section that has a solenoid wound around it and a ferromagnetic bar with a square cross section positioned axially within it. The longitudinal magnetic field produced when the solenoid is energized changes the permeability of the ferromagnetic bar, which can control the propagation constant of the passing RF energy. This permits phase shift to be varied incrementally by drive current so that the RF energy can be slowed in steps. These reciprocal phase shifters are widely used in phased-array radar antennas. See also "Radar Systems" in Sec. 23, "Military and Aerospace Electronics Systems."

## FARADAY ROTATOR PHASE SHIFTERS

A *Faraday rotator phase shifter* is a latching, reciprocal phase shifter housed in a rectangular waveguide section. An axially positioned Faraday rotator section consists of a small, square, ferrite-filled waveguide with nonreciprocal quarter-wave plates at each end. An axial coil is wound around the waveguide, and the magnetic circuit is completed externally to the thin waveguide wall with ferrimagnetic yokes. The nonreciprocal quarter-wave plates convert energy in the rectangular waveguide to either right- or left-hand circularly polarized energy, depending on the direction of propagation. The insertion phase of the energy is changed by the variable axial magnetic field supplied by the coil around the waveguide.

## TOROIDAL PHASE SHIFTERS

A *toroidal phase shifter* is a nonreciprocal phase shifter with a ferrimagnetic toroid located within a waveguide section. The toroid is wired to a drive amplifier that can supply either a positive or negative current pulse. These pulses induce a magnetic field that drives the toroid material into either positive or negative saturation. A complete digital phase shifter

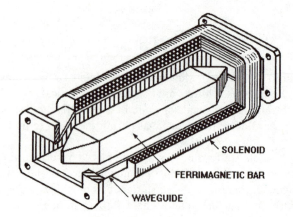

SOLENOID

FERRIMAGNETIC BAR

WAVEGUIDE

**Figure 7-18** Reggia-Spencer phase shifter.

contains several lengths of ferrite cores to give differential phase shifts of up to 180°. Analog versions must be capable of producing at least 360° of phase shift.

## SEMICONDUCTOR-DIODE PHASE SHIFTERS

A *semiconductor-diode phase shifter* is a digital phase shifter that depends on *PIN diodes* as control elements. The PIN diodes can be forward or reverse biased. The intrinsic regions in the diodes behave as poor dielectrics at microwave frequencies. As a result, the impedance presented between the network terminals can be varied with external reactive tuning elements. There can be from 10 to 16 PIN diodes per stripline module. This phase shifter is widely used in high-power radar phased-array antennas. See also "Radar Systems" in Sec. 23, "Military and Aerospace Electronics Systems."

# Microwave Attenuators

Microwave attenuators are microwave system components that can control the flow of RF energy passing in a waveguide. The three attenuator types are: (1) *flap,* (2) *fixed,* and (3) *rotary vane.*

## FLAP ATTENUATORS

A *flap attenuator,* as shown in Fig. 7-19, is a waveguide section with a hinged, tapered resistive card that can be moved in or out of a slot cut down the center of the wide wall. The hinge

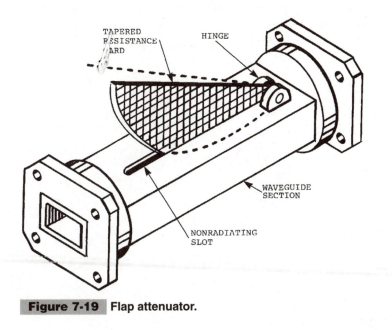

**Figure 7-19**  Flap attenuator.

permits card penetration to be varied so that attenuation can be changed from 0 to some maximum value, typically 30 dB. Because the longitudinal slot is centered on the broad wall, no microwave *energy* is radiated. It is also called a *variable resistive-card attenuator.*

### FIXED ATTENUATORS

A *fixed attenuator* is a waveguide section containing a resistive card tapered at both ends and attached along the centers of the broad inside walls. The card contours can be trimmed to obtain specific attenuation values. It is also called a *fixed resistive-card attenuator.*

### ROTARY-VANE ATTENUATORS

A *rotary-vane attenuator,* as shown in Fig. 7-20, is a variable attenuator consisting of three circular waveguide sections, each containing a resistive card. The two end sections are fixed in position, but the center section can be rotated. Input and output transitions at the ends permit the attenuator to be connected between standard waveguide sections. Attenuation is controlled by rotating the center section. Minimum attenuation occurs when all three cards lie in the same plane, and maximum attenuation occurs when the card in the center section is positioned at 90° with respect to the other two cards.

# Transmit-Receive (TR) Switches

A *transmit-receive* (*TR*) *switch* for UHF and microwave frequencies isolates the receiver from the transmitter when a powerful signal is being transmitted. Gas-filled *TR tubes* are used in high-power radar systems. Other TR switches are *ferrite circulators* and *PIN diodes.*

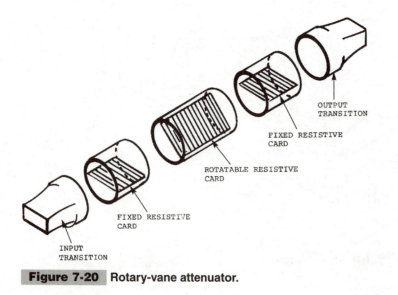

**Figure 7-20**    Rotary-vane attenuator.

PIN switches are easier to apply in coaxial transmission lines at the lower frequencies. Multiple diodes are used when a single diode cannot withstand the incident voltage or current.

## TR TUBES

A *TR tube*, shown in cutaway view Fig. 7-21, is a sealed, gas-filled waveguide section or glass envelope that protects a radar's receiver circuitry from damage or destruction that would be caused if it received high-power RF signals. It permits alternate use of the antenna for both transmitting and receiving. When the gas is ionized, the TR tube acts as a short circuit to protect the receiver; but when it is not ionized, it permits low-power radar returns or echoes to pass through it with minimal attenuation.

Energized electrodes projecting into the cavity form the "keep-alive" assembly that raises the gas ionization threshold to the level that will permit breakdown in the presence of all but the weakest incident RF input signals.

# Radio Telescopes

A *radio telescope* is a microwave receiver and antenna designed specifically for detecting and recording microwave-frequency signals from outer space. The antenna is analogous to the objective lens or mirror of an optical telescope, while the receiver-recorder is analogous to the eye-brain combination, photographic plate, or video recorder. The appearance of the sky at radio wavelengths differs from its optical appearance because the sun is a less significant RF source than the Milky Way, which radiates strong signals. The sky is dotted with radio sources that are almost entirely unrelated to celestial objects visible to the unaided eye. Because the radio-noise window covers wavelengths from a few millimeters to tens of meters, the receiving antennas take many different forms.

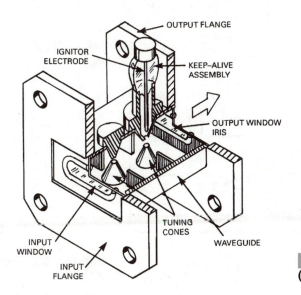

**Figure 7-21** Transmit-receive (TR) tube cutaway view.

# YIG-Tuned Oscillators

A *YIG-tuned oscillator* includes a high-$Q$ circuit formed by an yttrium-iron-garnet (YIG) sphere in a DC magnetic field which acts as a shunt-resonant tank. A simplified schematic of the oscillator is given in Fig. 7-22. RF signals are coupled from a transistor into the YIG sphere by a wire loop. The ferrimagnetic resonance frequency is a function of magnetic field strength, so these oscillators can be tuned linearly over several octaves, and they have been made to oscillate at frequencies from 2 to 40 GHz.

# Masers

A *maser* (microwave amplification by stimulated emission), as shown in the simplified section view Fig. 7-23, amplifies microwave frequencies by stimulating atoms or molecules to a higher, unstable energy level. A microwave input signal interacts with the atoms or molecules to stimulate the emission of excess energy at the same frequency and phase as

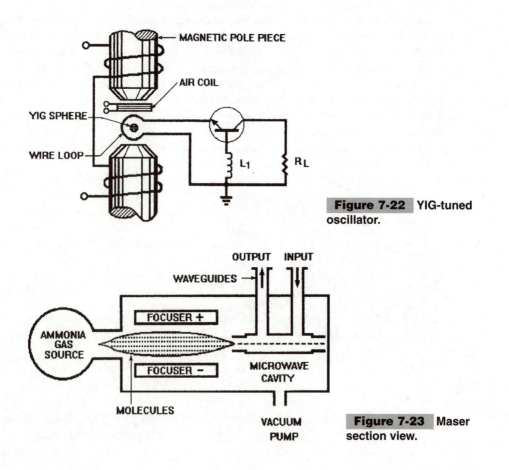

**Figure 7-22**  YIG-tuned oscillator.

**Figure 7-23**  Maser section view.

the stimulating wave. This response provides coherent amplification at a wavelength determined by the dimensions of the cavity or resonant structure.

The application of external energy required for amplification or oscillation in a maser is called *pumping*. The radiated energy greatly exceeds the energy level of the pumping signal. There are three types of masers: (1) *gas,* (2) *solid state,* and (3) *traveling wave.* Ammonia atoms or molecules are the parametric materials in beam-type gas maser oscillators. Solid-state masers depend on the electrons of parametric atoms or molecules. There are two- and three-level solid-state masers. Masers are low-noise preamplifiers for very weak signals received by radio telescopes and extremely long-distance radars, and are also used as time and frequency standards. The *laser* operates by the same stimulated-emission principles as does the maser in the visible-light and infrared regions.

# ANALOG AND LINEAR
# INTEGRATED CIRCUITS

# Overview

Many of the analog and linear integrated circuits that have been developed over the past 20 years are now recognized electronic building blocks. They continue to perform vital functions in circuits, products, and systems despite the significant gains made in digital circuitry. All circuits are either linear or digital, but some people distinguish between linear and analog circuits. *Linear circuits* are broadly defined as those whose outputs vary in direct proportion to their inputs, but *analog circuits* are considered to be a subset of linear circuits that represent physical quantities such as velocity, pressure, and temperature by variable values of voltage, current, or resistance. Many analog circuits were originally developed for use in analog computers.

Analog circuits generally include operational amplifiers, comparators, and both analog-to-digital and digital-to-analog converters. They were originally designed as vacuum-tube circuits, but they were subsequently redesigned with transistors. They were then offered commercially as discrete transistor modules, while others were made as proprietary hybrid circuits. The most popular analog circuits were then manufactured as monolithic ICs, which became available as standard catalog or off-the-shelf devices.

But discrete modules and hybrid ICs continued to be made long after the introduction of monolithic ICs because the commodity ICs were unable to achieve the higher speed or higher precision required for some applications. Later commodity ICs were included in modules and hybrid circuits to reduce the discrete device count, but they were still supplemented with off-chip precision resistors and capacitors for performance improvements.

Conventional low-frequency and high-frequency amplifiers, linear power supplies, solid-state relays, and phase-locked loop circuits are all classed as linear circuits. All or most of these circuits have been produced as ICs.

# Operational Amplifier ICs

The integrated circuit *operational amplifier* (op amp) is the most common analog IC. More than 10 manufacturers offer these devices in the world marketplace. It is possible to make a selection from hundreds of different part numbers because of the many combinations of performance classifications, temperature ratings, fabrication methods and package styles. The lowest-priced general-purpose IC op amps, typically under $1 in small purchase quantities, are considered to be commodity or "jelly bean" items.

Op-amp ICs are classified by performance as follows:

- General purpose
- Low offset, Low drift
- Low input current, high impedance
- Wide bandwidth
- High slew rate
- Low noise
- Low power
- High voltage
- High output current (power op amps)
- Programmable

Op amps are made to comply with commercial, industrial, and military temperature requirements and there can be from one to four op amps per chip (single, dual, and quad). For further information on the op amp and its circuit configurations see "Operational Amplifiers" in Sec. 4, "Basic Amplifier and Oscillator Circuits."

# Comparators

An IC *comparator* is an operational amplifier without feedback that detects changes in voltage level as required by analog-to-digital, digital-to-analog, and other types of converters. They are designed to interface analog signals with standard logic families such as transistor-transistor logic (TTL), complementary metal-oxide semiconductor (CMOS), and emitter-coupled logic (ECL). There are fewer IC voltage comparator part numbers and manufacturers than there are IC op amps. Response time is an important characteristic of comparators, so they can be conveniently classified by their speed as low, medium, high, and very high speed. They are also available in single, dual, and quad packages.

# Voltage Regulators

An IC voltage regulator is a circuit that includes a sensor capable of monitoring the load of a power supply and restoring its output within close tolerance limits despite changes in both load and input voltage. There are two basic types: *linear* and *switchmode.* Low-cost linear IC voltage regulators are capable of holding the DC output voltage levels of power supplies at levels from 3 to 30 V constant within ±2 percent. These ICs typically are packaged in standard tree-terminal or pin packages.

# Analog-to-Digital Converters (ADCs)

An *analog-to-digital converter* (A/D converter or ADC) continuously converts variable analog input signals into digital signals. These circuits are offered commercially as modules and monolithic ICs, but for some applications, particularly military, aerospace, and high-reliability (hi-rel) applications, they are made as high-performance hybrid circuits. Six types of analog-to-digital converters are:

1. Dual-slope integrating converters
2. Successive-approximation converters
3. Flash converters
4. Voltage-to-frequency converters
5. Synchro-to-digital converters
6. Resolver-to-digital converters

Dual-slope converter ICs combined with liquid-crystal digital displays made possible low-cost compact *digital panel meters* (DPMs), *digital multimeters* (DMMs), and other digital electronic instruments. The preferred A/D conversion circuit for communication and computer applications is the *successive-approximation converter* because it offers a favorable compromise between speed and accuracy. The faster high-speed *flash converter* ICs are widely used for video signal conversion. *Voltage-to-frequency converter* (VFC) ICs provide high-resolution conversion, and also permit long-term integration (from seconds to years), frequency modulation, voltage isolation, and arbitrary frequency division or multiplication. The time required for a complete measurement by an ADC is called *conversion time.* For most converters, this time is essentially identical with the inverse of conversion rate.

## DUAL-SLOPE INTEGRATING ADCs

A *dual-slope integrating ADC,* as shown in the block diagram Fig. 8-1, converts an unknown input voltage to a time value, which is then converted to a binary number. The unknown input voltage $V_{IN}$ is switched to the capacitor $C$ in the operational amplifier integrating circuit, charging it for the preset length of time $T$ (shown as the positive slope $V_{IN}$ in the triangular graphic). Then the reference input signal $V_{REF}$ is switched to the integrator, discharging capacitor $C$ from its integration value to zero (shown as the negative slope $V_{REF}$ in the triangular graphic). The output of the comparator is sent to the counter, where the time to discharge capacitor $C,$ shown as time value $N$ in the diagram, is converted by the counter to a binary number. The value of $N$ is proportional to the average value of the unknown signal. It can then be displayed on a digital readout.

## SUCCESSIVE-APPROXIMATION ADCs

A *successive-approximation ADC,* as shown in the simplified block diagram Fig. 8-2, converts analog-to-digital values by a method that is analogous to placing known weights in one tray of a scale in descending order of magnitude to balance an unknown weight in the other tray. The unknown input voltage $V_{IN}$ is on the positive terminal of the comparator. The internal digital-to-analog converter (DAC) switches a voltage one bit at a time to the negative comparator terminal under clock control, starting with the most significant bit (MSB) value.

The comparator then produces an output that indicates whether $V_{IN}$ is greater or less than the output of the DAC. If it is greater, the MSB will be reset to zero because that value will not contribute to a sum of values equal to $V_{IN}$; if it is less, the MSB is held in the successive-approximation register (SAR). The converter then continues to make comparisons with each

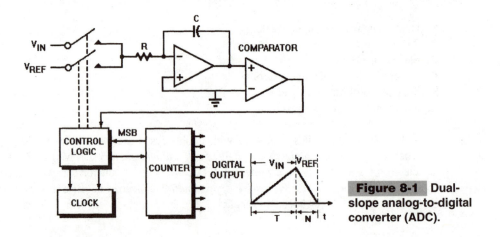

**Figure 8-1** Dual-slope analog-to-digital converter (ADC).

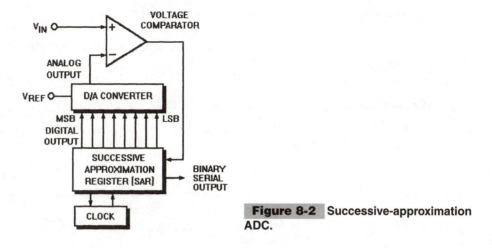

**Figure 8-2** Successive-approximation ADC.

bit in descending order of significance. It retains only those in its register that contribute to the digital approximation of $V_{IN}$, and resets those that do not. A serial binary output representing $V_{IN}$ can be taken from the SAR.

## FLASH CONVERTER ADCs

A *flash converter ADC,* as shown in the simplified block diagram Fig. 8-3, is an array of parallel comparators whose outputs are encoded as binary numbers by the encoding logic block. To convert an analog input to $n$ bits, $2^n$ comparators are needed to compare the input with $2^n$ different reference levels from a string of series resistors. For example, a 6-bit flash requires $2^6$ or 64 comparators. A string of 65 resistors across the voltage reference $V_{REF}$ divides the reference into 64 voltage levels, which are connected to the negative terminals of the 64 comparators. The unknown input voltage $V_{IN}$ is connected to the positive terminals. When triggered by latches in the encoding logic, the encoder converts the comparator

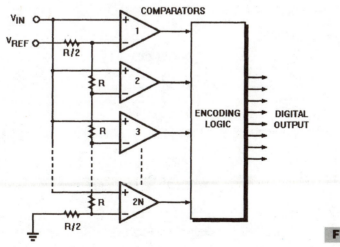

**Figure 8-3** Flash ADC.

outputs to binary code. The flash converter converts faster than a successive approximation converter because the digital output conversion is done in parallel instead of serially. Conversion speed is limited only by the propagation delay time of one comparator and by the encoding logic.

## VOLTAGE-TO-FREQUENCY CONVERTERS (VFCs)

A *voltage-to-frequency converter* (VFC), as shown in the block diagram Fig. 8-4, depends on a charge-balancing circuit. Conversion begins when a capacitor is charged from a current source which is proportional to the input voltage. The capacitor is then discharged with a precise current each time the charge on the capacitor reaches a preset level. VFCs are not effective for measuring low input voltages because of offset voltage errors. In addition, the slew rate and settling time of the amplifier limits the upper frequency. In typical VFCs, the comparator output pulses are fed to a counter for a fixed period of time. The accumulated count is proportional to the input voltage.

## SYNCHRO-TO-DIGITAL AND RESOLVER-TO-DIGITAL ADCs

*Synchro-to-digital* and *resolver-to digital ADCs* are used where angular or linear position must be measured precisely with high resolution in synchro systems before conversion into a digital readout. See "Synchro Systems" in Sec. 22, "Industrial Electronics Technology."

# Digital-to-Analog Converters (DACs)

A *digital-to-analog converter* (D/A converter or DAC) converts digital signals to time-varying analog signals. The earliest method for making this conversion was a *voltage-summing amplifier* consisting of a parallel array of precise resistors feeding an operational amplifier. Difficulties in matching resistors in ratios greater than about 20 to 1 led to the

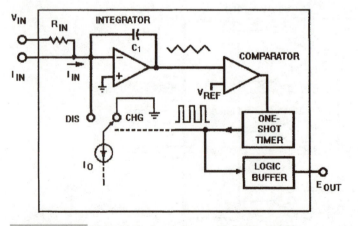

**Figure 8-4**  Voltage-to-frequency converter (VFC).

development of the *R-2R ladder network DAC,* shown in the simplified schematic Fig. 8-5. Only two different resistor values are needed in this converter. Switches connect each 2$R$-resistor leg to either the reference voltage $V_{REF}$ or ground. The switches are shown positioned for converting the binary number 0101 to an analog voltage. The inverted $R$-2$R$ ladder configuration made possible the fabrication of monolithic DACs because the IC process was unable to produce the precision resistors required for the voltage-summing amplifier. The $R$-2$R$ DAC produces an analog output voltage that is proportional to the product of the digital input and the reference voltage.

# Analog Function Circuits

Analog function circuits or *analog-to-analog converters* are computational and special-purpose circuits for conditioning analog signals. They can relieve a computer's central processing unit (CPU) of the burden of conditioning analog signals in data acquisition systems while also saving on the additional programming required. Among the more popular analog function circuits are those that perform multiplication, taking ratios, raising to powers, taking roots, and performing special-purpose nonlinear functions such as linearizing transducers. Analog function circuits can also make root-mean-square (RMS) measurements, compute trigonometric functions and vector sums, integrate, and differentiate. They can also transform current to voltage or voltage to current. Some functional circuits can be purchased as off-the-shelf multiplier/dividers or log/antilog amplifiers.

# Analog Switches

An *analog switch* is a semiconductor switch that opens or closes transmission paths for analog signals. The open-switch positions usually are digitally controlled. Analog switches

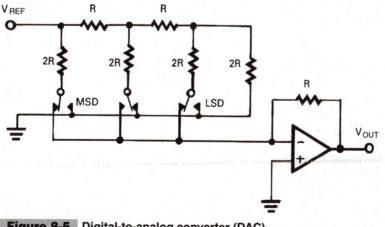

**Figure 8-5**   Digital-to-analog converter (DAC).

are fabricated with CMOS technology because of the superior performance of FET switches. Commercial analog switches are available in many styles and configurations. These include single, dual, and quad single-pole, single-throw (SPST) normally open (NO) and normally closed (NC) switches, single and dual single-pole, double-throw (SPDT), and double-pole, single-throw (DPST) monolithic devices. Analog switches are used in instruments, automatic test equipment, communication systems, telephone equipment, process control, and telemetry.

# Active Filters

An *active filter* is a filter that includes electronic devices such as operational amplifiers and transistors. These filters require a power source, but unlike passive filters, they provide gain and do not require inductors. They are also smaller and lighter than comparable passive filters. Some active filters are available as monolithic ICs, but some require external resistors and capacitors to give them desired responses. See also "Filters" in Sec. 1, "Passive Electronic Components."

## SWITCHED-CAPACITOR FILTERS (SCFs)

A *switched-capacitor filter* (SCF), as shown in the diagram Fig. 8-6, is an active IC filter that does not require supplemental discrete capacitors or inductors. They can replace passive *LC* filters in many circuits. It consists of a switching section and an operational amplifier integrator, all on the same chip. The switches are actually MOS transistors (identified as $S_1$ and $S_2$). Capacitor $C_1$ simulates a resistor. The output of this section is fed to the op-amp integrator with capacitor $C_2$ in the feedback circuit. A simple clock switches the transistor on and off and determines the filter's cutoff or center frequency.

The transistors are alternately switched on and off in a *break-before-make* fashion by the clock. When $S_1$ is open and $S_2$ is closed, the charge on $C_1$ flows to ground. With a constant input voltage, the faster the switching rate, the greater the flow of charge per unit time. Because current is the rate of charge flow, an average current flow can be determined. Thus an equivalent resistance value can be determined by dividing the known input voltage by

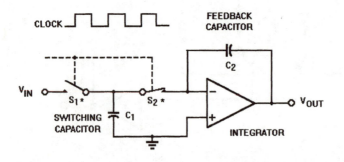

**Figure 8-6** Switched-capacitor filter (SCF).

\* Switching actually performed by MOSFETs

the average current. Because the equivalent value of resistance also equals the inverse of the product of the clock frequency and switching capacitor $C_1$, the equivalent input resistance to the integrator can be changed by adjusting clock frequency.

The cutoff frequency of the op-amp integrator can be determined from the value of the equivalent input resistor and capacitor $C_2$ in the integrator. The equivalent value of the cutoff frequency of the SCF is directly proportional to the product of the clock frequency and $C_1$, and is inversely proportional to the value of $C_2$.

The frequencies that are passed or rejected by the SCF (its response) and the shape of the gain versus frequency plot can be set by selecting the clock switching frequency, the capacitor ratio, and, in some instances, the value of the external resistors. A single SCF chip can be adapted to function as a bandpass, high-pass, low-pass, band-reject (notch), or all-pass filter in Butterworth, Chebychev, Bessel, or Cauer (elliptic) response formats. The analog frequency is typically 1 Hz to 20 kHz or more. SFC stability is 10 to 20 times better than conventional active filters. SCFs are now used in audio systems, electronic musical instruments, speech synthesis and recognition equipment, and test instruments.

# Sample-and-Hold Amplifiers (SHAs)

A *sample-and-hold amplifier* (SHA), as shown in Fig. 8-7, consists of an operational amplifier and capacitor. Essentially an analog switch, it can, on command, periodically sample the instantaneous level of the input signal and temporarily hold it as a DC value. That value represents the instantaneous voltage value in a continuously changing data stream from a sensor or transducer. The SHA holds the voltage long enough for an analog-to-digital converter to convert the sample and provide a stable output which can be digitally displayed. Also called *track-and-hold amplifiers* (THAs), SHAs are used in multimeters, peak detectors, and other instruments. The circuit has analog input, control input, and analog output pins, and it is designed to operate in either the sample or hold mode.

In the *sample mode,* the SHA rapidly acquires the input signal and tracks it until commanded to hold that value. In the *hold mode,* it retains the captured value of the input signal as a DC value. Available SHA ICs are compatible with transistor-transistor logic (TTL) or metal-oxide semiconductor (MOS) logic. The IC contains input and output operational amplifiers and a transistor switch, but it requires an external holding capacitor.

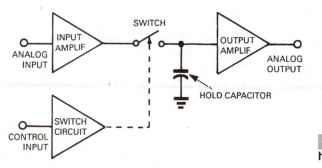

**Figure 8-7** Sample-and-hold amplifier (SHA).

The difference between SHA and THA operation depends on the duration of switch closure. The THA switch is closed for a relatively long period, permitting the output to change significantly. However, the output holds the level present at the instant the switch is opened. By contrast, the SHA switch is closed only long enough to charge the holding capacitor completely. There is no difference between the operation of the two circuits when the host data acquisition system updates data at rates greater than 1 MHz.

# Phase-Locked Loop (PLL) Circuits

A *phase-locked loop* (PLL), as shown in the block diagram Fig. 8-8, is an electronic circuit that can lock an oscillator in phase with an input signal. A PLL can demodulate a carrier frequency or track a carrier or synchronizing signal whose frequency varies with respect to time. The PLL circuit consists of a phase detector and low-pass filter with a feedback loop closed by a local voltage-controlled oscillator (VCO).

The phase detector detects and tracks small differences in phase and frequency between the incoming signal and the VCO signal, and it provides output pulses that are proportional to the difference. The low-pass filter removes the AC components, leaving a DC voltage signal to drive the VCO. This input voltage changes the output frequency of the VCO to that of the input signal. The phase detector and low-pass filter function as the mixer in a feedback loop and, in the loop, the output is driven in the direction that will minimize the error signal, a changed frequency. Thus the loop tends to drive the error signal back toward zero frequency. Once the two frequencies are made equal, the VCO will be *locked* to the input signal, and any phase difference between the two signals will be controlled.

The PLL was designed for demodulation in frequency-modulation (FM) receivers, but it now stabilizes many other circuits. For example, a PLL can synchronize the horizontal and vertical scanning signals in a TV receiver to remove the effects of Doppler shift in satellite tracking. It can also stabilize the frequency of klystron oscillators or filter out noise in communications circuits. PLLs are found in synchronous detection circuits, modems, tone decoders, and frequency-shift keying (FSK) receivers.

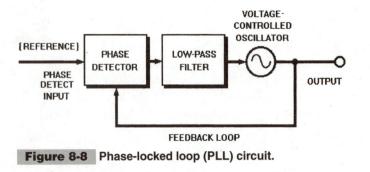

**Figure 8-8** Phase-locked loop (PLL) circuit.

# Power Integrated Circuits

A *power integrated circuit* is a monolithic IC that combines a signal-level analog or digital logic circuit with one or more power transistors on the same chip. Some power ICs are capable of handling 2 A or 2 W. They can save circuit-board space because the power switch and driver electronics are on the same chip. These devices were made possible because of special fabrication techniques that permit a power device to coexist on the same chip with small-signal transistors without destroying them or interfering with their operation.

The first power ICs were *power drivers* for high-voltage neon seven-segment displays. They combined bipolar digital logic with a bipolar power transistor on the same chip. Later power ICs mixed bipolar and MOS power devices with analog circuitry. CMOS logic was combined with bipolar transistors in BiMOS technology. Then CMOS logic was combined with DMOS (MOSFET) transistors in CMOS/DMOS. If lateral layout or topology is employed, the IC can contain two or more on-chip power devices. But with vertical topology, the IC is limited to only one power device.

BiMOS technology is suitable for medium voltage and currents; it has been used to make motor controllers, solenoid switchers, and printhead switching ICs. By contrast, CMOS/DMOS technology is suitable for either low-voltage, high-current, fast-switching or high-voltage, low-current fast-switching ICs. BiMOS has been used to make power supply pulse-width modulator, multiplexer, and voltage regulator ICs, and CMOS/DMOS has been used to manufacture AC plasma and electroluminescent display drivers.

Three different techniques are used to isolate control circuitry from the power device to prevent interference and breakdown on the monolithic IC:

1. *Self isolation,* an extension of CMOS technology in which a reverse-biased junction is located between the source and the drain region. This technique is usually limited to devices drawing less than 2 A, but voltage can be as high as 500 V.
2. *Dielectric isolation* (DI) employs single-crystal "islands" or "tubs" grown on a polycrystalline silicon substrate for the IC functions. Current must be brought out of the top of the chip within the tub, so the voltage levels are limited. DI is said to produce the lowest parasitic capacitance and to permit full isolation on the chip.
3. *Junction isolation* (JI) permits both lateral and vertical ICs. An epitaxial layer is formed on the substrate, and deep junctions are diffused to obtain isolated areas. Current flow is similar to that in discrete power devices.

# DIGITAL LOGIC AND
# INTEGRATED CIRCUITS

# Overview

Standard digital logic integrated circuits made by many different semiconductor technologies are circuit building blocks formed from logic gates. They can range from simple gate-level logic functions to very large scale integrated (VLSI) circuits. Standard digital logic ICs are manufactured for inventory and are listed in the manufacturers' catalogs. All ICs in the same family are mutually compatible, regardless of the manufacturer, and they do not require special interface circuits for system integration. The wide variety of products available in standard digital IC families is a consequence of constant improvements and innovations in manufacturing methods and ongoing success in boosting gate speed while reducing gate power dissipation.

Unfortunately there is no ideal digital logic family because each has its share of merits and drawbacks. So circuit designers rank device performance characteristics in the order of their importance in their intended applications. For example, in a new design, if low power consumption is of paramount importance, one of the latest CMOS families would be selected. On the other hand, if blinding speed is a prime requirement, one of the emitter-coupled logic (ECL) families would be chosen. But if neither of those factors is critical, one of the TTL families is the likely candidate, particularly if it is readily available from multiple sources, is a complete family, and is mature so that prices are dropping to their lowest levels.

Cost is an important consideration, and designers are willing to waive certain desirable features to keep costs down. Thus designers must be willing to accept tradeoffs of characteristics that rank lower on their priority lists. This has led to compromises by mixing different families within the same circuit to gain the benefits of both where they will be most effective. A prime example is BiCMOS, a combination of TTL and CMOS.

The rising popularity of battery-powered electronic products has elevated the importance of the low-voltage, low-power requirement despite a speed tradeoff. Nevertheless, the characteristics of some of the newer CMOS digital logic families are approaching the ideal, and they will work efficiently from a low 3-V supply. However, manufacturers are reluctant to redesign successful products that are still selling well in the marketplace just to gain minor or incremental performance improvements. Some manufacturers will continue to buy and use obsolete logic families that are no longer recommended for new products despite their rising unit costs because their use will still not justify the cost of reengineering the product.

# Truth Table

A *truth table* is a matrix of ones and zeros that lists all possible combinations of input values and output values of a logic gate or circuit. Examples of truth tables are given in Figs. 9-1, 9-2, and 9-3. In truth tables for simple gates, *A* and *B* are typical designations for inputs while *C* designates the output. In truth tables for more complex gates, *C* can indicate a carry and *S* can indicate a sum. A logic *low* is indicated as 0 and a logic *high* is indicated as 1. The actual values of the low and high voltages are determined by the circuit design. Both 0 and 1 can represent either negative or positive values provided there is a detectable difference in value between them.

SCHEMATIC SYMBOL          TRUTH TABLE

(a.) NOT

| A | B |
|---|---|
| 0 | 1 |
| 1 | 0 |

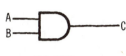
(b.) AND

| A | B | C |
|---|---|---|
| 0 | 0 | 0 |
| 0 | 1 | 0 |
| 1 | 0 | 0 |
| 1 | 1 | 1 |

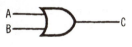
(c.) OR

| A | B | C |
|---|---|---|
| 0 | 0 | 0 |
| 0 | 1 | 1 |
| 1 | 0 | 1 |
| 1 | 1 | 1 |

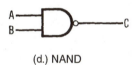
(d.) NAND

| A | B | C |
|---|---|---|
| 0 | 0 | 1 |
| 0 | 1 | 1 |
| 1 | 0 | 1 |
| 1 | 1 | 0 |

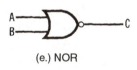
(e.) NOR

| A | B | C |
|---|---|---|
| 0 | 0 | 1 |
| 0 | 1 | 0 |
| 1 | 0 | 0 |
| 1 | 1 | 0 |

**Figure 9-1** Gate (logic) symbols and truth tables: (*a*) NOT, (*b*) AND, (*c*) OR, (*d*) NAND, and (*e*) NOR.

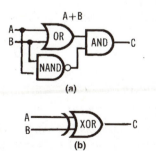

| A | B | C |
|---|---|---|
| 0 | 0 | 0 |
| 0 | 1 | 1 |
| 1 | 0 | 1 |
| 1 | 1 | 0 |

(c)

**Figure 9-2** Exclusive OR (XOR) gate: (*a*) logic, (*b*) symbol, and (*c*) truth table.

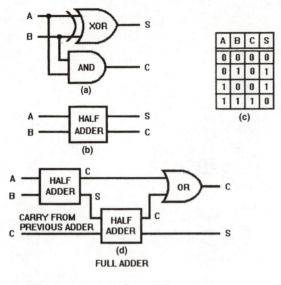

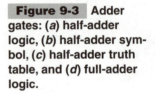

Figure 9-3 Adder gates: (*a*) half-adder logic, (*b*) half-adder symbol, (*c*) half-adder truth table, and (*d*) full-adder logic.

# Basic Logic Gates

There are five basic logic gates: AND, OR, and NOT are positive forms, and NAND and NOR are negative forms. The schematic symbols for each of these five logic gates are given in Fig. 9-1*a* through *e* with their truth tables. To interpret a truth table, consider the table for the AND gate in Fig. 9-1*b*. Reading horizontally across the top, *A* and *B* designate the input terminals, and *C* designates the output terminal. Reading down the columns, notice that if both *A* and *B* are 0 or *low,* the output *C* is also 0 or low, but if either *A* or *B* are 0 or low while the other is 1 or *high,* the output *C* will be 0 or low. However, if both *A* and *B* are 1 or high, the output at *C* will be 1 or high because that is the one condition in which the inputs add. These statements fully describe the AND function. The descriptions of the other basic logic gates can be determined by similar analysis.

More complex functions such as the *EXCLUSIVE OR (XOR) gate* require the interconnection of three gates, OR, AND, and NAND, as shown in its logic diagram in Fig. 9-2*a*. Its schematic symbol is shown in Fig. 9-2*b,* and its truth table is shown in Fig. 9-2*c*. It can be seen that if either *A* or *B* is high, the output at *C* will be high, but if both *A* and *B* are high, the output at *C* is low. These statements describe the XOR function.

Figure 9-3*a* shows the logic diagram for the *half adder,* a binary circuit with two input and two output terminals. Its schematic symbol is shown in Fig. 9-3*b,* and its truth table is shown in Fig. 9-3*c*. Figure 9-3*d* shows the logic diagram of a *full adder,* showing how it combines two half adders.

# Flip-Flop Circuit Variations

A *flip-flop circuit* is a basic two-stage multivibrator circuit that was described in "Flip-Flop Circuits" in Sec. 5, "Fundamental Electronic Circuits," and Fig. 5-1 is its schematic diagram. For counting and scaling purposes, a flip-flop can deliver one output pulse for each two input pulses. There are four different variations of the flip-flop circuit: (1) *D flip-flop,* (2) *J-K flip-flop,* (3) *R-S flip-flop* and (4) *T flip-flop.* See also "Combinational and Sequential Logic" in this section.

## DELAY (D) FLIP-FLOPS

The *delay (D) flip-flop,* as shown in the logic diagram in Fig. 9-4*a,* is a J-K flip-flop and a NOT circuit that delays its input by one clock pulse period. The output is therefore a function of the input that appeared one pulse earlier. Its schematic symbol is given in Fig. 9-4*b.*

## J-K FLIP-FLOPS

The *J-K flip-flop,* as shown in the logic diagram Fig. 9-5*a,* consists of four interconnected NAND logic circuits. The state of its output changes if both the *J* and *K* inputs are 1 (high) when a clock pulse arrives. If only the *J* input is 1, a clock pulse drives the output to 1, but if only the *K* input is 1, a clock pulse drives the output to 0. The choice of the letters *J* and *K* for this flip-flop was arbitrary. Its schematic symbol is given in Fig. 9-5*b.*

## RESET-SET (R-S) FLIP-FLOPS

The *reset-set (R-S) flip-flop,* as shown in the logic diagram Fig. 9-6*a,* consists of two NAND logic input circuits. One is designated *R* for reset and the other is designated *S* for

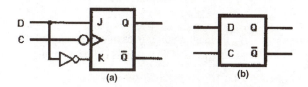

**Figure 9-4** Delay (D) flip-flop: (*a*) logic, and (*b*) symbol.

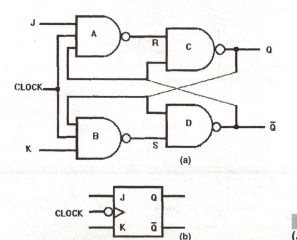

**Figure 9-5** J-K flip-flop: (*a*) logic, and (*b*) symbol.

set. Only one of the two inputs can be high, or logic 1, at a time. The flip-flop's operating speed is determined by the time it takes to charge the capacitors in the clock-steering network connected between the $R$ and $S$ inputs. If there is a logic 1 at the $S$ input when the clock pulse arrives, the flip-flop and its $Q$ output go to the 1, or on, state. A logic 1 at the $R$ input resets the circuit to the logic 0, or off, state. This flip-flop is also called the *set-reset flip-flop* or the *S-R flip flop*. Its schematic symbol is given in Fig. 9-6b.

## RST FLIP-FLOPS

An *RST flip-flop* has an additional trigger input $T$ which can be energized to make the flip-flop change state. If the flip-flop output is off, a pulse on the $S$ or $T$ inputs will turn it on, but a pulse on the $R$ input will not cause the state to change. However, if the flip-flop output is on, a pulse on either the $R$ or $T$ inputs will turn it off. In this situation, a pulse on the $S$ input will not cause the state to change. This flip-flop is also called a *set-reset trigger flip-flop* and an *SRT flip flop*.

## T FLIP-FLOPS

The *T flip-flop,* as shown in the logic diagram Fig. 9-7a, is a J-K flip-flop that changes state with each application of a trigger or clock pulse to its single input terminal $T$. Widely used in counter circuits, it is also called a *toggle*. Its schematic symbol is given in Fig. 9-7b.

# Binary Counters

A *binary counter* is a chain of flip-flops that can count pulses to control digital circuits. The basic counter formed from T flip-flops, as shown in Fig. 9-8, is called a *ripple counter*. The

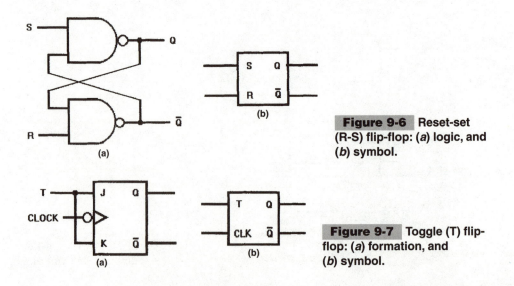

**Figure 9-6** Reset-set (R-S) flip-flop: (*a*) logic, and (*b*) symbol.

**Figure 9-7** Toggle (T) flip-flop: (*a*) formation, and (*b*) symbol.

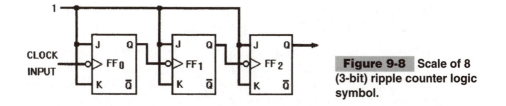

**Figure 9-8** Scale of 8 (3-bit) ripple counter logic symbol.

*J* and *K* terminals of each flip-flop are connected and returned to logic 1. This connection converts the J-K to a T flip-flop. The output *Q* of each flip-flop is at logic 0 because the *J-K* terminals are returned to a logic 1. The outputs of $FF_0$, $FF_1$, and $FF_2$ are designated $Q_0$, $Q_1$, and $Q_2$, respectively. The output at $Q_0$ is the least significant bit (LSB), and the output at $Q_2$ is the most significant bit (MSB).

The output of $FF_0$ changes its state once for every 2 clock pulses; the output of $FF_1$ changes its state for every 4 clock pulses; and the output of $FF_2$ changes its state for every 8 clock pulses. Because of this sequence, a counter with 3 cascaded flip-flops is called a *divide-by-8 circuit*. With 4 flip-flops, it would be called a *divide-by-16 circuit*.

The *ripple counter* is an *asynchronous* counter because the individual flip-flops are not clocked simultaneously. It is also known as an *up* or *forward counter* because it adds each input pulse to the count. All of the flip-flops of a *synchronous counter* are clocked and change their states simultaneously. Other types of counters are as follows:

■ A *down counter* is a reverse counter that subtracts 1 from a preset number during each input pulse.
■ An *up-down counter* is a bidirectional counter that can operate as either an up or down counter.
■ A *modulo counter* is a counter that counts other than binary multiples. An example called a *modulo-5 counter* counts 1, 2, 3, 4, and resets itself to 0 on the 5th pulse.

The diagram of a *modulo-10 counter* is given in Fig. 9-9. It is a decade counter that counts 1 to 9 and resets itself to 0 on the 10th pulse. However, a NAND gate is connected so that six of the counts will not be used.

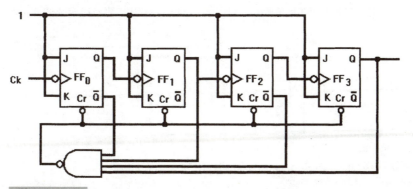

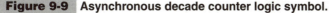
**Figure 9-9** Asynchronous decade counter logic symbol.

# Shift Registers

A *shift register* is a circuit that can store a binary number or an instruction. For example, if a binary number or instruction has $n$ bits, it is referred to as an $n$-bit word. An $n$-bit shift register consists of $n$ cascaded flip-flops. Flip-flop $FF_0$ contains the least significant bit (LSB) and $FF_{n-1}$ the most significant bit (MSB) of the word stored in the register. There are five general types of shift register:

1. *Serial registers* store a word and read it out serially. If the word is shifted to the right, it is called a *right-shift* register; if it is shifted to the left, it is a *left-shift* register.
2. *Serial-in, parallel-out registers* store words serially and read them out in parallel. They are also called *serial-to-parallel converters.*
3. *Parallel-in, serial-out registers* store words in parallel and read them out serially. They are also called *parallel-to-serial converters.*
4. *Parallel-in, parallel-out registers* store and read out words in parallel.
5. *Circulating registers* circulate words continuously. They are also called *dynamic shift registers* and *shift-ring read-only memories.* The 4-bit serial shift register, shown in Fig. 9-10, includes J-K flip-flops. During each clock pulse, each flip-flop assumes the state of the flip-flop that precedes it.

# Combinational and Sequential Logic

Standard commercial digital logic IC families include many devices that are classified as *combinational* or *sequential.* These circuits have no memory, so they make decisions based on the inputs they receive. For every combination of signals at the input terminals, there will be a definite signal combination at the output terminals. Examples of *combinational devices* include *NAND, NOR, AND, OR,* and *XOR gates* as well as *inverters* and *Schmitt-trigger inverters.*

By contrast, a *sequential-logic device* can store information, so its output signal depends not only on the most recent input signal, but also on input signals it has received in the past—a sequence of inputs. This memory capability adds complexity to these digital building blocks and the systems in which they are placed, both in design and operation.

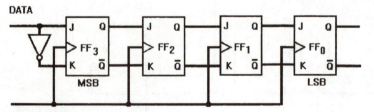

**Figure 9-10**   Serial-in, serial-out (SISO) shift register logic symbol.

Flip-flops, examples of sequential logic, are simple gate circuits capable of storing information. Modern digital logic families include D flop-flops, some with set and reset and others that are three-state, noninverting. Flip-flops are the basic circuits in other important standard digital logic ICs such as *transparent latches, registers,* and *counters,* all of which are memory devices.

# Standard Logic Families

Standard digital logic ICs are general-purpose digital building blocks suitable for many different applications. These devices can be interconnected to form complex logic systems. The many different configurations of logic gates available in each family allow the designer to assemble a system without redundant elements. However, if a desired function is not available as a catalog item, the designer has the choice of assembling as much of the system as possible with standard ICs and implementing the remainder of the circuit with discrete components or, alternatively, ordering an application-specific IC (ASIC) tailored to the system's needs. These could be either custom or semicustom ICs or a gate array.

Mature digital logic families include all of the logic gates discussed under "Basic Logic Gates," as well as such functions as *decoder/demultiplexers, multiplexers, voltage-controlled oscillators, multivibrators, comparators, first-in, first-out (FIFO) memory buffers,* and *memory controllers.* There are also three-state inverting and noninverting *buffers.* Standard logic ICs called *transceivers, receivers,* and *line drivers* are classed as *interface circuits.*

The ongoing process of reducing line width to submicron size has made it possible to integrate as many as eight functions on a single silicon chip and keep them electrically isolated from each other. Many different logic functions are available four (quad), six (hex), and eight (octal) to a package. This permits higher component density at a lower cost per function, and conserves circuit-board space by reducing the package population on the board, both attractive features to circuit designers.

In general, very complex IC functions are less likely than simple functions to become standard products. The designer's decision to use alternatives if a standard IC is not available will be influenced by available circuit-board space, power limitations, number of systems required, cost and availability of the alternatives, and urgency in getting the product to market.

Even if a product planned for high-volume production can be made from standard logic ICs, designers will look for higher-level integration to reduce the parts count, power requirement, and circuit-board size to improve reliability and reduce unit cost. The alternatives to standard logic today include *field-programmable logic devices* (PLDs), *gate arrays, standard cells,* or a *full custom IC.*

Standard digital logic ICs now include *small-scale integration* (SSI), limited to about 12 gates, *medium-scale integration* (MSI) with up to 100 gates, *large-scale integration* (LSI) with up to 1000 gates, and *very large scale integration* with more than 1000 gates. Many LSI and VLSI ICs available as commercial products perform specific functions in automotive, telecommunications, or consumer entertainment systems.

The introduction of the microprocessor (MPU) and the microcontroller (MCU) has reduced overall demand for standard digital logic ICs, but they will still be needed for "gluing" or completing circuits that contain custom and semicustom ICs.

# Bipolar Versus CMOS Families

Digital logic families are in three general categories: (1) *transistor-transistor logic* (TTL), (2) *emitter-coupled logic* (ECL), and (3) *complementary metal-oxide silicon* (CMOS) technology. The TTL and ECL families are based on bipolar junction transistors (BJTs), while CMOS is based on a combination of N- and P-channel MOSFET transistors. The basic CMOS NOT gate or inverter is shown in section view in Fig. 9-11*a* and schematically in Fig. 9-11*b*.

The term *bipolar* refers to NPN and PNP junction transistors used in TTL and ECL devices. Both the output drivers and input buffers use transistors, so there is a direct transistor-to-transistor connection. The earlier logic families *resistor-transistor logic* (RTL) and *diode-transistor logic* (DTL) were interconnected with resistors and diodes. Both of these families are now obsolete and are no longer used in circuit designs.

The original TTL devices from Texas Instruments, more than 40 years ago, were called *gold-doped TTLs*. Over the years improvements have been made in TTL devices to reduce their power consumption and increase their speed. A major improvement was the addition

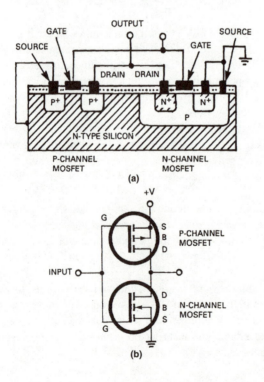

**Figure 9-11** (CMOS) NOT gate: (*a*) circuit, and (*b*) schematic.

of Schottky diodes to the bipolar transistors. This led to a number of new families with performance improvements over the original TTL. These include standard Schottky TTL (S-TTL), low-power Schottky TTL, (LS-TTL), advanced low-power Schottky TTL (ALS-TTL), and fast advanced Schottky TTL (FAST).

The principal advantages that TTL had over CMOS were higher speed, higher output drive, and higher transistor gain, but these advantages are disappearing.

The N- and P-channel MOSFETs in CMOS devices differ from the bipolar junction transistors in TTL and ECL in both structure and operation. CMOS devices were first introduced as CMOS 4000B by RCA. Its primary advantage was low power consumption and higher component density per chip. Recent performance improvements permitted CMOS to catch up with TTL. The original CMOS 4000B has been superseded by metal gate CMOS (MG-CMOS), high-speed silicon-gate CMOS (HC/HCT-CMOS), improved high-speed silicon-gate CMOS (HCS/HCTS-CMOS), advanced CMOS (ACL and FACT-CMOS), and low-voltage CMOS (LVX-CMOS).

Emitter-coupled logic (ECL) is derived from the common differential-amplifier configuration in which one side of the amplifier consists of multiple input bipolar transistors with their emitters tied together. An input bias on the opposite side of the amplifier causes it to operate continuously rather than saturate. As a result, ECL consumes more power in either state, but it still offers the fastest switching speeds of any logic family. Over the years there have been ECL improvements. Three of the newer improved families are 100K ECL, 10HECL, and ECLinPS.

Digital IC performance is evaluated with a two-input NAND gate. Speed is determined by measuring average gate propagation delay, measured in nanoseconds, and power consumption determined by measuring average gate power dissipation in milliwatts. Despite what might appear to be obvious choices in the selection of logic families based on speed and power, other factors must be evaluated. These include the available power source and environmental conditions in which the circuitry will operate (consumer, industrial, or military/aerospace). Other considerations are device unit cost and the availability of alternative- or second-source manufacturers.

# Speed-Power Graph

The *speed-power graph* is a plot of speed versus power performance characteristics for digital logic families. The vertical axis of the graph, *average gate propagation delay* (speed), is graduated in nanoseconds, and the horizontal axis, *average gate* dissipation (power) is graduated in milliwatts. The distribution of 12 points representing average values for 12 different digital logic families is given in Fig. 9-12.

CMOS families offer the lowest average gate power distribution, while ECL families offer the highest speeds. The many TTL families show the widest variations in both speed and power. The 10-ns × 10-mW point is the plot for standard gold-doped TTL, an important benchmark for indicating progress. The objective of all digital logic development over the years has been to move the plotted average value regions for each logic family down and to the left, toward the ideal 0-0 origin standing for infinite speed with zero power dissipation.

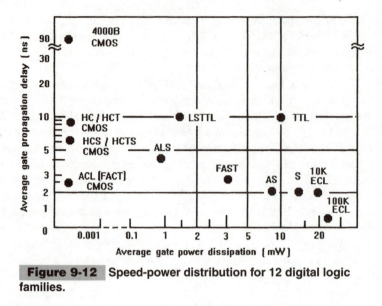

**Figure 9-12** Speed-power distribution for 12 digital logic families.

Improvements in such Schottky TTL families as ALS-TTL and FAST have dragged the TTL averages toward the origin, but CMOS has made faster progress. ACL CMOS logic offers average gate speed of less than 3 ns and average gate power of less than 0.001 mW.

The speed-power graph is based on averaged test data taken from a number of sample 2-input NAND gates fabricated by different manufacturers in each of the 12 technologies. Each point plotted is a centroid that lies within the envelope determined by encircling the plotted points of each of the devices sampled. Not all manufacturers agree that a single point truly represents their products, but most agree that a point representing just their own devices would be close by. The validity of the chart depends upon uniform test procedures performed under identical ambient temperature and environmental conditions.

# BiCMOS Integrated Circuits

Memories, microprocessors, and other integrated circuits made with both bipolar and CMOS circuitry on the same chip benefit from low CMOS power consumption and high bipolar speed. Bipolar ICs switch faster than CMOS ICs, offer greater current drive, and are more suitable for analog applications. CMOS ICs, on the other hand, consume less power, dissipate less heat, and permit higher gate density than bipolar ICs.

BiCMOS chips are more complex than either bipolar or CMOS ICs. BiCMOS technology has been used to manufacture analog ICs years longer than it has been used to make digital ICs. The three most important reasons for using BiCMOS technology in digital ICs are:

**1.** Compatibility with analog devices
**2.** Improved digital device performance
**3.** Reduced system cost

BiCMOS SRAM memories consuming the same power as bipolar memories offer faster access times or equal access times at lower power. In typical high-density, low-power BiCMOS memories, the control functions and memory cells are CMOS and the I/O circuitry is either TTL or ECL. However, low-density, high-power memories use ECL for decoding, sensing, and output drivers.

Signal delays in critical paths have been reduced and I/O throughput has been increased in BiCMOS microprocessors, and BiCMOS gate arrays consume less power and have higher gate density than bipolar gate arrays. BiCMOS gate arrays can be made with any combination of bipolar and CMOS that best fits their application. Initially CMOS gate arrays had only bipolar I/O, but more recently, bipolar transistors have been put in adders and registers to take advantage of the higher NPN transistor speed. Analog/digital ICs combine CMOS digital signal processing for higher density with bipolar analog signal processing.

# Digital Logic Characteristics

Speed is usually the most important specification in any logic family. But high speed can introduce problems such as noise generation, higher power consumption, higher costs for associated components, and more difficult board layout. Digital logic speed is measured as OR gate propagation delay (ns), D-type flip-flop toggle rate (MHz), and output switching time (ns). They are usually presented as an average of minimum and maximum values.

The important digital logic characteristics are:

- *Switching speed.*   The speed at which a gate switches, or changes its output from *high* to *low* or vice versa. It is specified as *average propagation delay* and measured in nanoseconds (billionths of a second). The fastest switching speed is desired in all logic circuits, so propagation delay should be as short as possible.
- *Power dissipation.*   A measure of the electrical energy that is converted into heat in powering the device. It is specified as *average gate-power dissipation,* and measured in milliwatts. This value should be as low as possible to conserve power and minimize cooling problems.
- *Speed-power product.*   The product of average gate propagation delay (ns) and average gate-power dissipation (mW). It is expressed in picojoules (pJ), a unit of energy.
- *Noise margin.*   A measure of how securely the digital logic IC transmits and receives information without errors in the presence of electrical noise. It is desirable to have noise margin be as *wide* as possible. The output voltage must be larger than the input voltage required to set the logic states correctly. Noise margin is measured in volts.
- *Fan-out.*   The number of inputs to other gates that can be driven successfully by the digital IC. High fan-out capability is important because it helps to reduce the number of ICs on the circuit board.
- *Circuit density.*   A measure of the effective use of silicon "real estate." It is important that active functions such as gates occupy a minimum amount of space on the chip, as measured in square mils (thousandths of an inch). The number of transistors and gates that can be integrated on a chip depends on its dimensions and such factors as required electrical isolation and ability to dissipate heat.

■ *Manufacturing cost.* Some digital logic IC families cost more to manufacture than others. Cost is directly related to the number of masking and processing steps required to produce the wafer, the test procedures required, the size of the chip, and the process yield.

# Programmable Logic Devices (PLDs)

A *programmable logic device* (PLD) is an unstructured array of AND and OR logic gates that can be organized to perform dedicated logic functions by selectively opening or otherwise altering the interconnections between the gates. This can be accomplished by blowing fusible links in some devices or reducing the conductivity of interconnections in others by the selective application of overvoltage. PLDs available as commercial off-the-shelf devices can meet the requirements for certain kinds of custom or semicustom logic. Because PLDs have many alternate sources and are made in high volume, they are relatively inexpensive.

PLD technology is an extension of *field-programmable read-only memory* (PROM) technology. Although PROMs are primarily used as memories, they can be programmed to perform some logic functions. PLDs expand that capability and are now made with TTL and ECL as well as three different CMOS technologies: (1) *fuse programmable,* (2) *ultraviolet erasable,* and (3) *electrically erasable.*

The term *PLD* avoids references to the proprietary names and acronyms by which they are better known. The PAL (for *programmable array logic*) was developed by Monolithic Memories (now Advanced Micro Devices, AMD), and the FPLA (for *field-programmable logic array*) was developed by Signetics (now Philips Semiconductor). Altera Corp. manufactures CMOS PLDs that it calls ELPDs (for *electrically programmable devices*). Other terms such as *integrated fuse logic* (IFL) and *fuse-programmable logic* (FPL) are also synonymous with PLD.

The first PLDs (PALs and PLAs) were intended as replacements for small-scale integration (SSI) logic devices and had from 4 to 100 gates. Subsequently those gate counts were increased from 100 to 500 gates. Either PALs or FPLAs can replace four or more conventional logic ICs, thus conserving circuit-board space and reducing power consumption.

Large-scale CMOS PLDs are available with 50,000 or more gates, but those with 1200 to 15,000 are most in demand. The decision to use a PLD or gate array (ASIC) hinges on the anticipated quantities, development cycle, electrical performance, and price. Cost and time-to-market considerations give PLDs the edge, if electrical performance goals can be met. However, for high volume, unit costs favor ASICs, but they have longer development time.

PLDs were designed for programming with commercial benchtop PROM programming equipment which can be modified with plug-in "personality cards." The cards contain the circuits needed to provide the voltages required for blowing fuses or altering junction conductivity. PLDs can also be programmed with desktop computers that run the required applications software.

The internal structures of PLAs and PALs are similar to those of PROMs. All of these devices have comparable AND/OR structures, but flexibility and programmability differ.

The AND/OR structure consists of an AND matrix that accepts inputs, performs the desired AND functions on the inputs, and then sends those results to an OR array which combines the AND functions to produce AND/OR outputs. This structure permits PLDs to perform boolean sum-of-product logic.

By contrast, the PROM has a fixed AND array and a programmable OR array, the PAL has a programmable AND array and a fixed OR array, and the PLA has both programmable AND and OR arrays.

The programmable AND array allows the PAL to solve equations with many inputs. It has a programmable input/output, internally registered feedback, and choice of output polarities. By contrast, FPLAs with both programmable AND and OR arrays are more versatile than PALs. However, the programmable AND arrays in both PALs and FPLAs permit them to overcome the programming logic limitations of PROMs.

# Gate Arrays

A *gate array* is a semicustom semiconductor integrated circuit prefabricated as a matrix of uncommitted identical cells with each cell containing transistors and resistors. Gate arrays are also known as *logic arrays, macrocell arrays,* and *undefined logic arrays* (ULAs). All gates, drains, sources, and channels are accessible, and all mask levels except the final one or two metal masks are predefined and fixed. The final metal masks uniquely define the interconnects for each application.

A gate array is considered to be an ASIC, as are custom-designed ICs, standard cells, and PLDs. The dedication or personalizing of a gate array starts with a fully diffused or ion-implanted semiconductor wafer with a matrix of identical primary cells arranged in columns. There are routing channels between the cell columns in the $x$ and $y$ directions and I/O devices around the periphery. The information available on each prefinished array includes the equivalent gate density, an indication of the number of functions that can be performed with the chip, and the size of each individual device.

Gate arrays can be fabricated in CMOS, ECL and TTL logic. They can also be made from gallium arsenide, silicon on sapphire, and various combinations of these technologies. Delivery or lead time is reduced because a gate array wafer is from 70 to 80 percent complete. Unit cost is lower than for other comparably sized ASICs because the wafers can be made in volume. Initial engineering cost is relatively low because only one or two of the masks must be custom designed and fabricated. Gate arrays are selected because they can:

- Reduce onboard standard logic and perhaps board size.
- Increase system reliability and performance.
- Reduce the number of interconnections and connectors.
- Lower power requirements.

The interconnection of the cells and interconnections between cells are customized with the aid of routing performed on a computer-aided design (CAD) workstation.

A digital gate array can contain only logic functions, or it can also include memory cells and analog circuits such as operational amplifiers. There are also 100 percent analog gate arrays. The functional blocks, referred to as *macrocells* (or *macros*) and *macro functions,* permit a wide selection of precisely characterized logic or other functions.

Gate array density normally is measured in *equivalent gates,* typically a two-input NAND gate. Macrocells, the basic gate array elements, include inverters, NANDs, NORs, latches, flip-flops, decoders, multiplexers, shift registers, and buffers. Macrocells have predefined metal interconnections.

Macrofunctions such as adders, arithmetic logic units (ALUs), comparators, decoders, flip-flop registers, and counters are integrated from macrocells. In contrast with macrocells, macrofunctions do not have predefined metal interconnections; redundant parts of macrofunctions can be deleted if required.

Gate arrays are made commercially in TTL, ECL, and CMOS, and gallium-arsenide arrays have been developed. They can be fabricated with different array and I/O technologies. The I/O circuitry is selected to be compatible with other kinds of logic. For example, the array could be ECL while the I/O section of it could be a CMOS array with a TTL I/O section.

ECL arrays provide subnanosecond delay times, while CMOS and TTL arrays with two-layer interconnections have delays between 1 and 5 ns. Geometries are generally in the 1.5- to 3-μm size range. Where delays in excess of 5 ns are acceptable, single-layer 5-μm CMOS arrays are specified.

As gate array technology progresses, improvements are being made in speed and number of equivalent gates. Increases in equivalent gate density result in higher performance in all of the technologies because of the shorter time delays. Feature size in commercial gate arrays dropped from 5 μm to less than 1.0 μm in less than 5 years. ECL arrays with more than 3500 equivalent gates and CMOS arrays with more than 10,000 equivalent gates have been announced.

# Field-Programmable Gate Arrays (FPGAs)

A *field-programmable gate array* is a gate array on a single chip based on *static RAM* (SRAM) *transistors* or *antifuse links* to control the array's signal paths. When the system is powered up, a program is downloaded into the FPGA to determine which transistors are on or off. Other FPGAs have antifuse links that melt together to form permanent interconnections when pulses from a programmer are applied.

# Standard Cells

A *standard cell* is a logic device that is completely fabricated by computer-aided methods from a blank wafer. The process produces a device from scratch that could be electrically equivalent to a gate array but occupies a smaller chip area because it contains only those gates, memory bits, and I/O pads that are actually needed for the specific application. Compare this with gate arrays that are 70 to 80 percent prefabricated but have redundant elements that take up valuable chip space. Both standard cells and gate arrays can use the same library of tested macrocells stored in CAD workstation memory. However, to gain this advantage, the standard cells will have higher unit prices, longer lead times, and higher up-front engineering costs.

A standard cell differs from a full custom IC because its design depends on the library of predesigned cells stored in the workstation memory. Skilled designers of full custom ICs can intercede and modify the macrocells in workstation memory to make them more efficient. If necessary, new cells more appropriate for the application can be introduced, further optimizing the design. This results in a more efficient IC that occupies a smaller chip, but up-front custom engineering charges are higher, and lead times are longer than for a standard cell.

# Semiconductor Memory Devices

A *semiconductor memory device* is an integrated circuit that retains information in the binary format of ones and zeros for storing programs and data for digital computers, microcontrollers, and other digital circuitry. These memories were developed to replace ferrite core memories as the main memories of digital computers more than 40 years ago. Core memories were made by weaving hundreds of tiny ferrite rings on frames with read and write wires. They remained in military computers long after nonmilitary computers had adopted semiconductor memories because ferrite cores were less susceptible to erasure by nuclear radiation bursts than were early semiconductor memories. Later, shielding and methods to protect semiconductor memories against this threat were developed.

The most basic semiconductor memory is a transistorized *flip-flop* that can store 1 bit in either state like a toggle switch. Eight flip-flops in series can, for example, store an 8-bit word or *byte* of information.

Semiconductor memories are classified by type, characteristics, and fabrication technology. An important distinction is how the information is accessed: *randomly* or *serially*. Another is how the memory retains data. *Read/write memories* permit new data to be entered by writing over existing data and permit the erasure of existing data. *Read-only memories* store data permanently or semipermanently, and *erasable read-only memories* are semipermanent memories that can be erased by exposure to ultraviolet light or the introduction of electrical signals and then reprogrammed.

*Volatile memories* retain their data only as long as they are powered, and *nonvolatile memories* retain data after the power is shut off. With power on, some memories retain their data permanently (*static*), while in others the data must be refreshed periodically (*dynamic*).

Semiconductor memories can also be grouped by fabrication process: *bipolar* or *metal-oxide semiconductor* (MOS). The bipolar memory technologies are *transistor-transistor logic* (TTL) and *emitter-coupled logic* (ECL). Three different MOS technologies have been developed: (1) *P-channel* (PMOS), (2) *N-channel* (N-MOS), and (3) *complementary* (CMOS).

## READ/WRITE VERSUS READ ONLY

The term *read/write* applies to semiconductor memory that permits data to be written in it and erased from it with logic-level voltages. This property is useful when the data and instructions change frequently, as in word processing or doing spread sheets. It is available only with the so-called (but inaccurately named) random-access memories (RAMs). By contrast, a *read-only memory* (ROM) without additional modifiers stores data permanently. ROMs are factory programmed by conductor depositions in the last stages of chip manufacture. Programmable ROMs are programmed permanently as a postmanufacturing operation by computer or specialized equipment.

## RANDOM ACCESS VERSUS SERIAL ACCESS

Read/write (RAM) and read-only (ROM) memories can be randomly addressed. That is, the data can be written or read without searching the entire file, as must be done with a serial or sequentially accessed memory. Examples of randomly accessed memories are all types of semiconductor random-access and read-only memories. However, other memory media such as hard-drive disks, diskettes, compact discs (CDs) and digital video discs (DVDs) can also be randomly accessed. Examples of serially accessed semiconductor memories are *shift registers* and *charge-coupled devices* (CCDs). But magnetic video- and audiotapes must be serially accessed. In general, it takes less time to access data in a random-access memory than in a serial-access memory.

## VOLATILE VERSUS NONVOLATILE

Two important classes of *volatile* semiconductor RAMs are dynamic RAMs (DRAMs) and static RAMs (SRAMs). As stated earlier, data stored in these memories will be lost if the power is shut off, but the data can be preserved if the memories are rapidly switched to standby or backup battery power. Most computers are designed so that the contents of their RAMs can be transferred to a magnetic hard-disk drive or other storage medium prior to a planned power shutdown. All ROMs, factory and field programmed, as well as electrical and ultraviolet erasable, are inherently nonvolatile, so data will be retained after power shutdown and no backup battery power is required.

## STATIC VERSUS DYNAMIC

The modifier *static* applies to the volatile static RAM (SRAM) and the modifier *dynamic* applies to the dynamic RAM (DRAM). Static RAM memory cells retain all written data as long as they remain powered. They are fast but complex and expensive devices. DRAMs, by contrast, have a simpler memory cell structure and higher bit density than comparable SRAMs, but they are slower and less expensive. The simpler structure and higher density are achieved at the cost of periodic data refreshing.

## PERMANENT VERSUS ERASABLE PROGRAMMED MEMORY

ROMs are permanently programmed at the factory during manufacture, and programmable ROMs (PROMs) are permanently programmed in the field by selectively blowing fuse connections between cells or by electrically altering the properties of the cells with overcurrent. The alternative to permanent data storage is semipermanent storage available only with erasable ROMs (EPROMs, EEPROMs, and flash memories). They can be programmed, erased and reprogrammed many times with special equipment. EPROMs, for example, can be erased with ultraviolet light, while both EEPROMs and flash memories can be erased electrically.

# Semiconductor Memory Families

Within the past 30 years, the design improvements in semiconductor memory devices have been driven by progress in microprocessors and personal computers. Microprocessors have become faster and more powerful, creating a demand for faster memory with higher storage density per chip, particularly SRAMs and DRAMs. At the same time, demand for better microprocessor performance has led to the doubling and quadrupling of the addressing capability of computers from 8 to 16 to 32 bits and to a seemingly insatiable appetite for more memory.

The ideal semiconductor memory device would have an access time less than 1 ns, high bit density per chip, low power dissipation, and be low in price. It would also be randomly accessible, nonvolatile, highly reliable, and standardized around the world. This ideal has yet to be realized because there is no known processing technology that will make this possible.

The selection of the appropriate memory device for any specific application remains an economic and engineering tradeoff, because of the differences between the characteristics of the semiconductor memory devices. The diversity of memory devices and the relationships between them are illustrated by the semiconductor memory family tree in Table 9-1.

CMOS technology has emerged as the preferred manufacturing technology for DRAMs. TTL and ECL memories account for only a small percentage of total memory sales. TTL

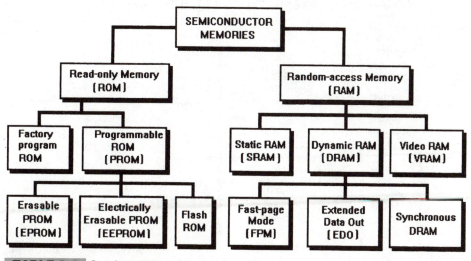

**TABLE 9-1**  Semiconductor Memory Family Tree

SRAMs offer high speed for cache, buffer, and scratchpad memories where short cycle time and fast access are critical.

*Access time* is the time interval between the instant that information is called for from memory and the instant when it is delivered (*read time*). It is also the time interval between the instant information is made ready for storage and the instant when storage is complete (*write time*).

*Cycle time* is the time between the instant that memory is accessed and the instant when it can be accessed again.

# Random-Access Memories (RAMs)

## DYNAMIC RAMS (DRAMs)

A *dynamic random-access memory* (DRAM) is built as a rectangular array of cells, as shown in Fig. 9-13, a simplified block diagram of a 4 Mb × 4-bit-wide DRAM. To read or write data, the CPU sends an address to the DRAM, which typically multiplexes it, supplying first the row address and then the column address. A row address can access a cell in 40 to 80 ns, the column address can access a cell in 20 to 40 ns, and the precharge time is 30 to 50 ns. Thus the cycle time can be from about 80 to 150 ns.

Each of the of data storage cells in the DRAM array consists of a single select transistor paired with a storage capacitor that can store 1 bit of information, as shown in the schematic in Fig. 9-14a. When one of the selection lines or rows in an array is actuated, it turns on all the transistor switches connected to it. The transistor serves as an on-off switch connecting the capacitor to its assigned data line, a column in the array. The simultaneous activation of a row and a column picks the cell for reading and writing.

The price paid for a memory consisting of simple cells formed from a single transistor and capacitor rather than a multitransistor flip-flop is the requirement that the data be periodically refreshed, hence the term *dynamic*. Refreshing regenerates the charge that is lost

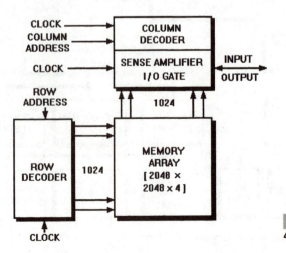

**Figure 9-13**  Block diagram of 4 Mb x 4-bit dynamic RAM (DRAM).

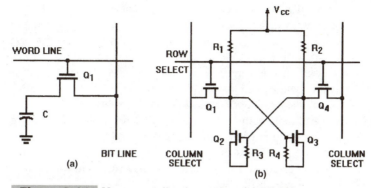

**Figure 9-14** Memory cell schematics: (*a*) DRAM, and (*b*) SRAM.

by reading and leakage. Refreshing, which occurs about once every 2 μs, is performed by the refresh amplifier when the switch in the data line is closed. The charge must always be large enough to permit the memory state to be read without ambiguity. A charge will remain on a continually refreshed DRAM cell capacitor indefinitely until new data is written over it or the power to the memory is shut off. The development of on-chip refresh circuitry reduced the cost of DRAM for personal computers by eliminating the need for external refresh circuitry.

The simple DRAM memory cell permits rapid access, draws less power than a SRAM cell, and permits large, dense arrays on a single chip. But reading the DRAM's capacitor cell drains current from it. It takes a finite amount of time to write data back into the cell after it has been read, and the bit lines must be charged before the cell can be read again.

By contrast the SRAM cell, as shown schematically in Fig. 9-14*b*, stores data in a transistor flip-flop and is ready for the next read cycle as soon as one is complete. While the SRAM is 3 to 5 times faster than a DRAM, its cells occupy more space than a comparable number of DRAM cells, and it consumes power continuously. The figure shows a two-transistor flip-flop with two passive pull-up transistors. See also "Static RAMs," following in this section.

The cost of memory is a function of silicon chip size, so large cells on a chip lower bit density and increase cost more per bit. Thus the smaller DRAM cell size permits higher bit density and lower cost per bit despite its refresh requirement.

The first DRAMs had a memory capacity of 1024 bits (1 kbit) and were made with the PMOS process. With the introduction of the NMOS process, rapid increases in DRAM density occurred. That technology reached its density limit at about the 16-bit level. The CMOS process made today's higher densities possible. The density of commercial DRAMs has now reached the 64-Mb level, typically organized in 4 to 8 major divisions and 32 subdivisions.

The demand for DRAMs is expected to surge dramatically with the introduction of high-definition television (HDTV) and other digital TV receivers because they will require a lot of digital memory. DRAM memory in some existing receivers can store a full frame of video in memory and display it twice, at twice the normal speed to reduce flicker and to store and display text on the screen.

While the density of DRAMs has been quadrupling about every three years, neither their access nor their cycle times have improved as rapidly. Computer manufacturers have continued to install slow but inexpensive DRAMs as main memory because they extended its useful storage methods, such as *page mode,* and they introduced caching and interleaving. *Caching* is the use of an intermediate faster memory between the registers of the microprocessor and the main memory and *interleaving* is the use of additional parallel arrays or banks of memory to increase memory bandwidth or throughput.

The first DRAMS had input/output organizations of 1-bit word widths, and these were acceptable until the introduction of the 64K DRAM created a requirement for 4-bit and 8-bit (byte-wide) word widths. DRAM organizations today are about 35 percent 4-bit, 25 percent each for 8-bit and 16-bit, and the remaining 15 percent 32-bit widths. It is expected that 64-bit widths will be available by 2003.

The clock rates or speed of DRAMs fell behind the clock rates of microprocessors, and this spurred the development of faster synchronous DRAMs including cache, Rambus, and Joint Electron Devices Council (Jedec) DRAMs. Until recently the disparity in speed between DRAM and the microprocessor was resolved by inserting small, fast SRAM cache memory between the processor and the main memory. The cache is then loaded with copies of those blocks of data stored in main memory that the microprocessor is most likely to request for the specific operation it is currently performing. However, this calls for additional circuitry to keep track of the data entering and leaving the cache. The alternative solutions are the inclusion of cache memories in the microprocessor and in the organization of some DRAMs.

## SYNCHRONOUS DRAMS (SDRAMs)

A *synchronous DRAM* (SDRAM or syncDRAM), or self-timed DRAM, operates at frequencies of 100 MHz to improve the match between the microprocessor and memory. These DRAMs offer 4 times the bandwidth of conventional DRAMs, and memory bandwidth is directly proportional to the clock frequency. All inputs and outputs are synchronized to the rising edge of a clock pulse. When a read operation is performed, more than one word is loaded into a high-speed shift register, and these words are shifted out, one word per clock cycle. However, access times are no faster than for conventional DRAMs. Synchronous DRAMs are now available at the 64-Mb level.

## JEDEC DRAMs

*Joint Electron Devices Council (Jedec) DRAMs* are SDRAMs that have a small SRAM cache inserted between their external pins and internal DRAM. This addition increases access speed.

## CACHED DRAMS (CDRAMs)

A *cached DRAM* (CDRAM), a proprietary development of Mitsubishi Corp., has a small SRAM cache inserted between its external pins and an internal DRAM. This addition increases access speed. In computer systems where main memory is connected directly to the microprocessor, CDRAM could replace external cache and eliminate one level in the memory hierarchy.

## RAMBUS DRAMs

*Rambus DRAM* is a proprietary DRAM system from Rambus Inc. Each chip has built-in decoding and hit-detection logic. The memory controller is simpler and fewer memory devices are required. Its output architecture permits rapid bursts of data from a wide internal bus before the entire access cycle is complete. It could eliminate external cache in some computer systems by connecting the microprocessor directly to memory. Rambus DRAM requires a special interface containing the I/O drivers, phase-locked loop, and logic.

## VIDEO DRAMs (VDRAMs)

The *video DRAM* (VDRAM) is a specialized DRAM that includes SRAM for a better match with the video application to maximize data transfer speed. It permits very fast graphics manipulation in personal computers and workstations. However, its chip is larger in a larger package, and its unit cost is higher than that of a DRAM of equal density.

## STATIC RAMs (SRAMs)

A *static random-access memory* (SRAM) consists of four- to six-transistor cells configured as flip-flops which do not depend on a capacitor for data storage. The flip-flops retain all data written into them indefinitely unless the data is overwritten or the power is turned off. Most SRAMs today are made with the CMOS process. SRAM speeds have kept up with microprocessor speeds because of the introduction of wider buses, alternative technologies, and special interfaces. As stated in "Dynamic RAMs (DRAMs)" earlier in this section, the SRAM has an inherently shorter cycle time than a DRAM because the flip-flop is ready for the next read cycle as soon as the previous cycle is complete. The cells in small CMOS SRAMs can be accessed every 8 ns, or about 5 times faster than DRAM cells, but SRAM access time nearly doubles for the larger devices, reducing their speed advantage by about 50 percent. SRAMs are available in 4-bit and 8-bit (byte-wide) architectures.

The highest-speed SRAMs are made with combined bipolar TTL and ECL processing. SRAMS can also be partly bipolar and partly CMOS (BiCMOS) with bipolar outputs and line drivers and CMOS cells. However, gains in speed are made at the expense of higher power consumption and lower cell density. Faster SRAMs have also been made from gallium-arsenide (GaAs) and silicon-on-sapphire processes. An SRAM's data can be retained on accidental or deliberate power shutdown if it has a backup battery.

# Read-Only Memories (ROMs)

A *read-only memory* (ROM) is a semiconductor memory that has been permanently programmed at the factory in the final stages of its manufacture. There are two kinds of ROMs. The *semicustom ROM* is factory programmed with final metal deposition that interconnects selected transistors while the chips are still part of the wafer. The alternative *full-custom ROM* is completely fabricated as a dedicated device, and all of the masks used in making its transistor connections are custom generated. Generally performed only for large-volume production orders, full-custom ROM fabrication makes more efficient use of the silicon in the chip than the semicustom ROM, so its chip can be smaller for the same bit density as the semicustom device. Most ROMs are now fabricated by the CMOS process, and they are available with bit densities of 1 Mb or more and 8-bit (1-byte) and 16-bit (2-byte) word widths.

## PROGRAMMABLE READ-ONLY MEMORIES (PROMs)

A *field-programmable read-only memory* (PROM) is a ROM that is permanently programmed as a postmanufacturing operation by specialized programming equipment or a computer with the proper software. The first PROMs available included fuse links between logic cells that could be selectively opened or blown with overcurrent so that the memory became a dedicated or application-specific matrix of ones and zeros. Later the alterable-transistor PROM was developed. In this technology, cell programming is done by short-circuiting base-to-emitter junctions in an array of transistors with overcurrent.

The fastest PROMs are made with ECL technology, the largest 64K PROMs are made with bipolar TTL technology, and the medium-speed PROMs are made with CMOS technology. In addition to performing memory functions, PROMs can be digital logic arrays because they contain a fixed AND array followed by a programmable OR array.

## ELECTRICALLY PROGRAMMABLE ROMs (EPROMs)

An *electrically programmable ROM* (EPROM) is a reprogrammable ROM packaged with a quartz window covering its memory chip so that previously programmed data can be erased by exposure to ultraviolet light so that it can be reprogrammed. To erase existing data, the EPROM must be removed from its circuit board and the opaque protective tape covering its quartz window must be removed to expose the chip to ultraviolet (UV) radiation. The chip is exposed to UV for up to 20 min for complete erasure or dissipation of the charge that defined its data. If the device is not removed from the board, the 12.5-V reprogramming voltage could damage or destroy adjacent sensitive components.

Standard EPROMs are packaged in ceramic cases, which are needed to support the transparent quartz windows. This package style accounts for a significant part of the cost of an EPROM. Most EPROMs are made with CMOS technology and they work from a 5-V supply. EPROMs have a useful life of about 100 erasure and rewrite cycles.

The *one-time-only* (OTO) *EPROM* is an EPROM chip in a smaller, lower-cost plastic package without a window. It is used where EPROM performance is better than that of other PROMs, but the expensive quartz-window package is not required because the program will not be changed. Typically the program, developed and debugged with conventional EPROMs, has proven to be successful in the host product.

## ELECTRICALLY ERASABLE PROGRAMMABLE ROMs (EEPROMs)

An *electrically erasable programmable ROM* (EEPROM) is a reprogrammable ROM that can be partially or completely erased by an overvoltage and reprogrammed while the device remains in its circuit board. This memory can be remotely reprogrammed provided that its host circuit board has the wiring necessary for erasure and reprogramming signals. The host circuit could be in a spacecraft, in an undersea instrument, or at the bottom of a deep well. Reprogramming can be done by a computer with a radio or cable link to the EEPROM. The ability to alter data in electronic equipment with a connection to a distant computer during routine maintenance procedures could be advantageous at an overseas military base where neither the trained personnel nor a suitable computer are available.

EEPROMs can be packaged in inexpensive plastic IC cases, but the EEPROM case contains more pins than an OTO EPROM case of comparable density because the EEPROM has the extra pins for erasure and reprogramming. EEPROM access times are slightly longer than those for EPROMs with comparable density, but they also operate from a 5-V power supply and most are made with CMOS technology.

## FLASH MEMORIES

A *flash memory* is an erasable, reprogrammable ROM whose data, unlike that an EEPROM, must be completely erased before reprogramming. Similar in structure to the EEPROM, it can also be reprogrammed without being removed from its host circuit. Flash memories can be based on either EPROM or EEPROM technology, and they offer a compromise between the two architectures. Flash memories are suitable for applications where frequent reprogramming is required. Depending on the process technology used to make it, the flash memory can be erased and rewritten from 100 to 10,000 or more times. These memories can be bulk erased in seconds, far faster than EPROMs. Moreover, a flash memory consumes less power than either an EPROM or an EEPROM. Some can store up to 1 Mb. Flash memories can be used to upgrade system software and replace disk drives and PROMs. Some have faster access times than EEPROMs. They offer performance that makes them competitive with SRAMS that have battery backup.

# Nonvolatile RAMs (NV-RAMs)

A *nonvolatile RAM* (NV-RAM) combines the characteristics of a volatile static RAM and a nonvolatile memory. During normal system operation, the RAM section functions as a read/write memory, but when it receives a store signal indicating a power shutdown, either unplanned or routine, the data in the RAM is transferred to the nonvolatile section for storage. Thus the nonvolatile section serves as the backup for the data. NV-RAMs are made with NMOS technology.

# Universal Receiver-Transmitters

*Universal receiver-transmitters* are large-scale monolithic IC receiver-transmitter circuits that combine many essential communications functions on a single chip, saving circuit-board space and assembly time. A single receiver-transmitter chip can perform such functions as framing, formatting, modem control, or microprocessor bus interfacing. The three kinds of receiver-transmitter LSI ICs available are:

1. Universal asynchronous receiver-transmitter (UART)
2. Universal synchronous receiver-transmitter (USRT)
3. Universal synchronous/asynchronous receiver-transmitter (USART).

*Universal asynchronous receiver-transmitter* (UART) ICs perform the following functions:

- Assemble and serialize asynchronous or isosynchronous formatted characters.
- Add start and stop bits for characters to be transmitted and delete them from received characters.
- Detect start bits and receive data sampling based on fast clocks.
- Detect errors and special conditions.
- Generate breaks.

Some UARTs also contain a bit-rate generator (BRG) and modem-control I/O pins. Single- and dual-channel UARTs can both be packaged in a single IC case.

*Universal synchronous receiver-transmitter* (USRT) ICs perform the following functions:

- Assemble and serialize synchronous formatted characters.
- Detect and generate synchronization characters.
- Frame and format one or more data link controls, a set of procedures followed by terminals, computers, and other devices to ensure orderly transfer of information on a single data link.
- Detect receive errors and special conditions.

*Universal synchronous/asynchronous receiver transmitter* (USART) ICs are receiver-transmitters that can support either synchronous or asynchronous transmission with the software programming of internal mode registers. In the synchronous mode, a USART can do the framing and formatting for character- or bit-oriented data link controls, or it might only perform the character synchronization of character-oriented data link controls. In asynchronous mode, a USART functions like a UART.

These ICs are available with one or two serial communication channels per package. Some UARTS and USARTs include bit-rate generators, and many USRTs and USARTs support character- and bit-oriented data-link controls. UARTs, USRTs, and USARTs are specified for printers, personal computers, modems, multiplexers, data PBXs, and communications test equipment.

# Gallium-Arsenide Digital ICs

A wide variety of digital and analog functions have been implemented in *gallium arsenide* (GaAs). These ICs are made by integrating transistors on semi-insulating GaAs substrates using processes that are similar to those used to fabricate silicon MOS ICs. The future of digital GaAs ICs is seen in custom and application-specific devices rather than in standard digital logic families. GaAs ICs are designed to be compatible with other logic families, particularly ECL. However, ongoing improvements in silicon ICs have maintained the narrow performance gap between silicon and GaAs ICs, and this has reduced the incentive for designers to switch to all GaAs IC circuits or even to specify more GaAs ICs.

GaAs ICs have been produced for two different markets: there is a commercial market for the higher-speed GaAs versions of certain types of silicon digital logic ICs and a military/aerospace market primarily for GaAs analog microwave monolithic ICs (MMICs).

GaAs SSI and MSI digital logic, memories, and gate arrays have been developed, but they lag silicon bipolar ICs in integration density. GaAs digital and analog interface chips are niche parts that supplement or complement slower silicon bipolar and CMOS linear and interface ICs in systems that require higher data rates or faster data conversion. These devices are also used to interface microwave and digital systems. For further information on MMICs see "Microwave Monolithic Integrated Circuits (MMICs)" in Sec. 7, "Microwave and UHF Technology."

# BATTERIES AND
# POWER SUPPLIES

# Overview

Modern electronic circuitry requires direct current for operating semiconductor devices. This direct current can be obtained from linear or switching power supplies, batteries, solar cells, or even fuel cells. This section covers the principal practical means for obtaining DC power at voltages that are typically less than 50 V. The technical definition of a battery is two or more power cells connected in series to provide an output voltage that is the sum of the voltages of the individual cells. But battery manufacturers' literature and consumer advertising, as well as the general public, typically refer to both 1.5-V C cells and 9-V alkaline batteries as *batteries*. This section distinguishes between the two electrochemical power sources.

# Batteries

A *battery* is an assembly of two or more electrochemical power-producing *cells*. The power cell consists of four principal components:

1. Anode or negative electrode (a reducing material or fuel)
2. Cathode or positive electrode
3. Oxidizing agent
4. Electrolyte which provides for internal ion transfer

Electrolytes are usually liquid, but some cells have solid electrolytes. In addition, there is a material separating the anode and cathode.

The cell converts the chemical energy within its active materials directly into electric energy by an oxidation-reduction process. The anode gives up electrons during cell discharge in an oxidation reaction and electrons are transferred from one material to another. The anode is separated from the cathode, which is capable of accepting electrons. Electrons flow in the external circuit connecting the anode and cathode through the electrolyte to provide power.

Cells can act as stand-alone power sources or they can be connected in series or parallel to give the desired output voltage or capacity. If the cells are connected in series, the output voltage is the sum of individual cell voltages, but if they are connected in parallel, the output current is the sum of the individual cell currents, but the voltage remains that of a single cell.

Batteries and cells are identified as *primary* or *secondary*. Primary (nonrechargeable) batteries or cells are capable of one continuous or intermittent discharge and are then discarded. Secondary (rechargeable) batteries or cells can be recharged a finite number of times following partial or complete discharge. They are recharged by passing direct current through them in a direction opposite to that of the discharge current.

*Reserve* cells and batteries are primary power sources that are activated by breaking a separate container filled with an electrolyte that has been isolated from the other parts of the cell or battery. This type of construction gives them a long shelf life in the inert condition. Another form of reserve power source is the *thermal battery,* which remains inactive until it is heated to melt a solid electrolyte.

If the cell electrolyte is contained in an absorbent or separator material, it is called a *dry cell,* but if it is liquid it is called a *wet cell.* There are also semiliquid *gelled-electrolyte cells.*

Electrochemical cells and batteries are important DC power sources for electronics and they have made possible many cordless electronic products, including portable radios, tape recorders, TV sets, cordless telephones, cellular mobile telephones, and field test instruments. Miniature *button cells* power watches, hearing aids, cameras and heart pacemakers. Cylindrical cells and round, flat *coin cells* provide standby power on CMOS memory circuit boards to preserve the data stored in volatile memories whenever power is shut down—unplanned or routinely. CMOS DRAMs or SRAMs backed up by power cells can function as nonvolatile ROM.

The discussion of cell and battery systems here is limited to those having direct application to electronics. Many different electrochemical processes produce the energy in commercially available cells and batteries suitable for powering electronics. The measure of performance of a power cell is determined by its *energy density* in terms of weight and volume. Comparisons can be made between systems based on watthours per kilogram (Wh/kg) or watthours per cubic inch (Wh/in$^3$). In most electronic applications where cells and batteries must be small, Wh/in$^3$ is a more meaningful indicator. But useful comparisons can be made only between cells or batteries in the same size packages or cases. *Rated capacity, nominal cell voltage, typical operating temperature,* and *shelf life* are other important specifications to be compared.

# Primary Electrochemical Systems

## ZINC-CHLORIDE (Zn/Cl) CELLS

The *zinc-chloride* (Zn/Cl) *cell* is an improved version of the earlier zinc-carbon cell. The cell has a nominal voltage of 1.5 V and an energy density of 85 Wh/kg or 165 Wh/L, but it has a 50 percent longer life than the zinc-carbon cell. These cells are inexpensive and have moderate shelf lives. They are still are used for powering flashlights and toys. The same electrochemical system is used to make flashlight batteries in two-terminal rectangular cases.

## ALKALINE–MANGANESE DIOXIDE (Zn/MnO$_2$) CELLS

The *alkaline manganese-dioxide* (Zn/MnO$_2$) *cell* is made from the same zinc–manganese dioxide chemical system as the zinc-chloride cells, but it has a highly conductive potassium-hydroxide electrolyte. This cell, as shown in cross-section view Fig. 10.1, functions at lower temperatures and higher drain rates than the zinc-chloride cell, and it has a shelf life in excess of three years. It can deliver from 2 to 10 times the ampere-hour capacity of the zinc-chloride cell. Open-circuit voltage is from 1.5 to 1.6 VDC, and its operating voltage is from 1.3 to 1.1 VDC. Specific energy is 130 Wh/kg and energy density is 375 Wh/L.

Alkaline cells are popular as factory installed and replacement power sources for electronic circuitry. They are recommended for powering such products as smoke and carbon monoxide (CO) detectors, handheld instruments, and pocket radios. Nine-volt batteries, formerly called *transistor batteries,* are made as six alkaline cells enclosed in a metal case.

## ZINC–SILVER OXIDE (Zn/AgO) CELLS

The *zinc–silver oxide* (Zn/AgO) *cell* is typically made as a button cell. The nominal output of this cell is 1.55V, and it is able to provide high capacity per unit weight, good shelf life,

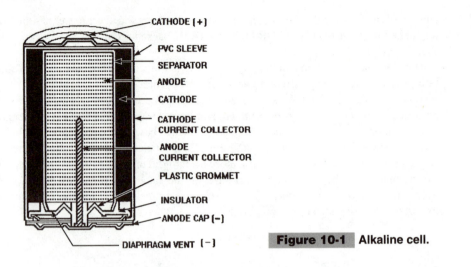

**Figure 10-1**  Alkaline cell.

and a flat discharge curve. Zinc–silver oxide cells are widely used to power hearing aids, watches, clocks, timers, calculators, and cameras.

## ZINC-AIR (Zn/O₂) CELLS

The *zinc-air* (Zn/O₂) *cell,* as shown in cross-section view Fig. 10-2, is typically packaged as a button cell. It takes oxygen from the air as its active cathode ingredient. The air cathode occupies only one-tenth of a cell's internal volume, so the anode can be larger than for other primary cells. This system offers the highest energy density per unit volume and weight of any primary cell. Nominal cell voltage is 1.5 V, specific energy is 146 Wh/Kg, and energy density is 146 Wh/L.

The zinc-air anode is an amalgam of powdered zinc mixed with a gelling agent, and the electrolyte is a water-based solution containing potassium hydroxide and zinc oxide. The air cathode assembly is a mixture of carbon, Teflon, and manganese dioxide pressed onto a nickel-plated screen. A semipermeable membrane of Teflon separates the electrodes to prevent moisture from entering and leaving the cell. The cell is activated when the seal is removed, and oxygen from the air is the active cathode ingredient. The cells are typically used to power hearing aids, medical instruments, and pagers.

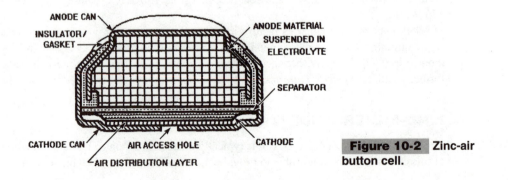

**Figure 10-2**  Zinc-air button cell.

# Lithium Primary Cells

*Lithium primary cells* are popular for powering electronic products because of their high energy (watthours) per unit volume. Some lithium systems offer twice the voltage output of alkaline cells. Although favored in military applications, about six different consumer lithium systems are in production. Nominal output voltage of a lithium cell is from 1.5 to 3.6 V.

Lithium cells are classified by their electrolytes, which can be organic or inorganic liquids or solids. Lithium chemistry is identified by the chemical abbreviation for lithium (Li) followed by abbreviations for each electrolyte. Some practical lithium chemistries are:

Lithium–sulfur dioxide (Li/$SO_2$)

Lithium–thionyl chloride (Li/$SOCl_2$)

Lithium–manganese dioxide (Li/$MnO_2$)

Lithium–carbon monofluoride (Li/$CF_x$)

Lithium–copper oxide (Li/CuO)

Lithium–solid electrolyte (iodine) (Li/$I_2$)

The advantages of lithium cells include high energy density (up to 250 Wh/kg), high power density, flat discharge characteristics, excellent service life over a wide temperature range, and excellent shelf life. Their higher cost is justified because a smaller number of cells is needed for equivalent energy. For example, one lithium cell can replace two alkaline cells in the same case size in certain applications. Lithium primary cells can be categorized on the basis of their electrolytes and cathode materials as *soluble cathode, solid cathode* and *solid electrolyte.* Some lithium electrochemistries can be recharged, so they can be classified as secondary systems. See "Lithium Ion (Li-ion) Cells," following.

## LITHIUM–SULFUR DIOXIDE (Li/$SO_2$) CELLS

The *lithium–sulfur dioxide* (Li/$SO_2$) *cell* offers excellent high-rate and low-temperature performance. Pressurized and hermetically sealed $SO_2$ cells offer excellent shelf life and a flat discharge characteristic. These cells have a nominal cell voltage of 3.0 V. They are being used in both military and special industrial applications, where their high capacity, high rate, and ability to operate in extreme temperatures is valued. Typical applications include powering appliances, telemetry, computers, clocks, CMOS memory, radio transceivers, emergency lights, medical instruments, security systems, and sonobuoys.

## LITHIUM–THIONYL CHLORIDE (Li/$SOCl_2$) CELLS

The *lithium–thionyl chloride* ($SOCl_2$) *cell,* as shown in cross-section view Fig. 10-3, is used for many general-purpose applications as well as MOS memory backup. These cells have a nominal output voltage of 3.6 V and a flat discharge curve. Packaged in hermetically sealed cases, they will operate satisfactorily at temperatures as high as 100°C. Batteries are made with this electrochemistry.

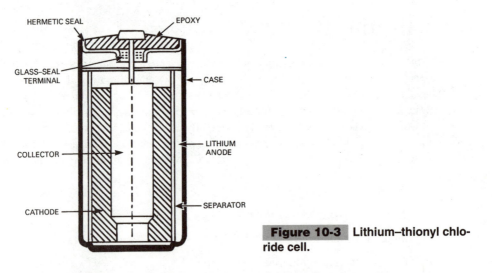

**Figure 10-3** Lithium–thionyl chloride cell.

## LITHIUM–MANGANESE DIOXIDE (Li/MnO₂) CELLS

The *lithium–manganese dioxide* (Li/MnO$_2$) *cell* offers good rate capability, low-temperature performance, excellent shelf life, and a flat discharge curve. It has a nominal voltage of 3.0 V. These cells typically are made as coin cells to power cameras, watches, calculators, CMOS memory, and instruments. They are also widely used for MOS memory backup. The cells are not hermetically sealed.

## LITHIUM–SOLID ELECTROLYTE (IODINE) (Li/I₂) CELLS

The *lithium–solid electrolyte (iodine)* (Li/I$_2$) *cell* can provide a nominal voltage of 2.8 V, and it is the principal power source in the heart pacemaker.

# Secondary Battery Systems

## NICKEL-CADMIUM (NiCd) CELLS

The *nickel-cadmium* (NiCd) cell is rechargeable and consists of a cadmium negative electrode, a nickel-hydroxide positive electrode, and an electrolyte of aqueous potassium hydroxide. It has a nominal voltage of 1.2 V, specific energy of 50 Wh/kg, and energy density of 150 Wh/L. Minimum charge time is 15 min, and operating temperature is −20 to 70°C. Portable sealed versions are available in case sizes comparable to those of alkaline cells. NiCd cells are found in batteries for powering handheld video cameras (camcorders), notebook computers, cellular mobile telephones, portable television sets, and cordless power tools.

## NICKEL–METAL HYDRIDE (Ni-MHD) CELLS

A *nickel–metal hydride* (Ni-MHD) *cell* is rechargeable and has a nominal voltage of 1.2 V, specific energy of 60 Wh/kg, and energy density of 250 Wh/L. Minimum charge time is 1 h, and operating temperature is −20 to 50°C. Batteries made from these cells have the same applications as NiCd batteries, but they offer higher energy density.

## LITHIUM-ION (LI-ION) CELLS

A *lithium-ion* (Li-ion) *cell* is made with a rechargeable lithium technology. It has a nominal output voltage of 3.6 V, specific energy of 120 Wh/kg, and energy density of 300 Wh/L. Minimum charge time is 1 to 2 h, and operating temperature is −20 to 60°C. The cathodes can be cobalt oxide or manganese dioxide, and the anodes are graphite. Batteries made from these cells are used in the same applications as NiCd and Ni-MHD batteries, but they offer higher output voltage and energy density.

## SEALED LEAD-ACID BATTERIES

The *sealed lead-acid battery* is based on a variation of the same electrochemistry that is used in the standard automotive battery. But it has been adapted for powering electronics by sealing the cells to prevent accidental spillage of sulfuric acid that could corrode and ruin electronic components. The sealed battery has cells with lead negative electrodes, lead-dioxide positive electrodes, and an electrolyte of liquid or gelled sulfuric acid. The nominal voltage of a lead-acid cell is 2.0 V, its specific energy is 35 Wh/kg, and its energy density is 70 Wh/L.

Two of these variations of the conventional lead-acid battery have either a *gelled* or *starved electrolyte.* The gelled cell has a polymeric gel electrolyte, and the cell is packaged in a rectangular case. The starved electrolyte cell contains a small amount of liquid electrolyte, and it is made in the form of a spiral-wound "jelly roll" and packaged in a cylindrical case. Both types of battery are made as proprietary products and are used to power such products as professional camcorders and portable television sets.

# Reserve Batteries

There are many military requirements for reliable primary *reserve batteries* capable of producing their full rated power output within seconds or minutes of activation after years of storage. After activation the cells and batteries function as conventional cells and batteries. Until recently the most widely procured reserve batteries for military use were based on zinc–silver oxide (Zn/AgO) cell electrochemistry. However, the lithium–thionyl chloride reserve cell, as shown in the cutaway view Fig. 10-4, has replaced them. The liquid electrolyte is held in a glass ampoule. After activation by a mechanical or explosive impulse that fractures the glass ampoule containing the electrolyte, the cell voltage rises rapidly to its nominal 3.65-V rated voltage. Reserve cells and batteries are used to power military survival radios, beacons, and life-support equipment.

Other military applications such as powering electronic circuitry within missiles, torpedoes, and submarine decoys require primary reserve batteries able to produce one-shot bursts of high electrical power output for periods of up to an hour within a fraction of a second of activation after 20 years of storage. They are built to withstand extreme centrifugal and linear acceleration forces.

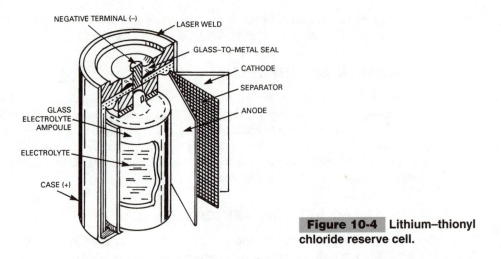

**Figure 10-4** Lithium–thionyl chloride reserve cell.

These requirements are being met by the *lithium alloy–iron disulfide thermal battery,* shown in the cutaway view Fig. 10-5. A solid electrolyte separates the lithium alloy anode and iron-disulfide cathode of each cell in the battery while it is on standby. When activated electrically, pyrotechnic sources within the battery initiate a chain reaction which raises the temperature within the case to 300 to 450°C, necessary to melt the electrolyte in each cell and produce electrical power. Output voltages can range from 1.5 to 100 V, depending on the battery size and internal connections.

# Battery and Cell Packaging

Standard commercial case sizes developed for zinc-carbon, zinc-chloride and alkaline cells and batteries are widely used for electronics. The cell case sizes include the cylindrical

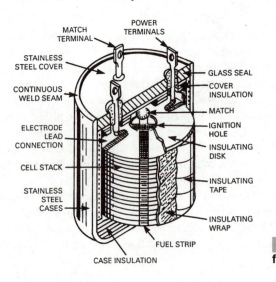

**Figure 10-5** Lithium–iron disulfide thermal battery.

AAA, AA, C, and D. The widely used 9-V alkaline battery is packaged in the rectangular case, as shown in Fig. 10-6. It can be seen that there are six cells within the metal case, which has snap-on terminals for positive contact connections.

A wide range of button cell sizes has been developed for alkaline, lithium, silver-oxide and zinc-air chemistries specifically for hearing aids, cameras, calculators, watches, and clocks. Some lithium systems have been adapted to the standard case sizes, but many are packaged in nonstandard cases for military and life-support equipment. They are also offered in *coin cell* packages.

Commercial/consumer-grade nickel-cadmium cells are packaged in the standard AAA, AA, C, and D cylindrical as well as the 9-V transistor battery cases as replacements for primary cells. Because of their lower voltage ratings, more nickel-cadmium cells are needed to provide the equivalent energy of primary cells. Proprietary NiCd batteries are also available for use in cordless telephones, and NiCd battery packs are available as replacements for cellular mobile telephones and notebook computers.

Gelled lead-acid batteries are available in many different rectangular packages, but some starved-electrolyte cells are packaged in D-size cases, which can be encased to form multicell batteries.

The *button cell* is a small, low-voltage power cell made as a metal disk with a diameter of about 0.4 in (11 mm) and a height of about 0.2 in (5 mm). One face is the *positive terminal* (anode), and the opposite face is the *negative terminal* (cathode). Examples include the lithium, silver-oxide, and zinc-air button cells.

# Power Supplies

A *power supply* is a circuit that converts alternating line current or unregulated direct current to regulated direct current. In the electronics industry the term implies a separately packaged AC-to-DC converter packaged as a potted module, an open-frame assembly, or an enclosed and shielded module. The two principal categories of AC-to-DC power supply are *linear regulated* and *switching regulated.*

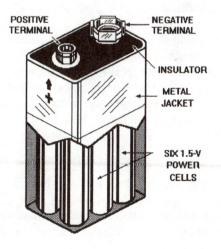

POSITIVE TERMINAL

NEGATIVE TERMINAL

INSULATOR

METAL JACKET

SIX 1.5-V POWER CELLS

**Figure 10-6** 9-V alkaline battery.

A DC power supply converts power from the AC line to DC, typically with a regulated output value. The AC input voltage is first rectified to provide a pulsating DC which is filtered to produce a smooth voltage. This voltage can then be regulated to assure that a constant output level is maintained despite variations in the AC power line voltage or circuit loading.

The switching power supply is now widely accepted for powering desktop and mainframe computers as well as network servers and computer workstations. It took years to overcome the resistance to switching or switchmode power supplies because of their inherent noise generation and poor regulation. They are still not accepted for powering sensitive instruments. The principal benefits of these supplies are their smaller size, lighter weight, and higher efficiency when compared with a comparably rated linear power supply. The regulation of most switching power supplies today is typically ±5 percent, and this is lower than for most linear power supplies. However, this regulation value is acceptable for powering most digital circuitry, including computers.

Many new components such as high-frequency bipolar and MOSFET power transistors, ferrite-core transformers, fast-recovery and Schottky rectifiers, and aluminum electrolytic capacitors were designed and manufactured specifically for switchmode applications.

Switching power supplies generate electromagnetic interference (EMI) that can adversely affect other electronic circuits in the vicinity. This problem is only partially solved with filters installed by the manufacturer. Many applications require additional filtering and shielding tailored to the sensitivity of the host product or system.

# Linear-Regulated DC Supplies

The *linear-regulated DC supply* is one whose regulator circuit provides a linear response to load changes, keeping the output voltage constant. There are three principal functional blocks in a voltage-regulated linear power supply, as shown in Fig. 10-7; (1) *rectifier,* (2) *filter network,* and (3) *regulator.* The *rectifier,* typically a bridge, delivers unregulated DC to the filter. Isolation from the AC line is provided by the transformer. The rectifier output is a steady DC voltage with an alternating voltage superimposed on it. The *filter network* removes the alternating voltage (ripple) and delivers it to the pass transistor $Q_1$ in series with the load.

Linear power supplies can have either a series or shunt *voltage regulator,* as determined by the location of the pass element (typically a transistor) with respect to the load. The position of transistor $Q_1$ in Fig. 10-7 indicates that it is in a *series-pass regulator,* the most common closed-loop regulator configuration. As either the load or the input line voltage changes, the voltage across $Q_1$ changes. The amplifier compares the voltage at the junction of resistors $RR_2$ and $R_3$ in a voltage divider with the reference voltage across zener diode $D_1$, and if it detects a change it produces an output that varies the bias on the base of $Q_1$ to regulate the current through the $Q_1$ to restore the output voltage to its intended value. This keeps the voltage delivered to the load essentially constant. The regulator forms a closed feedback loop.

Commercial linear power supplies are rated from 3 to about 1000 W with single or multiple outputs. Linear power supplies provide close regulation, are highly reliable, and are insensitive to minor shifts in frequency. Proven circuits, they are made in volume to sell at low prices, particularly in the low-power ratings where inexpensive transformers can be

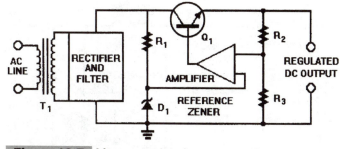

**Figure 10-7** Linear-regulated power supply.

used. The pass transistor acts as a variable resistor, and because it dissipates power continuously, it reduces the power supply's efficiency to values between 15 and 35 percent. Another drawback is its requirement for a large, heavy transformer for rectifying and filtering 50/60-Hz line power. Transformers are the largest and heaviest components of linear power supplies.

# Ferroresonant Power Supplies

A *ferroresonant power supply* is basically a tuned transformer that also functions as a filter. Both reliable and efficient, this supply is, however, sensitive to changes in line frequency. A ferroresonant transformer operating at a frequency of 50/60 Hz is large, heavy, and expensive—it is about three times as heavy and occupies about twice the volume of a 50/60-Hz transformer for a comparably rated linear power supply. Although a ferroresonant power supply's regulation is excellent and its efficiency can reach 80 percent, this value will depend on the stability of the AC line frequency. Another drawback is its loud annoying hum, making it unsuitable for use in rooms occupied by people. It also is known as a *constant-voltage supply.*

# Switching-Regulated Power Supplies

A *switching-regulated* or *switchmode power supply* is smaller in size, lighter in weight, and more efficient than comparably rated series-regulated or ferroresonant power supplies. The basic functional blocks of an offline switching-regulated power supply are shown in Fig. 10-8. They are the *rectifier, filter network, pass transistor, pulse-width modulator* (PWM), and *comparator.*

The comparator compares the output voltage taken at the junction of resistors $R_1$ and $R_2$ in a voltage divider to the reference voltage across zener diode $D_2$, and this difference determines the *on* time, or pulse width set by the PWM. If the DC output voltage falls below its reference value because of increasing load, the width of the drive pulses sent to switching transistor $Q_1$ will be increased, which will, in turn, increase the *on* time of $Q_1$ to offset the

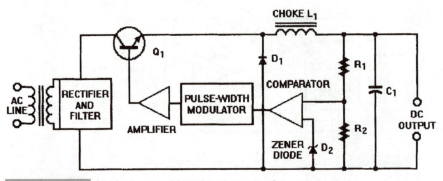

**Figure 10-8** Switching-regulated power supply.

voltage drop. Similarly, if the DC output voltage increases above its reference value due to decreasing load, pulse width will be decreased, reducing $Q_1$'s *on* time and reducing output to offset voltage increase.

The switching-regulated power supply is more efficient than a linear-regulated power supply because its regulator circuit conducts intermittently. Thus power is drawn only when the transistors are switching at frequencies typically from 20,000 to 100,000 times per second (20 to 100 kHz) or more. By varying the length of time the switch is on during each cycle, the amount of energy delivered to the filter can be controlled. Switching regulator efficiency can be as high as 85 percent in supplies with 12- to 15-VDC output—more than twice that of the most efficient linear supplies.

Because switching is done at a high frequency, the input transformer can be far smaller and lighter than the 50/60-Hz transformers required for series-regulated and ferroresonant power supplies. This also means that heat dissipation is lower. The reduction in transformer size and weight permits the switcher to be smaller and lighter.

A significant feature of the switching supply, not available from a linear supply, is *holdup time,* the length of time the supply's output voltage remains at a value high enough to sustain the load immediately after the loss of input power. The holdup times of switching power supplies are long enough for the circuitry being powered to make a safe transition to a standby power source. The length of holdup time depends on the energy stored in a high-voltage capacitor. This value can be increased by decreasing the output load or increasing the line voltage.

# Switching Regulator Topologies

There are many different configurations or topologies for switching power supplies. These include the following:

- Flyback or buck-boost converter
- Forward or buck converter
- Half-bridge regulator
- Push-pull regulator
- Full-bridge regulator

## FLYBACK CONVERTERS

A *flyback converter,* as shown in the schematic Fig. 10-9, is a *buck-boost switching power supply* that typically has a single switching transistor and no output inductor. Energy is stored in its transformer primary during the first half of the switching period (when the transformer is conducting), but it is transferred to the transformer secondary and load during the second half (the *flyback* period) when the transistor is off. The converter produces a negative output from a positive input, and is used in switching power supplies rated under 20 W.

## FORWARD CONVERTERS

A *forward converter,* as shown in the schematic Fig. 10-10, is a *buck-derived switching power supply* that typically has a single switching transistor. Energy is transferred to the transformer secondary while the transistor switch is conducting, and it is stored in the output inductor. The circuit is used in switching power supplies rated under 20 W.

## MULTITRANSISTOR CONVERTERS

A converter with two or more switching transistors is a *push-pull* or a *half-bridge* design. These designs are widely used in the 100- to 500-W power range. Four-transistor *full-bridge* topologies are typically used in switching power supplies rated at 500 W or higher.

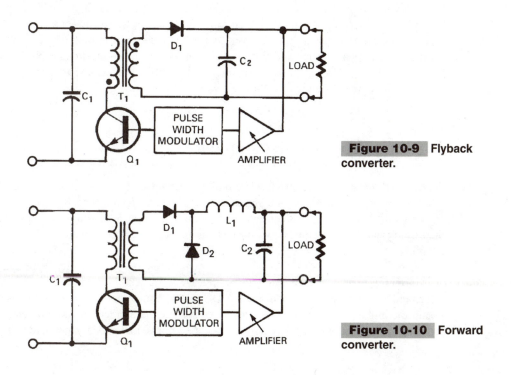

**Figure 10-9** Flyback converter.

**Figure 10-10** Forward converter.

# Laboratory Power Supplies

A *laboratory power supply* is a fully enclosed, versatile benchtop or rack-mounted power supply capable of providing a range of output voltages and currents for circuit testing, development, and maintenance. Considered to be electronic test instruments, most lab supplies are linear and they feature external controls for adjusting voltage and current and one or more panel meters for reading those values.

# Power Supply Packaging

Factory-made, stand-alone power supplies are offered in a wide range of sizes, styles, and output voltage ratings for powering equipment such as a computers, instruments, and test equipment. They are offered in the following package styles:

- *Encapsulated modules.* These are power supplies potted or encapsulated in small, rigid, rectangular plastic boxes. They are typically rated for 10-W output. They are typically used in instrumentation applications.
- *Open-frame packages.* This open-style packaging is common for factory-made linear and switching power supplies rated from 100 to 500 W. The power supply components are mounted on a circuit board for card cage mounting or on an open L-shaped metal chassis for installation within an enclosure of the host product or equipment.
- *Enclosed modules.* This packaging style, used for both linear and switching power supplies, is common for supplies rated for 500 W or greater. The power supply components are mounted on a metal chassis that is protected with a metal cover, typically formed from pierced sheet metal for ventilation. The openings are small enough to provide RF shielding. These units usually include EMI/RFI filters; some are equipped with fans or blowers to provide forced-air cooling.

# DC-to-DC Converters

A *DC-to-DC converter* is an electronic circuit that accepts a DC input at one voltage level and converts it to a DC output at a higher or lower voltage. This is typically accomplished by "chopping" the input direct current to convert it to a coarse alternating current and then amplifying and rectifying the alternating current. In the transistorized version, shown in the simplified schematic Fig. 10-11, the chopping is done by a *pulse-width modulator* (PWM) that modulates a power MOSFET $Q_1$ whose square-wave output is fed to the primary of transformer $T_1$. The coarse AC output that appears at the secondary of $T_1$ is rectified, filtered, and regulated by a zener diode $D_3$. DC-to-DC converters of this type are used to obtain the higher voltages necessary for powering such equipment as radio transmitters, radars, computers, and TV receivers from 12-V automotive batteries or shipboard batteries.

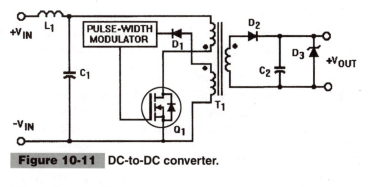

**Figure 10-11**  DC-to-DC converter.

# Fuel Cells

A *fuel cell* is a power source that converts chemical energy directly into electricity without combustion when it is supplied with hydrogen-rich fuel and oxygen. Fuel cells are alternatives to batteries as power sources that do not require recharging. They essentially reverse the process of electrolysis of water into hydrogen and oxygen. The fuel cell consists of a sandwich of two porous *flow-field plates* laced with tiny channels that are separated by a thin membrane, as shown in Fig. 10-12. Each field plate contains an electrolyte, usually phosphoric acid, inserted between two electrodes, the anode and the cathode. Both flow-field plates contain a small amount of a catalyst such as platinum to enhance the chemical reaction.

A hydrogen-rich fuel, which could be hydrazine, kerosene, or hydrogen gas, is fed into the plate on the anode side, where it circulates within the plate. Oxygen (air) is injected into the plate on the cathode side. Hydrogen molecules are separated by a chemical reaction into two positive hydrogen ions or nuclei—single protons—and two electrons within the anode plate. The electrons flow through an external electrical circuit, providing power for the load.

Hydrogen nuclei are drawn toward the cathode and seep through the membrane to combine with oxygen from the air, and returning electrons from the power circuit on the cathode plate form water as a byproduct. This water, which can be useful, is drained off. The cell gives off some excess heat, but it runs much cooler than a combustion engine. If necessary, the water byproduct can be used as a coolant.

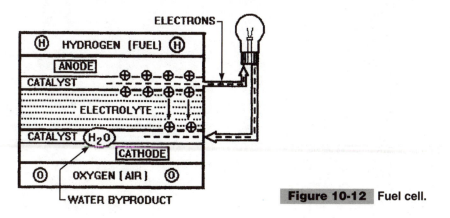

**Figure 10-12**  Fuel cell.

As a power source for electronics only a relatively small number of fuel cells is required. They are stacked and interconnected in series and parallel to provide the required power output at a useful voltage. Although their efficiency is low, fuel cells have proven to be practical substitutes for batteries in manned spacecraft where batteries and onboard generators would introduce contaminants.

If fuel cells are to provide standby power for commercial power plants, large numbers of cells are needed. It has been demonstrated that fuel cells can be more efficient than gas-powered turbine generators. Stacks of hundreds of fuel cells can be organized as alternative power sources for utilities. Stationary fuel-cell power plants now operating have produced 200 kW. However, the DC power output must be converted to AC if it is to be dispatched over a power network.

# ELECTRONIC TEST
# INSTRUMENTS

# Overview

Electronic test instruments are important tools for the design, development, manufacture, calibration, and repair of electronic circuits and equipment. The ancestors of today's instruments were invented by physicists and electrical engineers well before the start of the electronics era, and they proved to be essential in the development of the telegraph, the telephone, motors, and electrical power generators. The classical test instruments include the Wheatstone bridge and the many moving-coil meters descended from the galvanometer—the voltmeter, ammeter, ohmmeter, and wattmeter. These instruments were themselves the predecessors of the analog multimeter or volt-ohm milliammeter (VOM), the multipurpose instrument that made bench and field testing convenient and affordable.

The analog multimeter has since been replaced by the digital multimeter (DMM), now the multipurpose test instrument of choice. In addition to being able to measure the five basic electrical variables, it has consolidated a benchful of other instruments in a single case. DMMs can also measure capacitance, temperature, frequency, electrical continuity, and can even evaluate diodes and transistors.

These capabilities were made possible by the introduction of the monolithic analog-to-digital converter IC, the liquid-crystal display, and the decoder/driver IC. The convenience of being able to read a measurement directly from a meter instead of having to estimate its value by interpolating a pointer position made routine testing easier, faster, and more accurate. However, the analog "moving bar" display on meters has been preserved on digital instruments because it indicates trends and transients in real time.

The oscilloscope is another classical electronic test instrument whose origins can be traced back to the physics laboratories of more than a century ago. The cathode-ray tube (CRT), at first a laboratory curiosity, went on to become the key component in the oscilloscope and the TV receiver, and its development led to the invention of the video tube that made electronically scanned TV a reality.

Modern analog oscilloscopes can make simultaneous measurements of two variables over a wide range of frequencies and display them on the same CRT. A subsequent innovation, the digital oscilloscope, offers features that were not possible on analog oscilloscopes—the ability to freeze waveforms on the screen for instant study and store them in memory for later reproduction and computer analysis. Nevertheless, the analog oscilloscope still remains a valuable bench tool.

Other basic electronic test instruments include the function, signal, and sweep generators, and the frequency counter. More specialized instruments such as the spectrum analyzer permit RF measurements to be made well into the microwave frequency band. While on the one hand improved test instruments have helped to make many innovations in electronics possible, some of these innovations have been fed back to improve instruments. For example, the microcontroller, a byproduct of the development of the microprocessor for calculators and computers, has added to the versatility of many instruments and made them easier to use. Microcontrollers in test instruments such as oscilloscopes and multimeters have extended their capabilities and permitted them to communicate directly with computers. Plug-in software modules for the microcontrollers permit faster setups for a wider range of measurement tasks.

A recent innovation, the *virtual test instrument,* is created by inserting a circuit card and software into a personal computer. The computer monitor's screen is turned into a virtual instrument panel, complete with graphic display and working controls. A virtual digital oscilloscope might lack the versatility and precision of the real instrument, but its measurement accuracy can match that of many general-purpose oscilloscopes.

# Meter Movements

## PERMANENT-MAGNET MOVING-COIL (PMMC) METER MOVEMENTS

A *moving-coil meter* includes a *permanent-magnet moving-coil* (PMMC) *meter movement,* as shown in Fig. 11-1. This movement, a descendant of the *galvanometer,* is technically known as a *d'Arsonval meter movement.* The moving-coil mechanism is generally set in a jewel-and-pivot suspension system to reduce friction. Another suspension method called the *taut-band system* is more sensitive but is also a more expensive meter movement.

Regardless of its suspension, the movement operates on the DC motor principle. A C-shaped permanent magnet has soft iron pole pieces attached inside its north and south poles, and a cylindrical aluminum bobbin with a fine wire coil wound around it is mounted between the pole pieces. It can rotate within the constraints of a spring. A pointer attached to the bobbin deflects across the scale as the moving coil rotates.

Current from the circuit being measured passes through the bobbin coil, causing it to act as an electromagnet with its own north and south poles. These poles interact with the poles of the permanent magnet, causing the bobbin and pointer to rotate. Because the movement responds only to current, the scale must be calibrated for any measurement being made, such as volts, amperes, or ohms, in terms of its deflection caused by current.

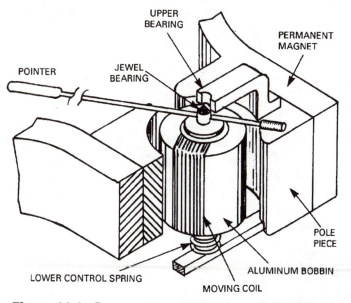

**Figure 11-1**   **Permanent-magnet moving-coil (PMMC) meter movement.**

## DC MOVING-COIL AMMETERS

The *DC moving-coil ammeter* is capable of measuring current flow when it is placed in series with a conductor. The DC ammeter, as shown in Fig. 11-2, has a low-resistance shunt placed in parallel with the coil circuit. The shunt diverts all but a fraction of the current flowing in the circuit from the sensitive meter movement. By adjusting the ratio of current flow between the moving coil and the shunt with an appropriate selection of resistors, the meter scale can be calibrated in microamperes, milliamperes, or amperes.

For example, if the resistance value of the shunt is selected so that it is one-tenth of the resistance value of the coil, the shunt will carry nine-tenths of the current and the coil will carry one-tenth of the current.

The DC moving-coil ammeter cannot measure AC because the alternation of the current will cause the pointer to oscillate and make an accurate reading impossible. However, by rectifying the AC, the AC is converted to a DC level and the dial can be calibrated to read the equivalent AC value.

## AC MOVING-VANE METER MOVEMENTS

The *moving-vane meter* can measure AC current and voltage directly because it works on a different principle than the moving-coil meter. In the moving-vane meter, there are two curved iron plates. One curved plate or vane is fastened to the central shaft of the moving element and is located concentrically inside a second curved plate or vane that is fixed in position. Both vanes are placed within a coil. When current passes through the coil the movable vane, repelled by the stationary vane, drives the moving element containing the pointer in a clockwise direction. Pointer movement depends on the amount of current flowing in the coil. A spiral spring restrains the motion of the moving vane.

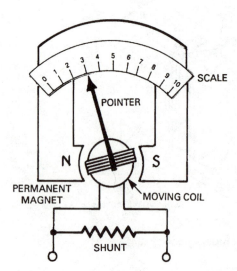

**Figure 11-2** DC moving-coil ammeter.

## AC ELECTRODYNAMOMETER MOVEMENTS

The *electrodynamometer* operates on the same principle as the PMMC meter except that it has two large fixed coils in place of the permanent magnet, as shown in Fig. 11-3. A smaller movable coil is fastened to the central shaft so that it can rotate inside the two fixed coils and a pointer is attached to the end of the shaft. It can be organized as either an AC or DC voltmeter or ammeter by connecting the fixed and movable coils in series. For voltage a dropping resistor is placed in series with the coils, and for current a shunt resistor is placed across the coils. Because the polarity of the fields produced by the coils reverses whenever current reverses, the deflection of the moving coil and pointer will always be in the same direction, regardless of the direction of current through the coils. However, this movement has been used primarily to measure AC power, the product of voltage and current. When the fixed coils are energized by current and the movable coil is energized by voltage, pointer deflection will be proportional to the true power expended in watts.

# Analog Panel Meters (APMs)

## MOVING-COIL ANALOG PANEL METERS

An *electromechanical analog panel meter* (APM) is a general-purpose PMMC meter, as shown in Fig. 11-2, packaged in a case that can be mounted on a panel of an instrument or console. The panel meter can be organized and calibrated so that its pointer movement is proportional to voltage, current, power, or some physical variable such as velocity, pressure, temperature, or revolutions per minute.

These instruments can be modified to measure the different variables by adding series resistors, shunts, or rectifiers. An APM can measure AC values if it includes a rectifier. A power source is not needed because these meters obtain their power from the input signal.

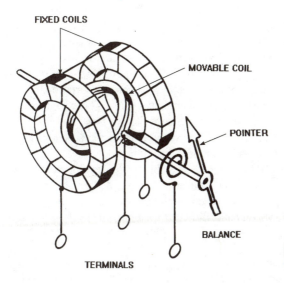

**Figure 11-3** Electrodynamometer meter movement.

Standard, off-the-shelf moving-coil panel meters have rated accuracies of $\pm$ 0.2 to 5 percent of full scale, but with custom graduation and calibration they can provide $\pm$0.1 percent accuracy. These APMs are packaged in industry-standard cases for mounting in panel cutouts. Standard meters have round faces with 2 in (5 cm) diameters and rectangular faces measuring $4 \times 4$ in ($10 \times 10$ cm).

Some electromechanical APMs have been made as moving-bar indicators. The moving coil moves a curved plate behind a window instead of a pointer. When viewed edge-on, a colored stripe on the curved plate appears as a moving bar.

## ELECTRONIC ANALOG PANEL METERS

An *electronic analog panel meter* is a panel meter based on an analog-to-digital converter (ADC) that converts input signals to values which can be displayed as moving bargraphs. These bars can be illuminated directly or indirectly by external illumination. An example of an illuminated electronic APM is one whose bar display is formed from a stack of discrete light-emitting diodes (LEDs) with rectangular end caps that appear as a seamless illuminated stripe. The length of the bar displayed is determined by the number of LEDs energized. As many as 100 LEDs have been stacked to form the bar in some APMs.

By contrast, the moving bar in a liquid-crystal display (LCD) is formed by energizing a closely spaced row of rectangular electrodes formed within the display. The length of the bar depends on the number of those elements energized. The visibility of the display depends on its level of external illumination or backlighting. The resolution of both bargraph panel meter displays depends on the number of active elements in the display.

# Digital Multimeters (DMMs)

A *digital multimeter* (DMM) is a versatile, multipurpose electronic test instrument capable of measuring electrical variables and displaying them on a single digital display. The general-purpose handheld DMM, shown in Fig. 11-4, is suitable for most amateur and professional users. It can make the five basic electrical measurements: AC and DC voltage, AC and DC current, and resistance. The measurement functions of this type of DMM are set with a rotary switch, a popular control for this class of instrument. Most general-purpose DMMs can also test diodes and transistors and perform audible continuity checks. These instruments can test circuits and equipment, calibrate instruments, perform factory tests, and make laboratory measurements. Some handheld models can also measure and display capacitance, temperature, or frequency, but not all of them with the same instrument.

The low-cost, multipurpose DMM was made possible by the development of monolithic dual-slope integrating analog-to-digital converter ICs and inexpensive low-power-consuming LCDs. Because LCDs draw less power than other digital displays, they made it practical to operate a 5- to 8-function handheld DMM with 0.5 in (13 mm) characters from a 9-V disposable alkaline battery. Momentary backlighting of the display improves its readability in reduced light.

The most popular DMMs have 3½- and 4½-digit, 7-segment displays. A 3½-digit DMM has a resolution of 1 part in 1999 or $\pm$0.05 percent, a 4½-digit DMM has a resolution of 1

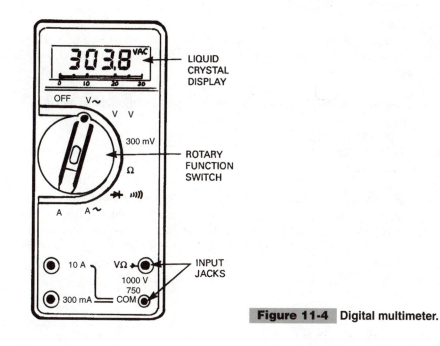

**Figure 11-4** Digital multimeter.

part in 19,999 or ±0.005 percent. Accuracy in DMMs is stated as DC voltage measurement accuracy. (A typical value for a 3½-digit DMM is ±[0.5 percent + 1 digit] or about 10 percent of the value of the resolution.)

The typical handheld DMM is packaged in a plastic case that measures about $7 \times 3.5 \times 2$ in ($18 \times 9 \times 5$ cm), and it weighs about 1 lb (450 g). The centrally located rotary function switch is common, but on many models additional pushbutton switches perform some of the routine functions.

The following measurement ranges are typical for a 3½-digit handheld portable DMM:

| | |
|---|---|
| DC voltage | 200 mV and 2, 20, 200, and 1000 V |
| AC voltage (45 to 500 Hz) | 200 mV and 2, 20, 200, and 750 V |
| Resistance | 200 ohm; 2, 20, and 200 kohm; and 2 megohm |
| DC current | 2, 20, 200, and 2000 mA and 10 A |
| AC current (45 Hz to 1 kHz) | 2, 20, 200 and 2000 mA and 10 A |

Benchtop DMMs are capable of greater accuracy and have other features not available in the handhelds. They are typically packaged in low, flat, rectangular cases with the display and controls on the front panel. A long depth dimension gives the instrument stability on the bench. Most are equipped with retractable bails to raise the front face off the bench. Some AC-line-powered bench DMMs have LED numeric displays which are easier to read in reduced light, and rechargeable batteries, making them more suitable for use as field-test instruments.

The standard features included in many DMMs are automatic polarity, automatic zeroing, high input impedance for voltage measurements, overload protection, and low battery indication. Other features that are available on both handheld and bench DMMs include the following:

- Autoranging—the ability to select the range and position the decimal point automatically on voltage and resistance scales
- True root mean square (RMS) AC measurement
- Peak-hold circuitry
- Level detection

Microprocessors have enhanced the capabilities of some DMMs by providing automatic range selection to remind users of that selection. Analog bargraph displays combined with the numerical display on the LCD display permit users to observe peaking or dipping trends. Some premium-priced DMMs now have large LCDs that plot waveforms, and others can transmit digitized measurement data to a personal computer for processing or logging.

# Oscilloscopes

An *oscilloscope* is an electronic test instrument with a cathode-ray tube (CRT) display for viewing waveforms as plots of instantaneous voltage versus time. This display draws the graphs of input waveforms and permits variables such as frequency and phase to be measured. Figure 11-5 is a block diagram for a single-channel oscilloscope that shows its four

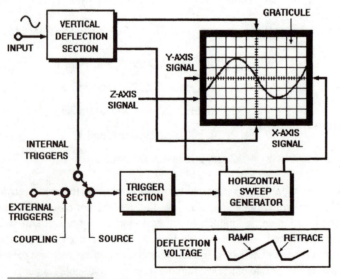

**Figure 11-5**  Oscilloscope block diagram.

principal functional blocks: (1) *vertical deflection section,* (2) *horizontal sweep generator,* (3) *trigger section,* and (4) *display* or CRT. Unless modified by another term, *oscilloscope* means an analog oscilloscope.

The *graticule* is a grid of lines, typically etched or silk-screened on the inside of the glass CRT faceplate, to eliminate measurement inaccuracies called *parallax errors* that would be present if it were a separate grid mounted in front of the faceplate. Graticules are usually $8 \times 10$ patterns. Each of the 8 vertical and 10 horizontal lines block off *divisions* on the screen. The labeling on oscilloscope controls always refers to major divisions which help the user to estimate the amplitude values at any point of the waveform being displayed.

The *vertical deflection section* provides the *y*-axis or vertical information for driving the electron beam as it draws the waveforms on the CRT screen. This section receives input signals and develops the vertical deflection voltages that the CRT needs to control or deflect the electron beam. This section also permits a choice of input signals and internal triggers.

An oscilloscope CRT needs horizontal drive signals to draw a graph. The *horizontal sweep generator* supplies this second dimension by generating the deflection voltage which moves the electron beam horizontally. It produces the sawtooth waveform, as shown in Fig. 11-5, that controls the oscilloscope's sweep rates. The *ramp* is the rising part of the sawtooth, and the falling edge is the *retrace*. The sweep of the electron beam across the screen is controlled by the ramp, and the beam returns to the left side of the screen during the retrace.

The horizontal sweep generator provides a linear rate of rise in the ramp, permitting horizontal beam movement to be calibrated directly in units of time. Because of this property, the sweep generator is also called the *timebase*. Time units can be selected so that any waveforms displayed can be observed for either very short periods, measured in nanoseconds or microseconds, or relatively long periods of several seconds. The *z* axis of a CRT determines the brightness of the electron beam and whether it is on or off.

The *trigger section* provides a stable waveform display because the sweep is started each time at the same deflection voltage. The oscilloscope has both internal and external triggers. Coupling controls the connection of an external trigger to the trigger circuit and the mode control determines the operations of the trigger circuit. The source switches are used to select the trigger signal.

For an external signal, the trigger signal is connected to the trigger circuit with the external coupling controls. Slope and level controls define the trigger points on the trigger signal. The *slope control* specifies either a positive or *rising* edge or a negative or *falling* edge on the waveform. The *level control* allows the user to select the deflection voltage when the trigger event will occur. When the trigger circuit recognizes that the selected voltage level on the trigger signal has been reached, the horizontal sweep generator will be turned on each time that level recurs.

Thus the waveform on the screen is the composite of all sweeps overlaid into what appears to be one graph. For example, if the 0.05 μs/div setting is made, the oscilloscope draws one waveform every 0.5 μs (0.05 μs division × 10 screen divisions). This means that about 2 million graphs are drawn every minute.

Today most analog oscilloscopes are *two-channel* or *dual-trace oscilloscopes*. They have two separate vertical input circuits that permit two waveforms to be observed simultaneously on one partitioned CRT display. A general-purpose oscilloscope is capable of measuring frequencies up to 100 MHz. The key specifications for a typical 100-MHz analog oscilloscope are:

| | |
|---|---|
| Bandwidth (3 dB) | 100 MHz |
| Rise time | 3.5 ns |
| Input impedance | 1 megohm in parallel with 22 pF |
| DC gain accuracy | $\pm 3$ percent |
| Deflection factor | 5 mV/div to 5 V/div |
| Sweep time | 2 ns/div to 0.5 s/div |

# Digital Oscilloscopes

A *digital oscilloscope* displays voltage waveforms by converting the original signal to a digital format that can be displayed or stored in memory. Because it can store the waveform digitally, it is also called a *digital storage oscilloscope* (DSO). This instrument is ideal for viewing very slow events, capturing and comparing waveforms, capturing and viewing one-time events, and viewing pretrigger events. Automatic measurements and calculations can be performed on the waveform, and the waveforms can be printed out on a printer.

The input circuitry of a DSO is similar to that of an analog oscilloscope. After preamplification, the input signal is sampled by a sample-and-hold (S/H) circuit and digitized by an analog-to-digital converter (ADC). The sample is then stored in a digital memory. A crystal oscillator in the timebase circuit measures the time difference between the trigger signal and the sample clock. A microprocessor then determines where the waveform is to be positioned on the screen. The timebase moves across the display from left to right as in an analog oscilloscope, and the waveform is refreshed each time a trigger occurs. Timebase linearity and accuracy of the DSO are superior to those of an analog oscilloscope.

# Function Generators

The *function generator* is the most widely used general-purpose signal source. As shown in the simplified block diagram Fig. 11-6, it can supply both sine waves and nonsinusoidal

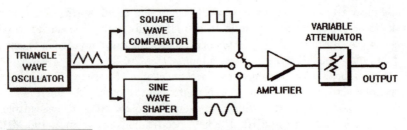

**Figure 11-6** Function generator block diagram.

waveforms such as square waves, triangle waves, and pulse trains. Some function generators also provide modulation and swept frequencies. These instruments include circuitry that is similar to that in a sine-wave oscillator. They typically produce a primary waveform with a free-running oscillator and then derive the other waveforms from it.

In Fig. 11-6, a triangle wave is generated by the oscillator. A square wave is then derived from the triangle wave by passing it through a square-wave comparator. Similarly, a sine wave is derived from the triangle wave by passing it through a sine-wave shaper. The desired waveform is selected, amplified, and made available at the output. Function generators typically include a variable attenuator to adjust the output, and a DC offset adjustment for adding a positive or negative DC level to the output.

The square wave output can function as a clock for digital circuits, and the triangle wave can act as a sweep generator. Function generators can also perform audio sine-wave testing. A laboratory-quality instrument will typically be able to generate sine, square, triangle, ramp, and pulse waveforms over a frequency range of 0.001 Hz to 20 MHz. The output voltage is 30 V peak-to-peak in an open circuit and 50 V peak-to-peak across a 50-ohm load.

# Radio-Frequency Signal Generators

A *radio-frequency signal generator*, as shown in the simplified block diagram Fig. 11-7, is a general-purpose source of RF sine waves useful for testing radio receivers. Most signal generators are also capable of generating audio-modulated signals from an audio oscillator with frequency- and amplitude-modulation circuitry, and some also provide phase and pulse modulation. Precise, wide-range attenuation and low RF leakage are typical requirements for signal generators that perform full-range receiver tests.

The key circuit of an RF signal generator is the voltage-controlled oscillator (VCO), whose frequency is determined by the voltage at the control input. Increases or decreases in the control voltage cause corresponding changes in VCO output frequency. Thus, the signal applied to the control voltage sets the oscillator frequency, a frequency-modulation requirement. The audio-modulating signal drives the control input of the VCO to produce a frequency-modulated carrier. As shown in Fig. 11-7, a modulator follows the VCO. This

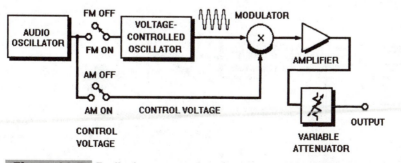

**Figure 11-7** Radio-frequency signal generator block diagram.

circuit varies the amplitude of the VCO output without altering its frequency, making the output an amplitude-modulated signal.

A laboratory-quality RF signal generator can cover the frequency range from 500 kHz to more than 500 MHz in 10 bands. The calibrated and metered output will be adjustable from about 0.02 µV to 2 V. Internal AM and FM modulation as well as external AM, FM, and pulse modulation are provided. The metered and calibrated modulated signal is variable from 20 to 600 Hz.

# Sweep Generators

A *sweep generator,* as shown in the simplified block diagram Fig. 11-8, is a sine-wave source whose frequency can be changed in a controlled manner. A swept frequency is useful for testing circuits over a wide frequency range. The sweep generator can produce fixed-frequency sine waves if the sweep feature is turned off. The instruments typically sweep linearly in frequency, but some can also sweep logarithmically. The sweep output voltage is proportional to sweep frequency. This output can drive the horizontal axes of other instruments such as oscilloscopes. The VCO is driven by the ramp generator to provide the frequency sweep.

# Frequency Counters

A *frequency counter,* as shown in the simplified block diagram Fig. 11-9, measures frequency. A crystal oscillator and digital counting circuit can measure both period and frequencies from low to RF. Some frequency counters can also measure waveform rise time and pulse width. The signal being measured is amplified and changed into a digital pulse train that passes through a logic AND gate to drive a series of digital counters. The gate is opened for a finite length of time, and the number of waveform cycles, representing waveform frequency, that occur during that time is measured. To obtain the next measurement, the digital counter is reset, and the gate is reopened.

Frequency counters typically include frequency dividers that reduce the frequency of the digital signal. They can be switched in and out of the circuit to increase the counter's

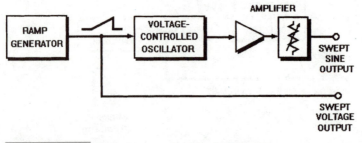

**Figure 11-8**  Sweep generator block diagram.

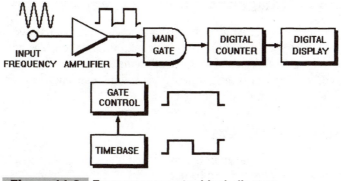

**Figure 11-9** Frequency counter block diagram.

frequency range. At the input, the divider increases the instrument's frequency limit by reducing incoming signals to a usable range. In this position the divider is known as a *prescaler.* If the divider is added to the timebase circuit, it increases the *on* time of the gate, permitting lower frequencies to be measured while also increasing frequency resolution.

# Spectrum Analyzers

A *spectrum analyzer* is a real-time analyzer that simultaneously displays the amplitude of all the signals in the frequency range of the analyzer. It can provide information about the voltage or energy of a signal as a function of frequency. In effect, it permits viewing the frequency domain just as an oscilloscope permits viewing the time domain.

Spectrum analyzers measure the amplitudes of the frequencies of the components of a complex waveform throughout the frequency range of the waveform. The waveforms are displayed as vertical voltage amplitude bands representing a very narrow band of frequencies positioned along the horizontal $x$ axis. A pure frequency will be displayed as a single vertical trace. The instrument can, for example, measure the amplitude and frequency of a single sine wave or the energy at each frequency in the output of a microwave generator.

Spectrum analyzers are designed and built on the basis of different concepts. Three of the most common types are: (1) *bank of filters,* (2) *fast Fourier transform* (FFT), and (3) *swept spectrum.*

## BANK-OF-FILTERS ANALYZERS

A *bank-of-filters analyzer* includes many bandpass filters that are tuned to very narrow bands within the total spectrum. Each filter removes only those specific signal frequencies from the spectrum, rejecting all others. The output of each filter is an AC voltage whose amplitude is a measure of the energy contained within its narrow band. The outputs of all filters are then detected and displayed on a CRT screen as a row of vertical bars of varying amplitudes, providing frequency domain information.

## FAST FOURIER TRANSFORM (FFT) ANALYZERS

A *fast Fourier transform* (FFT) *analyzer* is based on the Fourier transform technique for computing the waveform spectrum from its time domain information. The analog waveform is converted to digital format and sampled at regular intervals. The voltage at each sample is digitized, and the data is then processed by a digital signal processor (DSP) to provide the frequency spectrum for display on a CRT.

## SWEPT-SPECTRUM ANALYZERS

A *swept-spectrum analyzer,* as shown in the simplified block diagram Fig. 11-10, has a single tunable filter channel instead of a bank of filters. The single channel is automatically swept in frequency and the results are displayed on the CRT. It is able to provide the same frequency domain display as the other spectrum analyzers. The output of its detector is converted by an analog-to-digital converter. Some models have the ADC in the intermediate frequency section so that resolution bandwidth filtering is done digitally. The microprocessor receives the data in a digital format and processes it for the CRT display.

# Logic Analyzers

A *logic analyzer* permits the state of logic signals in many channels of a digital circuit to be measured simultaneously. It stores the state of logic inputs in memory for display on a CRT. A conventional logic analyzer functions both as a timing and state analyzer. When functioning as a timing analyzer it samples the waveform at each edge of its internal clock, but it can only determine whether the signal is a logic high or logic low without providing any information on voltage resolution. When acting as a state analyzer, it uses a signal from the circuit being tested as the clock for determining when the logic signals are stored in memory.

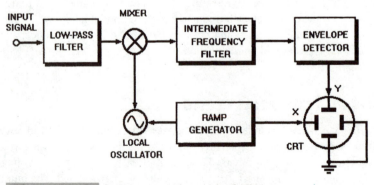

**Figure 11-10**  Spectrum analyzer block diagram.

# Magnetic Field-Strength Meters

A *Hall-effect gaussmeter* measures magnetic field strength. It includes a *Hall-effect transducer* (HET) on the end of a pair of flexible wires which can be inserted between the poles of the magnet whose field strength is to be measured. A battery or other DC source provides current through the transducer, and a voltmeter in the instrument's case is calibrated to give a voltage that is proportional to magnetic field strength. See also "Hall-Effect Sensors" in Sec. 17, "Electronic Sensors and Transducers."

# DC Measurement Bridges

## WHEATSTONE BRIDGE

The *Wheatstone bridge* is the simplest form of comparison measurement bridge circuit. Bridge circuits are widely used to measure resistance, inductance, capacitance, and impedance. They operate on the *null detection* or *balance principle,* meaning that the indication is independent of the calibration of the indicating device.

The schematic for a Wheatstone bridge Fig. 11-11 shows that it has two parallel resistance branches with each branch containing two series resistors, $R_1$ and $R_2$, and $R_3$ and $R_4$. A DC voltage source connected across this network provides current through the network. A null detector, typically a meter ($M_1$), is connected between the parallel branches to detect a balance or null condition.

Meter $M_1$ indicates the difference between the output voltages of the voltage dividers. If $R_1$ equals $R_2$ and $R_3$ equals $R_4$, the meter will read 0 and indicate a null or balance. This can be written as

$$\frac{R_1}{R_2} = \frac{R_3}{R_4} \qquad \text{or} \qquad R_1 \times R_4 = R_2 \times R_3$$

If $R_1$ equals $R_2$ and $R_4$ is an unknown resistance value, the meter will indicate a null when $R_4$ equals $R_3$. If $R_3$ is a variable resistor, its value can be adjusted until the meter reads 0. At

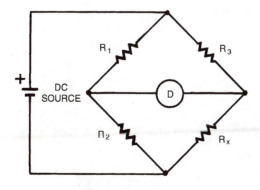

**Figure 11-11** Wheatstone bridge.

this setting, the value of the unknown resistor $R_4$ is the same as the value of $R_3$. The Wheatstone bridge has probably been in use longer than any other electrical measuring instrument. Measurement accuracy of 0.1 percent can be obtained with this bridge.

## KELVIN BRIDGE

The *Kelvin bridge,* shown in Fig. 11-12, is a seven-arm bridge that is a modified version of the Wheatstone bridge. It can eliminate the effects of contact and lead resistance when low values of unknown resistance values are being measured. Its second set of ratio arms compensates for low lead-contact resistance values. Resistors in the range of 1 ohm to about 1 microhm can be measured accurately with this bridge. It is also called a *double bridge* or a *Thomson bridge.*

# AC Measurement Bridges

The concept of the Wheatstone bridge has been extended to permit the measurement of capacitor and inductor values. These impedance measurements require a bridge in which the DC voltage source has been replaced by an AC source. Many different AC bridge designs have been developed, but all are variations of the Wheatstone bridge. One or more resistors in the Wheatstone bridge are either replaced or supplemented by inductors or capacitors. The unknown impedance is one arm of the bridge circuit, and one or more of the other arms must be adjusted until the null condition is reached.

The best known AC bridge is the *Maxwell bridge* for the measurement of inductance by capacitance. Other well-known AC bridges are the *Hay* and *Owen bridges,* both modifications of the Maxwell bridge for the measurement of inductance. All three bridges have a pure capacitance in the arm opposite the pure inductance, and the other two arms become pure resistances. The differences between the three bridges lie in the methods used to balance the resistance component of the coil. The *Schering bridge* is one of the most important AC bridges for the measurement of capacitance. It has generally replaced the *Wien bridge.* Other AC inductance bridges include the *Anderson, Campbell,* and *Carey-Foster bridges.*

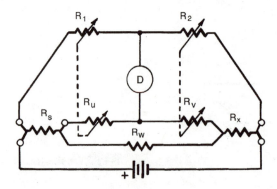

**Figure 11-12** Kelvin bridge.

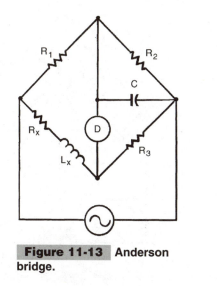

**Figure 11-13** Anderson bridge.

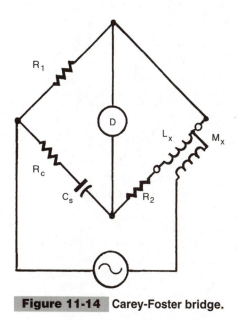

**Figure 11-14** Carey-Foster bridge.

# AC Measurement Bridge Examples

The *Anderson bridge,* shown in Fig. 11-13, is a six-branch modification of the Maxwell-Wien inductance bridge for measuring a wide range of inductances with reasonable values of fixed capacitance. This bridge can also measure the residuals of resistors by a substitution method that eliminates the effects of residuals in the bridge elements. Bridge balance is independent of frequency.

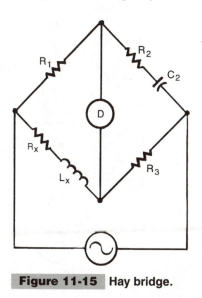

**Figure 11-15** Hay bridge.

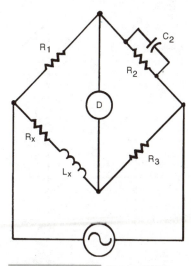

**Figure 11-16** Maxwell-Wien bridge.

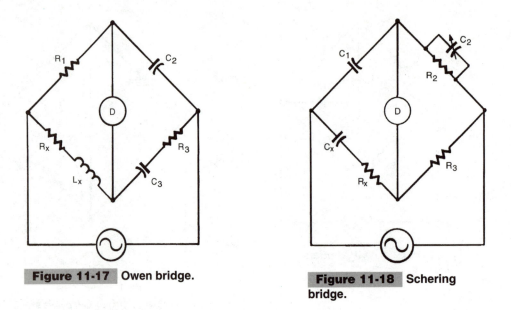

**Figure 11-17** Owen bridge.

**Figure 11-18** Schering bridge.

The *Campbell bridge* is an AC bridge for measuring mutual inductances. It can compare unknown and standard mutual inductances with different values.

The *Campbell-Colpitts bridge* is an AC bridge for measuring capacitance by the substitution method.

The *Carey-Foster bridge,* shown in Fig. 11-14, is an AC bridge for measuring the mutual inductance of inductors in terms of capacitance and capacitance in terms of mutual inductance. It is also called the *Heydweiller bridge.*

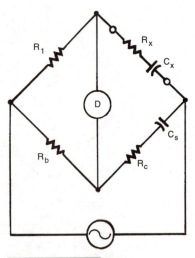

**Figure 11-19** Universal bridge.

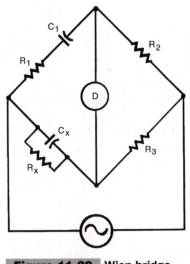

**Figure 11-20** Wien bridge.

The *Hay bridge,* shown in Fig. 11-15, is a four-arm bridge similar to the Maxwell-Wien bridge. It can measure inductances with large $Q$ values and also determine the incremental inductance of iron-cored reactors. Its balance is frequency dependent.

The *Maxwell-Wien bridge,* shown in Fig. 11-16, is a four-arm AC bridge for making accurate inductance measurements. It depends on a capacitance standard rather than a resistor. Bridge balance is independent of frequency.

The *Owen bridge,* shown in Fig. 11-17, is a four-arm AC bridge for measuring self-inductance in terms of capacitance and resistance. Bridge balance is independent of frequency.

The *Schering bridge,* shown in Fig. 11-18, is a four-arm AC bridge for measuring capacitance and dissipation factor. Bridge balance is independent of frequency.

The *universal bridge,* shown in Fig. 11-19, is a four-arm AC bridge for measuring capacitance. There are two versions: (1) the *series resistance bridge* for measuring equivalent series capacitance, and (2) the *parallel resistance bridge* for measuring equivalent parallel resistance.

The *Wien bridge,* shown in Fig. 11-20, is a four-arm AC bridge for measuring the equivalent capacitance and parallel loss resistance of an imperfect capacitor such as a sample of insulation or a length of cable. Bridge balance depends on frequency. It is used as a frequency-determining network in RC oscillators. See "Wien-Bridge Oscillators" in Sec. 4, "Basic Amplifier and Oscillator Circuits."

# OPTOELECTRONIC COMPONENTS AND COMMUNICATION

# Overview

The benefits of the marriage between optics and electronics are readily apparent in television cameras, camcorders, digital cameras, and night-vision telescopes, to mention but a few common examples. But optoelectronics can be considered as having two realms: the *visible* and the *nonvisible*. Operating in the visible realm are devices, circuits, and systems whose outputs are signals, indications, text, or images that can be seen by the human eye. These include displays that either emit visible light or are intended to be seen in visible light. By contrast, devices, circuits, and systems that emit nonvisible infrared (IR) and ultraviolet (UV) energy, or those whose functions depend on that energy, are considered to be in the nonvisible realm. This section focuses primarily on the nonvisible realm.

The development of simple, reliable, low-cost semiconductor IR-emitting diodes and lasers with matched photodetectors has made possible such devices as optocouplers and such products as digital remote controls, laser printers, scanners, code readers, and CD (compact disk) players. It has also made long- and short-haul optoelectronic communication practical.

Fiberoptic cables are now competing with wire and coaxial cables as conductors of telephone conversations, television programming, computer data, and still pictures. These cables can transmit information in the form of modulated light over long distances without additional amplification. Complete families of IR photoemitters and photodetectors have evolved, making it easy to convert IR energy back and forth to electrical signals.

Night-vision tubes permit scenes to be viewed at night that are illuminated only by natural background light sources, and converter tubes make it possible to view scenes in total darkness that are illuminated by IR sources. This equipment permits soldiers, sailors, and airmen to observe enemy targets at night and aim weapons at them while remaining unseen by the enemy. The same equipment also permits scientists to study nocturnal animals without alarming their subjects or view weather patterns observed from satellite IR cameras at night. Night-vision equipment also gives the police and other law-enforcement personnel an opportunity to observe criminal acts taking place under the cover of darkness, again without being seen by the perpetrators.

While devices and products based on the transmission and reception of IR are more common, UV energy has been harnessed to perform such tasks as erasing semiconductor memories and registering masks on photoresist-covered semiconductor wafers or circuit boards.

Section 13, "Optoelectronic Display Technology," focuses on devices, products, and systems that emit or respond to visible light, or are best seen when illuminated by visible light.

# Visible and Infrared Energy

The diagram of the electromagnetic spectrum Fig. 12-1 shows visible light in a thin band at the center extending from the shortest violet wavelength of 380 nm (0.38 µm) to the longest red wavelength of 770 nm (0.77 µm). The infrared (IR) band extends from the red end of the visible band to 1 mm ($10^6$ nm), the shortest-wavelength end of the RF band. The ultraviolet (UV) region extends from the violet end of the visible band to 10 nm (0.01 µm), the longest-wavelength end of the X-ray region.

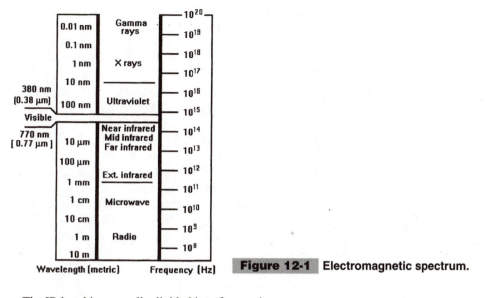

**Figure 12-1** Electromagnetic spectrum.

The IR band is generally divided into four regions:

**1.** Near IR (NIR), 0.77 to 3.0 μm (770 to 3000 nm)
**2.** Middle IR (MIR), 3.0 to 6.0 μm (3000 to 6000 nm)
**3.** Far IR (FIR), 6.0 to 15 μm (6000 to 15,000 nm)
**4.** Extreme IR (XIR), 15 to 1000 μm (15,000 nm to 1 mm)

IR is considered to be heat radiation, but technically it is not. IR feels warm because the skin transforms its radiant energy into heat as it is absorbed. IR is produced by such varied sources as candles, fires, incandescent lamps, and most living animals. But inanimate moving objects such as electrical motors and internal combustion engines are also IR sources.

When illuminated by an invisible IR source such as a laser, scenes in total darkness can be viewed with special telescopes and binoculars that include image-converter tubes. These images can be transferred to CRT or LCD monitors for ease of viewing.

Another class of products, called *night-vision systems,* are based on the light-multiplier or image-intensifier tube. While these night-vision aids are designed to function effectively in reduced natural light and do not require supplementary IR illumination, their tubes are sensitive to and therefore capable of viewing objects that emit IR radiation.

The IR energy can be modulated for nonvisible communication. Handheld, battery-powered remote controllers are familiar appliances that can turn TV sets off and on, change channels, and adjust the picture or sound. These actions are carried out when the IR receiver in the TV set receives the proper coded pulses. These controllers can also adjust stereo systems, VCRs, CD players, and other entertainment products. Cordless mice and keyboards can communicate with a host computer over IR links, and computers can be linked in networks within a room by IR signals.

Military weapons systems have been designed for operation in the IR region. These include IR-emitting lasers and IR sensors mounted on vehicles and aircraft for target illumination and tracking. The addition of IR guidance kits to "dumb" bombs makes them "smart" so that they can home in on IR illuminated surfaces. Medical instruments based on IR detection are able to monitor patients' temperatures noninvasively in intensive-care units, and industrial instruments can display temperature gradients in various colors on video screens to detect faults in metal components or locate unwanted hot spots in moving machinery.

IR-sensitive instruments in earth-orbiting satellites now sense and monitor the nighttime temperatures of ocean water and cloud-tops, providing valuable information on the intensity and movement of storms. The satellites convert this information to radio frequencies for transmission back to earth, where they can be received by private, commercial, and government receivers. When displayed on a computer monitor, the images appear as photographic views of the earth from space. They supplement the visible light images obtained from weather satellites made during daylight hours for weather forecasting.

However, fog and rain in the atmosphere reduce the useful range of IR transmission. The most severe attenuation occurs between 4.5 and 8.0 μm in the MIR and FIR regions. In addition, atmospheric carbon dioxide attenuates IR transmission between 14 and 16 μm in the FIR and XIR regions.

# Infrared-Emitting Diodes

An *infrared-emitting diode* (IRED) converts electrical energy into nonvisible IR. IREDs are similar in construction to *visible-light-emitting diodes* (LEDs) and the *diode injection lasers*. They are made as PN junctions from gallium arsenide (GaAs) or aluminum gallium arsenide (AlGaAs).

An N-type epitaxial layer is grown on the GaAs or AlGaAs wafer, and P-type diffusion forms the upper layer. IREDs can also be made with alternate layers of GaAs and AlGaAs. When an IRED is *forward biased* (positive lead connected to the P-type material and negative lead connected to the N-type material), charge carrier recombination occurs. Electrons cross from the N side to recombine with holes on the P side of the junction. Because electrons are at a higher energy level than the holes, phonons (heat energy) and photons (IR energy) are emitted. The output wavelength of the diode depends on the *bandgap* of the materials used to construct it. The peak energy from a GaAs IRED occurs at about 900 nm and from an AlGaAs IRED at about 830 nm, both beyond the visible band.

Both GaAs and AlGaAs are transparent to IR emission, so the energy is emitted from the face of the upper P-type layer. However, IREDs that emit from both end surfaces are made by different epitaxial growth techniques. A forward voltage on an IRED of about 1.2 V results in milliwatt-level IR output with the value depending on drive current.

IREDs matched to silicon photodetectors such as photodiodes, phototransistors, or photodarlingtons are used in short-range data communication. Peak photodetector response occurs at about 850 nm, closely matching peak IRED output. Special packages have been developed to adapt IREDs for short-range fiberoptic systems. These packages might include short stubs of optical fiber called *pigtails* to simplify the task of matching the IRED to the optical fiber. IRED dies also are included in many different optoelectronic components such as *optocouplers, optical interrupters, optical encoders,* and *optical reflector modules.*

# Photomultiplier Tubes

A *photomultiplier tube,* as shown in the diagram Fig. 12-2, is a vacuum phototube containing a *photocathode,* an *anode,* and many additional electrodes called *dynodes* positioned between the photocathode and anode. The tube illustrated has a reentrant configuration with curved-plate-type dynodes facing each other. A voltage-divider network applies successively larger voltages in steps of about 100 V to the dynodes so that the dynode nearest the anode has the highest voltage.

When light strikes the photocathode, photons drive off free electrons within the cathode surface material. The liberated electrons are accelerated toward dynode 1 by the positive voltage on that dynode with respect to the photocathode. All dynode surfaces are coated with a material that promotes *secondary emission.* At dynode 1 some electrons are liberated by each cathode electron that strikes its surface, and they are accelerated towards the more positive dynode 2. The electrons emitted from the dynode surface are called *secondary electrons* to distinguish them from the *primary* or *incident electrons.* This process continues, with the original photoemission current being amplified incrementally. Stated in another way, the number of secondary electrons is multiplied by each successive dynode until the increased number of electrons is finally collected by the anode.

Electron multiplication generates photoemission currents initially measurable in microamperes and they are boosted to milliampere level. Current can be amplified 10 million or more times, depending on the number of dynodes in the tube. Cathode photocurrent amplification (DC current gain) is typically between $10^5$ and $10^7$ when the photomultiplier has between 9 and 14 dynodes.

Anode voltages of 500 to 5000 V are required to operate a photomultiplier. *Dark current,* a current which flows when the cathode is not illuminated, results from *thermal emission* and the effects of the high-voltage electrodes. For incident illumination at a specific wavelength, the number of emitted electrons is directly proportional to the illumination inten-

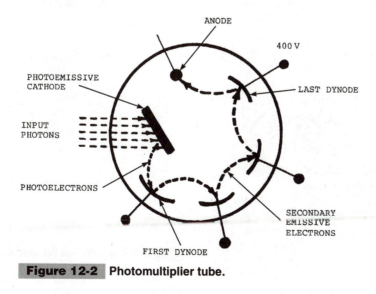

**Figure 12-2**  **Photomultiplier tube.**

sity. Thus, for a specific illumination intensity, the anode current in a photomultiplier remains constant as the anode voltage is increased. However, dark current adds to the anode current produced by illumination, and secondary emission improves with applied voltage. Consequently, anode current increases with anode voltage.

Illumination levels in photomultiplier tubes are measured in *microlumens.* These tubes are so sensitive that if they are exposed to ordinary daylight levels, destructively large current could flow when voltage is applied to the electrodes.

Dynodes are typically made from silver-magnesium and beryllium-copper. Some photomultipliers have as many as 16 dynodes. A typical photocathode has a diameter of 0.25 to 2 in (6 to 50 mm). Some photomultipliers are made in a linear configuration with an arrangement of grid-type "venetian blind" dynodes distributed along the axis of the envelope. The photomultiplier is also called a *multiplier phototube* and an *electron-multiplier phototube.*

# Photoconductive Cells

A *photoconductive cell,* or *photoresistive detector,* as shown in the cutaway view Fig. 12-3, is a photoconductor whose electrical resistance changes in response to incident light wavelength and intensity. Its resistance decreases with light intensity. Because the device requires a fixed external voltage source, current output is proportional to light intensity. The light-sensitive stripe is deposited in a serpentine pattern across the face of an insulating wafer to maximize its length over the wafer diameter. The ends of the stripe are connected to terminal pins in the base. A glass or plastic cover protects the photosensitive material from abrasion and contamination.

Electrons absorb the energy of incident photons and are excited into free states. Electric conduction occurs either by electrons in the conduction band or by positive holes vacated in the valance band. The light provides enough energy to pull electrons from their atoms

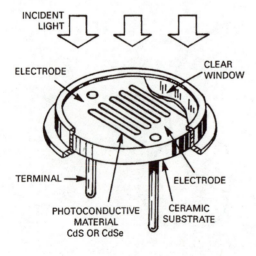

**Figure 12-3** Photoconductive cell.

within a semiconductor film, so the number of holes (charge carriers) created in the stripe increases with light intensity. This has the effect of decreasing the resistance of the stripe. However, as light intensity is dimmed, the process is reversed and resistance increases.

Resistance declines along a negative slope with increasing light intensity. When the cell is not illuminated, its *dark resistance* can exceed 100 kilohms, but when exposed to high light intensity, the minimum resistance value falls to a few hundred ohms. Cell *sensitivity* can be expressed in terms of cell current for a given voltage and given level of illumination.

The two most effective conductive materials for photoconductive cells are *cadmium sulfate* (CdS) and *cadmium selenide* (CdSe). Both respond slowly to changes in light intensity. CdSe has a response time of about 10 ms, while CdS has a response time of as much as 100 ms. The resistance of CdSe changes significantly with ambient temperature, but the resistance of CdS remains relatively stable.

The *spectral response* of a CdS cell is similar to that of the human eye, peaking at about 555 nm. By contrast, the spectral response of the CdSe cell starts at the violet 450 nm end of the visible region and peaks at about 700 nm in the visible red region, but it extends into the infrared region. These characteristics are evaluated when selecting a photoconductive cell for a specific application.

# Photodiode Detectors

## PN JUNCTION PHOTODIODES

A silicon *PN junction photodiode* has a positively doped P and negatively doped N region. When operated in the *photovoltaic mode* with no external bias, an area of neutral charge, called the *depletion region,* exists between these two regions. When light falls on the diode junction, electrons in the diode become excited. If the light energy exceeds the bandgap energy, electrons will flow into the conduction band. This creates holes in the valence band vacated by the electrons. These electron-hole pairs form throughout the diode.

The drift of electron-hole pairs results in a positive charge buildup in the N region and a negative charge buildup in the P region. The value of the charge is directly proportional to the light intensity on the diode. Current will then flow in an external circuit that connects the N and P regions.

In the *photoconductive mode* a reverse bias is applied to the PN junction photodiode. This bias increases both the electric field strength between regions and the depth of the depletion region. When light is incident on the diode, its response time is shorter and more linear than when the diode is in the photovoltaic mode. However, an effect called *dark* or *leakage current* occurs in this mode, and its value depends on reverse bias voltage. PIN photodiodes and avalanche photodiodes are generally operated in the photoconductive mode.

## PIN PHOTODIODES

A *PIN photodiode* is a depletion-layer junction detector that is operated in the photoconductive mode. Its sensitivity range and frequency response depend on the thickness of the intrinsic layer between the P layer and the N layer. Light must pass through the P region before it reaches the depletion region, where it excites electron-hole pairs which are quickly dissipated by a large electric field.

## METAL-SEMICONDUCTOR PHOTODIODES

A *metal-semiconductor photodiode* is a depletion-layer Schottky-barrier photodiode made as a thin metal film over an N or N on N⁺ layer. Light passes through the thin film, and the photogenerated electron-hole pairs in the semiconductor device produce an output current.

## POINT-CONTACT PHOTODIODES

A *point-contact photodiode* is a depletion-layer Schottky-barrier point-contact detector. Light falls on its Schottky barrier through an etched cavity in the semiconductor.

## HETEROJUNCTION PHOTODIODES

A *heterojunction photodiode* is depletion-layer photodiode made by forming a junction between two different semiconductors. It is also called the *depletion-layer photodiode*.

## AVALANCHE PHOTODIODES (APDs)

An *avalanche photodiode* (APD), shown in section view Fig. 12-4, is a photodiode operated at higher reverse-biased voltages than other photodiodes. Large reverse voltages create a high electric field at its PN junction. Electrons moving in the high electric field are accelerated and they collide with other electrons, creating additional electron-hole pairs. These pairs, when accelerated, produce other electron-hole pairs in a process called *impact ionization*. This is an electron multiplication process that produces internal gain. It is the

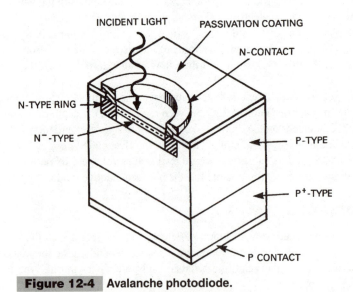

**Figure 12-4**  Avalanche photodiode.

semiconductor counterpart of the photomultiplier tube. There are both *planar* and *mesa* avalanche photodiodes.

# Solar Cells

A *solar cell*, as shown in section view Fig. 12-5*a,* is a semiconductor PN junction that absorbs energy from the sun and converts it directly into electric energy. The surface layer of P-type silicon is thin so that light can penetrate to the junction. The metallized contact on the P-type silicon is the positive output terminal, and the metallized surface on the bottom of the N-type silicon is the negative output terminal. Low-current solar cells for powering light loads such as calculators are typically packaged in clear plastic. However, devices intended to operate as solar energy converters require large surface or detector areas to maximize current capacity, low series resistance to provide maximum power transfer to the load, and very narrow depletion regions to provide a higher open-circuit output voltage.

Panels formed from cells connected in series and parallel can provide enough power to maintain the charge on yacht batteries, and power night-lights. Far larger arrays for powering satellites and spacecraft are assembled as folding panels or "sails" that are deployed after the spacecraft has reached its assigned altitude.

Solar cells are about 10 to 12 percent efficient, and their output power and current are proportional to their surface area. Small solar cells now function as light detectors in cameras. The schematic symbol for a solar cell is given in Fig. 12-5*b*. It is also known as a *photovoltaic diode* or a *solar energy converter.*

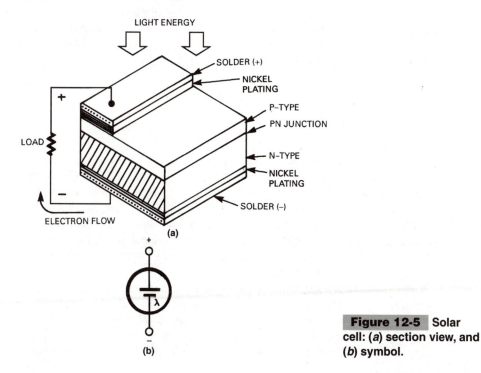

**Figure 12-5**   Solar cell: (*a*) section view, and (*b*) symbol.

# Phototransistors

A *phototransistor* is a photodetector made either as an NPN or PNP bipolar junction transistor (BJT). However, it has no base terminal. An NPN phototransistor is operated as a two-terminal device with the base floating and the collector positively biased. Collector-base leakage current acts as base current, and this is enhanced by incident light. As in the photodiode, *reverse saturation current* is increased by incident light energy on the junction of the phototransistor.

The phototransistor provides much more output current than a photodiode when exposed to the same amount of light. Phototransistors are packaged in transparent plastic cases to admit incident light to the base or base-collector junction. Although they are more sensitive than photodiodes, photodiodes can switch much faster. The circuit symbol for a phototransistor is the same as for a conventional BJT, except it has no base connection.

# Photofets

A *photofet* or *photosensitive field-effect transistor* is a junction field-effect transistor (JFET) that functions as a light-sensitive photodiode and high-impedance amplifier. Light focused on the gate-channel junction of a JFET produces a change in its drain current. For example, in an N-channel JFET the gate-source leakage current is the reverse saturation current at a reverse-biased PN junction. As happens in the phototransistor, this reverse current in a JFET is also susceptible to light. Illumination of the junction causes more charge carriers to be generated, causing the reverse current to increase. When the illumination exceeds a threshold value, the increase in reverse current causes the gate voltage to increase, turning the FET on. Thus light falling on the junction controls photofet drain current.

# Optocouplers

An *optocoupler,* as shown in the cutaway view Fig. 12-6, includes a photoemitter, a photodetector, and a short optical transmission path. It is actually a short fiberoptic transmission system. The photoemitter, typically an IRED, converts an electrical input signal into modulated light and transmits that light over a short optical transmission path to the photodetector, typically a phototransistor, that restores the light to an electrical signal. A plastic or glass bead, the transmission path, isolates the input signal from the output signal electrically to protect the output circuit from damage or destruction from voltage transients or surge currents in the input circuit.

Optocouplers can also prevent low-level noise from degrading signal transmission. They permit two circuits operating with different voltage levels and ground points to be coupled without a transformer. Most optocouplers have IRED photoemitters, but photodetectors can be photodiodes, phototransistors, or other photosensitive ICs. All photodetectors respond to the IR output of the IRED and provide an electrical output. However, unlike

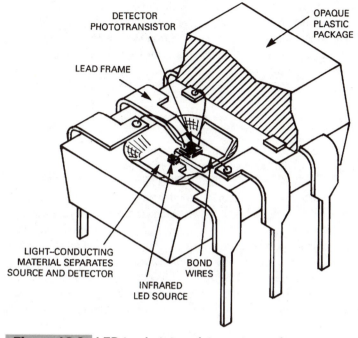

DETECTOR
PHOTOTRANSISTOR

OPAQUE
PLASTIC
PACKAGE

LEAD FRAME

LIGHT–CONDUCTING
MATERIAL SEPARATES
SOURCE AND DETECTOR

BOND
WIRES

INFRARED
LED SOURCE

**Figure 12-6**  **LED-to-phototransistor optocoupler.**

photodiodes, phototransistors and photo-ICs provide signal gain. The output of an opto-coupler can be an analog signal or train of digital pulses. IREDs can also be biased to estab-lish a zero level for the transmission of AC signals.

The *current transfer ratio* (CTR) of an optocoupler indicates its coupling efficiency. That value depends on the radiative efficiency of the IRED, the spatial separation between the IRED and photodetector, and both the sensitivity and the amplifying gain of the detector.

Optocouplers are classified by their photodetectors, and their photodetectors determine their characteristics. The characteristics are as follows:

- *Photodiode* couplers offer the highest speed but do not amplify.
- *Phototransistor* couplers are the most versatile and also provide signal gain.
- *Photodarlington* couplers offer higher CTRs and output currents than phototransistor couplers, but their switching speeds are slower.
- *Photo-SCR* couplers have the ability to switch AC and DC with logic-level inputs, but they require separate gate pins.
- *Phototriac* couplers can switch full-wave AC with logic-level inputs, but they also require separate gate pins.
- *Photo–Schmitt trigger IC* couplers provide logic-level square-wave outputs and they can be interfaced with computers, power supplies, motors, and other actuators.

Optocouplers, also called *optoisolators,* must be packaged in opaque cases to keep out light which could falsely trigger them. The common packages are four- and six-pin dual-in-line package (DIP) cases. Two related devices, the *optointerruptor* and the *optoreflector,* also provide electrical isolation, but visible light or IR beams from their photoemitters pass through the air to their photodetectors. For information on optocoupler packaging, see "Optocoupler Packaging" in Sec. 27, "Semiconductor Device Packaging."

# Lasers

A *laser* is a device capable of emitting coherent energy in some region of the electromagnetic spectrum when pumped with external energy. Laser action can be produced in the visible, infrared, ultraviolet, and X-ray regions. Gases and liquids in gas and liquid lasers are confined by suitable optically transparent containers. Solid-state lasers, as distinguished from semiconductor lasers, have crystalline lasing structures made from such materials as rubies and neodymium: YAG shaped as rods. Gallium-arsenide (GaAs) and gallium-aluminum-arsenide (GaAlAs) semiconductor lasers, known as *laser diodes* or *injection lasers,* are fabricated on semiconductor wafers by techniques similar to those used in fabricating LEDs.

Gas, liquid, and solid-state lasers have full mirrors at one end of their lasing structure and partially reflective mirrors on the other end to allow high levels of coherent energy to pass out of the structure. They are energized by external energy from one or more of the following sources:

- High-intensity flash lamps
- Electric current or pulses
- Chemical reactions
- Nuclear radiation
- Free electron streams

The laser is a higher-frequency version of the *maser,* a device that produces coherent energy principally in the microwave region. The maser is used as a detector of microwave signals. See "Masers" in Sec. 7, "Microwave and UHF Technology."

The coherent radiation wavelengths emitted from gas, liquid, and solid-state lasers is determined by the composition of the medium, so changes in ingredients result in shifting of the output wavelength.

As atoms within the medium are stimulated by external energy, their electrons are driven to higher orbits. As the electrons fall back to their original orbits, they emit energy in the form of *photons.* If the medium is a molecule, the molecule is distorted by external energy and in the process of returning to its original form, photons are also emitted.

When a photon interacts with an atom, the energy exchange stimulates the emission of another photon of the same wavelength in the same direction. Multiple contacts between photons and atoms create avalanches of photons within the medium. The photons shuttle back and forth until they gain sufficient energy (at a resonance condition) to escape from the medium through the partially reflective mirror as a narrow beam of coherent light.

Laser output is intense, directional, pure, and coherent, with a very narrow bandwidth. Visible-light-emitting lasers emit coherent light that is predominately of one color, and IR-emitting lasers emit energy that is concentrated at one wavelength. All coherent wave travel is parallel in the same direction. This behavior differs from the incoherent light emitted by the sun, a fire, or an incandescent lamp because it is a jumbled mixture of wavelengths radiating in many different directions.

The coherence of a laser's parallel beam permits it to travel long distances without dispersion. This property has been exploited in free-space communication, distance measurement, fiberoptic communication, and even the laser light from the small battery-powered pointers used in lectures and audiovisual presentations. Laser emission can heat the substances it illuminates, so lasers can perform welding and cutting in industry, science, and medicine. But that same property of laser light makes it dangerous to persons and animals. It can burn spots on skin and destroy parts of the retina in the eye. Consequently the use of lasers must be carefully controlled.

Glass fibers conducting laser energy can act as sensors if they are exposed to heat or pressure which affect the ability of the fibers to transmit the light energy. This means that a change in light transmission can be correlated with a specific physical influence which can be detected by a remote monitor. Laser beams can be continuous wave (CW) or pulse modulated. Fluorescent dyes in some lasers permit visible light to be modulated into an infinite number of wavelengths.

Laser power, like incandescent lamp power, is stated in watts. The choice of a laser for a specific application is determined by evaluating the advantages and disadvantages of each laser type that functions in the desired frequency region. For high peak power output solid-state or glass lasers are suitable because of their energy storage capacity. These and dye lasers are also suitable for generating ultrashort pulses.

# Solid-State Lasers

## RUBY LASERS

The *ruby laser,* as shown in Fig. 12-7, is a simple solid-state laser. A ruby-rod lasing structure is mounted axially within a helical flashlamp. Separate mirrors are located at each end of the rod. One is silvered for total reflection and the other is silvered to allow about 4 to 6 percent of the light to pass out (94 to 96 percent of the incident light is reflected back into the rod). When photon resonance occurs within the ruby rod, a coher-

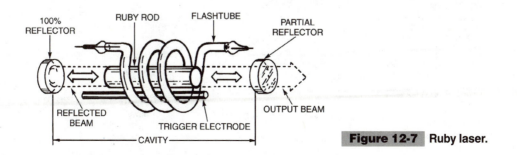

**Figure 12-7**  Ruby laser.

ent red visible-light beam with a wavelength of about 0.694 μm (694 nm) is formed. This type of laser can machine materials, but its output is particularly hazardous if directed into the eyes.

## NEODYMIUM: YTTRIUM-ALUMINUM-GARNET (Ne:YAG) LASERS

The *neodymium: yttrium-aluminum-garnet* (Ne:YAG) *laser* is a solid-state laser that has military applications because it can produce moderate to high output power and can be operated in either a continuous wave or pulsed mode. The Ne:YAG laser emits energy at 0.69 and 1.06 μm (690 and 1060 nm). At 1.06 μm, its emission wavelength is compatible with many NIR optical components such as frequency multipliers, harmonic generators, modulators and detectors. Ne:YAG lasers are useful for range finding, target designation and tracking, weapon fire control, missile and projectile guidance, reconnaissance and surveillance, and secure communications. Its output frequencies are also hazardous to the eyes.

# Gas Lasers

## HELIUM-NEON (He:Ne) GAS LASERS

The *helium-neon* (He:Ne) *gas laser,* as shown in the diagram Fig. 12-8, is a glass tube sealed at both ends by windows set at Brewster's angle. Two concave mirrors (one partially transmitting) form the optical cavity containing a mixture of helium and neon gas. The external voltage is applied to the cathode at one end—the source of electrons. The electrons excite the helium atoms, which in turn collide with neon atoms, driving their electrons into higher-energy orbits. As the electrons drop down from these higher orbits, they emit bursts of energy in the form of photons, which collide with other neon atoms to generate visible light at a resonance wavelength of 632 nm. The coherent light leaves the tube through the partially transmitting mirror. He:Ne gas lasers have been used in measurement instruments, printers, and optical-disk readers, but they have been replaced by semiconductor laser diodes in many of these applications because they are smaller, lighter, less fragile, and more reliable.

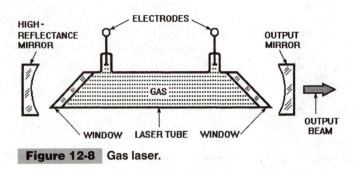

**Figure 12-8** Gas laser.

## ARGON-ION LASERS

The *argon-ion laser* is a gas laser that emits at wavelengths of 488 and 515 nm in the visible-light region; it has been used in printers, machine tools, and medical equipment.

## CARBON-DIOXIDE (CO₂) LASERS

The *carbon-dioxide* ($CO_2$) *laser* is a gas laser developed for industrial and military applications. Its principal active gas is carbon dioxide, but the addition of other gases such as helium and nitrogen can increase its output power. This laser generates IR energy at a wavelength of 10.6 μm in the FIR region that can penetrate smoke and fog. Because it can operate continuously at kilowatt output levels, it can perform such applications as cutting, drilling, heating, and welding. These lasers can produce output power as high as 10 GW for cutting plastic and fabric, welding metals, precision drilling, and heat treating.

The $CO_2$ laser is also used for long-range free-space communications and as an IR illuminator for directing laser-guided bombs and missiles. It is also effective in laser detection and ranging (ladar) systems. The laser can be pumped by chemical, electric, or optical energy, and its highest average output power is obtained by pulsing.

## CARBON-MONOXIDE (CO) LASERS

The *carbon-monoxide* (CO) *laser* is a gas laser with carbon monoxide as its active gas. Its highest power output is achieved at wavelengths of 4.9 to 5.7 μm in the MIR band.

## EXCIMER LASERS

The *excimer laser,* named for the abbreviation of "excited dimers" or "excited diatomic gas molecules," generates UV-light pulses when those molecules are excited. These lasers are used in IC photolithographic equipment to expose IC photomasks on semiconductor wafers during wafer fabrication. The blue-green light from modified excimer lasers penetrates sea water to depths of several thousand feet. These lasers can communicate more efficiently with submarines than can very low frequency (VLF) radio waves.

# Chemical Lasers

A *chemically-pumped laser* depends on chemical reactions rather than electric energy to produce the pulses of light required for pumping the laser. In one method, the light is emitted by a chemical reaction. In another, shock waves from an explosion generate light in a flash tube that triggers the laser. The shock wave from a pyrotechnic squib compresses argon gas so that it emits intense UV radiation over a wide band.

The *Alpha laser* is an orbiting space-based chemical laser intended for antimissile defense that operates at 2.7 μm in the NIR region. Its lasting action is produced by the combustion of hydrogen and fluorine. The fluorine is obtained on demand by a reaction between nitrogen trifluoride, deuterium, and helium. The laser is designed to produce 2.2 MW, and it can melt metal at long distances in space. The energy intensity at the laser's core exceeds that of the surface of the sun. Made largely of aluminum, it contains a system of mirrors known as the *Large Advanced Mirror Program* (LAMP) *system.*

# Liquid Lasers

The *dye laser* obtains lasing from a strongly fluorescing organic compound dissolved in an appropriate solvent as its active medium. The medium is pumped with a flash lamp or another laser, usually in a pulsed mode. The dye laser emits at wavelengths from 0.32 to 1.2 μm, across the visible-light band and into the NIR region. It is used in measuring instruments.

# X-ray Lasers

The *X-ray laser* has been demonstrated in feasibility tests. Its X-ray laser emissions are generated by focusing the energy of a nuclear explosion. Research in the United States has been focused on the development of an X-ray laser capable of shattering incoming ballistic missiles. The X-ray laser's radiation has an extremely short wavelength measurable in atomic units. This permits the rays to scatter from objects of atomic size or larger. Excimer lasers operating in the UV region have been used to pump the X-ray laser by pulsing it with millions of watts of power. In other research, an experimental low-power X-ray laser has been developed for biological and medical research. It might be able to provide superior pictures of DNA molecules, enzymes, viruses, or human cells. X-ray laser holograms could record these entities in three dimensions.

# Semiconductor Laser Diodes

Both the light-emitting diode (LED) and the laser diode are semiconductor PN junctions. When current is applied across them, they both emit light as positively charged holes join with negatively charged electrons. The LED emits light spontaneously in many directions over a wavelength distribution wide enough to result in incoherent light emission. By contrast, the laser diode emits light primarily by stimulated emission when the applied current rises above a threshold level, and its narrow wavelength distribution and highly directional output produce coherent light.

The laser diode, as shown in Fig. 12-9, is made as a sandwiched double-heterojunction structure with a light-emitting active layer set between two other cladding layers that provide larger energy level differences. The inner active layer has a greater refractive index than the outer cladding layers, effectively confining the light. The die is cleaved so that it has mirror-like reflecting surfaces or facets on both ends at right angles to the longitudinal mode, the path of the oscillation spectrum. These facets give the emitted light high directionality. To

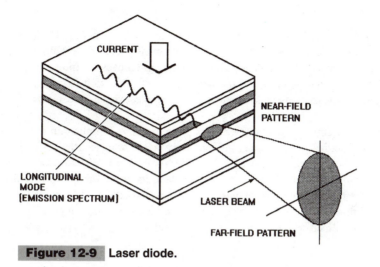

**Figure 12-9** Laser diode.

encourage stimulated emission from the laser diode, electrons must initially be held at high energy levels, and to conserve power, currents are concentrated in narrow current channels.

Two different laser diode structures, *gain guided* and *index guided,* are being manufactured commercially. As shown in Fig. 12-10, both structures have active layers sandwiched in parallel between outer layers with larger energy level differences that confine light in a direction that is parallel with the active layer. The index-guided laser provides a more stable horizontal transverse mode than the gain-guided laser.

As current increases through the diode, incoherent light is emitted. This light increases gradually with current until the *knee* is reached. At that threshold the laser goes into oscillation and starts to emit coherent light. However, the rated optical power is obtained only when the forward *operating current* is beyond the level at the knee.

An oscillating laser diode emits many wavelengths that differ slightly from each other in *longitudinal mode,* causing many standing waves between the reflective facets. Among those wavelengths, the one with maximum intensity is the *peak lasing wavelength.*

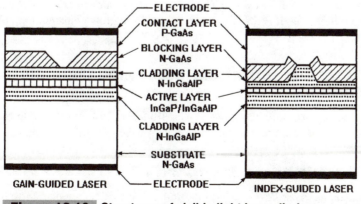

**Figure 12-10** Structures of visible-light laser diodes.

Laser diodes are called *multimode lasers* if they oscillate in many longitudinal modes, and *single-mode lasers* if they oscillate in a single longitudinal mode. The light intensity distribution at a facet is the *near-field pattern*. It has an elliptical shape, with its major axis parallel with or aligned to the die's lateral axis. As the light moves out into space, it fans out into an elliptical *far-field pattern* and its major axis shifts 90° so it is vertical or aligned with the die's transverse axis. The vertical spread angle exceeds the horizontal spread angle.

Red visible-light-emitting laser diodes made from indium gallium aluminum phosphide (InGaAlP) can be either gain or index guided. They lase at wavelengths between 630 and 690 nm, and their optical output power is typically from 3 to 5 mW.

When a laser diode oscillates at a constant current, its optical output power decreases with temperature rise. To hold optical output power constant, the diode is placed in an automatic power-control circuit with a heatsink attached. The optical output power of the main beam can be determined by measuring the photocurrent.

# Laser Diode Wavelengths

A typical semiconductor laser diode is about the size of a grain of table salt. It requires a power supply capable of 100 to 200 mW. These lasers have been widely used as emitters in short- and long-range fiberoptic communication lines and for reading CDs and CD-ROMs. With typical optical output powers of 3 to 5 mW, these lasers can produce a single stable beam and are cheaper, more compact, and more reliable than gas lasers. Moreover, laser diodes are easily modulated by switching the input current on and off. Single-mode diode lasers that produce 20 to 50 mW of optical output power are used for optical recording, high-speed printing, analog signal transmission, long-distance optical communication at high data rates, and communication between satellites in orbit.

## INDIUM GALLIUM ARSENIDE PHOSPHIDE (InGaAsP) LASER DIODES

An *indium gallium arsenide phosphide* (InGaAsP) *laser diode* emits IR energy of 1.30 to 2.1 μm in the NIR region. It is made as layers of InGaAsP grown on an indium phosphide (InP) substrate.

## INDIUM GALLIUM ALUMINUM PHOSPHIDE (InGaAlP) LASER DIODES

An *indium gallium aluminum phosphide* (InGaAlP) *laser diode* emits visible red light in the 630- to 690-nm region. It is made of InGaAlP grown on an InP substrate.

## GALLIUM ALUMINUM ARSENIDE (GaAlAs) LASER DIODES

A *gallium aluminum arsenide* (GaAlAs) *laser diode* emits IR energy of 780- to 870-nm in the NIR region.

## INDIUM GALLIUM ARSENIDE (InGaAs) LASER DIODES

An *indium gallium arsenide* (InGaAs) *laser diode* emits IR energy of 900- to 1020-nm in the NIR region.

## GALLIUM NITRIDE (GaN) LASER DIODES

A *gallium nitride* (GaN) *laser diode* emits blue coherent light in the visible band. These diodes are more efficient blue-light emitters than silicon-carbide (SiC) laser diodes. They are being used in laser diode printers. Higher resolution can be obtained with blue laser light because it has a shorter wavelength than either IR or visible red light. GaN laser diodes are also expected to make possible higher density data storage on CDs because the shorter blue wavelength allows each bit of data to be stored in a smaller spot, as much as a tenfold improvement over IR-emitting diodes.

# Fiberoptic Communications

In fiberoptic communication, modulated light is transmitted through a thin, transparent optical fiber that connects a photoemitter to a photodetector in a transmission system, as shown in Fig. 12-11. The transparent fiber core surrounded by a transparent cladding with a lower *index of refraction* confines most of the light to the fiber. (The index of refraction is the ratio of the velocity of light in a vacuum to that of light in a medium.)

The most efficient fiber cores and cladding are made of silica glass doped with impurities to give them different refraction properties. The clad fibers are protected from damage by the same jacketing materials that are used to protect wire cable. The jacket might include a steel wire or plastic fibers to impart additional tensile strength. Special connectors are available for interconnecting fiberoptic cables to active emitters and detectors. Cable is the costliest item in most medium- and long-distance fiberoptic communication systems. For additional information on optical cables and connectors, see Sec. 29, "Electronic Hardware, Wire, Cable, and Connectors."

Fiberoptic communications systems under 1 km in length are considered to be short-haul. They are typically used to connect computers and peripherals within a building or cluster of buildings. These systems are usually privately owned and maintained, and they are usually assembled from standard commercial hardware. A local-area network (LAN) equipped with fiberoptic cable would be a short-haul system. Medium- and long-haul sys-

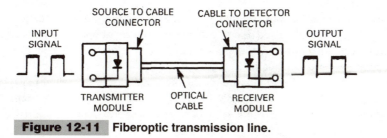

**Figure 12-11** Fiberoptic transmission line.

tems also connect computers and peripherals, but they can be hundreds of kilometers apart. Most of these systems are owned by the government or public-service telecommunications utilities. Hardware for these systems is more likely to be custom fabricated to meet their more demanding requirements.

A new form of optical transmission called *dense wavelength division multiplexing* (DWDM) increases communications traffic by projecting multiple light beams of different wavelengths or colors onto a single glass fiber. The different colors are combined into a single "white" beam by the fiber. At the termination, the DWDM system acts like a prism, separating the white light of the fiber into a rainbow of colors, each carrying its own voice, data, and video information.

# Optical Fibers

*Optical fibers* conduct modulated light in fiberoptic transmission systems. A basic fiberoptic transmission system is illustrated in Fig. 12-11. There are two types of optical fibers: *step index* and *graded index*. Step-index fibers transmit data at speeds of up to about 30 MHz. By contrast, graded-index fibers have an upper limit of 500 MHz. There are also *single-mode* and *multimode* optical fibers. Single-mode fibers have core diameters small enough to restrict transmission to a single, low-order mode of propagation.

*Multimode graded fiber,* an optical fiber with a graded core, supports more than one optical transmission mode, as does *multimode step-index fiber,* an optical fiber with a stepped core. Made of ultrapure quartz, clad with a polymer of lower refractive index or with a halide-doped, low-refractive-index glass, it has an upper limit of about 30 MHz.

The three alternatives available in optical fiber include:

**1.** All-glass step- and graded-index fibers
**2.** Plastic and glass combined in step-index, plastic-clad silica (PCS) fiber
**3.** All-plastic step-index fibers

The four major variables to be considered when selecting fiberoptic cable are:

**1.** Attenuation
**2.** Bandwidth
**3.** Numerical aperture (NA)
**4.** Core diameter

*Attenuation* is the loss in optical signal power due to the absorption and scattering of the transmitted energy. It is expressed as a rate of loss in decibels of optical power per kilometer (dB/km). Glass fibers made by depositing doped silica glass from high-purity gases exhibit the lowest optical attenuation.

*Bandwidth* or *dispersion* is a measure of the highest sinusoidal modulation frequency that can be transmitted through a length of fiber at a specified optical wavelength without losing more than 50 percent of the signal power. It is expressed in megahertz per kilometer of length (MHz/km). Bandwidths of 200 to 1000 MHz/km are obtained with graded glass 50-μm cores, but they are reduced to 20 MHz/km with 300-μm step-index glass cores.

The diagram Fig. 12-12 illustrates some of the terms used in describing optical fibers.

*Numerical aperture* (NA) is the measure of the ability of the fiber to accept incident light. In effect, it is the degree of openness of the input acceptance cone. Numerical aperture is defined mathematically as the sine of the half-angle of the acceptance cone. It is stated as a dimensionless number, and represents an angle that contains most of the output optical power. Typical values range from 0.20 to 0.27.

*Core diameter* is the diameter of the central region of an optical fiber that has a refractive index higher than that of its cladding. Core diameters for commercial fibers range from 50 to 300 μm. The cladding diameter of a 50-μm glass core is typically 125 μm; for a 300-μm glass core it is typically 440 μm. Commercial PCS fiber is available with core diameters of from 125 μm to 1 mm.

Most light rays from an emitter that strike the interface between the core and the cladding are reflected back into the core. Unless the fiber is bent sharply, the light rays will rebound off the sides indefinitely. Only rays that enter the fiber at a large angle escape. If the core has a uniform index of refraction, rays that make many reflections follow a longer path and arrive behind the generally axial rays that make fewer reflections. This is called *modal dispersion.*

Fiber with a graded refractive index overcomes modal dispersion. The fiber is made so that its index of refraction decreases with the radius distance from the axis. The grading is done during the fabrication process by varying the densities of the internal doping of the glass. The graded index causes light rays that deviate from the axis to travel faster and thus move at the same speed as those rays moving parallel to the axis.

Optical signals with only a few milliwatts of power can be detected after traveling dozens of kilometers in silica optical fiber. However, the signal is attenuated as it travels through the fiber because of light scattering. This requires that repeaters be installed to amplify and regenerate the signal to overcome losses over long distances.

At the shorter IR wavelengths of about 0.85 μm (850 nm), the typical attenuation of a silica fiber is 2.5 to 3 dB/km. Attenuation declines to 0.3 to 1 dB/km at 1.3 μm (1300 nm). However, it rises sharply again at about 1.390 μm (1390 nm) because of the presence of the oxygen-hydrogen (OH) impurity. The greatest transparency of silica is reached at around

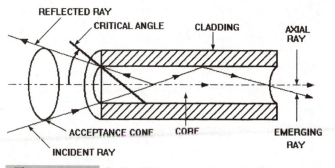

**Figure 12-12** Optical fiber response to light rays.

1.55 µm (1550 nm) where attenuation is less than 0.15 dB/km. Thus light at that wavelength can travel 3 to 10 times as far as a 0.85-µm (850-nm) signal with the same power.

*Chromatic dispersion* also affects repeater spacing. The index of refraction varies for different wavelengths of light. Therefore chromatic dispersion introduced by the fibers causes signals with slightly different wavelengths to travel through the fiber at different speeds. This becomes a problem in digital transmission. In extreme cases, each pulse becomes broad enough to interfere with neighboring pulses, which increases the bit-error rate. Thus chromatic dispersion forces a tradeoff between repeater spacing and effective bit rate.

# Image-Intensifier Tubes

A *night-vision light-multiplier* or *image-intensifier tube* magnifies low-level visible light to permit viewing an image without the need for supplemental illumination, either visible or IR. The most common example is the *night-vision tube,* the key component in passive night-vision telescopes and binoculars or goggles. Its operation depends on the presence of normal ambient nighttime light from such sources as the moon, the stars, or reflections from clouds or other light. The image-intensifier tube will not work in total darkness.

A third-generation single-stage image-intensifier tube is shown in the cutaway view Fig. 12-13. It provides a visible image that is brighter than its input image because of electron multiplication performed by an internal light-multiplier plate called the *microchannel plate.* It is the principal component in night-vision imaging systems such as the monocular *starlight scope* or binocular *night-vision goggles.* It requires a high-voltage power supply, typically a battery-powered voltage-multiplier circuit.

Image intensification occurs when electrons emitted by the photocathode strike the phosphor screen after being accelerated by the high voltage on the anode and focused inter-

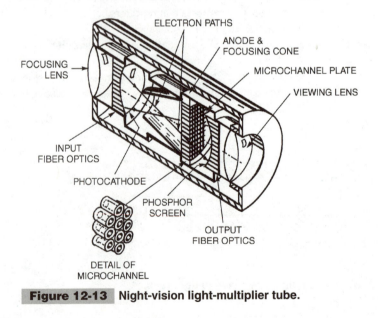

**Figure 12-13** Night-vision light-multiplier tube.

nally by an electron lens. Luminance gains in a single-stage image tube can be as high as 50,000. The latest image-intensifier tubes are smaller and lighter than earlier multistage tubes, and they can reduce blooming and smearing effects caused by nearby flashes of light such as those caused by flares or tracer bullets.

The microchannel plate (MCP), as shown in the detail in Fig. 12-13, acts as a multi-channel photomultiplier. It is a secondary-emission current amplifier that works on a pho-tomultiplier principle and is located between the photocathode at the input end and the phosphor screen near the viewing lens. The MCP is a matrix of tiny parallel glass tubes that have been bonded together to form a 1-mm-thick disk. The inner surfaces of each tube are coated with phosphor to multiply and accelerate electrons that have been converted from image photons by the tube's photocathode and electrostatically focused on the MCP. Elec-trons leaving each tiny tube of the MCP impact on a phosphor screen to form the image on the viewing lens.

Image-intensifier tubes are classified by the useful diameter of their photocathodes, typ-ically 18, 25, and 40 mm. The spectral response of the image tube is selected on the basis of application. Yellow-green phosphor is most frequently used for direct viewing because its spectral output peak closely matches the maximum sensitivity of the human eye. Other tubes can be matched for photographic or television applications.

For further information on the use of image-intensifier and image-converter tubes see Sec. 23, "Military and Aerospace Electronics Systems."

# Image-Converter Tubes

An *image-converter tube* converts an image obtained in one frequency band to another fre-quency band in the electromagnetic spectrum. An IR image converter can function in com-plete darkness and provide a visible-light output for direct viewing or display on a CRT or LCD monitor. However, an active IR source is required to illuminate the scene. For exam-ple, the tube can be used in military night-vision systems for aiming guns or missiles where IR illumination is provided by a laser. All image-converter tubes are similar in construction to image-intensifier tubes. For further information on the use of image-intensifier and image-converter tubes see Sec. 23, "Military and Aerospace Electronics Systems."

# OPTOELECTRONIC DISPLAY
# TECHNOLOGY

# Overview

Optoelectronic displays are acquiring increasing importance as interfaces between people and computers, cellular telephones, pagers, and other communications devices. These displays, ranging from cathode-ray tubes to liquid-crystal, light-emitting diode, and electroluminescent panels, have been steadily improved. They now present larger, brighter images than earlier versions, are more reliable, and draw less power. These displays are now integral parts of a wide variety of products from desktop and notebook computer monitors to test equipment, TV receivers, automotive and aircraft instruments, and electronic games. They can be found in homes, offices, factories, stores, travel agencies, financial institutions, and airline terminals, and even in the passenger sections of airliners. Some optoelectronic displays have reached monumental billboard size in cities, where they provide news bulletins as colorful, moving "zippers." The latest automobiles now include three or four different kinds of indicators and displays, and more are in the offing.

Of all the display technologies, liquid-crystal displays have shown the most conspicuous improvement within the past 10 years. They first appeared as black segmented digits on white backgrounds in watches and calculators, but they have now morphed into multicolor flat panels. With increasing resolution and wider color palettes, LCDs are now challenging CRTs as the leading computer monitor displays and they are now in TV receivers, camcorders and digital cameras.

The charge-coupled device (CCD) camera is also replacing traditional video camera tubes inside and outside of TV studios and in industrial and consumer still and video cameras.

The standard observer curve, shown in Fig. 13-1, is helpful for understanding the relations between wavelength and color distribution in the visible-light band and its effect on human vision. The curve shows that the human eye is most sensitive to yellow-green at 555 nm but acuity trails off in the violet and deep red regions. This curve indicates that dim sources of yellow and green light can be more eye-catching than bright sources of red or violet light.

This section focuses on devices, circuits, and systems that function in the visible-light realm while Sec. 12, "Optoelectronic Components and Communication," concentrates on nonvisible optoelectronics.

# Cathode-Ray Tubes (CRTs)

The *cathode-ray tube* (CRT) is a glass vacuum tube containing one or more electron guns, a phosphor screen to convert the energy of the electron beam into visible light, and provision for the electrostatic or electrodynamic control of one or more electron beams. It remains the most popular and cost-effective display today for desktop computers, TV

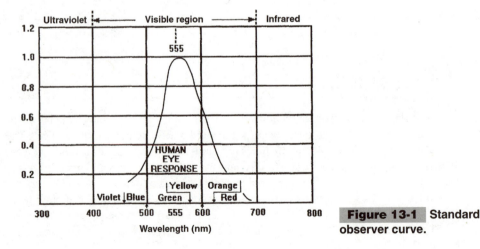

**Figure 13-1** Standard observer curve.

receivers, and electronic test instruments despite the fact that it was invented well over a century ago.

There are two basic kinds of CRT: *electromagnetic* and *electrostatic*. These tubes are further classified by color as black and white or *monochrome* and *multicolor,* and by the number of electron guns they contain. Computer monitors, TV receivers, and many kinds of scientific and medical apparatus include multicolor electromagnetic CRTs, while monocolor electromagnetic CRTs are still prevalent in radar and sonar systems. However, such electronic test equipment as oscilloscopes and spectrum analyzers contain monocolor electrostatic CRTs because of their higher resolution and suitability for presenting rapidly changing waveforms.

## ELECTROMAGNETIC-DEFLECTION CRTs

The *electromagnetic-* (EM-) *deflection CRT,* as shown in the diagram of a monocolor version Fig. 13-2, is designed for the installation of external electromagnetic coils around its neck for the control of the electron beam. Electrons are driven from the cathode, which has been heated by a filament, and they are formed into a beam by the control and screen grids of the *electron gun.* The beam is focused at the back of the phosphor-coated faceplate and is drawn to it by a potential as high as 2500 V on the anode. While transiting the length of the envelope, the beam passes through the magnetic fields of the focus and deflection coils that together form a magnetic lens capable of deflecting the beam in the *x* and *y* directions. The beam can be scanned in a raster sweep format for receiving TV images or it can be formed into a radial scan for a radar plan-position indicator (PPI) display. The phosphors on the faceplate cause the beam trace to persist long enough to form a visual image.

Color CRTs, operating on the same principles, typically contain three electron guns as shown in Fig. 13-3*a*. One is assigned to each of the colors (red, green, and blue). In a multicolor tube, the phosphors are grouped as dots or bars arranged in patterns to form *pixels*. A single beam impacting on a single phosphor dot, for example, creates a single color spot. Two dots activated simultaneously result in a mixed color, and the simultaneous illumination of all three dots produces white. A black spot appears if none of the dots in the triad is illuminated.

The three electron beams are coordinated to move synchronously and, because of their spatial separation, the beams pass through a *shadow mask* at slightly different angles to

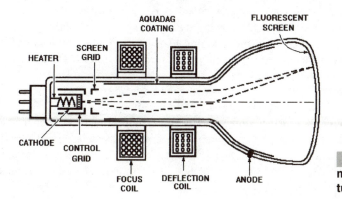

**Figure 13-2** Electromagnetic cathode-ray tube (EM CRT).

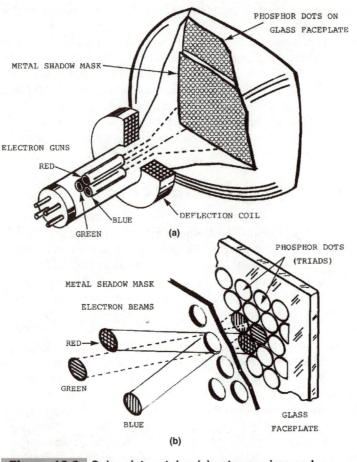

PHOSPHOR DOTS ON
GLASS FACEPLATE

METAL SHADOW MASK

ELECTRON GUNS

RED

BLUE

GREEN

DEFLECTION COIL

(a)

PHOSPHOR DOTS
(TRIADS)

METAL SHADOW MASK

ELECTRON BEAMS

RED

GREEN

BLUE

GLASS
FACEPLATE

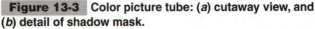

(b)

**Figure 13-3** Color picture tube: (*a*) cutaway view, and
(*b*) detail of shadow mask.

illuminate their assigned phosphor dots. The mask is a thin sheet of metal mounted just behind the phosphor-coated back of the faceplate. Holes in the metal sheet correspond to each phosphor dot on the faceplate, allowing the assigned electron beam (red, green, or blue) to strike the correct color phosphor. Figure 13-3*b* shows the phosphors deposited on the faceplate in triangular groups of three dots, or *triads,* but other CRTs have the phosphors arranged in vertical stripes.

For further information on how the multicolor electrodynamic CRT works in a television receiver, see Sec. 19, "Television Broadcasting and Receiving Technology." The main difference between color CRTs for computer monitors and those for TV receivers is the RGB phosphor pitch. At present the 0.28-mm pitch is becoming the standard for personal computer monitors, but computer workstation monitors feature 0.20- to 0.26-mm dot pitch. For further information on computer monitors see Sec. 16, "Computer Peripheral Devices and Equipment."

## ELECTROSTATIC-DEFLECTION CRTs

The *electrostatic-* (ES-) *deflection CRT* includes four electrostatic plates positioned 90° apart to control the position of the electron beam, shown in the diagram Fig. 13-4. These CRTs are well suited for displaying rapidly changing complex waveforms. Precise electrical measurements can be made with a properly calibrated ES CRT. Some have two electron guns to provide a dual trace on a divided graticule.

# Light-Emitting Diodes (LEDs)

A *light-emitting diode* (LED), as shown in Fig. 13-5, is a visible-light source that produces different colored light at low DC currents and voltages. LEDs can be switched at high speed and have working lives that are far longer than those of incandescent lamps. Moreover, they can withstand high shock and vibration. LEDs function as lamps, indicators, and segments in alphanumeric displays. They have been made to emit light across the visible spectrum from dark red to blue, but with wide variations in efficiency. Many different processes are used to make LEDs and some are more efficient light producers than others.

Most commercial LEDs are made from elements in Groups III and V of the periodic table of elements. Aluminum (Al), gallium (Ga), and indium (In) are in Group III and nitrogen (N), phosphorus (P), arsenic (As), and antimony (Sb) are in Group V. However, other

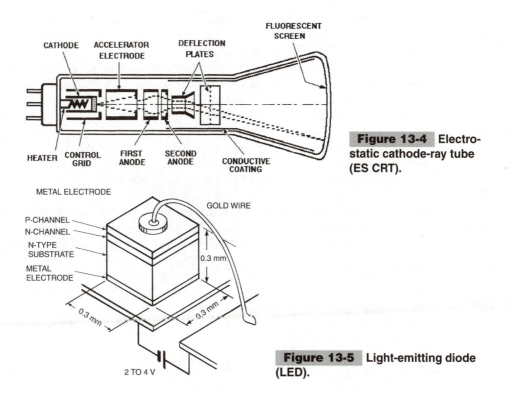

**Figure 13-4**  Electrostatic cathode-ray tube (ES CRT).

**Figure 13-5**  Light-emitting diode (LED).

compounds and elements such as silicon (Si) and carbon (C) from Group IV and zinc (Zn) from Group II have been used.

Most commercial LEDs are made as Group III-V two-element alloys such as GaAs and GaP, three-element alloys such as GaAsP, and four-element alloys such as InGaAlP. The simplest LED, as shown in Fig. 13-5, is formed by growing an *epitaxial* N-type layer on a N-type wafer of substrate and then forming a PN junction by diffusing a P-type dopant into the N-type layer. This is called a *homojunction LED* because both sides of the junction are formed from the same material. A metal layer deposited on the bottom of the N-type wafer forms one electrode and a wire bond on the P-type channel forms the upper electrode.

When the LED is forward biased (positive terminal at the P electrode and negative terminal to the N-type electrode), electrons and holes migrate into the active layer around the PN junction. As electrons are injected into the N-type region, recombination takes place at the PN junction. Photons created by recombination emerge as visible light. The wavelength or color of the light emission is a function of the *bandgap* of the materials from which the junction is made. LED light output or luminous intensity is generally proportional to current, so switching the bias voltage can modulate light output.

The photons generated are emitted in all directions. If the diode substrate is an opaque material such as GaAs, only those photons that are emitted upward within a critical angle will be emitted as useful light. All other photons emitted into or reflected into the GaAs crystal will be absorbed. When compared to GaAs, GaP is nearly transparent. Thus LEDs formed by growing an epitaxial layer on a GaP substrate are more efficient because most of the photons are emitted.

It was found that the performance of homojunction LEDs could be improved by using different materials on each side of the PN junction. The *single heterostructure* is fabricated by growing a window layer of N-type AlGaAs on the active layer of P-type GaAs. The window layer is transparent to the photons generated in the active layer. A later development called the *double heterostructure* further improved the efficiency of the LED. An active layer is sandwiched between two layers made from a material that differs from that of the active layer. The upper layer is still called the *window layer* and the lower layer is the *confining layer*. This structure permits a thinner active region, confines the injected electrons within the active layer, and minimizes photon absorption.

The luminous performance of LEDs is measured in lumens per watt (lm/W) and it has risen by a factor of 100 over the past 30 years. For example, homojunction LEDs emit less than 1 lm/W, single heterostructure LEDs emit 2 lm/W, but double heterostructures have emitted more than 10 lm/W.

The first commercial LEDs were GaAsP on GaAs substrates that emitted red light at 660 nm with a luminous efficiency of 0.15 lm/W. Red and yellow diodes made with GaAsP:N and green diodes made with GaP:N produced about 1 lm/W. Single heterostructure AlGaAs on GaAs substrates increased that value to 2 lm/W. In the late 1980s, double heterostructure AlGaAs on GaAs substrates achieved performances of 6 lm/W, more light than red-filtered incandescent lamps. Also, AlGaAs on AlGaAs substrate LEDs produced 8 lm/W, nearly the level of yellow-filtered incandescent lamps. Red AlGaAs, green GaP:N, and red, orange, and yellow GaAsP on GaP or GaAsP:N LEDs are in demand for interior and exterior displays.

LEDs that emit red light at about 646 nm are being made from aluminum gallium arsenide (AlGaAs) by a double-heterojunction process. They can emit at lower currents than earlier high-efficiency red materials. An N-type AlGaAs confining layer, a P-type AlGaAs active layer, and a P-type AlGaAs confining layer are grown on an $N^+$ GaAs substrate.

Both high-efficiency and AlGaAs LEDs can be multiplexed in alphanumeric and dot-matrix displays because they offer the higher brightness of GaP with the linear light versus the more limited diode current characteristics of GaAsP.

Commercial red and yellow LEDs made from double heterostructure AlGaInP produce luminous intensities of 2000 millicandelas (mcd) and devices made with this technology are expected to have luminous performance that exceeds that of unfiltered incandescent lamps.

There have been many introductions of blue LEDs over the past 10 years, but their luminous efficiency has been low. For example, commercial blue LEDs made from silicon carbide (SiC) offer a luminous efficiency of about 0.04 lm/W. They are homojunction LEDs that emit at 480 nm. More recently, commercial blue LEDs made as double heterostructures of P-type zinc-doped indium gallium nitride (InGaN) and N-type aluminum gallium nitride (AlGaN) have produced luminous intensities of 1000 mcd.

# LED Lamps

The most common LED lamp package is the bullet-shaped epoxy lens molded over a radial-leaded assembly, as shown in Fig. 13-6. The LED die is bonded in a conical cavity on the upper end of the cathode leads and a fine wire is bonded from an electrode on the die to the anode lead to complete the circuit. The lens focuses and distributes the light. It can be transparent, filled with a powdered diffusing material, or dyed red, green, or yellow.

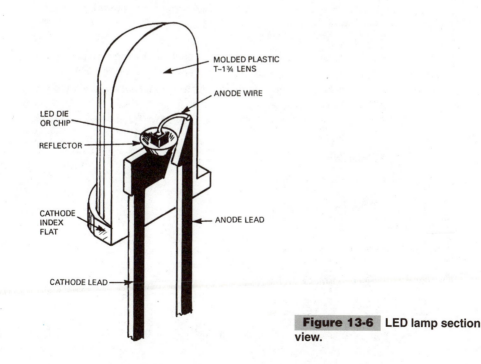

MOLDED PLASTIC
T-1¾ LENS

ANODE WIRE

LED DIE
OR CHIP

REFLECTOR

CATHODE
INDEX
FLAT

ANODE LEAD

CATHODE LEAD

**Figure 13-6** **LED lamp section view.**

The most popular case styles are the T-1 and T-1¾ radial-leaded plastic packages. Other package styles include flattop and surface-mount styles as well as rectangular molded cases which produce a rectangle of light when viewed end-on. For military and high-reliability applications, LEDs are packaged in hermetically sealed metal cases with glass lenses.

The forward bias voltages for most LEDs are from 1 to 5 V. Luminous intensity is stated in millicandelas. The *viewing angle,* the angle in which the luminous intensity is at least half of its axial value, varies from 18 to 150°. This angle is important in LED applications because LEDs with wide viewing angles can easily be seen when viewed from large offset angles. However, LEDs with narrow viewing angles appear brighter because the light is more concentrated.

LEDs continue to replace incandescent lamps, particularly where their high mechanical stability, low operating voltage, compatibility with semiconductor drive circuits, low operating temperature, and long service life compensate for their generally lower luminous performance. Examples include instrument readouts, light bars, and indicators for avionics applications; brake lights for automobiles; and outdoor displays.

# LED Alphanumeric Displays

## ALPHANUMERIC DISPLAY MODULES

Seven-segment digital displays have become commonplace because, as shown in Fig. 13-7, selective illumination of the segments can form all numbers and the letters from A through E. This concept is used in the seven-segment LED, as shown in Fig. 13-8, which is made by bonding individual LED dies to a substrate and enclosing them in molded plastic cases that define the segments with light pipes. The desired character is formed by switching on the appropriate segments.

The LED dies are bonded to a metal leadframe by cementing with conductive epoxy. Then fine wire is bonded from the individual dies to pin connections on the leadframe. After all the wire bonding has been completed, a molded form is placed over them and each of the triangular-shaped cavities is filled with a translucent colored plastic to form a light pipe. A colored filter diffuses the light output from each segment to offset any brightness variations.

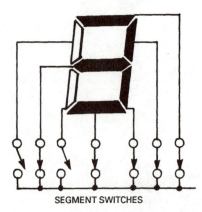

SEGMENT SWITCHES

**Figure 13-7** Seven-segment digital display.

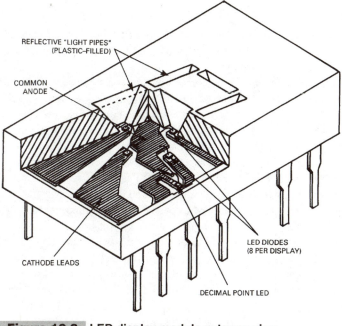

**Figure 13-8**  LED display module cutaway view.

Character heights of standard commercial LED displays range from 3 mm (0.15 in) to 20 mm (0.8 in), and a wide choice of colors is available. These displays are used in test instruments, sales terminals, clocks, and automobile instrument panels.

## MULTIDIGIT LED ALPHANUMERIC DISPLAYS

Multidigit LED alphanumeric displays are formed from end-stacked modules, each capable of forming a single character (with or without a decimal point). Modules are available with one or two seven-segment characters and dot-matrix formats.

Multidigit "smart" monolithic 16-segment LED display modules have onboard CMOS ICs that contain RAM memory, ASCII decoders, multiplexing circuitry, and drivers. They are used in displays for medical and process-control equipment and test instruments.

# Liquid-Crystal Displays (LCDs)

An *alphanumeric liquid-crystal display* (LCD) is an electronically switched display panel that must be viewed in ambient reflected light unless it is backlit by an electroluminescent panel. Characters are formed by the response of *liquid crystals* in the presence of an electric field. These crystals are viscous substances containing rodlike molecules that respond to electrical fields by reorienting themselves along electrical field lines to transmit or block the light that creates the image.

A film of liquid crystals is sandwiched between glass plates that have been imprinted with transparent metal electrodes which form diodes, as shown in Fig. 13-9. When voltage is applied selectively across the electrodes, the liquid-crystal molecules between them are rearranged so that light is either reflected or absorbed. LCDs offer higher contrast in direct sunlight than do CRT or LED displays.

All numbers and some letters can be formed by screening transparent metallized segments or dots on the inside of the upper glass plate and a metallized interconnection pattern on the inside of the lower plate. This permits the formation of multisegment and matrix displays capable of forming ASCII characters. The metal-film deposition process permits characters to be formed in different fonts, and any symbol or graphic that can be screened on the glass cover plate can be displayed as a visible positive or negative image.

The glass plates are separated by spacers to keep the gap and the thickness of the liquid-crystal layer uniform. The panels then are sealed around their edges. Most LCDs include a light filter and a protective transparent plastic cover. They are commonly packaged in dual-in-line package (DIP) cases for convenient insertion in circuit-board sockets.

Because LCDs have low voltage and power requirements they are popular displays for battery-powered calculators, watches, clocks, and instruments. More complex monocolor LCDs can form bargraphs, maps, diagrams, and waveforms for test instruments, calculators, and marine instruments such as depth finders, fishfinders, radar receivers, and GPS receivers.

*Twisted-nematic field-effect* (TNFE) *liquid crystals* provide either dark or black characters on a light (gray-white) field or the inverse. In the absence of an electrical field, TNFE molecules align themselves on a helical axis and typically twist polarized light up to 90°. This channels the light on a new axis through an exit polarizer so that the viewer sees a lighted pixel.

When an electric field is applied, the crystals align themselves parallel to the field so that the entering polarized light is blocked by the exit polarizer. The viewer then sees a dark pixel. But the polarizers can be arranged so that a dark pixel occurs when no electric field is applied and a lighted pixel occurs when it is.

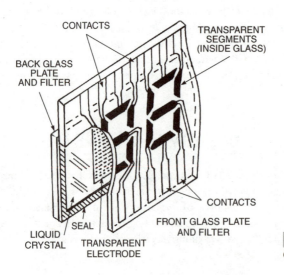

**Figure 13-9** Digital liquid-crystal display (LCD).

Recovery time of an LCD following a change in input signal is called *response time*. LCDs are driven by AC or pulses because electrolysis caused by DC will destroy the crystals. MOS ICs perform the decoding and provide the required drive pulses for LCDs.

TNFE displays have threshold voltages of 1.6 V, but 3 to 15 V is required to provide enough contrast for easy reading. "Guest host" LCD materials, introduced after TNFE, can add colors to the display.

# Liquid-Crystal Color Video Displays

Multicolor flat-panel LCD video displays are capable of displaying moving, full-color images. They now are installed in battery-powered notebook computers and are an option for AC-line-powered desktop computer monitors. They also are used as camcorder and still-camera viewfinders. The two principal multicolor LCD video display technologies are *dual-scan supertwisted-nematic* (DSTN) *display* and *active-matrix LCD* (AMLCD).

Full-color LCD video displays include red, green, and blue filters on their subpixels to produce colors. Filter and pixel-area losses reduce the light that passes through the crystals so that strong backlighting is required. Although more than twice as expensive as comparably sized CRTs, their performance is challenging the CRT. Moreover, they do not need high-voltage power supplies, and they generate less heat.

But LCD displays typically have bad pixels, including some that are stuck in either the *on* or *off* positions, causing imperfections called *freckles*. Another criticism of these displays is their narrow viewing angle. They must be viewed nearly straight-on because moving the head to the side, or up or down more than a few inches, can make images change color.

The displays are now being used in LCD monitors for desktop computers. They are being placed in thin cases with stands for desktop use or in frames for hanging them on the wall.

## PASSIVE-MATRIX LCDs (PMLCDs)

The multicolor *passive-matrix LCD* (PMLCD) was developed as a monitor for battery-powered notebook computers because a CRT was not practical. The pixel address information is applied by the computer processor row by row. The first models contained *twisted-nematic* (TN) liquid crystals, but the displays were slow because the crystals took 100 to 200 ms to respond to the address signals. This slow response minimized flicker but introduced ghosting and blurring when the displayed images changed rapidly.

TNLCDs were made with screen sizes of 10 to 12 in (25 to 31 cm), measured diagonally, but they lacked the contrast, were darker, and had more limited viewing angles than later displays that replaced them. The substitution of supertwisted-nematic (STN) liquid-crystal materials provided wider viewing angles, better contrast, and better resolution than TN. (*Supertwist* refers to added twist in the liquid-crystal molecule.) However, STN displays have now generally been replaced by dual-scan or double-layer supertwisted-nematic LCDs (DSTNLCDs). Two layers of liquid crystal improve contrast and viewing angle, but increase cost, complexity and weight. This class of display is now widely used in lower-cost notebook computers.

## THIN-FILM TRANSISTOR ACTIVE-MATRIX LCDs (TFT-AMLCDs)

The *thin-film transistor active-matrix LCD* (TFT-AMLCD) is an improved multicolor display suitable for TV receivers and computer monitors, particularly the notebook models. The parts of this display are shown in Fig. 13-10. Developed as an improvement over the PMLCD, it offers higher brightness and contrast, and it is capable of displaying rapidly moving images in natural color. Flicker is eliminated by keeping the pixels energized continuously.

This display includes a diffusion plate that spreads the fluorescent backlighting uniformly over the back of the entire display. The light passes through a polarizer and then moves through the array of thin-film amorphous-silicon transistors. Each transistor is addressed by row (address) and column (data) lines. The panel contains millions of these transistor switches deposited on a glass plate. The liquid-crystal material is sandwiched between transparent electrodes that form a capacitor which stores voltage until it is refreshed or changed in the following frame.

When no voltage is applied, the liquid crystals twist the polarized light to make it either parallel or perpendicular to a second polarizer, forming either a lighted or dark pixel. The application of voltage to the electrodes disrupts the orientation of the liquid crystals, leaving the polarized light unaffected. Pixels are divided into four parts, each with its own amorphous silicon transistor and a red, green, blue, and white filter. These displays are manufactured by photolithographic processes that are similar to those used in manufacturing ICs.

Active-matrix displays use an analog RGB interface to display National Television Systems Committee– (NTSC-) standard TV pictures or computer graphics of 1024 × 768 pixels. Commercial modules now measure approximately 14 × 9 in (36 × 23 cm) and are made to conform to Motion Picture Expert Group standard MPEG-1. The analog interface

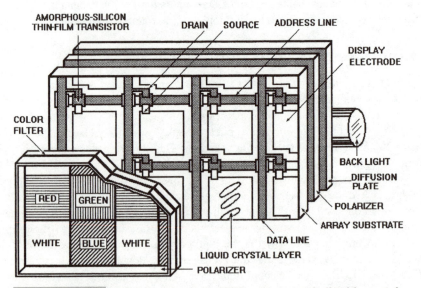

**Figure 13-10** Thin-film transistor (TFT) active-matrix liquid-crystal display (AMLCD).

consists of three ICs: (1) analog interface, (2) data inverter, and (3) analog driver. To achieve low power consumption and high speed, the devices are made with CMOS and BiCMOS technologies.

For more information on LCD monitors, see "Liquid-Crystal Display Computer Monitors" in Sec. 16, "Computer Peripheral Devices and Equipment."

# Electroluminescent (EL) Displays

An *electroluminescent* (EL) *display,* as shown in the cutaway view Fig. 13-11, generates light when an electric field is applied to an electroluminescent phosphor. Most electroluminescent materials require activators, impurities in the material that determine the characteristics of the emitted radiation. A typical electroluminescent phosphor is manganese-doped zinc sulfide.

EL displays are made by depositing the transparent thin-film metal electrodes, typically indium tin oxide, onto a glass faceplate through screens to form columns. Next the first dielectric insulator layer, typically silicon dioxide, is deposited, and this is followed by the polycrystalline EL phosphor compound layer which can be a powder or a 4000-Å (400-nm) thin film. Then the second dielectric layer is deposited, and opaque aluminum electrodes are deposited on it. Finally, a protective layer of insulation is applied. The transparent electrode is the column electrode, and the opaque electrode is the row electrode. A dot or pixel is the smallest independently addressable display area in the display.

The phosphor glows yellow, but colors from red to green can be obtained with filters. Most EL displays are AC powered to obtain their high luminescence and long life. If about 290 VAC is applied between the two electrodes, light is emitted through the transparent electrodes and glass substrate. EL displays can be configured as segmented alphanumerical or dot-matrix displays capable of producing all of the ASCII characters, the principal use of the displays today.

EL panels have been made for use as computer monitors. Some have provided up to $512 \times 256$ lines or 130,000 pixels and measure about $10 \times 5$ in ($25 \times 13$ cm). An EL system

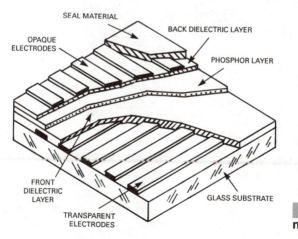

**Figure 13-11** Electroluminescent (EL) display.

consists of an EL display panel and a separate circuit board containing all of the necessary drive and interface electronics. The electronic circuitry converts serial signals to a video image on the display.

Because an insulating material is used between the transparent front column electrodes and the phosphor, EL display elements behave electrically like capacitors. As a result, the scanning rate and size of an EL panel are limited by the effective RC time constant set by the electrode's resistance and the panel's capacitance.

# Electrophoritic Displays

An *electrophoritic display,* as shown in the diagram Fig. 13-12, is a liquid-crystal display that has had light-absorbing dye added to the crystals to improve both color and contrast. Individual electrically charged dye particles move when an electric field is applied. If white dye particles are suspended in a black fluid between transparent electrodes, a DC voltage deposits the particles on one electrode, and the display appears white when viewed through that side, but when the voltage polarity is reversed, the particles move to the other electrode, and the display appears dark.

# Plasma Display Panels (PDPs)

A *plasma display panel* (PDP), as shown in the cutaway diagram Fig. 13-13, is a cold-cathode discharge tube that can display alphanumeric characters or images. When a DC voltage is applied across the anode and cathode electrodes, the gas, typically a mixture of neon and other gases, breaks down, forming a plasma discharge between the electrodes near the cathode. The electric discharge continues while voltage is applied to obtain a stable discharge.

The panel is made by sealing together two glass plates over the separators at their edges. Cell-separation walls and electrodes are printed in thick film on the plates and then baked

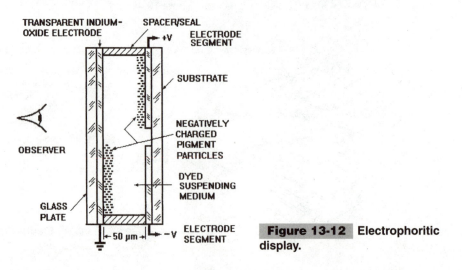

**Figure 13-12** Electrophoritic display.

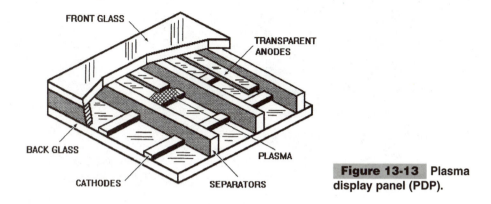

FRONT GLASS

TRANSPARENT
ANODES

BACK GLASS

PLASMA

CATHODES

SEPARATORS

**Figure 13-13** Plasma
display panel (PDP).

in the configuration shown in Fig. 13-13. The glass plates must withstand the printing and high-temperature baking processes, so they are thicker than the glass plates used in LCDs.

Monochrome PDPs display as many as 16 levels of shading. The luminance of PDPs can be controlled by changing the pulse width of anode data. Monochrome PDPs serve in bank terminals, point-of-sale terminals, and other business machines, as well as in medical and test instruments. They stand up well to vibration and shock, and they can withstand higher temperatures than do other display panels. A prime drawback of PDPs has been their high-voltage power-supply requirement. More than 50 V is needed to initiate gas breakdown, but actual requirements depend on the gas mixture.

A high-voltage driver IC as well as shift register and tone control ICs are required to drive the PDP. These functions are available in flatpacks with 60 to 80 pins. Interfaces have been developed to connect a PDP to a conventional CRT controller or to computers.

The color PDP is built in the same way as the monochrome PDP except that the anodes are transparent thin films. The cathode electrodes and separation walls are deposited by thick-film printing. Thin films of phosphors give the PDPs their color. In addition, a greater percentage of xenon is mixed with the neon in the mixed sealed gases.

UV radiation emitted by the electric discharge through xenon excites the phosphors applied as thin films to the inside of the glass panel, causing them to emit visible light. The red, green, and blue phosphors are contained in each cell, and are arranged so that selected phosphors emit light. The monochrome dots of each cell are divided into three parts in the cathode direction, and each cell corresponds to an anode. In a 10-in (25 cm) 640- × 400-dot color PDP cell, dot pitch is 0.1 to 0.12 mm. Separation walls provide 8 to 10 walls/mm cell sizes. Multicolor PDP wall-hanging TV monitors with 50-in-wide (1.27 m) screens can display 983,040 pixels. These monitors are less than 4 in (10 cm) deep and weigh about 100 lb (45 kg).

# Vacuum-Fluorescent Displays (VFDs)

A *vacuum-fluorescent display* (VFD), as shown in the cutaway view Fig. 13-14, is an alphanumeric or dot-matrix light-emitting display panel. Its dot or segmented elements are

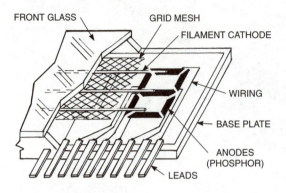

FRONT GLASS

GRID MESH

FILAMENT CATHODE

WIRING

BASE PLATE

ANODES
(PHOSPHOR)

LEADS

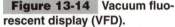

**Figure 13-14** Vacuum fluorescent display (VFD).

illuminated when a phosphor is excited by accelerated electrons in a hard-vacuum panel.
The VFD has a triode structure with combination heater and oxide-coated cathode, mesh
grid, and phosphor-coated anodes. The flat glass faceplate encloses the display, and it is
separated from the wiring printed on an insulating base plate.

The anodes are a conductive material surrounding phosphor depositions shaped as segments, and the grid is a metal mesh formed by photoetching a thin sheet of stainless steel.
The cathodes are fine tungsten wires that act as filaments, thinly coated with oxides of barium, strontium, and calcium so they will not obstruct the display segments. Heater voltage
applied to electrodes at both ends of the display raises the filament temperature to about
600°C, causing the oxide coating to emit electrons. When a positive voltage is applied, the
mesh grid accelerates the electrons from the filament toward the anode. They bombard the
anode, exciting the phosphor segments, which emit light to form the characters. Negative
voltage on the grid cuts off electrons moving toward the anode segments, controlling the
display.

The first VFD phosphors were green, but seven colors are now available, from reddish
orange to blue. There are two kinds of VFDs, the *dynamic drive* and *static drive* versions.
The dynamic drive is more suitable for VFDs with a large number of segments because it
can be operated by time-division pulse multiplexing. Microcontrollers with VFD drive
ports are available for that purpose. The static drive is more suitable for displays with a
small number of segments, and is recommended for VFDs in such applications as clocks
and automobile instrument panels.

The VFD offering the highest brightness level emits green light. VFDs can serve as multicolor displays with seven or more colors, and the designer is offered many options for patterns and type faces. The displays are reliable because they are essentially vacuum tubes
with a flat panel form factor. They can be driven by ICs and they respond rapidly. Pulse
driving permits the luminance to be adjusted easily.

Three types of VFDs are now in production: (1) *fixed segment,* (2) *character display,* and
(3) *graphic display.* Fixed-segment VFDs are widely used in automobiles, some for head-
up display (HUD) systems for projecting information on windshields. Dot-matrix alphanumeric character VFDs are used for displays in business machines, point-of-sale (POS)
terminals, and test instruments. Color graphic-display VFDs with 320 × 200 pixels have
been made to compete with other graphic displays.

# Holography

A *hologram* is a record of the interference pattern produced by the interaction of two electromagnetic waves, one of which is reflected from a subject. Holograms are most commonly produced by photographic methods with visible laser light, as shown in Fig. 13-15.

Light from the laser is split by a prism to create an *object beam* and a *reference beam.* The object beam is reflected by one mirror, passes through a dispersing lens, and is reflected by a second mirror to illuminate the subject. Light waves reflected from the subject radiate out in the direction of the photographic plate or film. Simultaneously, the ref-

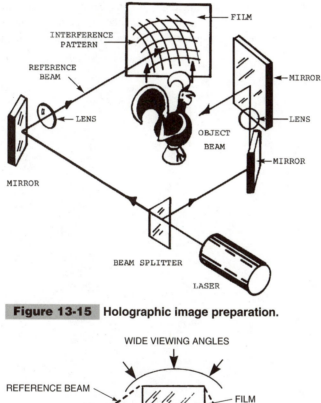

**Figure 13-15**  Holographic image preparation.

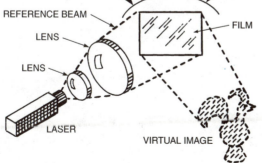

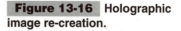

**Figure 13-16**  Holographic image re-creation.

erence beam is reflected by a mirror and dispersed by a lens in the direction of the film, but it does not illuminate the subject. Light from every part of the subject interacts with the reference beam to create an *interference pattern* which is recorded on the film. The developed film is the hologram.

The interference pattern is a three-dimensional record of the infinitely complex light waves that were reflected from the subject. When the hologram is illuminated by laser light, as shown in Fig. 13-16, the interference pattern defracts the light to re-create a virtual image. Viewers looking at the virtual image from different locations perceive it in three dimensions as if the light waves were coming from the original subject.

The term *hologram* is derived from the Greek words *holos* and *gramma,* meaning "the whole message." *Holography,* the practice of forming holograms, was developed by Dennis Gabor in 1947. Holograms have been adapted for inspecting of three-dimensional objects, scanning and reading universal product codes at checkout counters, and recording three-dimensional images of subjects for artistic, industrial, commercial, scientific, and educational applications.

# MICROPROCESSORS AND MICROCONTROLLERS

# Overview

The microprocessor, a computer's central processing unit on a single silicon chip, dramatically changed the course of electronic technology more than 30 years ago. It made the computer affordable for the average person and introduced the benefits of stored-program computing to virtually all sectors of human enterprise. Far from just eliminating the need for dozens of hard-wired components on a circuit board, it introduced new and previously unattainable versatility into communications and data processing equipment and accelerated the acceptance of digital electronics. Now that circuit capabilities are no longer limited to the capabilities of the components wired to the circuit board, software has become the controlling factor in determining many circuit applications.

The microprocessor made the personal computer a reality and turned word processing, spread-sheet preparation, programming, and database preparation into activities that can be done in the privacy of one's home. It has also introduced a new era in personal communications, information retrieval, and entertainment. With the development of the Internet and the World Wide Web, the personal computer is now competing with the library as a source of information, with the telephone and postal service for personal communication, and with the TV set as a home entertainment provider.

The microprocessor paces computer technology, and the performance of microprocessors has improved so rapidly that most personal computers have become technically obsolete in about two years. This has led to a lively computer replacement business, to say nothing of the growing market for new peripheral hardware and software to keep up with the performance improvements in the new models being introduced.

Having satisfied themselves that personal computer sales are nearing saturation, the microprocessor manufacturers and personal computer software publishers have been looking beyond PCs to network file servers, engineering workstations, and other more powerful computers, reversing the former downward spiral of technology transfer to an upwardly mobile one.

The names of microprocessor manufacturers, computer manufacturers, and software publishers, as well as their executives, have become highly visible in the financial and business community and are even well known in homes around the world. Because of concentrated advertising and promotion, most personal computer users are aware of the brand names of their microprocessors and the operating systems that run them. But this is not true for an important spin-off of the microprocessor, the *microcontroller*.

This device has brought the benefits of programmed control to a wide range of products, from toys, entertainment products, and appliances to automobiles and instruments. The less visible microcontroller is now managing the operation of TV sets, CD players, VCRs, camcorders, digital cameras, microwave ovens, and even power tools. It has also made its way into many commercial and industrial machines and processes. It performs its duties without user intercession and, indeed, is virtually invisible to the user.

The need for specialized microprocessors optimized for processing digital data has led to the development of yet another microprocessor spin-off, the *digital signal processor* (DSP) chip. These devices process sampled data at high rates and perform such operations as accumulating the sums of multiple products faster than can be done by a general-purpose microprocessor. DSPs are designed to take advantage of the repetitive nature of signal processing by "pipelining" data flow or starting to execute other tasks before completing those in progress to gain extra speed.

For further information on microprocessors in computer systems see Secs. 15, "Computer Technology," and 16, "Computer Peripheral Devices and Equipment."

# Microprocessors (MPUs)

A *microprocessor* (MPU) is a very large scale integrated circuit that performs the functions of a computer central processing unit (CPU). These functions were previously carried out by dozens of discrete components on a circuit board. Miniaturization and consolidation of these functions has reduced the cost, size, and power consumption of CPUs while increasing their speed and reliability. The MPU performs arithmetic, logic, and control operations. Now that the CPU is in all desktop and notebook personal computers as well as in many different kinds of subnotebook personal communication and scheduling products, it is also displacing CPU circuit boards in engineering workstations, network servers, mainframes, and supercomputers.

MPUs carry out their duties faster and more reliably while occupying far less circuit-board space and consuming far less power than their board-level predecessors. They have also reduced cooling requirements for computers. Moreover, they are capable of performing many functions that were simply beyond the capabilities of hard-wired circuits.

The MPU responds to inputs and produces outputs that are directed by the program or sequence of instructions that are stored in computer memory. Considered to be the "engine" in most computers, it can, in addition to computing, accept inputs from many different kinds of external devices and also send control signals to them through input/output (I/O) ports. But the MPU requires a large amount of off-chip memory to handle instructions, read and write data, and provide data to I/O buffers.

The CPU, as shown in the block diagram Fig. 14-1, is the core of the microprocessor. Its principal parts are the *arithmetic logic unit* (ALU); *address, status,* and *data registers;* and control and timing functions, which include the *instruction decoder, controller-sequencer, accumulator,* and *program counter.* The I/O ports or buffers, read-only memory (ROM), and random-access memory (RAM) are not part of the CPU, but a microprocessor can contain some of these functions.

Figure 14-2 is a simplified block diagram of an Intel Pentium microprocessor. MPU operations are synchronized by internally or externally generated clock pulses from a free-running oscillator or crystal. The Intel Pentium microprocessor can access external memory, its I/O interface, I/O devices, and peripherals through parallel conductors which carry data. It can control and service peripheral devices that share an external bus. The MPU can process large quantities of data quickly. Instructions from the program are stored in memory, typically in the computer's hard-disk drive. The principal functional blocks of a microprocessor are as follows.

## ARITHMETIC LOGIC UNIT (ALU)

The *arithmetic logic unit* (ALU) performs arithmetic and logic operations on data selected by the program that have been transmitted through various control lines. The ALU performs binary addition, subtraction, multiplication, and division, as well as many different

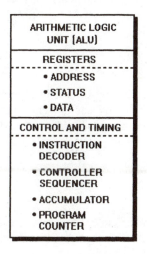

**Figure 14-1** Central processing unit (CPU) block diagram.

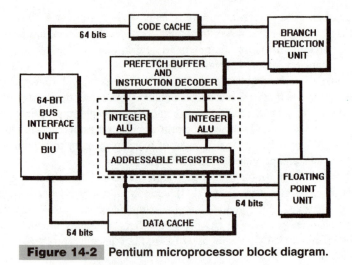

**Figure 14-2**  Pentium microprocessor block diagram.

logical comparisons by means of boolean operations. For example, it complements words by inverting and shifting them either right or left, 1 bit at a time.

## STATUS REGISTER

The *status register* contains bits resulting from the execution of an instruction. These bits, determined by the MPU design, include *carry, zero, not zero,* and *sign.*

## CONTROLLER-SEQUENCER

The *controller-sequencer* manages and supervises MPU activities by generating control signals for executing an instruction which are sent to the appropriate input pins. The control logic generates clock signals, reads and writes data to and from the memory and I/O port, and performs address latching (holding the location of data contents), bus control, and reset functions. These are directed by a *microprogram* stored in ROM on the MPU to control the digital gates and sequence the instruction set.

## ACCUMULATOR

The *accumulator* is a special-purpose functional register that performs a dual function during arithmetic and logic operations. Before an operation it holds an operand, and after an operation it holds the resulting sum, difference, or logical answer. It acts as a transfer focal point and binary adder. The accumulator can be complemented, tested, and shifted to perform basic MPU functions such as arithmetic or logic.

## DATA REGISTER

The *data register* is a temporary location for data received from the *data bus.* Addressed by the program, it is a general-purpose register. A register can hold data for both arithmetic and memory-addressing operations. The instruction to be decoded is held by a special register called the *instruction register* attached to the instruction decoder.

## INSTRUCTION DECODER

The *instruction decoder* analyzes the contents of the data register to determine the next operation to be performed. The controller-sequencer then supplies the appropriate preprogrammed signals for execution. Each instruction consists of a series of microinstructions stored in a microcode ROM on the MPU for performing such tasks as opening and closing logic gates, opening or closing circuits, or resetting status flags.

## PROGRAM COUNTER

The *program counter* is a special register for controlling the sequence of instructions, subroutines of a computer program. These instructions are stored in sequentially accessed memory locations. As each instruction is carried out, the register is incremented. It then points to the location in memory where the next instruction or data word is located.

## ADDRESS REGISTER

The *address register* temporarily stores the location or address of a data word in the memory or I/O being processed. The address is decoded by a logic circuit at the selected memory or I/O port, and that decoding logic translates the contents of the address register into specific memory locations. Address registers can be 32 bits wide. The different kinds of address registers are classified by the methods they use for addressing the register. These include *indexing, paging, stack pointing,* and *indirect.*

## MICROPROGRAM

A *microprogram* is a program contained in an on-chip ROM that cannot be accessed by the user. It contains the sequencing for the entire CPU instruction set. MPUs carry out their instructions by repeating three steps: (1) *I-fetch,* (2) *decode,* and (3) *execute.*

## MICROPROCESSOR CHARACTERISTICS

The microprocessor computes and processes digital data. The binary digits or bits are grouped in *words.* The 8-bit word, known as a *byte,* is the basic computer word length. MPUs are optimized for software-intensive applications that require many numerical computations and extensive data processing. Commercial, off-the-shelf microprocessors are available with 8-, 16-, and 32-bit architecture and they have clock speeds of up to 450 MHz. MPUs differ from microcontroller units (MCUs), which are optimized for real-time, hardware-intensive control applications. Their programs are permanently stored in some form of memory that might or might not be on the MCU chip. See "Microcontrollers (MCUs)," following.

The most advanced MPUs have a 32-bit internal architecture (32-bit internal address bus and 32-bit internal data bus) and a 32-bit external data bus. Driven by intense competition, manufacturers have been continually improving their MPUs because of customer demand for even higher clock speeds and more features. This trend has been encouraged by the simultaneous development of higher-density and faster semiconductor memories.

The latest 32-bit microprocessors include instruction and data caches, high-speed on-chip memory for holding the most recently used instructions and data that are most likely to be reused, thus reducing data storage and retrieval time. A memory-management unit (MMU) allows programmers to use system resources without considering the actual size of the memory in megabytes. The MMU also allows multiple programs and operating systems to be used simultaneously.

The first successful 8-bit MPUs were the Intel 8080, the Zilog Z80, and the Motorola 6800 and 6809. Motorola then introduced a 6000 line with 6000 transistors that included the 68008, 68010, 68012, 68020, and 68040. During the same time period Intel produced the 80186, 80286, 386, and 486 before introducing the Pentium series in 1993.

Table 14-1 summarizes the progress made by Intel Corporation, maker of the world's most popular microprocessors over a period of 18 years. The Celeron family is not listed because these MPUs are intended for use in low-priced computers, and are not optimized for highest performance. Truncated versions of the Pentium II, the first Celerons lacked their 128 kB cache memories, but they were restored in the later Celeron 300 A.

MPUs have been developed by other manufacturers in the United States and Japan. Sun Microsystems introduced the SPARC RISC microprocessors, and Hewlett-Packard produced the Precision Architecture-RISC processors for workstations and file servers. IBM, Motorola, and Apple Computer devised the Power PCs; Silicon Graphics introduced the MIPS processors; and Digital Equipment Corporation designed the Alpha MPU. None of these devices has been as successful as the Intel MPUs in terms of market acceptance and quantities produced.

The first MPUs were made with the NMOS process because it permitted smaller transistors than the PMOS process. The 8086 and 80286 were NMOS MPUs, but in 1985 Intel switched to CMOS for the 386 and 486. Then in 1992 Intel introduced its Pentium processor made with the BiCMOS process, a lower-power-consuming version of CMOS. The first Pentium ran about 4 times faster than the 486. Successive Pentium improvements, the Pentium Pro, Pentium MMX, and Pentium II, were also made by the BiCMOS process.

| Model | Introduction Date | Number Transistors | Design Rule (µm) | Bus Size (bits) |
|---|---|---|---|---|
| 4004 | 1971 | 2.3k | 10.0 | 4 |
| 8080 | 1974 | 2.3 k | 6.0 | 8 |
| 8086 | 1978 | 6.0 k | 3.0 | 16 |
| 8088 | 1979 | 12.0 k | 3.0 | 16 |
| 80286 | 1982 | 134.0 k | 1.5 | 16 |
| 80386 | 1985 | 275.0 k | 1.0 | 32 |
| 80486 | 1989 | 1.6 M | 0.6 | 32 |
| Pentium | 1993 | 3.3 M | 0.35 | 32/64 * |
| Pentium Pro | 1995 | 5.5 M | 0.35 | 32/64 * |
| Pentium MMX | 1997 | 4.5 M | 0.35 | 32/64 * |
| Pentium II | 1997 | 7.5 M | 0.35 | 64 |
| IA-64 Merced | 1998 | ~10.0 M | N/A | 64 |

* 64 bits external

**TABLE 14-1**  Microprocessor Progress

Line thickness (*design rule*) on microprocessors declined from 3.0 μm on the 8086 in 1978 to 0.356 μm in the Pentium II MPU in 1997. Over that same period, the number of transistors has increased from 29,000 to 7.53 million for the Pentium II. Intel has introduced variations of its Pentium microprocessor to address different performance requirements: Celeron at the low end, Pentium II in the midrange, and Pentium XEON and Pentium III at the high end. Pentium IIIs offer speeds of 450, 500, and 600 MHz.

Intel and Hewlett-Packard have introduced the Merced as the successor to the Pentium line for high-end computing applications. Known as the IA64 because it uses a 64-bit design, it will contain more than 10 million transistors and have a clock speed in excess of 800 MHz. With more than twice the speed of the Pentium processors being produced in 1999, it was designed for performing parallel computing. Industry experts have predicted that MPUs could have 100 million transistors before the year 2005.

# Microcontrollers (MCUs)

A *microcontroller* (MCU), as shown in the block diagram Fig. 14-3, is a monolithic IC that contains a CPU, program memory, data memory, I/O ports, and other timing, counting, and management functions on a single silicon chip. Its CPU performs the same functions as a microprocessor CPU, but it is less powerful and occupies less on-chip space to allow room for the other on-chip functions. The limited amount of on-chip ROM is typically programmed to carry out application-specific control functions. Thus the MCU is the closest approximation to a microcomputer on a chip that exists today because it contains many of the functions that are not included on microprocessor chips.

The MCU, optimized for control rather than computation, now controls products from toys, home appliances, and test equipment to computer keyboards, monitors, and printers. MCUs perform many different duties in automobiles, from engine control to managing the entertainment system and antilock brakes. MCUs are also embedded in TV-receiver chan-

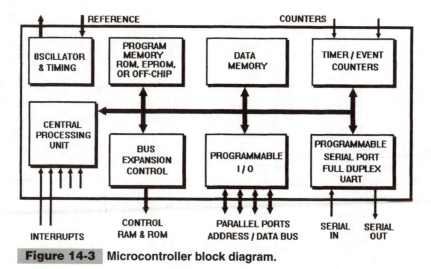

**Figure 14-3** Microcontroller block diagram.

nel switches and tuners, stereo-system volume and tone controls, camcorders, and digital cameras. All of these applications are directed by a dedicated program stored in some form of on-chip ROM. If the program is larger than the capacity of the on-chip ROM, it can be supplemented with an off-chip ROM.

The data memory is RAM, and the program memory can be factory-programmed ROM or field-programmed EPROM or EEPROM. Progress in MCUs closely follows that of microprocessors, taking advantage of improvements in wafer processing and line-width or feature-size reductions that increase its speed and memory density without increasing chip size.

Advanced MCUs can perform the same tasks as earlier generation MPUs with five external peripheral devices. They are now available with a wide range of performance capabilities in 4- to 16-bit architectures. MCU are designed in families to give the designer performance options. For example, standard MCUs can have different memory capacities to fit the end-use application. The lead part in each MCU family has a factory-programmed ROM for storing a proven program. When purchased in quantity, the unit price of a factory-programmed MCU will be lower than for other versions that let the customer program the memory. Some customers want to store a trial program in an erasable memory, and they purchase factory-programmed MCUs only after the trial program has been debugged.

ROM-less MCUs for product development are also available from some manufacturers to give designers the option of developing and verifying the program with off-chip EPROM or EEPROM. Other MCUs are offered only with fixed amounts of EPROM or EEPROM because some host products will never reach the volume levels needed to justify the up-front engineering costs for factory-programmed MCUs, so customers will not be able to obtain the lower unit pricing after the engineering costs are prorated.

Some MCUs now include on-chip analog-to-digital converters (ADCs) to permit the direct interfacing of MCUs with analog sensors for measuring physical variables such as pressure, temperature, or rpm. This feature, particularly attractive in industrial, scientific, and automotive applications, saves circuit-board space and assembly costs. Some of the more popular microcontrollers are the Intel 8048, 8051, and 8052; the Zilog Z8; and the Motorola 6801, 6805, and 68H C11.

# Digital Signal Processors (DSPs)

A *digital signal processor* (DSP) is a microprocessor optimized to process sampled data at high rates. It can accumulate the sum of multiple products faster than a microprocessor. The architecture of the DSP is specifically designed to take advantage of the repetitive nature of signal processing by "pipelining" data flow, or starting on another task before the one in progress is complete to gain extra speed. DSPs are used in wireless base stations, cellular telephones, pagers, modems, and other communications products.

Some DSPs include stand-alone microprocessors, and some are designed to work with a computer as host. In digital signal processing many repetitive mathematical operations such as fast Fourier transforms must be performed, and these require many multiplications and summations.

The general-purpose DSPs available include 16- and 24-bit fixed-point and 32-bit floating-point devices. A 16-bit fixed-point general-purpose DSP is a single-chip device that typically includes a microprocessor, on-chip RAM, serial ports, parallel port, on-chip codec, and a phase-locked loop (PLL). Some of them can perform 20 million instructions per second (MIPS) at 40 MHz. A 32-bit floating-point version can act as a host to support interfacing to other processors in multiprocessor applications.

In addition to image processing, DSPs can perform filtering and data processing in high-speed telecommunications lines and instrumentation, spectrum analysis, speech recognition and compression, radar return analysis, and pattern recognition.

DSPs are either general purpose (for digital signal processing) or application specific. Application-specific DSP ICs perform such tasks as digital filtering and Fourier transforms more accurately, faster, or more cost-effectively than general-purpose DSPs.

The DSP's fast array multiplier and accumulator distinguish it from a microprocessor. These functions allow the multiplication of two numbers and the addition of the product to previously accumulated results in a single clock period. In a typical fixed-point multiplier-accumulator two 16-bit numbers can be multiplied and the 32-bit product can be added to a 32-bit accumulator register in a single instruction cycle. A typical microprocessor would require about 25 clock cycles to carry out this task.

Most DSP chips include cache memories to service program and data pipelines. A *cache* is a small, fast memory located between a larger, slower memory and a processor to improve access to data and instructions. The DSP chips that carry out block floating-point arithmetic most efficiently have *barrel shifters,* circuits that allow arbitrary shifting of data, as well as other circuitry to detect the presence of exponents.

A typical DSP chip has two data memories and two data buses. It can deliver two operands required for a single-cycle execution of the multiply-accumulate function. In contrast with microprocessors that store both instructions and data in the same memory, most DSP processors have separate program and data memories so that instructions and data can be retrieved simultaneously. However, some DSPs store certain kinds of static data in the program memory for transfer to the faster, smaller data memories when needed. The time required for an instruction cycle or multiply-accumulate per operation ranges from 60 to 200 ns. Word lengths are from 8 to 32 bits long, but 16-bit lengths are typical.

DSP chips have multiple memories and corresponding buses to enhance system throughput by increasing the rate of data access. Some DSPs have dual internal data memories, and can access external data memory. Typical internal memories have 129 to 512 16-bit words. DSPs can also increase throughput with external program and data memories that are fast enough to supply operands and instructions in a single cycle.

DSP applications, such as image processing, require that the chips address at least 64,000 words of external data or program memory. Analog input data is converted for processing and reconverted to an analog output. In some applications, the DSP chip acts as a slave or peripheral to a microprocessor with program and data memory.

DSPs are packaged in dual-in-line packages (DIPs), ceramic and plastic leadless chip carriers (LCCs and PLCCs), and pin-grid array (PGA) packages.

# COMPUTER TECHNOLOGY

## Overview

The term *computer* today implies a stored-program electronic digital computer, but it has not always had that meaning. Fifty years ago, before the introduction of the modern solid-state computer, the word could mean any of many different kinds of electromechanical and electronic analog machines capable of computation. Application-specific analog computers, built from gears, levers, clutches, and motors, played an important role in World War II for setting fuses and gun elevations for artillery while accounting for such factors as range, elevation angle, propellant, wind, weather, and even the spin of the earth. Data was entered in these computers with knobs and hand cranks, and the useful output data appeared on dials or it controlled mechanical movements. It is significant that the first use of the large digital computers built by the U.S. government during the war was the preparation of tables for calculating artillery trajectories.

As the first vacuum-tube digital computers were being developed, engineers and scientists were using analog computers, consisting largely of operational amplifiers and precision potentiometers, to solve mathematical problems and perform simulations. Data was entered by setting banks of potentiometers, and the output could be in the form of meter readings, oscilloscope waveforms, or graphs on x-y plotters. The early analog computers had vacuum-tube circuitry, but these were later transistorized. Analog computers played a significant role in the design and testing of NASA equipment and the simulation of space vehicle behavior in the early days of the space program. They were also widely used in wind-tunnel and water-tank testing of scale models.

The basic ideas behind today's stored-program computer, as distinguished from the mechanical calculators and punched-card data processors, were proposed by Charles Babbage, a professor of mathematics at Cambridge University, in the 1830s. He designed and built a machine with decimal counting wheels that could add and print the results, but was never able to finish his "analytical engine." The assemblage of gears and levers was just too complicated for the technology of the day. Nevertheless, he proposed methods for automating computation by storing both data and instructions in a form that could be fed rapidly into the machine whenever they were needed. More than 100 years would pass before stored memories were included in computers.

The first programmable digital computers consisted of banks of electromagnetic relays which were later superseded by hundreds of vacuum-tube switches. The invention of the transistor made it possible to build computers that did not fill an entire air-conditioned room. Integrated circuits eventually reduced the size and power requirements for computers while dramatically improving their reliability.

Nevertheless, computers called *mainframes* remained large, cumbersome, and expensive for many years, affordable only by government agencies, scientific laboratories, corporations, and financial institutions. Although they changed the way both business and science were done, the computing process was tedious and slow, and it required the services of trained operators. Time sharing was introduced so that organizations and individuals who could not afford to own their own computers could make use of them on a part-time basis.

Minicomputers, the first of the benchtop computers, brought computer power to the factory floor and introduced digital control of machine tools and processes to industry. But it took the invention of the microprocessor to make private ownership of computers affordable. Many different desktop computers with microprocessors appeared, but the IBM PC caught on and captured the market. Since that time there have been consistent improvements in both hardware and software, making personal computing more attractive to ever-larger numbers of people. No longer limited to home office tasks such as word processing and spread-sheet preparation, they are now being purchased for playing electronic games, surfing the Internet, and sending and receiving E-mail. It is now possible to view movies and listen to stereo music on personal computers.

Efforts to make computers more user-friendly have not yet been entirely successful, even with the introduction of the graphical user interface and the mouse pointing device. Nevertheless, more than 40 percent of all homes have at least one computer, and they are being purchased in quantity for schools, where they are supplementing conventional teaching methods. It remains to be seen, however, if the computer will ever displace the TV set as the preferred entertainment medium. The next stage in computer evolution will be the introduction of voice recognition to eliminate many of the problems associated with keyboard data entry.

The world's fastest supercomputer in 1998 was built by Intel Corporation for the Sandia National Laboratory. It includes 9152 Intel Pentium P6 microprocessors, and it has achieved a peak speed of more than 1.3 trillion mathematical operations per second. The goal of the Accelerated Strategic Computing Initiative (ASCI) sponsored by the Department of Energy (DOE) is to have a supercomputer capable of performing 100 trillion calculations per second by 2004. This computer would be about 100 times faster than existing supercomputers and 100 million times faster than today's fastest supercomputer.

The commercial success of the desktop PC has led to the development of the battery-powered portable notebook models and a host of even smaller, more specialized sub-notebook-sized personal communicators and pocket-sized daily planners.

This section is focused on personal computers because of the rapid technological changes occurring in that sector that are now leading the way for the rest of the industry. For more information about computer-related technology, see Sec. 14, "Microprocessors and Microcontrollers," and Sec. 16, "Computer Peripheral Devices and Equipment."

# Digital Computers

A *digital computer* is a stored-program electronic computer with internal memory that depends on software for system management and the automatic execution of stored programs. The simplified block diagram Fig. 15-1 applies to all digital computers. Its internal

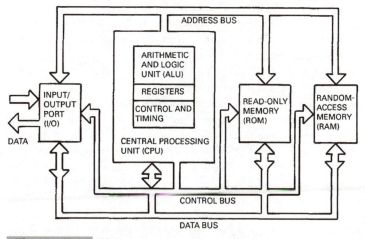

**Figure 15-1** Digital computer block diagram.

memory stores both data and instructions entered either by keyboard or from different sources such as diskettes, compact disk (CD) ROMs, backup disk cartridges, or magnetic tape. Its instruction sequence for input, processing, and output is performed automatically.

The central processing unit (CPU) has evolved from circuit boards full of active and passive discrete components to the microprocessor chip, while main memories have evolved from ferrite beads strung on a wire to semiconductor chips capable of storing millions of bits.

Computer monitors, direct descendants of color TV receivers, are now able to display transactions, act as interactive devices, and reproduce more than 65,000 colors. They have replaced the awkward teleprinters initially used for input and output functions into the 1970s. Desktop printers are now able to produce letterpress-quality printing and graphics at rates of more than 10 pages per minute, and they now can reproduce color photographs that are nearly indistinguishable from the originals. Data can now be introduced and manipulated in human-computer interactions by a host of different pointing devices including keyboards, mice, trackballs, joysticks, and data-entry tablets.

Most modern computer CPUs are based on the *complex instruction set concept* (CISC) but a rival *reduced instruction set concept* (RISC) was also developed. Some semiconductor manufacturers developed commercial RISC microprocessors but they were not successful in the marketplace. However, CISC CPU designers have incorporated RISC principles in the latest microprocessors to take advantage of some of the improvements they offered.

*Microcomputer* is a generic term still used to identify any computer whose CPU is a microprocessor, but these computers are now more commonly identified by more specific descriptions such as *desktop personal computer, notebook computer,* or *network server.*

*Personal computer* or *PC* is now a generally accepted generic term to identify all computers in the desktop and notebook class, but it was coined by International Business Machines (IBM) Corporation to identify its brand of small computer that included a microprocessor from Intel Corporation and an operating system from Microsoft Corporation. But the term *PC compatible* has a more definite meaning. It refers to computers made by manufacturers other than IBM that are compatible with or can run the same applications software as IBM machines. A compatible PC includes a Microsoft operating system (DOS, Windows 95 or 98) but the microprocessor could be made by Intel or be a clone made by other manufacturers. The term *PC compatible* distinguishes a computer from any models made by Apple Computer, Inc., such as *Macintosh* computers that use a proprietary Apple operating system and Power PC microprocessor. These computers are not compatible with Intel microprocessors and Microsoft operating systems.

A *workstation* or *engineering workstation* is a computer optimized for computer-aided design (CAD) and computer-aided engineering (CAE) that places heavy emphasis on graphics capability. Present generation workstations are based on 32-bit microprocessors.

A *network server* is a specialized computer intended as a common source for software in a computer network. These computers have memory capacities that far exceed those of standard desktop computers so that they can store and download selected specialized applications programs on demand to any computer in the network.

The term *minicomputer,* rarely used today, refers to a computer generally used as an industrial machine or process controller. Most minicomputers today have 32-bit custom-designed central processors.

The term *mainframe computer* still applies to powerful computers used in industry, government agencies, and banking, insurance, and securities organizations for large-scale data processing, accounting, and the maintenance of large databases.

# Computer Organization

Most digital computer systems are organized as shown in the block diagram Fig. 15-2. The computer reads or inputs data from such sources as keyboards, modems, diskette drives, or hard-disk drives, and it writes or outputs to the CRT monitor and other devices such as printers or *x-y* plotters. Computer memory stores program instructions as well as the data being processed. The computer's "brain" is its CPU, today typically a microprocessor. For more information on the CPU see Sec. 14, "Microprocessors and Microcontrollers."

# Computer Memory

The *central processing unit* (CPU) is connected directly to the memory as well as to input/output ports, as shown in the block diagram of computer memory hierarchy Fig. 15-3. In level 1 there are fast, low-density internal registers within the CPU, and some CPUs contain a *cache memory* consisting or fast, expensive, low-density static RAMs (SRAMs). There might also be a separate cache memory at level 2 consisting of more SRAMs. The main memory, level 3, consists of slow, inexpensive, high-density dynamic RAMs (DRAMs). The hard-disk drive is considered to be level 4. It is a high-density, low-cost-per-bit memory that serves as mass storage. The diagram shows that the costly fast memory is concentrated near the CPU, while the inexpensive slow memory is located further away.

The cache memory stores copies of data blocks from the main memory that the processor is most likely to request while performing its present task. Caches provide shortcuts to get around the slower response of main memory. The SRAMs run at about the same speed

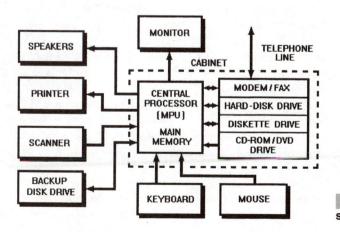

**Figure 15-2** Computer system organization.

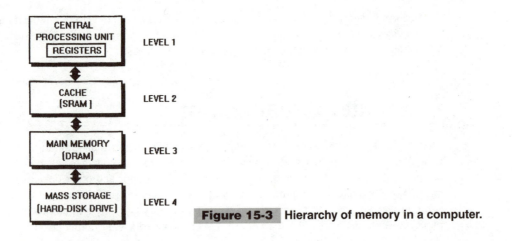

**Figure 15-3**  Hierarchy of memory in a computer.

as the processor, but they are expensive. Moreover, the cache memory must include circuitry that keeps track of which memory blocks are in the cache and what their status is. Microprocessors now contain on-chip cache memory along with registers. But both registers and caches must be small because of space restraints on the chip. Consequently, these caches are usually supplemented by another off-chip SRAM cache at level 2.

Main memory stores programs and data, and it is a source of input and the destination for output. Main memories consist of DRAMs, which are slower than the SRAMs but also are less expensive. Data and instructions are stored in areas called *locations*. Each location in main memory has an address so that data can be located. The capacity of main memory is determined by the size and application of the computer. The computer's CPU or processor can only operate on the data from the main memory and only instructions from the main memory can control the computer. Some SRAM cache memory is on the same chip as DRAM main memory, making it difficult to assign a definite location for all cache memory.

Level 4, mass storage, today is a magnetic disk drive that can be supplemented by a floppy-disk or diskette drive or a backup magnetic-disk cassette drive. Computers now include a scheme called *virtual memory* that fools the processor into thinking that main memory is much larger than it is. Pages of data are called from the hard disk and placed in main memory when they are needed or returned to disk when they are not. A *memory-management unit* (MMU) keeps track of which pages are in main memory and what their status is.

Hard-disk storage is inexpensive in terms of cost per bit. Commercial hard-disk drives that are factory installed or can be purchased as plug-in replacements for upgrading a personal computer can store more than 16 GB.

*Demand-paged memory systems* divide both disk and RAM memories into fixed-sized *pages*. In moving from disk to RAM, blocks of information are switched into the same number of pages. Demand-paged virtual memory permits multiple users and multiprocessing. *Segmented memory systems* impose partitions or segmentation on RAM so that it accommodates the longest program or data construct needed.

Erasable optical disk drives such as compact disk (CD) and digital video disk (DVD) ROMs are expected to increase computer storage significantly. Diskettes, by contrast, are limited to 1.44 MB, a small fraction of the capacity of typical hard-disk drives today.

Auxiliary offline memory devices are available for backup, long-term, archival data. or program storage. Among these memories are magnetic-tape drives, and magnetic-disk drives with removable disk cartridges. Tape drives are serial-access memories with relatively long access time, but removable magnetic-disk cartridges are essentially hard-disk drives with removable platters. CD- and DVD-ROMs can also provide archival data storage.

For more information on semiconductor memory see Sec. 9, "Digital Logic and Integrated Circuits," and for further information on memory systems see Sec. 16, "Computer Peripheral Devices and Equipment."

# Bus Structures

A *bus* is the signal and power interconnection that interconnects the internal circuitry of the computer. The amount of data that can simultaneously be transported along a bus is determined by the number of connections for moving binary numbers. A 16-bit bus can transport up to 16 binary digits, and a 32-bit bus can transport up to 32 binary digits.

*Bus structures* have been standardized so that computers with different microprocessors can communicate with each other to share and exchange instructions and data. The wiring for data transfer must conform to mutually acceptable rules. Because a computer's internal and external devices do not all operate simultaneously, the bus structure simplifies internal communication and eliminates the need for separate wiring between parts. All devices share one or more common buses. A bus can also interconnect computers that operate at different speeds and are separated by long distances.

Many different bus standards exist, and none has been found to be suitable for all applications. Some standards were designed to be proprietary by microprocessor manufacturers and special-interest groups, and this led to designs that duplicate the functions of existing standards. The industry now recognizes the benefits of *open* independent standards. All components connected to a standard bus must be able to work together.

A bus structure becomes a standard when it is accepted by many users and other manufacturers see the benefits of making products that are compatible with it. Allegiance to a bus structure is typically based on its success in the marketplace rather than its technical superiority. Examples of successful bus designs are Motorola's VME bus and IBM's PC bus.

Some manufacturers have submitted their proprietary bus structures for endorsement by national and international standards groups, and they have been accepted. Examples include the General Purpose Instrumentation Bus (GPIB), the Standard (STD) 100 bus, and Multibus.

Bus designs are not static, and some have been improved after technical review. Special-interest groups, independent of any manufacturer, have worked with standards organizations to design and develop standards that meet their particular needs. As part of the standardization process, the IEEE has given identification numbers to standards and has proposed standards.

Bus architecture dictates the dimensions, grouping, and cooling of components, as well as the power distribution and connector arrangements on the PC board supporting it. The

Eurocard system allows several different buses to share the same mechanical specifications. It specifies a range of standard sizes for interface cards and connectors.

Some bus systems, such as standard (STD) bus, were designed for simple single-processor computer systems, while VME bus and Multibus II are families of related buses for supporting multiple high-end 32-bit microprocessors. A complete computer system often includes a hierarchical bus structure, a backbone or system bus, and several application-specific subsystem buses.

Adapters attached to the system bus allow access to specialized peripheral interface buses such as *small computer system interface* (SCSI) for disk and tape drives and *general purpose interface bus* (GPIB) for connecting measurement and control equipment.

Five classes of signal on a bus transfer data: (1) address, (2) data, (3) control, (4) response, and (5) timing. Together they form a *transaction bus*. Address and data signals can flow sequentially over the same conductors, and this *multiplexing* reduces bus size. An *arbitration bus* assures competing devices orderly access to a bus, while the *interrupt bus* accepts requests or attention from devices.

# Standard Bus Structures

### STANDARD (STD) BUS

The *Standard (STD) bus,* developed for computers with Intel 8080A processors in 1976, has become the *IEEE-689 bus.* A 100-pin parallel-circuit bus, it has three power supplies: +8 V, −8 V, and −18 V. It plugs into 100-pin connectors which have 8- or 16-pin connections, up to 16 DMA functions, and 8 interrupt lines. It is still used in industrial applications and can support Intel 8-bit 8086/88 and Motorola 8- and 16-bit MC6800/68000 microprocessors.

### PC BUS

The *PC bus,* first developed by IBM for its personal computer line, was expanded into the *PC/AT bus* with 28 address lines, 16 data lines, 11 interrupt levels, and provision for memory and DMA channels. Later it was upgraded into the *extended AT bus.* Its power supply voltages are +5 V, −5 V, +12 V, and −12 V. Circuit cards, CRT controllers, memory, modems, and disk-drive controllers can be plugged into the expansion slots of this bus.

### GENERAL PURPOSE INTERFACE BUS (GPIB)

The *general purpose interface bus* (GPIB), first developed by Hewlett-Packard, was adopted as the IEEE-488 standard in 1975. It is widely used to interface personal computers with electronic test instruments for displaying and storing data.

### MULTIBUS

*Multibus I and II* were developed by Intel. Version I supported the 8-bit 8088 microprocessor and version II was developed to accommodate 16-bit and 32-bit 386, 486, and Pentium microprocessors. Version II has five different interconnecting buses.

## VME BUS

The *VME* and *Multibus II* buses were developed by Motorola to support both 16- and 32-bit microprocessors. VME has four individual buses and uses asynchronous protocol, and Multibus II has five individual buses and uses synchronous protocol.

## INDUSTRY STANDARD ARCHITECTURE (ISA) BUS

The *industry standard architecture* (ISA) *bus* structure is the IBM PC/AT bus with added features to accommodate the 80286 16-bit data bus and 24-bit address bus.

## MICROCHANNEL ARCHITECTURE (MCA)

The *microchannel architecture* (MCA) is a bus structure introduced by IBM in 1987. It is a 32-bit address and data bus with provision for adding 8-, 16-, and 32-bit data paths.

## EXTENDED INDUSTRY STANDARD ARCHITECTURE (EISA) BUS

The *extended industry standard architecture* (EISA) *bus* is a modified ISA bus designed to accommodate 32-bit microprocessors.

## VESA

The *Vesa* or *VL* bus is a local 32-bit bus for personal computers that permits system buses to be accessed at speeds close to those of the processor. It can have a frequency of 33 MHz and has three expansion slots.

## PERIPHERAL COMPONENT INTERFACE (PCI) BUS.

The *peripheral component interface* (PCI) *bus* is a local bus with both 32- and 64-bit data paths. It is suitable for high-performance processors such as Pentium II, allows burst mode data transfer and bus mastering, and is compatible with ISA, EISA, and MCA cards.

## IEEE 1394 BUS

Popularly known as *Firewire,* this bus system is expected to become the high-speed standard for connecting digital video cameras, VCRs, digital VCRs, cable modems, network cards, and other peripherals that require the highest bit-transfer rates.

# Disk Interface Standards

## ENHANCED SMALL DEVICE INTERFACE (ESDI)

The *enhanced small device interface* (ESDI) is an improved version of disk controller standard ST412 that permits data rates of 10 to 5 Mb/s and higher disk data density. It was developed to assure hard-disk drive compatibility.

### INTEGRATED DEVICE ELECTRONICS (IDE)

The *integrated device electronics* (IDE) standard made the controller part of the hard disk, thus eliminating a separate controller. However, it limits disk capacity to 504 MB.

### SMALL COMPUTER SYSTEM INTERFACE (SCSI)

The *small computer system interface* (SCSI), originally intended as a hard-disk drive interface standard, is also a suitable interface for peripherals such as CD-ROM drives.

# Computer Ports

Computer ports are connection points on a computer that are terminated by connector jacks for accepting plugs from peripherals. *Serial ports,* also called *communications ports,* are suitable for interfacing devices that do not require fast transmission rates, such as mice, modems, and keyboards. They use 9-pin connectors. *Parallel ports* permit the high-speed transfer of data 8 bits (1 byte) at a time, and are suitable for interfacing printers and removable-media backup drives. They use 25-pin D-type connectors. These ports are labeled LPT1 or LPT2.

# Universal Serial Bus (USB)

The *universal serial bus* (USB) port on newer computers is a computer interface that makes the setup and configuration of computer peripherals easier. It permits daisy chaining as many as 127 computer peripherals to one port. Its installation will eliminate serial and parallel ports and the need for computers to interrupt requests. Peripherals such as keyboards and monitors can act as additional plug-in sites. USB ports can handle bandwidth of 12 MB/s so that they can accommodate any equipment that supports the Motion Picture Experts Group 2 (MPEG-2) video compression standard. It has become standard on computers that run Windows software.

# Personal Computer Systems

A large circuit board called a *motherboard* is the heart of a personal computer. It contains the microprocessor, main memory, video memory, audio and video ICs, a network controller IC, and many other active and passive circuit components. In addition, it contains connectors for exterior peripherals such as the keyboard, printer, and mouse, as well as the interior peripherals such as the bus, hard-disk, diskette and CD-ROM drives, and memory expansion cards.

The microprocessor is packaged as a removable and replaceable component, so it is installed in its own connector which transmits all of the information from the processor to other parts of the computer. A connector for a Pentium MPU, for example, has two parallel rows of contacts for a total of 242 pin connections.

The size of the motherboard and the arrangement of its parts are not standardized, and will vary according to manufacturer, model number, and case style. It might be located in a flat metal case that will permit the monitor to be mounted on top of it, or it might be in a tall, upright case called a *tower,* as shown in Fig. 15-4.

Microprocessors now being factory installed in personal computers have speeds ranging from 166 to 450 MHz. A minimum system includes a monitor (CRT or LCD), keyboard, and mouse. While all systems include a hard-disk drive there are wide variations in data storage capability, from about 2 to 9 GB. Newer models are substituting magnetic-disk cassette drives for diskette drives and DVD drives for CD-ROM drives. Optional accessories include stereo speakers, printers, and scanners. A trackball can be used in place of a mouse, and a joystick is needed for playing many electronic games. A digital entry pad is also useful for entering sketches and written messages. Compatible computer printers, now typically inkjet or laser, can print out the text or graphics that appear on the monitor in color. Dual speakers are needed for reproducing stereo sound from CD-ROMs, audio CDs, DVDs, and audio and video data from the World Wide Web.

## DESKTOP PERSONAL COMPUTER SPECIFICATIONS

In 1999 a high-end desktop personal computer had the following technical specifications:

Microprocessor      Intel Pentium II, 450 MHz

Main memory      128 MB of 100-MHz SDRAM, expandable to 384 MB

Hard Drive      17.2 GB Ultra ATA

Monitor      CRT 19 in (18 in VIS) 0.26 dp, or flat panel LCD 14.1 in VIS.

Graphics card      3D AGP with 16-MB SGRAM

DVD-ROM/CD-ROM drive      5X DVD-ROM/CD-ROM and decoder card

Sound      A3D 64 voice sound card

Speakers      Dual digital speakers with subwoofer

Modem      Data 56 kb/s ITU V.90, fax 14.4 kb/s

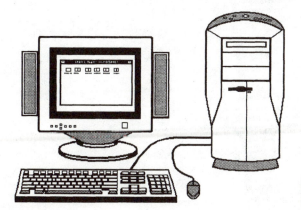

**Figure 15-4** Desktop personal computer.

Secondary disk drive      100-MB magnetic-disk cartridge internal.

Keyboard      104 keys with Internet access key

Most new personal computers now include USB ports for simplifying the connection of modems, printers, scanners, and other peripherals.

## NOTEBOOK COMPUTER SPECIFICATIONS

A *notebook computer* is a battery-powered portable computer in a briefcase-style hinged case. The keyboard is its lower half and a flat-panel LCD display is inside the cover, as shown in Fig. 15-5. In 1999 high-end models included:

Microprocessor      Intel Pentium II 300 MHz

Main memory      128 MB of SDRAM, expandable to 160 MB

Cache memory      512 kB integrated L2 pipeline burst

Monitor      Flat panel LCD 15 in XGA active matrix TFT

Hard drive      8.0 GB

CD-ROM, DVD-ROM and diskette drive      Removable 2X DVD-ROM/3.5-in diskette

Video      2X AGP with 8 MB of VRAM

Battery      Lithium ion

Modem      56 kb/s ITU V.90

Graphics card      2XAGP with 8 MB VRAM

Slots in the case accept either the CD-ROM/DVD-ROM drive, a 3.5-in diskette drive, or a second battery. Some models include stereo speakers. These computers are usually powered by lithium-ion or nickel–metal hydroxide (NiMH) batteries. Their weight is between 5 and 8 lb (2.3 and 3.6 kg). The keyboard typically contains 85 keys. It is also called a *laptop computer.*

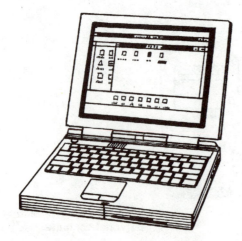

**Figure 15-5** Notebook personal computer.

## PALMTOP PERSONAL COMPUTERS

A *palmtop personal computer* is any of many different limited-function personal computers typically weighing less than 2 lb (900 g). It is smaller than a notebook computer so it has a smaller keyboard, as shown in Fig. 15-6. These computers are now available with 256-color, high-resolution CSTN LCD displays. Computers in this class might contain 16 MB of RAM, a voice recorder, and playback ability. The operating system is a modified version of standard operating systems. Its features could include a clock, a calculator, an information manager provision for accepting E-mail, slots for additional memory modules, and short-form versions of standard applications software.

These small computers can keep notes, schedule, save lists of addresses, calculate, and perform some communications functions. Cradles are available for some of these computers that permit them to "dock" with a larger compatible computer to transfer data. These products are also called *subnotebook computers.*

## PERSONAL DIGITAL ASSISTANTS (PDAS)

A *personal digital assistant* (PDA) is any of many different brands of small, battery-powered, limited-duty computers intended primarily for personal scheduling and note taking. Most will fit in a user's hand, but a stylus is required to enter data such as phone numbers and notes. Some PDAs offer E-mail and fax capability. They typically include from 512K to 1 MB of RAM and individual function keys in place of a keyboard. The display is typically a small monocolor LCD panel, and the units have slots for memory cards. Some models include cradles for docking the PDA to a larger computer for transferring data. Computers in this class are also called *personal communications devices* (PCDs) or *handheld personal computers* (HPCs).

# Reduced Instruction Set Computers (RISCs)

A *reduced instruction set computer* (RISC) is one that contains a RISC microprocessor with a simpler instruction set than that used by *complex instruction set computers* (CISCs). The RISC design is based on the finding that 80 percent of the functions of a software

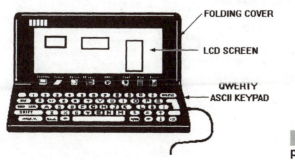

**Figure 15-6** Subnotebook personal computer.

package in a CISC computer are controlled by 20 percent of the computer's CPU. By placing the burden on software and simplifying the hardware, it was expected that the speed and performance of a RISC processor would be superior to that of a CISC processor. The Power PC developed jointly by International Business Machines Corporation, Motorola Inc., and Apple Computer, Inc. is a RISC microprocessor. However the Intel Pentium, a CISC MPU, has adopted many RISC ideas to improve its performance.

# Explicitly Parallel Instruction Computing (EPIC)

*Explicitly parallel instruction computing* (EPIC) is a computing concept that depends on parallel instructions to overcome delays inherent in CISC and RISC computing. The IA-64 Merced microprocessor from Hewlett-Packard Corporation and Intel is an EPIC MPU.

# Computer Board-Level Modules

*Computer board-level modules* are pretested, populated ready-to-plug-in circuit boards that contain the essential components for a stand-alone computer. They are typically purchased for installation in host equipment that will be computer controlled. Figure 15-7 is a simplified block diagram for a single-board computer. It includes a microprocessor and a selection of peripheral devices such as parallel and serial I/O ports and memory, as well as empty sockets for add-on memory cards. These modules permit manufacturers to assemble their own proprietary computer systems, embed computers in industrial equipment or scientific instruments, or add capability to their existing computer-based systems. The modules are a value-added product that allows manufacturers to avoid the expense of owning or

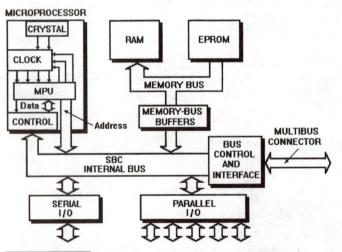

**Figure 15-7** Single-board computer block diagram.

leasing plant facilities and hiring personnel to assemble the computer boards. They are also called *single-board computers* (SBCs).

Other commercial plug-in circuit boards for computers include:

■ *Memory modules* with arrays of memory ICs to augment existing computer system main memory. These can be DRAMs, SRAMs, EPROMs, EEPROMs, NVRAMs, ROMs or combinations of these memory devices.
■ *Mass storage controllers* for controlling diskette, hard-disk and tape drives.
■ *Digital input/output and timer modules* for increasing system I/O.
■ *Analog I/O modules* for instrumentation and process control.
■ *Communications controllers* meeting such standards as RS-232 and IEEE-488.

# Programming and Software

Computers need instructions to carry out their hardware functions, and these are provided by a *program,* a list of instructions in a logical order that can solve a problem in a finite number of steps. Programs that are either stored in a magnetic disk drive or are entered into the computer by diskette or CD-ROM are called *software.* There are two kinds of software: *applications programs* and *systems programs.*

An *applications program* is software written to solve a specific problem or set of problems, perform certain tasks, or to accomplish some kind of repetitive task. It can be as simple as 10 statements in BASIC language for repetitively solving a simple mathematical formula, or it can consist of thousands of statements for the design of an aircraft wing. Common examples are, however, word processing, database management, and spreadsheet preparation and drawing. The two basic types are *custom* and *package.*

*Custom programs* are proprietary programs written to solve problems or perform tasks that are unique to a particular user. Some are developed by the user, particularly if they involve proprietary products or processes. Others are developed under contract by outside software vendors. An example might be a program for directing an automatic pick-and-place machine in placing electronic components on a circuit board.

*Package programs* are prepared by commercial software vendors for common applications and sold competitively on the open market. These programs are written to be compatible with specific operating systems such as MS-DOS, Windows 95, or Windows NT. Most can be used to perform many tasks. They are sold through retail stores or by direct mail. Examples include programs for word processing, preparing spread sheets, generating databases, drawing, desktop publishing, circuit analysis, tax preparation, bookkeeping, playing video games, and even the simulation of driving an automobile or flying an airplane.

# Operating Systems

An *operating system* (OS) is software that controls a computer's operation. It sets priorities, determines the locations of data in main and disk memories, and lists all of the files and their status. It also integrates peripheral equipment functions, controls input/output activities, and manages data. The OS relays information entered at the keyboard, instructs the printer how and when to print, and manages information storage on disks. The operating systems for IBM-compatible personal computers are DOS, Windows 3.1, Windows 95, Windows 98, Windows CE, Windows NT, a short-form Windows for handheld or palmtop computers, and OS/2 Warp. Mac OS #8 is an operating system for Macintosh computers made by Apple Computer, Inc. Windows NT and Unix are operating systems for engineering workstations and network connection.

# Basic Input/Output System (BIOS)

A computer's *basic input/output system* (BIOS) is a set of coded computer startup and self-test instructions that are stored on a ROM chip. These instructions control the startup process of the computer and its basic peripherals such as keyboards, monitors, and hard-disk and diskette drives. The instructions can be stored in a factory-programmed ROM or a field-programmed EPROM. A 128-kB ROM is required to contain a BIOS for a desktop personal computer, but a 256-kB ROM is required for a notebook computer. The BIOS program identifies the computer's processor, performs self-testing, and collects information about the operating conditions of the peripherals and the status of the memory. It evaluates and tests them so that it can enable, turn on, and configure all of the hardware. Finally, it supplies user services and acts as an interface between the user and the computer's operating system to resolve conflicts.

# Glossary of Common Computer Terms

**A3D 64 voice sound**    A technique for creating a lifelike three-dimensional listening experience permitting one to hear sounds from all sides and above when using speakers or headphones.

**access time**    The length of time taken by a computer to process a data request and retrieve that data from main memory or a storage device. It can range from a few nanoseconds when accessing a file in the computer's memory to hours when retrieving data from the Internet. Individual memory devices also have access times.

**advanced graphics port (AGP)**    A high-performance graphics interface.

**assembly language** or **assembler**    A computer program that converts or translates assembly language source code instructions into *machine language*.

**audio IC**    A circuit that accepts data from CD-ROMs, microphones, and other sources and digitizes it for the computer, which converts the data into a signal reproducible by speakers.

**bundling**    The inclusion of popular applications programs and options like graphics cards with factory-manufactured computers, generally as a sales incentive.

**bus connector**   A connector on a computer motherboard designed to act as a socket for an expansion bus.

**cache**   Static random access memory (SRAM) that keeps a copy of frequently accessed information for quicker data retrieval.

**compiler**   A computer program or circuitry that translates a high-level language into an executable program in a single operation.

**daughterboards** or **daughtercards**   Small circuit boards or cards that plug into connectors on motherboards. Semiconductor memory modules containing from 16 to 128 MB are examples.

**expansion slots**   Spaces for plug-in cards and other add-on peripherals. A combination of PCI and ISA slots is desirable on a personal computer.

**firmware**   A read-only memory (ROM) that permanently stores computer code which can include data conversion tables, applications programs, subroutines, and high-level language programs.

**graphics adapter**   A circuit board that determines the resolution, color range, and speed at which images can be displayed on a monitor. It typically has 4 MB of its own memory (DRAM or VRAM) that will display millions of colors at a resolution of $800 \times 600$ pixels.

**high-level language**   An applications-oriented programming language, as distinguished from a machine-oriented programming language, It is also called a *computer language.* Examples include Basic, C, C++, Cobol, Java, and Pascal.

**Internet**   A worldwide network of computers that can be accessed by private citizens to obtain a wide range of information and to send and receive E-mail. Access can be gained with a computer, software, and a telephone or cable connection.

**interpreter**   A computer executive routine that translates a program in high-level language or code into machine language or code. Unlike a compiler, the interpreter translates and executes one line at a time. See also *assembler* and *compiler.*

**I/O ports**   An abbreviation for *input/output ports,* sockets for connecting external peripheral devices such as printers or scanners to the computer.

**machine instruction**   An instruction written in a *machine language* that a compiler can recognize and execute without translation.

**Machine language**   Instructions that can be executed by a computer processor. It is also called *machine code.*

**Mac OS**   A proprietary operating system for Macintosh and compatible personal computers made by Apple Computer, Inc. See also *Windows.*

**motherboard**   A large circuit board into which smaller *daughterboards* or *daughtercards* can be plugged. In a computer, it is the large board that contains the microprocessor, peripheral ICs dedicated to specific functions, and connectors for both internal and external peripheral equipment and add-on memory cards. Most motherboards are designed by the computer manufacturer.

**network controller IC**    A circuit that sets up communications with a local-area network (LAN) and manages incoming and outgoing data to the host computer.

**object program or code**    A program or instructions in the language of the processor that can be executed directly by the computer. It is also called *machine code* or *machine language*.

**SDRAM** or **SyncDRAM**    Synchronous dynamic random-access memory.

**source language**    A language in which a problem is programmed for a computer. It must be translated into an object program in machine language by an *assembler, compiler,* or *interpreter* for use by the computer.

**source program**    A program that is written in source language or code.

**synchronous dynamic random-access memory** (SDRAM)    Dynamic random-access memory offering fast burst-mode data transfer. It is also called *syncDRAM.*

**video IC**    A circuit that draws images on the monitor screen. It keeps that data in the video memory and converts it into a signal that can be interpreted by the monitor. It also processes requests from the MPU to draw on the screen by altering the video memory.

**VIS or v.i.s.**    Visible image size, the size of the image on a monitor screen.

**video memory**    4 MB of memory for holding an image on the monitor screen.

**Windows**    A reference to proprietary operating systems from Microsoft, Inc. Versions of it are the operating systems in all recent IBM-compatible personal computers. See *operating systems* in this section.

**World Wide Web (WWW)**    An extensive network of pages of data and programs on the Internet that is interconnected by *hyperlinks* or keywords. Personal computer users can switch from one item to another by clicking a mouse on a word or icon representing the item on the monitor. Most electronic commerce and publishing is done on the Web.

# COMPUTER PERIPHERAL DEVICES AND EQUIPMENT

# Overview

Computer peripherals are devices or equipment added to the basic processor and main memory to facilitate human-machine interface and permit the computer to carry out control functions or perform useful applications. Certain peripherals active in most computer operations such as hard-disk, diskette, and CD-ROM drives are mounted within the computer case for user convenience, while other peripherals such as the monitor, keyboard, and mouse are connected by cable or infrared link to the computer to permit their positions to be personalized on the desktop for user comfort. Other computer peripherals such as printers, scanners, and backup drives with cable connections can be located away from the computer in positions that will be convenient for their effective use.

The microprocessor, the most expensive and critical component in any personal computer, is the "engine" that determines the speed and versatility of the system, and it also dictates the operating system and applications software that can be used. Computers are sold with a basic main memory, but it can be supplemented if needed with add-in memory on plug-in daughter cards, which themselves can be considered as peripherals.

Hard-disk drives for personal computers for sale as retail products have now exceeded capacities of 16 MB for a 3.5-in (90-mm) drive, and drives have matched the rate of progress in IC chips, increasing storage density at the rate of about 60 percent annually. Recently drives for subnotebook computers with diameters of 1 in (25 mm) and capacities of 340 MB have been introduced, 1.5 in (38 mm) smaller than the 2.5-in (64-mm) drives for notebook computers.

A modem, required for communication over a conventional dial-up telephone line for such activities as sending and receiving E-mail and connection to online services or the Internet, is now usually factory installed. But modems are also available as separate free-standing peripherals. Printers, needed for printing out text and graphic files or downloading information from online sources, have always been stand-alone, cable-connected equipment.

The most popular printers for personal computers are the inkjet and laser models. Both are available in a wide range of print speeds and resolution. The choice depends on the user's preference and budget. Laser printers do the best job of printing out text in near-letterpress quality, but are most economical for the kind of high-volume printing needed in business and professional offices, particularly where they will be shared with other networked computers. On the other hand, the inkjet printer is more economical for most home use, and it offers the added benefit of economical color printing, not available on low-cost laser printers. The dot-matrix printer, formerly the most popular printer for personal computers, is now used primarily in business offices where it is capable of printing multipart business forms, spread sheets, telephone lists, and address labels. It is the only printer that can print multiple copies simultaneously on chemically treated paper or with carbon-paper inserts.

The mouse is the standard pointing device, but the trackball is an option, and the joystick is required for playing many computer action games or simulations. Scanners save the time and effort of keying text or data into the computer, although some editing might be required. They can also reproduce graphics and copy color photographs, whose quality will depend on the resolutions of both scanner and printer as well as the printer's ability to reproduce many different colors.

The fax machine is a useful adjunct, especially for home or business offices. Some multipurpose computer peripheral machines now perform inkjet printing, faxing, copying, and scanning. Sending and receiving E-mail is a feature just introduced on some of these machines. Hard drives with removable disk cartridges were developed as stand-alone products to back up computer hard drives, but modular versions are now being factory installed in computer slots as replacements for diskette drives. With memory capacities of 100 MB or more these drives can store the data equivalent of more than 65 diskettes.

Recently the hardware and software have been introduced that will permit cordless peripherals such as mice, keyboards, joysticks, and game pads to be connected to computers by digital wireless infrared links. The proposed IrBus bus has been designed to take advantage of the universal serial bus (USB) interface.

# Computer Monitors

A *computer monitor,* as shown in Fig. 16-1, is a video input/output peripheral device. It can display text and graphics either entered by the user or obtained from computer memory. Most monitors for desktop computers include three-gun multicolor *electromagnetic cathode-ray tubes* (EM CRTs), but some are now available with active matrix TFT or dual-scan liquid-crystal flat-panel color displays. Monitors for battery-powered portable notebook PCs are typically active-matrix TFT or dual-scan LCDs, but most subnotebook or palmtop computers have monocolor LCD displays. Multicolor CRTs in computer monitors are similar to those in TV receivers. The block diagram of a computer monitor, Fig. 16.2, shows its principal functional circuits.

The first monitors were built to support the color graphics adapter (CGA) and enhanced graphics adapter (EGA) graphics standards. These have been superseded by the video graphics adapter (VGA) and super video graphics adapter (SVGA) graphic standards.

Monitors are rated in viewable image size (VIS), and for personal computers these range from 13.7 to 19 in (35 to 48 cm) VIS. The specifications that distinguish monitors are their scan rate, resolution, bandwidth, and whether or not they are capable of multiscanning.

## CATHODE-RAY TUBE MONITORS

A computer monitor cathode-ray tube (CRT) typically includes three electron guns, with one gun to each of the three colored phosphors, red, green, and blue. Electrons driven from the heated cathodes are formed into beams and move toward the faceplate under the influ-

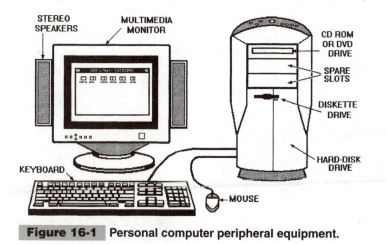

**Figure 16-1** Personal computer peripheral equipment.

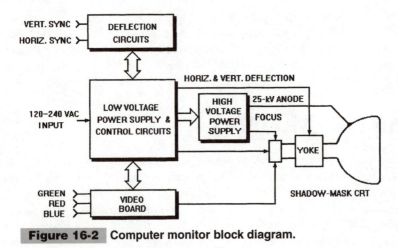

**Figure 16-2** Computer monitor block diagram.

ence of the high voltage on the anode. The back surface of the faceplate is coated with phosphor dots that respond to the incident electron beam by glowing red, green, or blue. The phosphor dots are typically deposited in triangular groups of three dots (triads), as shown in Fig. 16-3. (Some CRTs have the phosphors arranged in alternating vertical red, green, and blue stripes.)

The electrons from each of the three guns (or single gun in some tubes) are swept in synchronization under the control of the magnetic lens formed by the focus and deflection coils around the neck of the tube so they strike their assigned phosphor dots. However, the beams in transit to the faceplate pass through a perforated, thin metal shadow mask that regulates the beams so they strike the assigned phosphor dots without undesirable fringing effects.

As in the CRTs described in Sec. 13, "Optoelectronic Display Technology," the electron beams are turned on and off as they scan the phosphor dots, which glow long enough to be perceived by the human eye as integrated images.

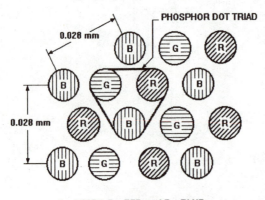

G = GREEN, R = RED, and B = BLUE.

**Figure 16-3** CRT phosphor-dot pitch defined.

*Pixel* (from *picture cell*) means the smallest pictorial element in a CRT image. Pixel size is determined by the diameter of the electron beam, and it can include more than one illuminated phosphor triad.

A *raster* is a pattern for deflecting electron beams on a CRT screen, and it can be seen when no modulation is present. See Fig. 19.2 and "U.S. National Television Systems Standard Committee (NTSC)" in Sec. 19, "Television Broadcasting and Receiving Technology." The vertical scan rate is the time taken to fill a screen with lines from top to bottom. The latest multiscan or multifrequency monitors have variable vertical scanning or refresh rates of from 1/40 to 1/100 s.

The horizontal scanning frequency of an NSTC standard TV receiver and a GCA monitor is 15.75 kHz, but for the EGA system this rises to about 22 kHz. For VGA systems it is 31.5 kHz or higher. Multiscan monitors can have scanning frequencies of 15.5 to 100 kHz.

*Resolution* is the number of pixels that can be displayed on the screen. The Video Electronics Standards Association (VESA) has $800 \times 600$ as the SVGA standard. The newest systems are multiples of this standard. Examples are $1024 \times 768$, $1152 \times 870$, and $1280 \times 1024$. Thus, the higher the resolution, the greater the number of lines and the closer they are together. This requires that they be scanned faster on the screen. As a result, more pixels are lit up by the electron beam as its sweeps across the faceplate. Higher resolution means sharper images. Unlike the conventional interlaced TV scanning raster, most computer monitors feature noninterlaced scanning.

*Dot pitch* is the center-to-center distance between like-colored phosphor dots, as shown in Fig. 16-3. High-resolution monitor CRTs have dot pitches of 0.28 mm or less. The smaller the dot pitch, the higher the resolution, but smaller-dot screens are difficult and expensive to manufacture. The pitch in CRTs with vertical colored phosphor stripes is the center-to-center distance between like-colored stripes.

*Refresh rate* is the number of times per second that a full image is redrawn or updated on the CRT screen. A monitor must have a refresh rate of 70 Hz to eliminate noticeable flicker.

*Video bandwidth* refers to a monitor's useful range of frequencies. A multiscan monitor can accept horizontal frequencies of from 31 to 82 kHz and vertical frequencies of from 55 to 100 Hz. An SVGA monitor requires a bandwidth of at least 30 MHz, and many high-resolution monitors require bandwidths of 180 to 200 MHz.

The VGA cards were capable of writing 16 colors, but to obtain true color, a computer must include a video board with many fast memory devices. It must be able to drive the monitor so it will display a large number of shades in separate distinct hues or pure colors. At least 1 MB of video memory is required to show 65,536 colors at $800 \times 600$ resolution on a 15-in (381-mm) monitor. More memory is needed for higher resolution or more colors.

## LIQUID-CRYSTAL DISPLAY COMPUTER MONITORS

A liquid-crystal display (LCD) computer monitor is based on the multicolor active and passive LCD panel widely used as the only practical monitor for notebook computers. The latest LCD flat-panel monitors offer big, bright images and are suitable replacements for desktop computer CRT monitors. They consume less power, weigh just a few pounds, and have less flicker. Moreover, they are intrinsically digital displays and they can receive digital signals directly from the computer, as they do in notebook computers. Unfortunately, they typically cost 2 to 3 times the price of an equivalent-size CRT monitor, and this has inhibited their acceptance in the marketplace.

The stand-alone LCD monitors now available for desktop computers offer 14- or 15-in (36- or 38-cm) VIS and are made with thin-film transistor (TFT) active-matrix liquid-crystal display (AMLCD) technology. Their typical resolution is 1024 × 176 and they are capable of producing 16 million colors. Manufacturers are offering a choice of wall-mounting picture frames or desktop stands, some of which can swivel. Unlike CRT monitors, LCD monitors do not emit signals that could interfere with nearby sensitive equipment.

Desktop computers typically include a video graphics card that converts the computer's digital signal into an analog format which the monitor can interpret. This means that for LCD monitors to work with most desktop computers not designed to be compatible with an LCD display, the analog signals must be converted back to digital signals, an undesirable and expensive reconversion step that might not always be successful because of variations in timing on analog video cards. The LCD monitors offered as CRT monitor replacements include analog-to-digital converters. Digital video cards installed in desktop computers would eliminate video-card compatibility problems but add to expense.

For further information on multicolor LCD panels used in LCD monitors, see "Liquid-Crystal Color Video Displays" in Sec. 13, "Optoelectronic Display Technology."

# Hard-Disk Drives

A *hard-disk drive* or *hard drive* is a mechanical mass-storage subsystem that reads and writes digital data on the surfaces of magnetic disks which rotate under electromagnetic *read/write heads*. Figure 16-4 shows a hard-disk drive with 3.5-in (90-mm) disks or *platters* for desktop personal computers. Circuitry in the drive takes digital data from the computer's main memory and writes it simultaneously on a stack of moving disks or platters or reads the data from and returns it to main memory. The data can then be scanned

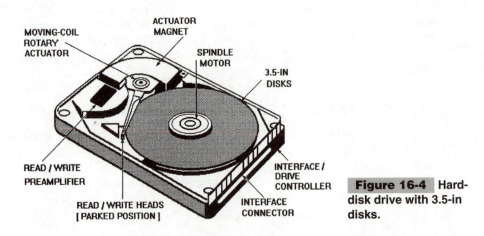

MOVING-COIL ROTARY ACTUATOR

ACTUATOR MAGNET

SPINDLE MOTOR

3.5-IN DISKS

READ / WRITE PREAMPLIFIER

READ / WRITE HEADS [PARKED POSITION]

INTERFACE CONNECTOR

INTERFACE / DRIVE CONTROLLER

**Figure 16-4** Hard-disk drive with 3.5-in disks.

to originate or restore a video image on a monitor's screen. The operating principles and magnetic storage techniques of both hard drives and diskette (floppy-disk) drives are similar.

Hard drives are secondary or mass-storage devices for computers at level 4 of the computer memory hierarchy. As many as 10 platters are stacked on a single spindle so that they all turn at the same speed. The number of read/write heads is related to the number of platters in the stack. All hard drives can be randomly accessed, meaning that their read/write heads can be positioned rapidly at the exact location of the desired data for reading and writing. Digital data is stored on the coated surfaces of the spinning disks in concentric rings called *tracks* and angular pie-shaped slices called *sectors*. Hard drives typically have 17 to 36 sectors per track and 512 bytes per sector. The tracks are stacked to form *cylinders*.

Information is stored on the disks when a pulse of current causes the read/write heads to magnetize a spot on the track beneath the head. When the tracks are read, the head is positioned over the track and the lines of force from each magnetized spot cause a pulse of energy to be induced in the head. During a precise block of time, an induced voltage from the magnetized spot represents a binary 1, and the absence of a pulse, meaning that no magnetized spot is present, represents a binary 0. The heads, which are mechanically linked, can move in two degrees of freedom radially over the spinning media and vertically to position them close to the spinning media. The address of any given byte must include a reference to the cylinder, the actual track (designated by the read/write head), and the sector of that track. The platter stack spins in a sealed case pressurized above room air pressure to keep out dust and moisture.

The mirror-finished nonmagnetic aluminum platters are thin and stiff. They are coated on both sides with a thin film of magnetic metal alloy in a vacuum chamber, and as many as 10 are stacked vertically and fastened to an axial shaft or spindle. Enough space is left between them to allow for read/write head movement.

Lightweight read/write heads are positioned on both top and bottom sides of each platter, so there can be as many as 20 heads all moving in unison on a 10-disk drive. One platter surface and one read/write head are dedicated to the closed-loop servo control for the multiple heads. The platter stores data and track locations. The heads are moved by a voice-coil actuator from a rest or neutral position near the center of the disk to the exact location for reading or writing, where they "fly" microinches over the surface of the platter, never at any time touching it. Any unevenness in the platter surfaces could cause a head crash. When the drive is shut down, the heads return to their neutral positions.

The spindle-platter assembly is driven by a motor under computer control at speeds from 3600 to 6400 rpm. The drive motor starts the assembly spinning when the computer is turned on, and it takes several seconds to get it up to speed. The motor continues to spin it at constant speed, drawing power continuously until the computer is shut down, unless the computer's power-saving circuitry shuts it down following a period of user inactivity.

The two most popular types of drives are the small computer system interface (SCSI) and integrated device electronics (IDE). Most of the disk-controlling functions are integrated into the SCSI and IDE drives. An interface electronics card is still required to transmit the data in 8-bit parallel format to the disks. The SCSI and IDE interfaces allow for recording up to 54 sectors per track.

A 3.5-in hard drive will have from 300 to 3000 tracks per inch (tpi). This compares with 135 tpi and 18 sectors per track for a 3.5-in double-sided, high-density diskette. However, both disk styles have 512 bytes per sector. SCSI and ESDI systems have transfer rates of 10 to 15 MB/s or more. Hard drives with capacities of 12 GB are now being produced for personal computers. In addition to capacity, access time is a key specification.

# Diskette Drives

A *diskette* or *floppy-disk drive* is a magnetic memory drive that uses removable diskette or floppy-disk magnetic media. The term *floppy disk* is a misnomer because the popular double-sided 3.5-in (90-mm) diskettes that replaced the 5.25-in (133-mm) floppy disks have their disks mounted inside rigid plastic jackets. Diskette drives function as secondary mass storage. The drive motor spins the media at 300 rpm, and metal hubs assure high tracking accuracy. The heads gain access to the magnetic media through a spring-loaded shutter that snaps shut to protect the disk when it is not in use.

## DISKETTES

A *3.5-in diskette,* as shown in Fig. 16-5, can store or back up 1.44 MB of data in a high-density (2HD) format. The 3.4-in. (85-mm) disks are made of polyethylene terephthalate plastic coated with iron- or cobalt-oxide magnetic compounds. It is spun within the rigid plastic jacket by a spindle drive motor, and read/write heads, one on each side, are moved in unison by a stepper motor across both faces of the disk. Each disk side is configured with 80 concentric circles called *tracks,* numbered 0 to 79, and the tracks are divided into pie-wedge-shaped sectors. The top surface is side 0 and the bottom surface is side 1.

When one head is over track 0 on the top, another is also over track 1 on the bottom. A head positioner moves both heads as a single unit to the various tracks. The heads are electronically switched so that after one writes on track 0, the second starts to write on the same track on side 1 because it takes less time to switch the heads electronically than to move them to another track. The 80 tracks of a 3.5-in diskette have a density of 135 tpi.

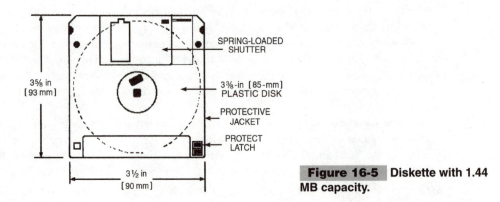

**Figure 16-5** Diskette with 1.44 MB capacity.

The tracks are divided into 18 wedge-shaped sectors, each containing 512 bytes. There are 1440 track sectors per disk side and 737,280 bytes per side. Each disk contains 1,474,560 bytes. This number is reduced by about 15,000 bytes for formatting, leaving about 1,459,000 bytes or, rounding off, 1.44 MB. A controller IC manages the reading and writing of data on each disk. The data can be stored on the diskette for years without degradation, and diskettes can be exchanged, easily carried, copied, and mailed. They can store data files, graphics, and short programs, and can be removed from the computer for security reasons.

Diskette technology is now considered to be obsolete because of low diskette capacity. Internal magnetic-disk cartridge drives with capacities of at least 100 MB per disk are now being installed in advanced desktop computers to replace the diskette drives.

# Magnetic-Disk Cartridge Backup Drives

Magnetic-disk backup drives with removable magnetic-disk cartridges are now available in two versions. Stand-alone versions with different storage capacities are providing backup for hard drives, and internal drives with 100-MB capacities are being factory installed in new computers to replace diskette drives. The cartridges, like the diskettes, can be removed for storage or distribution. These disks can store the equivalent of more than 60 1.44-MB diskettes. Disk cartridge drives are now available with capacities of 100, 135, 200, and 230 MB and 1 and 1.5 GB. These drives are being offered by different manufacturers and they are not compatible with the diskettes, but they meet EIDE standards. The higher capacity 1- and 1.5-GB cartridges can be used for editing and playing back video files as well as mixing audio. Connections are made to these units by SCSI or parallel ports.

# Compact-Disk Read-Only Memory (CD-ROM) Drives

The *compact-disk read-only memory* (CD-ROM) *drive* for computer systems is capable of playing back either CD-ROMs or audio CDs. The CD-ROM drive for computers is a refinement of the audio *compact-disk* (CD) *drive* that can accept and play audio compact disks. The CD-ROM, like the audio CD, is a plastic optical disk with a 4.72-in (120-mm) diameter on which data is permanently recorded in the form of microminiature pits. To read the data on the disk, a laser beam is directed at the disk surface, and the intensity of the light reflected reveals the presence or absence of a pit, as shown in Fig. 16-6.

A CD-ROM can store 650 MB of data, the equivalent of more than 300,000 pages of text, or 450 times the storage capacity of a 3.5-in diskette. A CD-ROM can store multimedia applications containing audio, video, and color graphics, such as encyclopedias. It can also be used to store software applications programs, such as graphics preparation or accounting programs. Because it is a read-only memory, it cannot be erased and reused.

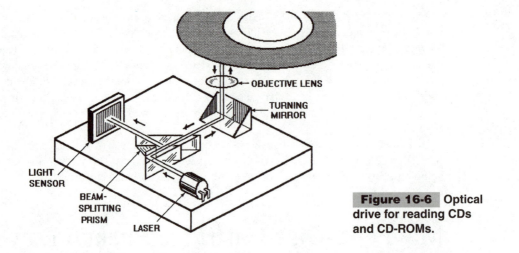

**Figure 16-6** Optical drive for reading CDs and CD-ROMs.

Audio CDs can be played on a computer's CD-ROM drive, but computer CD-ROMs cannot be played on a stereo CD player because it can read only audio data. Computer manufacturers now routinely offer their programs on CD-ROM when they exceed 5 MB, the equivalent of four diskettes. Although an application program's operating instructions might have been downloaded to the computer's hard-disk drive, many programs require that the disk remain in the drive to provide most of the required reference information, such as databases, telephone directories, and photograph and graphics libraries.

CD-ROMs are classified by speed ratings, measured as data transfer rates. These refer to the time required for the drive to transfer data from the CD-ROM to the computer. The first CD-ROM drives had transfer rates of 150 kB/s so all faster drives are designated in multiples of that rate. The fastest drives available in 1999 were designated 32X, meaning that they are 32 times faster than the first drives introduced in 1984. CD-ROM drives are still being factory installed on some personal computers, but the trend is toward installing digital video disk (DVD) drives that can also read both audio CDs and CD-ROMs. See "Digital Video Disks (DVDs) and Drives," following in this section.

# Alternative Compact Disks and Drives

In addition to the audio CD and CD-ROM there are several other derivatives of compact-disk technology. These include the recordable CD (CD-R), also called the write-once, read-many CD (CD-WORM) and the rewritable or read/write CDs (CD-RW) also called erasable CDs (CD-E). The disks are made from an organic polymer on which pits can be formed by a phase-change process. It is possible that many of these disks and compatible drives will not succeed in the marketplace because higher-performance versions of digital video disks (DVDs) are now becoming available.

## RECORDABLE CDs (CD-Rs)

A drive for the recordable CD (CD-R) can both read and record data, and is useful for keeping records, transferring large files, and distributing databases such as catalogs and directories. Speeds for these drives are significantly lower than those of CD-ROM drives, ranging from 2X to about 6X for reading data and 2X to about 4X for recording data. A data transfer rate of 900 kB/s is attained by CD-R drives in the 6X read mode. CD-R drives can read data from CD-ROMs, but they can write only to the CD-R disks. However, CD-ROM drives can read data from CD-R disks. Removable CD-R disks have pregrooved tracks and are available in 550- and 650-MB formats. Mistakes cannot be erased or changed. Some digital cameras use variations of the CD-R disk for storing digitized photographic images.

## REWRITABLE (ERASABLE) CDs (CD-RWs)

A drive for the rewritable CD (CD-RW) permits both writing and erasing data from CD-RWs. Available drives offer 2X writing and 6X reading speeds and a SCSI. The drive directly overwrites on the CD-RW, and the disks can be read by DVD drives. The maximum transfer rate is 5 MB/s, and average access time is 350 ms. The CD-RW drive allows multisession writing and writes as files rather than as tracks. CD-RWs are similar to CD-ROMs, but they cannot be read by all CD-ROM drives.

# Digital Video Disks (DVDs) and Drives

There are seven different formats for *digital video disks* (DVDs), also called *digital versatile disks.* Some are intended for playing on personal computers and others are intended for playing through television sets. DVDs are the same size as CDs, but they have higher data density. As on CDs, data is stored as millions of tiny pits or craters that have been cut into the disk by a modulated write laser beam. There are three methods for increasing that pit density:

1. Forming smaller pits closer together
2. Forming two layers of pits with the lower pitted surface semitransparent, allowing the read laser to read both its outer surface and the surface above it
3. Making a double-sided disk by bonding two thin DVDs back-to-back

DVD technology is rapidly developing, and the existing disk storage capacities stated for each of the DVD formats will probably change for technical and economic reasons. DVD drives are replacing CD-ROM drives in new computers, but many CD-ROM drives and players are still in use. Some form of DVD player will be able to play all of the existing CDs. Moreover, DVD-video and Divx are intended as replacements for the videocassette, particularly as a media for full-length movies. While DVD drives can read CD-ROMs as well as DVD-ROMs, audio CDs, and some DVD video disks, CD-ROM drives cannot read either DVD-ROMs or DVD videos. DVD-ROM drives were rated to 5X in 1999.

The DVD-ROM, intended primarily for computers, can contain theater-quality, full-length movies with extras such as multiple language tracks, as well as interactive games and reference and utility programs. A DVD-ROM can store up to 17 GB, the equivalent of about 6.25 h of movies. A computer DVD-ROM drive can also play CD-ROMs, audio CDs, CD-Rs, and CD-WRs.

*DVD-video* is a DVD format intended for use with TV sets. With a storage capacity of 9 GB, it can store two 2-h movies. DVD-videos are home entertainment products for storing movies, TV shows, and movie-type video presentations—making them a replacement for the videocassette. They can be read on a stand-alone DVD player connected to a television receiver. Players for these DVDs will also play audio CDs.

*Divx* is a DVD format that will record full-length movies. Disks in this format can be played back only on stand-alone Divx players. A variation of the DVD-video format, they are intended for rental. They have capacities of up to 17 GB. *DVD-audio* is a DVD format for playing high-fidelity sound. These disks also have capacities up to 17 GB, equal to about 25 audio CDs. See "DVD Players" in Sec. 21, "Consumer Electronics Products."

## DVD-R

DVD-R (for DVD record once) is a DVD format intended for recording information, including music and video. DVD-Rs require special recording equipment not normally considered to be computer peripheral equipment. DVD-Rs have a storage capacity of 3.95 GB on each side. Recorded information can be played back in players that will also play DVD-videos, DVD ROMs, CD-ROMs, and audio CDs.

## DVD-RAM

DVD-RAM is a DVD format that allows for re-recording information, including music and video. These disks can be erased and new data can be recorded on them. They have a storage capacity of 2.6 GB on each side. Players for DVD-RAMs will also play DVD-videos, DVD-ROMs, CD-ROMs, and audio CDs. The DVD-RW (read/write) is a format that is similar to DVD-RAM. It will have a storage capacity of 3.0 GB.

A double-layer DVD with a storage capacity of about 8 GB has the equivalent storage capacity of about 6000 3.5-in high-density diskettes. By comparison, a CD holds about 650 MB, equal to about 450 diskettes. Also, a 7-GB hard-disk drive has the equivalent data storage of about 5000 diskettes.

# Magneto-Optical and Phase-Change Disk Drives

Two data-disk recording technologies under development today are the *magneto-optical* (also known as *thermomagnetic*) and *phase-change* methods.

## MAGNETO-OPTICAL RECORDING

In *magneto-optical recording,* the energy in the laser beam heats a spot on the disk beyond the disk material's Curie point (about 200°C). The direction of polarization of each magnetic domain that has been heated above the Curie point can be altered by an external magnetic field. When the material cools below its Curie point, the direction of polarization is frozen, and this records the data until it is heated again. This is a reversible technology, and more than a million overwrite cycles are possible. The polarization of the laser light is altered by reflection from the magnetic domain, making it possible for the binary data recorded on the disk to be read. The drawback to this technology is its requirement for a bias magnet in the disk drive, which adds to the power consumption.

## PHASE-CHANGE RECORDING

In *phase-change recording,* pits are formed on an alloy disk that has two phases with differing optical properties. To mark the disk, a laser melts a spot of the material. When cooled, the spot assumes the amorphous phase, giving it different reflective properties than the surrounding crystalline-phase material. Marks that have been recorded can be erased by an annealing process in which a continuous-wave (CW) laser beam heats the material to a temperature just below its melting point, so that marked regions return to their crystalline state. An alloy of tellurium, selenium, and tin has proven to be a suitable recording medium because it has both crystalline and amorphous phases within a reasonable temperature range.

The drawback to phase-change technology is its lower number of overwrite cycles (estimated to be about 50,000), and the time required for it to change phase, which reduces its data-writing rate.

# Data/Fax Modems

A *modem,* an abbreviation for *modulator-demodulator,* is a circuit that converts digital data to an audio signal for transmission over telephone lines and also converts received audio signals back to digital data that can be read by a computer. This process calls for modulating and demodulating the computer's digital signals. Modems connect computers at both ends of a telephone line and both transmit and receive data, as shown in the block diagram Fig. 16-7. Modems are important computer peripherals, now typically bundled as standard equipment.

The digital signals from the computer are converted to audible tones, with the higher-pitched tone representing the binary digit 1 and the lower-pitched tone representing 0. Modems send and receive data simultaneously in blocks, and they distinguish between incoming and outgoing data by their tones.

Modems also exchange an added mathematical code called a *checksum* to let both computers know that the data blocks are being transmitted properly. If the checksums do not match, the modems resend the missing segments of data. Modems also have special circuits

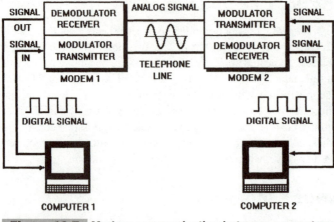

**Figure 16-7** Modem communication between computers.

for compressing digital signals before modulating them and then decompressing them after demodulating the signals. Compression/decompression speeds up transmission.

Modems are made both as stand-alone products and as modules for installation within computer cases. Internal modems for desktop computers are plug-in circuit boards, but for notebook computers and personal communicators they take the form of card-style modules. Factory-installed internal modems are integrated into the computer, so they need no special attention, but card modems for notebook and smaller computers normally must be purchased as add-on accessories. They can be inserted into the PC card slot or removed as needed.

Modems include the capability for sending and receiving facsimile (fax) transmissions compatible with Type II and Type III fax machines, so they are actually *fax/modems*. But special software is needed to send and receive faxes from a computer.

Modem speeds are measured in kilobits per second (kb/s), the number of information bits that can be transferred in 1s. The lowest standard data transmission speed for modem communication is 28.8 kb/s, and the highest rate for transmission over conventional telephone lines is 56 kb/s. Most modems will automatically adjust to the optimum rate that is acceptable by both sending and receiving modems. Three common modem speeds are 14.4, 28.8, and 33.6 kbps, but the latest modems for personal computers can receive at 56 kb/s and send at 33.6 kb/s. Facsimile is transmitted at 14.4 kb/s. The standard for modems is V.91.

# Computer Printers

The *computer printer* is a machine for printing out data, calculations, text, and graphics on paper. It is required for word processing, spread sheets, desktop publishing, and many kinds of computer-aided design tasks. It is also needed for downloading hard copy from the Internet and printing out E-mail.

High-quality printing, referred to as *near-letterpress quality* (NLQ), equivalent to that of the IBM Selectric typewriter, was obtained from printwheel or "daisy-wheel" impact printers designed as computer peripherals. But they were expensive, large, complex, and slow.

Dot-matrix printers were developed to print at higher speeds, but even in their letter-quality mode, their print quality is generally inferior to that of NLQ. But they offer very high speed draft-quality printing for informal communications, and they can print multiple copies in one pass with carbon-paper inserts or chemically treated reproducing paper. They can also print elementary line drawings and diagrams and, under software control, they can print nonstandard fonts. Some can do limited color printing. They are still being sold for printing multiple copies of business forms, labels, and address lists.

Thermal printers were once popular in locations such as offices where the noise of impact printing was annoying, but they required costly chemically treated paper, their printing rate was slow, and the copies deteriorated in time.

Inkjet and laser printers have proven to be far more versatile because they can form a wide selection of character fonts and sizes under computer control. Affordable laser printers are available for personal use but are more suitable for business and professional offices. But color laser printers are not affordable for personal use and are primarily purchased by corporate or commercial printing facilities. For quality color printing at reasonable prices as well as black printing, the inkjet printer has proven to be most cost-effective.

Both of these printers are rated by print speed in *characters per second* (cps), *words per minute* (wpm), or *pages per minute* (ppm). Page definition is defined in *dots per inch* (dpi). They contain buffer memories that temporarily store some of the data to be printed so that the computer can continue its work without interruption. A typical buffer holds 64 kB of data. Printers also contain microcontrollers to supervise the printing process.

## INKJET PRINTERS

An *inkjet printer* is a nonimpact printer for use with personal computers that prints by spraying fine ink droplets from a liquid-ink storage cartridge to form characters and graphics on plain paper. The printheads for personal computer printers are usually part of the replaceable (or refillable) ink cartridge. The printhead contains many tiny nozzles that convey ink from the reservoir to the paper as tiny spots. When the ink supply runs out, a replacement unit is snapped into position on a carriage mechanism that traverses the width of the paper. Roller feed mechanisms within the printer feed sheets of cut paper into the proper position under the printhead and scroll the paper precisely to keep pace with the moving printhead.

The individually controlled nozzles in the printheads have openings with diameters smaller than those of human hair. There are two basically different impulse methods for controlling the firing of the minute ink droplets from the nozzles—*piezoelectric* and *thermal*—and there are many proprietary variations on those methods.

The nozzles can be arranged in the printheads in vertical, horizontal, diagonal, or zigzag formats. Regardless of projection method, the dots overlap to form alphanumeric characters and determine line width. In color printing, the more complex hues can be obtained by overprinting various combinations of the available colored inks in the cartridges.

In *thermal inkjet printers,* resistors in close proximity to each nozzle are heated rapidly by current under computer control. The heat vaporizes the ink in the nozzle to form a vapor bubble. The collapsing bubble forces an ink droplet from the nozzle and sprays it onto the paper as a small spot. The collapse of the bubble draws new ink into the nozzle by capillary action. The ink must cool for a fraction of a second before a new bubble can be formed.

In *piezoelectric inkjet printers,* piezoelectric material that contacts the ink within the nozzle receives an electrical pulse under computer control. The material then contracts and expands, squeezing an ink droplet from the nozzle by electrostriction to spray it onto the paper as a small spot. Ink feeds into the nozzle array manifold during printing, and it is automatically retracted back into the ink reservoir at other times to prevent it from drying and clogging the nozzles. This method is said to be faster because the ink does not have to cool between pulses.

Inkjet printers are available with either three or four ink colors. A three-color printer can accept either a black cartridge or a tricolor cartridge containing cyan (blue), magenta (red), and yellow inks. It can also print a composite black by blending all three colors together. But this is uneconomical for printing pages with high black content (text or graphics) because the mixture is more expensive than true black ink and is not a true black.

Four-color inkjet printers accept cyan, magenta, yellow, and black ink cartridges, permitting simultaneous printing in true black and in color. Some inkjet printers have cartridges containing six colors for more precise rendition of color photographs. These printers typically print 600 horizontal by 600 vertical dpi in both black and colored ink, but $1440 \times 720$ dpi models are available. Black printing speed is 5 to 10 ppm and color printing speed is up to 7 ppm.

## LASER PRINTERS

A *laser printer* is a nonimpact printer for personal computers that prints text and graphics by an electrostatic process similar to xerography. The internal assembly of electronics and optics of the laser printer is called the *engine.* The light beam from the semiconductor laser (or LED in lower-cost versions) can be directly modulated, or the light beam can be modulated by an electro-optical modulator.

The principles of laser printing are illustrated in Fig. 16-8. A low-power semiconductor *injection laser* or *LED* receives digital signals serially that represent the scanned elements of a page (characters or image) from the host computer or internal printer memory. It converts the signals to light pulses, which are directed at a rotating scanning-prism mirror. The light is then focused through a cylindrical lens onto the charged surface of a rotating photoconductor drum.

Where light strikes the drum, charge is removed, creating a *latent image* on the drum's surface. A dry powder called *toner* is attracted only to the charged areas. Transfer- and detach-charge coronas apply an electrostatic charge to plain paper driven by rollers across the surface of the drum. After the toner pattern is transferred to the paper, heated rollers fuse it to the paper. Any remaining toner is removed from the drum before it is recharged to begin the cycle again.

Some laser printers can print at rates of up to 24 ppm, but rates of 8 to 12 ppm are more typical. Resolutions can be up to $1200 \times 1200$ dpi, but $600 \times 600$ dpi is more typical. A print memory of 2 MB is common in these printers, and they typically have paper trays with capacities from 100 to 350 sheets. The printers are compatible with MS-DOS and Windows 95 and 98 operating systems. Laser printers are offered with serial RS-232C and parallel

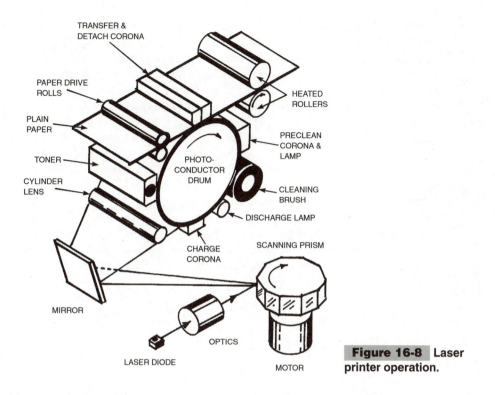

**Figure 16-8** Laser printer operation.

Centronics interfaces. Multipurpose machines that combine the functions of printing with fax/modem, copying, and scanning are available.

# Full-Travel Keyboards

A *keyboard* is a matrix of momentary keyswitches, as shown in Fig. 16-1. It is the primary input device for a computer that permits the entry of text, numbers, codes, and computer commands. For personal computers the keyboard is typically a separate *full-travel keyboard* connected by cable or infrared link to the computer. The standard keyboard for desktop personal computers has 104 keyswitches, but keyboards with from 85 to 107 keyswitches are available. The keyswitches have square keycaps labeled with letters, numbers, punctuation marks, symbols, and computer functions.

Depressing a keyswitch switches the path of an electrical signal to the computer rather than generating one because signals are being sent to the computer continually, even when no keys are depressed. A grid within the keyboard housing consists of wires assigned to each row and column of keys.

Most keyswitches are simple mechanical devices mounted on the keyboard's internal base. A plastic or rubber spring imparts a tactile response similar to that of an electric typewriter during the full travel of the keyswitch. The keyswitch's resistance to finger pressure is nonlinear, increasing as the keycap is depressed until contact is made and the keyed signal is sent. When released, a metal or plastic spring returns the keycap to its initial position. The keycaps of a full-travel keyboard must be depressed from 0.120 to 0.150 in (3 to 4 mm) to make the positive closure needed for generating an electrical code.

The familiar QWERTY keyboard originally developed for mechanical typewriters (named after the first six letter keys in the upper row) has been adopted by computer manufacturers because of its general acceptance despite its awkward layout. It was believed that changing to a more efficient format would call for time-consuming and costly retraining for skilled typists, so it was retained. This reason might no longer be valid. To overcome some of the complaints against the existing keyboard style some unusual semicircular and split "ergonomically correct" keyboards have been developed, but the QWERTY layout has been retained.

Keyboards generate digital code in accordance with the American Standard Code for Information Interchange (ASCII). A computer keyboard can generate 128 numbers, letters, symbols, and special control codes with its double-function keys. An internal *microcontroller* decodes the keys and organizes the data for serial transmission to the host computer. The standardization of keyboard interfaces by manufacturers permits them to be interchanged.

# Pointing Devices

The *mouse* is the most common, readily recognized pointing device. Other less familiar examples include the *trackball, pointing stick, touchpad, digitizer, light pen,* and *touch screen.*

## MICE

A *mouse,* as shown in Fig. 16-9, is a small, rounded pointing device that can move in any direction to cause cursor movement on the monitor screen. It contains a small flexible ball that contacts, two horizontal shafts which are 90° apart. When the ball moves, one shaft turns in response to the $y$ or vertical motion, while the other tracks $x$ or horizontal motion. Each shaft is terminated with a slotted optical encoder wheel that rotates with the shaft. The moving blades of the wheels interrupt or "chop" a beam of light from an internal LED that shines on two phototransistors for both the $x$- and $y$-encoder wheels.

The phototransistors convert the chopped light patterns falling on them into two non-synchronized pulse trains for each encoder wheel because of their spatial separation. The phase differences between the pulse trains can be interpreted as up or down vertical move-

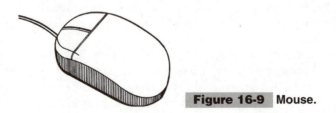

**Figure 16-9**  Mouse.

ments or left or right horizontal movements. The one to three buttons on the mouse are actually small switches. Pressing the button (clicking) closes a circuit, which is interpreted by the computer as a command.

An internal IC accepts the movement pulses from the phototransistors and the closure signals from the buttons and combines them into a data stream which is sent over the flexible cord to the computer for processing. The motion pulses are combined and translated by the computer into cursor position and speed of mouse movement, and the switch-closure commands are obeyed. The cursor on the monitor screen appears as a visible vertical line or an arrow in response to mouse motion. Clicking once or twice is interpreted as different commands. Clicking permits the user to select an item from a menu, position the cursor precisely on the screen, or even highlight text for deletion or repositioning. Cordless mice transmit commands by infrared IR link.

## TRACKBALLS

A *trackball* is an alternative device for positioning the cursor on a monitor screen. Operating on the same principle as a mouse, it is actually an inverted mouse with a rigid ball on top of the housing rather than a flexible ball on the bottom. The ball is rotated by the computer user's thumb or fingertips while the housing remains stationary. The trackball takes up less space than a mouse and eliminates the need for leaving about a square foot of horizontal space on the desktop for mouse operation. Trackballs are available in many different sizes and shapes, and balls vary from marble to baseball size. However, the most popular one is ping-pong-ball size. These positioning devices are favored by the handicapped and those who just do not like mouse manipulation. Cordless IR-linked trackballs are also available.

## JOYSTICKS

A *joystick* is a pointing device for controlling computer games. As shown in Fig. 16-10, it is designed to simulate the vertically mounted aircraft control stick. The handgrip is made to fit comfortably in the hand, and it is equipped with a trigger for finger control of action video games, while pushbuttons pressed by the thumb control playing conditions. Joysticks are mounted on rectangular housings intended to stand on a desktop. A typical unit contains a trigger, four- and eight-way switches, and sometimes control knobs.

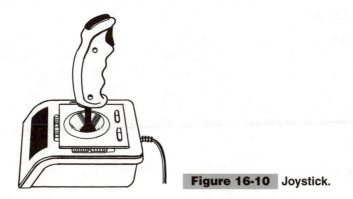

**Figure 16-10**  Joystick.

## POINTING STICKS

A *pointing stick* is a computer cursor-control device with a small lever that is manipulated by finger pressure. Finger movements direct the cursor on the monitor screen and the force applied influences their response speed. Because they conserve space, pointing sticks are used on notebook computers. However, tests show that these sticks take more time to move the cursor than does a mouse, and they lack the flexibility of control offered by a mouse.

## TOUCHPADS

A *touchpad* is a computer cursor control in the form of a small, rectangular pad that translates finger or stylus tapping on the pad into cursor movements. Tapping and dragging causes the cursor to move on the screen. Most touchpads are based on capacitive technology. The fingertip or stylus disturbs an electric field in the pad's surface, and voltage measurements made in each direction locate the intended cursor position. The touchpad will work even if its surface is not actually touched because the finger can alter the electric field when it is only near the pad's surface. Alternatives to mice, touchpads conserve keyboard space on notebook computers. Switches surrounding the touchpad program its functions. Touchpads are designed for serial or PS/2 ports.

## LIGHT PENS

A *light pen* is an optoelectronic pointing device that interacts directly with the computer monitor's screen when touched to the faceplate or when the user actuates a switch on the pen. Both of these actions have the same effect as clicking a mouse switch. The pen, which is connected to the computer by an electrical cord, has a photoreceptor at its tip that is focused on a small area of the screen. As the monitor CRT's electron beam scans the display, the light pen detects the passing increase in light intensity and responds by generating an electrical signal that is sent to the computer.

Software translates the position of the passing light pulse into screen coordinates to pinpoint the position of the pen tip. The light pen is easy to use, and it can address individual pixels on the monitor screen. A drawback is that holding the pen so that its tip is close to the screen for long periods can be tiring. Moreover, the hand and light pen obstruct the user's view, making it necessary to lean over to view the blocked parts of the screen.

## TOUCH SCREENS

A *touch screen* is a feature built into some computer monitors that permits a fingertip pressed on the monitor screen to act as a pointing device. The fingertip interrupts an ultrasonic field formed across the surface of the screen, and this interruption of the field causes it to send electrical signals to the computer. Software translates the $x$ and $y$ ultrasonic signals into coordinates that pinpoint the position of the fingertip within specified areas on the screen. This technology permits people who are intimidated by computers to enter data simply by pressing a finger against a labeled box in a menu appearing on the screen. The touch screen is suitable for initiating preprogrammed computer presentations, making choices of available items, controlling machines, or responding to surveys. It could also be used for voting.

## DIGITIZERS

A *digitizer* is a flat, rectangular tablet that interacts with a pen or stylus to enter handwritten or -lettered commands or sketches into computer memory. When the pen is touched to the surface of the tablet, the cursor appears at the same relative position on the monitor screen, and it acts like the primary switch of a mouse.

There are three digitizer technologies: (1) *capacitive,* (2) *resistive,* and (3) *electromagnetic.* In a capacitive digitizer, the pen alters an electrical field across the surface of the tablet, and in a resistive digitizer, pen pressure causes two electrically conductive layers to touch. But in an electromagnetic digitizer, an electromagnetic field set up on the surface of the tablet induces a resonance in a pen circuit as the pen is moved over the tablet surface.

The tablets of all three kinds of digitizers scan for all surface discontinuities caused by the pen and translate them into cursor motion. Digitizers can directly enter writing, printing, and drawing into the computer. Drawing software within the computer permits line thickness and color saturation to be altered. Both cord-connected and cordless digitizers are available.

# Scanners

A *scanner* is a computer peripheral that scans pictures, text, and documents optically, and digitizes the information for computer entry. It can save the time and effort required to retype text, and it can reproduce photographs and graphics with high resolution. All scanners have scanning heads that include a charge-coupled device (CCD) camera similar to those in camcorders and digital cameras. Some scanners also include an illumination source. The head scans the reflected light from the original document optoelectronically and detects it as shades of gray or color. These digital signals are converted into digital signals corresponding to dots per inch (dpi) and sent to the computer for hard-drive storage.

Some color scanners contain filters that block certain colors, allowing others to be digitized more readily. Other scanners use prisms, called *beam splitters,* to separate the light into colors. A few scanner models shine their built-in lights through the original document or graphic and measure the light that has passed through.

The *flatbed scanner* and the *sheet-fed scanner* are general-purpose units, while the *photo scanner* and the *handheld scanner* are specialized units. Flatbed scanners move their CCD optical reading heads past original documents which have been placed facedown on a fixed transparent surface; sheet-fed and photo scanners move the documents past a fixed CCD optical reading head with drive rollers. Photo scanners are small machines optimized for scanning photographs. Handheld scanners require that the scanner be passed manually over the document like a wand.

Some scanners are connected by cable to the computer at a SCSI port, and others use a parallel port. Some high-end color scanners can scan up to a maximum enhanced resolution of $9600 \times 9600$ dpi, but for most applications 300- to 600-dpi optical resolution is satisfactory. High-end scanners provide 30-bit color and 10-bit gray scale, but 24-bit color and 8-bit gray scale are sufficient for most applications.

Scanners are usually sold with two or three different kinds of software: the first kind controls or drives the scanning, the second adjusts the image's appearance after it has been scanned, and the third recognizes text. Image-editing software permits the image to be altered after the final scan to change its brightness or contrast or to reset its borders. Optical character recognition (OCR) software scans text directly into a word-processing file.

# Memory Cards or Modules

A *memory card* is an assembly of four to six semiconductor integrated-circuit memory devices mounted on a circuit card, as shown in Fig. 16-11. These cards can be inserted into connectors on the computer's motherboard to supplement main memory. The cards typically have edge contacts, and the board edge acts as a plug for a card-edge receptacle connector. With contacts on one side they are single-in-line memory modules (SIMMs); those with contacts on both sides of the card are dual-in-line memory modules (DIMMs).

Three kinds of memory are packaged as modules: (1) *flash,* (2) *dynamic random access* (DRAM), and (3) *static random access* (SRAM). Both DRAM and SRAM modules require power sources, but SRAM modules can be battery powered. Flash-memory cards do not require power. Add-in memory modules for notebook computers are in PCMCIA standard Type I, II, and III cases. They are used in digital cameras, personal digital assistants (PDAs), pagers, and cellular telephones.

# Video Graphics Cards

Video graphics cards are plug-in circuit cards that translate output instructions from the microprocessor into data for display on the computer's monitor. If the monitor is a CRT monitor, the card must perform digital-to-analog conversion because the CRT is an analog device. These cards include dedicated onboard memory to supplement the host computer's main memory for making the calculations necessary to display complex graphics. A card operating in the *text mode* can produce uniform characters on the monitor screen for the presentation of control and guidance messages. But when the card is under the control of the computer's operating system, both text and graphics are formed by pixels.

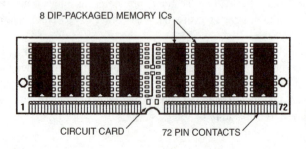

8 DIP-PACKAGED MEMORY ICs

CIRCUIT CARD     72 PIN CONTACTS

**Figure 16-11**  Single-in-line memory module (SIMM).

The video graphics card determines the size of the pixels that will appear on the monitor screen. (For more information on pixels and their formation see "Computer Monitors," preceding in this section.) The standard arrangements for monitor pixels are $640 \times 480$, $800 \times 600$, $1024 \times 768$, $1280 \times 960$, $1280 \times 1024$, and $1600 \times 1200$. These numbers are called the *resolution* because they indicate the number of pixels that can be resolved on the screen at one time, but *pixel addressability* is more accurate because it tells how many pixels the computer can address at one time.

Video graphics cards are organized by bits: 8, 16, and 24. An 8-bit card can provide 256 colors, a 16-bit card can provide 32,768 colors, and a 24-bit card can provide about 16.7 million colors. A monitor with a $1280 \times 1024$ resolution with a 24-bit color VGC can produce about 220 trillion colors (effectively unlimited).

The memory buffer for a video graphics card is rated in megabytes. Most video graphics cards contain at least 2 MB of video memory. Video data is read from the memory buffer by the digital-to-analog converter (DAC), which converts it to an analog signal. Specialized three-dimensional (3-D) video cards include circuitry that enhances the perceived depth and color. The cards are also known as *graphics controllers* or *video adapters*. There are also digital video graphics cards.

# Sound Cards

*Sound cards* are plug-in circuit boards containing ICs that permit computers to produce sounds through their speakers. There are two kinds of sound cards: *frequency-modulation* (FM) *synthesis* and *wavetable synthesis*. The FM synthesis cards, introduced first, produce music and other effects by modulating the frequency of analog sounds to simulate different sound sources and musical instruments. By contrast, wavetable synthesis cards, developed later, include memory that stores digitized samples of actual instruments being played. The card alters the pitch of the sample to produce different notes. As a result, wavetable synthesis cards can produce the sounds of musical instruments more realistically during playback than can FM synthesis cards. A *digital signal processor* (DSP) is used to record the actual sounds of instruments and the audio memory built into the wavetable card plays them back. This is done with Musical Instrument Digital Interface (MIDI) software.

# ELECTRONIC SENSORS AND TRANSDUCERS

# Overview

The terms *sensor* and *transducer* are generally considered to be interchangeable. They refer to devices that convert physical or chemical variables into electrical signals or, in some instances, convert an electrical signal back to a physical quantity. However, instrumentation specialists consider the term *sensor* to be more comprehensive and not limited to the conversion of physical quantities. Examples include the conversion of chemical concentrations and composition into a numerical value.

In this section a *sensor* is defined as a device capable of sensing changes in magnitude of some physical or chemical variable such as frequency, radiation, heat, pressure, or salinity and responding with a proportional electrical output. By contrast, a *transducer* is any device that converts energy in one form into energy in another form.

A sensor can transmit information, initiate a change, or actuate a switch. That output can control some function directly or be converted and measured to provide accurate measurements of a variable. Sensors can act as interfaces between the physical world and electronics, making it possible for electronic circuitry to "see," "hear," "smell," "taste," and "touch." They are essential for many process and machine controls and measurement systems.

Sensors are in home appliances, office equipment, food and chemical process lines, machine tools, and instruments in ships, aircraft, and automobiles. They are required in telecommunications equipment, computers, robots, and many different scientific and medical instruments. Demand for sensors has grown because of the increasing sophistication of existing instruments and their applications to an ever-expanding range of human activities.

Sensors can be classified by the physical or chemical changes they are designed to sense, and those changes have been divided into six domains:

1. *Thermal*—temperature, heat, and heat flow
2. *Mechanical*—force, pressure, velocity, acceleration, and position
3. *Chemical*—concentration and composition
4. *Magnetic*—field intensity and flux density
5. *Radiant*—electromagnetic wave intensity, wavelength, polarization, and phase
6. *Electrical*—voltage, current, and charge

The sensors being manufactured in the largest quantities are those used for measuring acceleration, position, pressure, speed, and temperature, and the worldwide automobile industry is now one of the largest customers.

Sensors are also classified by power requirement: those that can generate an electrical output without a power source are called *self-generating* or *passive*. An example is the thermocouple. On the other hand, sensors that will generate an output only when energy is applied are called *modulating* or *active*. Examples are photodiodes and phototransistors.

Three different transducers and sensors important to electronics are discussed in this section, although they could also be covered under the sections on radio transmission and reception. The classification of microphones, speakers, and encoders as sensors or transducers might not be obvious, but microphones are mechanical sensors because they convert acoustic pressure into electrical signals; speakers convert electrical signals to acoustic pressure; and encoders convert pulsating light signals into distance or rotation or counting measurements.

# Thermal Sensors

## THERMOCOUPLES

A *thermocouple* is a temperature sensor consisting of two dissimilar wire conductors welded together at their ends to form a *sensing junction*. When this junction is heated, a voltage is developed between the junction and the wire ends. The magnitude of this voltage depends on the temperature difference between the junction and the dissimilar alloy materials in the wires. It can be used to measure a heat source by direct contact or it can measure radiant energy. Chromel and Constantan wire pairs are widely used in thermocouples.

The organization of a practical thermocouple measuring circuit is shown schematically in Fig. 17-1. The ends of the thermocouple wires are kept at a constant cold reference temperature. Extension wires to a data transmitter or recorder are attached at the cold reference junction. The thermocouple voltage, which is proportional to temperature, is measured across the extension wire terminals at the transmitter or recorder. That value can be converted to a digital value for local display or be sent to a data acquisition system.

## RESISTANCE-TEMPERATURE DETECTORS (RTDs)

A *resistance-temperature detector* (RTD) is a solid conductor, usually in wire form, whose resistance increases with temperature (*positive* temperature coefficient [PTC] of resistivity). Its resistance change can be sensed directly as a voltage change across a current-driven resistor, or by the output of a resistance bridge. The most common RTDs are made of platinum because of their useful $-200$ to $850°C$ temperature range, but they are also made of nickel, copper, and nickel-iron alloy. RTDs are made in many different shapes and sizes as shielded or open units for both wet and dry applications.

## BARRETERS

A *barreter* or *bolometer* is a temperature sensor made from fine resistive wire or metal film whose resistance increases with increasing temperature (*PTC* of resistivity). Barreters are used to measure output power in microwave systems.

## THERMISTORS

A *thermistor* (for thermally sensitive resistor), as shown in section view Fig. 17-2, is a resistor that can sense heat changes. It is a common form of axial-leaded bead-type thermistor that is encapsulated in glass. Thermistor sensing elements are made by sintering combinations of metallic oxides into beads, rods, and other shapes. Oxides of manganese, nickel, cobalt, copper, iron, and titanium are commonly used. The resistance of these materials decreases as the temperature increases along an exponential curve, so they exhibit *negative* temperature coefficients (NTCs) of resistance.

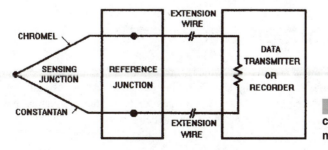

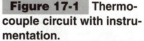

**Figure 17-1** Thermocouple circuit with instrumentation.

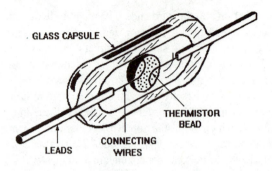

GLASS CAPSULE

THERMISTOR BEAD

LEADS   CONNECTING WIRES

**Figure 17-2** Thermistor section view.

Thermistors are used in both temperature control and measurement circuits, and they can determine the temperature of surfaces as well as fluids. Because of their nonlinear resistance-versus-temperature characteristics, they can measure a large resistance change over a narrow temperature range. Available commercial thermistors offer a wide choice of power rating, size and shape, resistance tolerance, and thermal time constant. At 25°C, thermistor resistance can be from about 100 ohms to more than 10 megohms. Typical measurements can be made within a −100 to 450°C range.

Glass-encapsulated thermistor beads with diameters as small as 0.03 mm suspended on precious-metal alloy leads can respond in milliseconds. Thermistors made as disks, washers, and rods can dissipate higher power, and some have positive temperature coefficients.

## SEMICONDUCTOR TEMPERATURE SENSORS

Temperature sensors can be made from semiconductor materials such as silicon. They are made as bulk resistors, diodes, and monolithic ICs. The simplest are bulk resistors with positive temperature coefficients that can provide temperature readings between −65 and 200°C, with resistive values of 10 ohms to 10 kohms. They are typically used in bridge circuits. Diodes and transistors can measure temperatures between −40 and 150°C when calibrated or when they are matched pairs in bridge circuits. Some semiconductor sensors are transistors connected as diodes with their collector and base leads shorted so that the temperature measurements are made across their base and emitter leads. Two-terminal ICs optimized for temperature sensing are current sources capable of measuring temperatures between −55 and 150°C.

# Mechanical Sensors

Many different mechanical sensors can convert force, pressure, velocity, acceleration, and position or motion into electrical signals. These include such sensors as strain gages for force or pressure measurements, strain-gage bridges for force measurements, and piezoelectric devices for measuring force, pressure, and acceleration.

## CAPACITIVE PRESSURE SENSORS

A *capacitive pressure sensor,* as shown in the section view Fig. 17-3, senses changes in pressure and converts them to electrical signals that are proportional to pressure. Pressure applied to the flexible quartz diaphragm deflects it so that an annular electrode on its underside makes contact with a common electrode. The vent hole allows air to escape the chamber when the diaphragm is deflected.

## PIEZOELECTRIC PRESSURE SENSORS

A *piezoelectric pressure sensor* measures pressure by application of the *piezoelectric effect,* in which a mechanical force applied to piezoelectric material produces voltage that is proportional to the applied pressure. The most accurate pressure measurements can be made with these sensors by placing four of them in a Wheatstone bridge or by positioning a single sensor so that it is most sensitive to shear stress.

## PIEZOELECTRIC ACCELEROMETERS

A *piezoelectric accelerometer,* as shown in the cutaway view Fig. 17-4, makes use of the piezoelectric effect to measure acceleration. When the accelerometer is accelerated, the internal inertial mass opposes the axial force of the spring and compresses the piezoelectric crystal. That mechanical compression generates voltages that are converted into practical measurement units. Piezoelectric accelerometers provide continuous acceleration information in *inertial guidance systems.*

## LINEAR VARIABLE DIFFERENTIAL TRANSFORMERS (LVDTs)

A *linear variable differential transformer* (LVDT) is a sensor that can measure position, acceleration, force, or pressure, depending on how it is installed. Figure 17-5*a* is a section view of an LVDT, and Fig. 17-5*b* is its schematic diagram. It responds to axial movements of its iron core within a transformer wound as a solenoid. Secondary windings $S_1$ and $S_2$ are wound on a common core with primary winding $P$. The output voltage of the LVDT

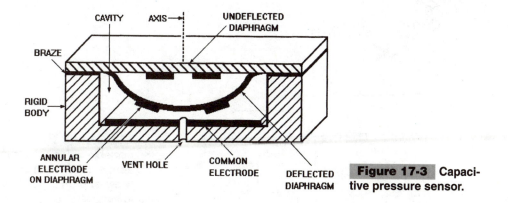

**Figure 17-3** Capacitive pressure sensor.

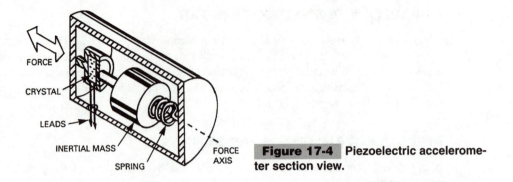

**Figure 17-4** Piezoelectric accelerometer section view.

changes as the inductances of coils $S_1$ and $S_2$ are changed in equal but opposite amounts by linear core movement.

## STRAIN GAGES

A *strain gage* is a resistive sensor that can be used for measuring weight, pressure, and mechanical force or displacement. Figure 17-6 shows a wire strain gage bonded onto a common plastic base. Strain gages are cemented to the structure whose strain is to be measured. They are usually mounted in a *strain-gage Wheatstone bridge,* as shown in Fig. 17-7.

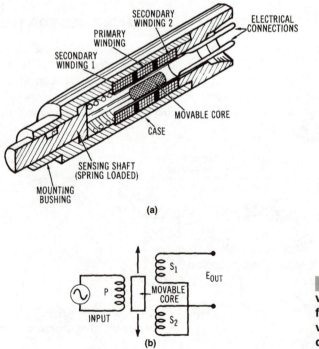

**Figure 17-5** Linear variable differential transformer (LVDT): (*a*) section view, and (*b*) schematic diagram.

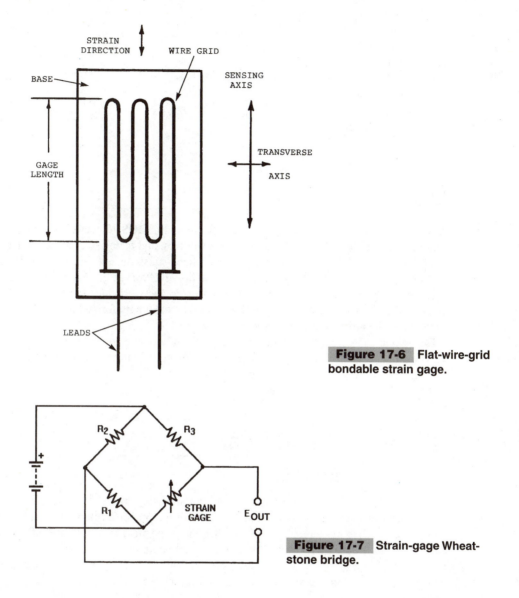

**Figure 17-6** Flat-wire-grid bondable strain gage.

**Figure 17-7** Strain-gage Wheatstone bridge.

When the bridge is energized, the gage elongates or contracts with the surface on which it is mounted, and its deformation alters the gage's resistance, which is converted into a voltage for display or recording. A single strain gage can be used if a precise measurement is not required or where temperature variations will not affect accuracy. However, a matched gage can be mounted in a second arm of the bridge to compensate for temperature changes. It is mounted at right angles to the first gage so that its sensitive axis will not participate in the measurement, and it is not exposed to the same temperature excursions as the first gage. In a four-gage bridge, two gages show increases in resistance and two gages show decreases, thus giving a larger reading than a single gage.

There are five different kinds of strain gage:

1. Bondable bare wire
2. Bondable wire mounted on a paper or plastic base
3. Bondable metal foil
4. Semiconductor
5. Deposited thin metal film

# Gas Sensors

A *gas sensor* is a chemical sensor made with thin-film MOS IC technology to provide warnings of dangerous levels of gas in homes, vehicles, and factories. A heater embedded in the silicon dioxide ($SiO_2$) layer between the silicon substrate and the metal-oxide sensing film raises the film temperature to its sensitivity level for the gas to be detected. Conductivity in the external circuit across the sensor contacts provides the warning signal. Some of these sensors have a micromachined silicon diaphragm formed in the silicon substrate under the heater to reduce power consumption.

Depending on its composition, the sensing film can sense the presence of carbon monoxide (CO) or methane ($CH_4$). A matching control integrated circuit provides signal conditioning, output driving, and interfacing.

# Magnetic Field Sensors

## FLUX-GATE COMPASSES

A *flux-gate compass* is an electronic compass that measures the relative strength of magnetic fields passing through two coils of wire. With the proper electronic circuitry, these sensors can deduce the direction of the earth's magnetic field. They are still affected by local magnetic fields from nearby steel objects or magnets, but they can be adjusted to compensate if those fields do not change in strength or location. The sensor can be mounted remotely from the readout instrument. External power is required for the coils and electronics.

## HALL-EFFECT SENSORS

A *Hall-effect sensor* is a semiconductor magnetic field sensor, as shown in the diagram Fig. 17-8. It produces a Hall-effect voltage across the opposing faces of a doped semiconductor chip when it is placed in a magnetic field and a current is passed through it perpendicular to the magnetic field. The Hall voltage developed across the faces of the chip is proportional to the strength of the magnetic field. The voltage from the sensor can be converted

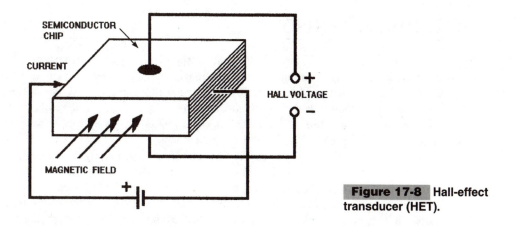

**Figure 17-8** Hall-effect transducer (HET).

into a linear or digital output for display or recording. It can also detect the speed and position of moving objects. It is also called a *Hall-effect transducer* (HET).

## MAGNETORESISTORS

A *magnetoresistor,* as shown in Fig. 17-9, is a magnetic field sensor whose resistance changes with the strength of the applied magnetic field. These sensors are generally mounted in pairs in a Wheatstone bridge circuit, as shown in Fig. 17-7, with a permanent magnet providing a magnetic bias. Current flow through series-opposing coils wound on the magnet poles reduces the flux at one resistor while increasing it at the other. This unbalances the bridge and provides an output current which can be converted to a voltage measurement.

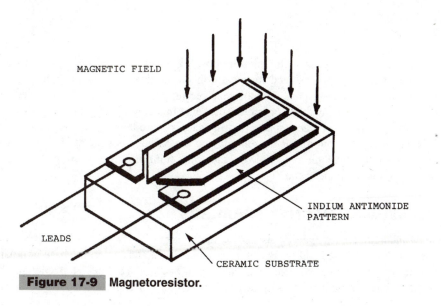

**Figure 17-9** Magnetoresistor.

# Radiant Sensors

A *radiant sensor* is a sensor that can detect and respond to radiated electromagnetic energy and produce an electrical output that is proportional to radiated power. A common example is a *photodetector* that can detect visible or infrared energy and convert it into an electrical signal, with or without signal gain. *Photoconductive cells, photodiode detectors,* and *solar cells* are all photodetectors. *Phototransistors* and *photodarlingtons* can perform photodetection, but they are not usually used as primary sensors.

All of these sensors can detect visible light and IR energy directly or as transmitted to them through optical fibers. The radiation response of a silicon photodetector is a function of the characteristics of the silicon and the diffusion depth of the light-sensitive PN junction. Photoconductive cells can be made so that their peak frequency responses are in the visible-light region, but all silicon photosensors exhibit peak responses in the near-IR region at about 900 nm. For more information on photosensitive devices see "Photoconductive Cells," "Photodiode Detectors," "Solar Cells," and "Phototransistors" in Sec. 12, "Optoelectronic Components and Communications."

# Speakers

A *speaker*, also called a *loudspeaker*, is an electroacoustic transducer that converts audio-frequency currents from an amplifier into sound waves, generally within the frequency range of human hearing (10 to 20 kHz). There are three different basic speaker designs:

1. Permanent magnet (PM)
2. Electrodynamic
3. Electrostatic

General-purpose speakers for most radios, TV receivers, and other communications equipment are usually designed to operate in the middle of the audio range, but high-fidelity speakers are designed to emphasize certain parts of that range. High-fidelity systems typically include three or more speakers. In addition to a midrange speaker, it typically includes a *woofer*, a speaker optimized for reproducing the lower audio frequencies, and a *tweeter*, optimized for reproducing the higher frequencies. These sounds can extend into the ultrasonic range beyond the upper limit of human hearing to enhance the realism of the reproduced sound.

## PERMANENT-MAGNET SPEAKERS

A *permanent-magnet speaker*, as shown in the section view Fig. 17-10, is a dynamic speaker. It has a voice coil wound on a movable sleeve attached to the base of a paper or plastic cone and held in position by a diaphragm. The sleeve is free to move longitudinally in the magnetic field that exists in the gap between the north and south poles of the permanent magnet. When the audio signal current from the audio-frequency amplifier flows in the voice coil, the electromagnetic field around the coil alternates at the audio frequency. This changing field interacts with the permanent-magnet field, causing the voice coil and the base of the speaker cone to oscillate longitudinally. This vibration of the cone produces alternating compressions and rarefactions of the air, which cause the sensation of sound in the ear.

The voice coil is wound from a few turns of low-resistance wire, and it has relatively low impedance throughout the audio-frequency range. At the low audio frequencies this impedance value is between about 2 and 8 ohms. Consequently, an impedance-matching transformer must be inserted between the power amplifier and the speaker.

## ELECTRODYNAMIC SPEAKERS

An *electrodynamic speaker,* as shown in the section view Fig. 17-11, is similar to the PM speaker except that an electromagnet produces the constant magnetic field rather than a permanent magnet. This requires additional circuitry, making it more expensive than a PM speaker, and a DC power source is needed to energize the electromagnet. However, the electromagnet of the electrodynamic speaker is lighter than the permanent magnet of the PM speaker, and it occupies less space.

## ELECTROSTATIC SPEAKERS

An *electrostatic* or *capacitor speaker,* as shown in the section view Fig. 17-12, is based on the principle of electrostatic attraction and repulsion. Its operation depends on the force of a high-voltage electrostatic field between two plates, one of which is directly connected to the base of the speaker cone. The high voltage varies with the audio frequency, and the flexible plate is repelled from the fixed plate when the voltage increases, and it is attracted to it when the voltage decreases. The vibration of the movable plate causes the speaker to produce sound.

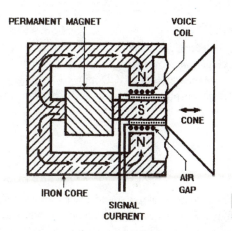

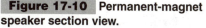

**Figure 17-10** Permanent-magnet speaker section view.

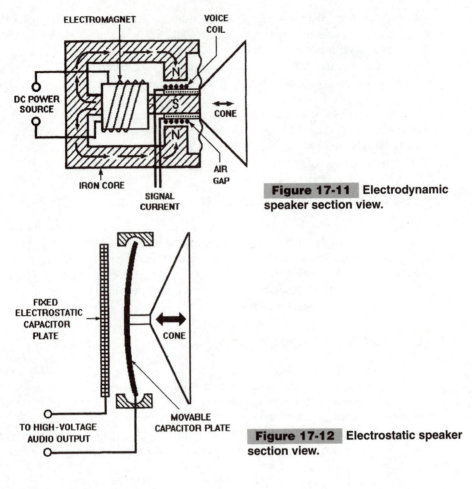

**Figure 17-11** Electrodynamic speaker section view.

**Figure 17-12** Electrostatic speaker section view.

# Headphones

*Headphones* operate on basically the same principles as speakers: moving-iron and moving-coil magnetic circuits, and piezoelectric and electrostatic principles. Conventional telephones and hearing aids have moving-iron movements, while high-fidelity headsets and battery-powered portable radio–tape cassette and -CD players typically include electrostatic, moving-coil, and piezoelectric earphones. Military and aircraft headsets typically include moving-coil earphones.

# Microphones

A *microphone* is an electroacoustic sensor that converts sound waves into AC signals at audio frequencies. It can be classified as a mechanical sensor because sound waves exert pressure. Every microphone consists of a diaphragm (which vibrates in the presence of a sound wave) and some form of transducer that is capable of converting the mechanical vibration into electrical signals. The three basic microphone designs are: (1) *carbon,* (2) *permanent magnet,* and (3) *piezoelectric.*

The selection of the optimum microphone for an application is determined by power requirements, size constraints, weight, and ability to stand up to rough treatment. Other selection criteria are frequency response, impedance, sensitivity, and directivity. Microphones vary in size from the inconspicuous lapel microphones used to amplify the voices of radio and TV announcers, news readers, and persons being interviewed to those large enough and sensitive enough to pick up sounds from the distant sections of an orchestra. The microphones in telephones, home-entertainment and public-address systems, and radio transceivers are typically custom designed for the product.

## CARBON MICROPHONES

A *carbon microphone,* as shown in the section view Fig. 17-13, was one of the first microphones to be invented in the nineteenth century. It is an assembly of a covered cup or *button* filled with carbon granules that act like a variable resistor. The button cover is linked by a stiff strut or piston to a flexible metal diaphragm. In the presence of incident sound waves the diaphragm vibrates, and those vibrations are transferred as variable pressure to the cover, which acts as a piston, keeping the carbon granules agitated. The movement of the granules changes the resistance of the button in response to the sound waves.

The variations in the resistance of the button in series with a low-voltage power source alternates the current flowing through the primary of the audio transformer, and the output across the secondary turns is an audio signal. In the absence of sound waves, steady DC flows in the circuit, so the transformer produces no audio output.

The carbon microphone offers high sensitivity at a high output level, low impedance (11 ohms or less), and it is inexpensive. However, its applications are limited to use in a controlled environment because its response is degraded by high humidity. It also has a poor frequency response and produces a noticeable background noise.

## PERMANENT-MAGNET (PM) MICROPHONES

A *permanent-magnet (PM) microphone* is a dynamic microphone that is similar in construction to a PM speaker. (It has a moving voice coil and a permanent magnet to provide a

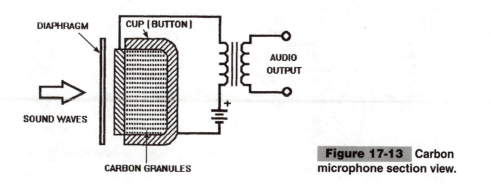

**Figure 17-13** Carbon microphone section view.

constant magnetic field). As shown in the section view Fig. 17-14, a diaphragm replaces the cone of the speaker to provide the inverse function—that of converting sound energy into electrical energy. When sound waves strike the diaphragm, it vibrates, causing the attached voice coil to vibrate. This vibration cuts magnetic flux lines and induces an output that is an audio-frequency signal corresponding to the incident sound waves. PM microphones are commonly used in broadcast studios.

The PM microphone offers low output impedance and wide frequency response, and it does not need a supplementary DC power source. It is a rugged movement, and its performance is not degraded by exposure to high humidity and wide temperature swings.

## CRYSTAL MICROPHONES

The *crystal* and *ceramic microphone* both depend on the piezoelectric effect to produce an audio output. Figure 17-15 is a section view of a crystal microphone. Sound waves impinging on the diaphragm of this microphone vibrate the crystal wafer sandwiched between a diaphragm and an electrode at a frequency within the audio range. This vibration causes an AC audio signal to appear across the wires from the diaphragm and electrode.

The piezoelectric crystal in a crystal microphone can be Rochelle salt or ammonium dihydrogen phosphate (ADP). Typically, two crystals are cemented together. The output signal is then amplified. The crystal microphone offers good frequency response and it introduces no background noise. It does not require a supplementary power supply. However, its performance is degraded by exposure to excessive heat, shock, and high humidity.

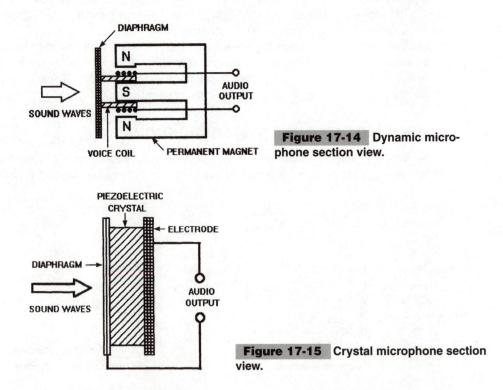

**Figure 17-14**  Dynamic microphone section view.

**Figure 17-15**  Crystal microphone section view.

## CERAMIC MICROPHONES

The *ceramic microphone* operates on the same piezoelectric principle as the crystal microphone, and its construction is similar to that of the crystal microphone shown in Fig. 17-15. It has characteristics that are similar to those of a crystal microphone, except that the crystal is replaced by a ceramic material such as barium titanate. A more rugged ceramic microphone can withstand wider temperature and humidity variations than a crystal microphone.

## VELOCITY (RIBBON) MICROPHONES

The *velocity (ribbon) microphone* has a diaphragm that is a very light aluminum ribbon which can move freely in a magnetic field. When sound waves cause a pressure differential on opposite sides of the ribbon, a velocity gradient causes the ribbon to vibrate. The corresponding audio signal is then induced into the ribbon. The output impedance of this microphone is extremely low, so its output must be fed to a step-up transformer. This increases the output level while also matching the low microphone impedance to the surge impedance of a standard audio line. It is widely used in broadcasting, but because wind can affect the delicate ribbon, it is not normally used outside of the studio. The velocity microphone provides good frequency response and high output level. Its background noise is low, it is unaffected by temperature and humidity, and no supplementary power supply is required.

# Shaft-Angle Optical Encoders

*Encoders* are mechanical and electromechanical transducers that convert shaft rotation into output pulses that can be counted to determine shaft revolutions or shaft angle. A common type of encoder is the rotary optical shaft-angle encoder, shown in the cutaway view Fig. 17-16. It includes light emitters, photodetectors, and a shaft-mounted disk with alternating transparent and opaque patterns. When rotating, the disk "chops" the light beam or beams to produce output pulses. The two basic types of optical encoder are the *absolute optical shaft-angle encoder* and the *incremental optical shaft-angle encoder*. Both can serve as feedback sensors in closed-loop control systems. There are also *direct contact* or *brush-type* and *magnetic* encoders, but they are not widely used in electronic control systems.

## ABSOLUTE SHAFT-ANGLE OPTICAL ENCODERS

An *absolute shaft-angle optical encoder* is an encoder whose digital word output uniquely defines each shaft angle. It contains a linear array of light-emitting diodes (LEDs) facing a matched linear array of photodetectors separated by an absolute encoder disk that rotates on the input shaft. When the shaft rotates, a unique digital word is generated as the encoder disk's opaque sectors chop the light beams between the emitters and detectors.

An absolute shaft-angle encoder disk is shown in Fig. 17-17. The output code word is read radially. The most significant bit is read from the inner coded track and the least significant bit is read from the outer coded track. The disk will retain the last angular position of the encoder shaft if it is stopped accidentally because of power interruption. These disks,

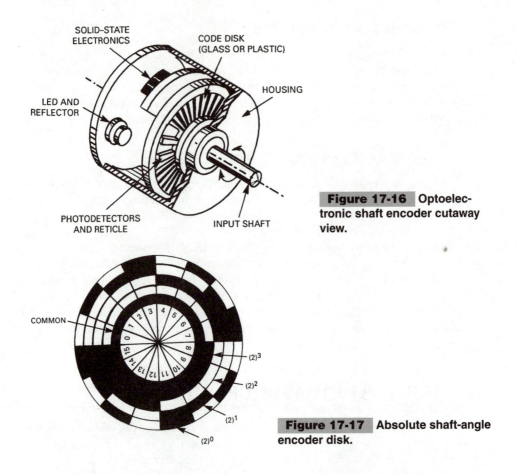

**Figure 17-16** Optoelectronic shaft encoder cutaway view.

**Figure 17-17** Absolute shaft-angle encoder disk.

made of thin glass or plastic, can produce either the natural binary or the Gray code. Shaft position accuracy is proportional to the number of annular rings or channels on the disk.

## INCREMENTAL SHAFT-ANGLE OPTICAL ENCODERS

An *incremental shaft-angle encoder* has an internal LED, and matching photodetectors with an incremental shaft-mounted code disk rotating between them. Its encoder disk is a thin, transparent plastic or glass disk with a pattern of equally spaced opaque spokes radiating out from the center, as shown in Fig. 17-18. When the code disk rotates, the light beam is chopped, and the pattern is converted by photodetectors into two or three square-wave output pulses equal to the number of radial lines on the disk. Shaft speed and position can be determined with respect to a previous reference angle by counting pulses.

# Surface Acoustic Wave (SAW) Devices

A *surface acoustic wave* (SAW) device can delay the passage of radio-frequency signals between its input and output terminals by converting them to a slower acoustic wave dur-

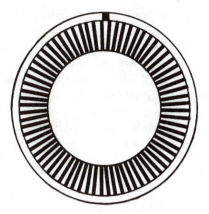

**Figure 17-18** Incremental shaft-angle encoder disk.

ing their transit across its surface. At the output terminals the RF signals can be restored to their original format. The SAW depends on the ability of certain crystals to delay signal propagation, permitting the devices to function as delay lines, bandpass or dispersive filters, pulse-compression filters, resonators, oscillators, and signal generators and decoders.

The substrate for a SAW device can be piezoelectric materials such as quartz, lithium niobate, or bismuth silicon oxide. These materials permit the efficient exchange of energy between mechanical and electric forms, and the uniformity of their crystal lattices supports the propagation of wave motion through the crystal in different modes. The surface mode propagates along the polished surface of the SAW as do ripples across the surface of water.

In a typical application, such as the delay line shown in Fig. 17-19, the RF input signal applied to the SAW is converted to an elastic acoustic or Rayleigh wave that propagates across the surface of the substrate as a composite of longitudinal and transverse waves. This acoustic energy is efficiently coupled to the output end by matching transducers deposited at each end of the substrate surface. The transducers, as shown in Fig. 17-19, are typically formed as quarter-wavelength interdigital slow-wave structures by photolithographic and thin-film deposition techniques. The line widths and spacing of the digits determine the frequency response of the SAW device and its efficiency.

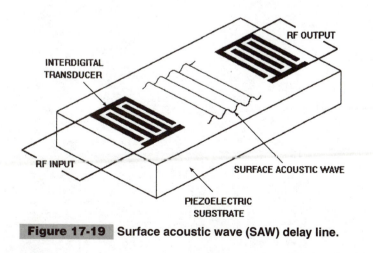

**Figure 17-19** Surface acoustic wave (SAW) delay line.

Elastic acoustic waves created by electromagnetic frequencies travel five orders of magnitude slower in the solid ceramic than they do in air (a ratio of 1 to 10,000). An acoustic path length of 1 cm gives a delay of 3 $\mu$s. A SAW delay line can be shorter by that ratio than a cable or waveguide delay line, saving considerable weight and space in communication and radar systems. The mass and spacing of the crystal ions in quartz and lithium niobate results in a propagation velocity of about 3 mm/$\mu$s. Acoustic energy injected into one transducer on the surface of the substrate can be collected at intermediate transducers on the same substrate, permitting the addition, subtraction, or division of input frequencies. See also "Surface Acoustic Wave (SAW) Filters" in Sec. 1, "Passive Electronic Components."

# RADIO TRANSMITTERS AND RECEIVERS

# Overview

Radio broadcasting has made a significant contribution to humankind by shrinking the time needed to communicate between distant places. It has provided the timely distribution of news and public service messages, and offered mass entertainment at no cost to the listener. It has also helped to save lives by speeding up responses to natural and humanmade emergencies. The first radio communications were commercial message traffic transmitted across oceans or between shore stations and ships at sea. When regular sponsored radio broadcasts began, radio became the first worldwide mass media. Millions of people at all levels of society could hear the news, be entertained, hear storm warnings, and even get the right time.

Private citizens soon became radio amateurs who were able to communicate with other amateurs elsewhere in the world. These amateur links proved helpful in informing the rest of the world about wars and disasters that cut off other forms of communication.

Despite the popularity of broadcast and cable television, radio remains popular with many people, especially those engaged in other activities such as driving automobiles or working. Radio talk shows became the first interactive media when listeners were able to telephone the radio station with comments and criticisms. Music broadcasts programmed to appeal to different groups of listeners have also helped to keep radio alive and competitive.

The first radio stations broadcast with amplitude-modulated (AM) signals, which can be heard over wide listening areas, but they are subject to humanmade and natural interference. The later development of frequency modulation (FM) eliminated the annoying static and permitted higher quality of sound reception, but FM has a more limited listening range. Despite its technical advantages over AM broadcasting, FM never drove out AM, and both coexist amicably today.

There are, of course, many other radio broadcasting services, some unknown or barely known to the general public. They are government weather broadcasts, citizens band radio, and bands set aside specifically for marine and vehicular mobile communications, aviation safety and navigation, and police and other emergency services. Computers and digital electronics have made possible the radio broadcast networking needed for cellular telephony and paging.

# Radio Signal Propagation

The three different RF transmission modes are illustrated in Fig. 18-1. (1) *Sky waves* are transmitted signals that are reflected from the ionosphere before they reach the receiver hundreds or even thousands of miles away. They might actually be returned to earth and be reflected again two or more times before reaching the distant receiver. (2) *Direct waves* are received along a visual line of sight, as between two relay towers or from the ground to an aircraft. (3) *Reflected waves* are received after they have been reflected from the ground but not from the ionosphere, and (4) *surface waves* are radio waves that have been received over the horizon without reflections from the ground.

The range of high-frequency UHF and microwave transmissions is restricted to the *radio horizon,* as shown in Fig. 18-2. Signal range is a function of antenna height $h,$ and it is limited to the radio horizon distance $d_h$ if the receiver is close to ground level. That distance is approximately the product of 1.4 times the square root of the antenna height $h$.

The *ionosphere* is a region in the earth's outer atmosphere where the air is so thin that the quantities of free ions and electrons will affect radio-wave propagation. The lowest part of this region starts about 30 mi (50 km) above the earth, and it extends to about 250 mi (400 km) above the earth, with its height depending on the season of the year and the time of day.

During daylight hours the $D$ layer, which exists at 30 to 60 mi (50 to 100 km) above the earth, causes most of the RF attenuation between 1 and 100 MHz, and it absorbs energy from radio waves below about 10 MHz. Moreover, the daylight $E$ layer at 60 to 90 mi (100 to 150 km), reflects signals along paths 1 and 2, as shown in Fig. 18-3. The $F_1$ and $F_2$ lay-

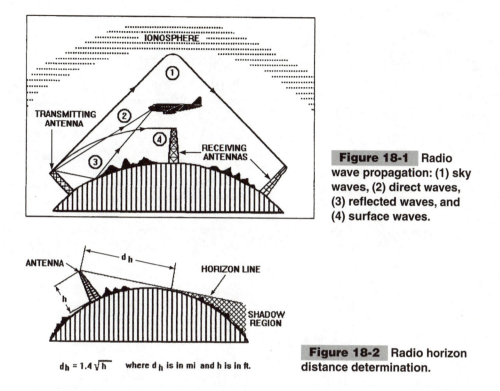

**Figure 18-1** Radio wave propagation: (1) sky waves, (2) direct waves, (3) reflected waves, and (4) surface waves.

**Figure 18-2** Radio horizon distance determination.

$d_h = 1.4\sqrt{h}$   where $d_h$ is in mi and h is in ft.

ers together occur during daylight at heights of 90 to 250 mi (150 to 400 km) and reflect signals along paths 3 and 4. The $F_2$ layer, extending from 150 to 250 mi (250 to 400 km), remains intact at night. Signals will follow path 5 through all of the ionospheric layers into outer space if they are directed at an angle higher than the *critical angle.*

# Amplitude Modulation (AM)

*Modulation* is the modification of an RF carrier by an audio or coded signal to introduce intelligence to the transmission. The modulated signal must be *demodulated* at the receiver to reproduce the modulating speech or music. The RF carrier is a high-frequency sine wave that has three features which can be modulated: (1) *amplitude,* (2) *frequency,* and (3) *phase.*

In *amplitude modulation* (AM), the instantaneous voltage amplitude of the RF wave is linearly related to the instantaneous voltage magnitude of the audio-frequency (AF) signal, and the rate of amplitude variation equals the modulating frequency.

The unmodulated carrier, shown in Fig. 18-4a, is a high-frequency sine wave, and the modulating signal, shown Fig. 18-4b, is a low-frequency sine wave that modulates the RF carrier, as shown Fig. 18-4c. The percentage of modulation is measured by two different expressions. The one for positive peak is $(E_{MAX} - E_o) \times 100/E_o$ percent, where the amplitude of the unmodulated carrier is $E_o$, and the expression for the percentage of modulation on negative peaks is $(E_o - E_{MIN}) \times 100/E_o$ percent.

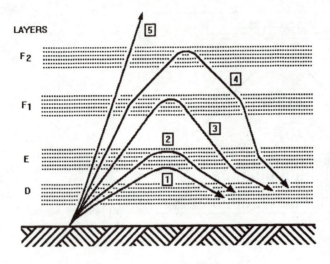

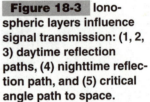

**Figure 18-3** Ionospheric layers influence signal transmission: (1, 2, 3) daytime reflection paths, (4) nighttime reflection path, and (5) critical angle path to space.

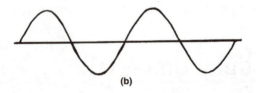

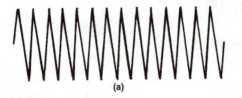

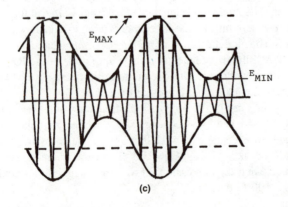

**Figure 18-4** Amplitude modulation (AM): (*a*) carrier signal, (*b*) modulating audio waveform, and (*c*) resulting modulated envelope.

An increase in the amplitude of the modulating signal causes the modulation percentage to rise. This improves the signal-to-noise ratio at the receiver and increases the audio power output. However, if the modulation percentage reaches 100 percent or more, overmodulation occurs, and the modulation envelope becomes severely distorted. This distortion can be heard at the receiver. Overmodulation also generates unwanted sidebands that interfere with adjacent channels by mixing with the sidebands in those channels.

The AM voltage signal is an RF signal because no audio voltage is present. It contains the following three components which, when added together, result in the AM waveform:

**1.** An *upper sideband* whose amplitude is half the depth of modulation and whose frequency is the sum of the modulating and carrier frequencies
**2.** A *carrier component* whose amplitude is that of the unmodulated carrier and whose frequency is that of the carrier.
**3.** A *lower sideband* whose amplitude is half the depth of modulation and whose frequency is the difference between the carrier and modulating frequencies

The AM signal is contained within a *bandwidth* defined as the difference in frequency between the upper and lower sidebands. When speech or music contains many instantaneous frequencies of varying amplitudes, each audio component will produce a pair of sidebands, and the bandwidth occupied by the AM signal will be the frequency difference between the highest upper sideband and the lowest lower sideband transmitted. Thus *bandwidth* equals twice the highest audio modulating frequency.

# AM Radio Transmitters

An *AM radio transmitter* is a transmitter whose carrier is amplitude modulated. A simplified block diagram of an AM transmitter capable of double-sideband voice communication is shown in Fig. 18-5. The RF section generates the required carrier power, and the AF section provides the audio power necessary to modulate the amplitude of the carrier.

The RF section is divided into the following blocks:

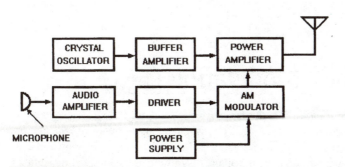

**Figure 18-5** Amplitude modulation (AM) transmitter block diagram.

- Crystal-controlled oscillator
- Buffer amplifier stage
- Final RF power amplifier

The Federal Communications Commission (FCC) assigns a specific carrier frequency to each AM station, and it also requires that the output frequency be contained within narrow tolerance limits. For example, if an AM broadcast station has been assigned a carrier frequency of 880 kHz with an allowed tolerance of $\pm 20$ Hz, the output frequency must not rise above 880 kHz + 20 Hz = 880.02 kHz, nor fall below 880 kHz − 20 Hz = 879.98 kHz. A crystal-controlled oscillator in a thermostatically controlled oven can maintain a stable output frequency within these narrow tolerances. In addition, the oscillator is normally operated from a well-regulated and filtered power supply. An oscillator stage is typically operated between Class AB and Class C, and can provide several watts of power output.

The oscillator is followed by a *buffer amplifier stage* that assures frequency stability. This stage has a high input impedance and is typically operated under Class AB conditions. The power gain of the buffer stage is low, but it presents a constant load to the oscillator, an important factor in stabilizing the frequency. The intermediate power amplifiers increase the power level of the carrier. These stages are typically operated under Class C conditions for the highest efficiency, and a tank circuit is usually the collector load. Frequency multipliers or doubler stages are used in the intermediate amplifiers to obtain the necessary operating band.

The final *RF power amplifier stage* is the most important stage in the entire RF section because it delivers the rated power output of the transmitter to the antenna system. The FCC assigns an authorized power output to AM standard broadcast stations, but it makes allowance for limited output power tolerance. For example, the power tolerance for these stations extends from 5 percent above to 10 percent below the authorized power output. Thus a station with an assigned power of 20 kW can actually broadcast with power as high as 21 kW or as low as 18 kW.

Because the final RF stage produces the highest level of RF power, the amplifier must be operated under Class C conditions to obtain an efficiency as high as 90 percent. This is typically accomplished with a push-pull RF circuit that eliminates all of the even harmonics of the carrier frequency. Shielding between the tank coil and the coupling coil reduces the capacitive coupling between the coils to eliminate the transfer of high-frequency harmonics to the antenna. In addition, wave traps are inserted in series and in parallel with the coupling coil to prevent any harmonics that might appear across the coupling coil from reaching the antenna.

# Frequency Modulation (FM)

In *frequency modulation* (FM) the RF carrier is frequency modulated by an audio modulating signal, as shown in Fig. 18-6a. The instantaneous amount of *frequency shift* or *deviation* away from the average unmodulated value of the FM wave, shown in Fig. 18-6b, is linearly related to the instantaneous voltage magnitude of the modulating signal. Even after modulation, the amplitude of the RF carrier remains constant. As a result of this form of modulation, the RF power output and the antenna current of the FM transmitter remain independent of the modulation.

In addition, the rate of frequency deviation ($f_d$) equals the modulating frequency. However, the amount of frequency shift in the RF carrier is independent of the modulating frequency. If both the frequency shift in the RF carrier and modulating frequency have the same amplitude, modulating tones of 200 and 400 Hz will provide the same amount of frequency shift in the FM wave.

The higher audio frequencies, generally above 800 Hz, are progressively accentuated (preemphasized) by FM and TV transmitters to improve their signal-to-noise ratios at the receiver. Preemphasis is measured by the time constant of an *RC* circuit whose audio output increases with frequency. The FCC specifies this time constant as 75 µs. To restore the tonal balance, the receiver discriminator output is fed to an *RC* deemphasis circuit with the same time constant.

At 100 percent modulation of an FM carrier, the frequency shift reaches the maximum value allowed for the system, called the *highest frequency deviation.* In commercial FM broadcasting, it is a shift of 75 kHz on either side of the unmodulated carrier frequency. Consequently, the output frequency swing for 100 percent modulation is $\pm 7$ kHz. By contrast, in a TV transmitter, 100 percent modulation corresponds to a frequency deviation of 25 kHz.

The *percentage modulation* and the amount of frequency shift are directly proportional, so that 40 percent modulation in commercial FM broadcasting corresponds to an output frequency swing of $\pm 40 \times 75/100 = \pm 30$ kHz.

In FM broadcasting, the degree of modulation is defined by the *modulation index.* In music and speech transmission, the instantaneous value of the modulation index can vary from less than 1 to over 100. However, for any individual system, there is a specific modulation index value called the *deviation ratio.*

In commercial FM broadcasting, bandwidth is 200 kHz, and the transmitted audio range is 50 Hz to 15 kHz. Because the frequency deviation value for 100 percent modulation is 75 kHz, the deviation ratio is 75 kHz/15 kHz = 5. This is a typical value for a wideband FM system. By contrast, marine FM transmitters use a narrow-band system that has a bandwidth of only 20 kHz. In these systems, frequency deviation for 100 percent modulation is

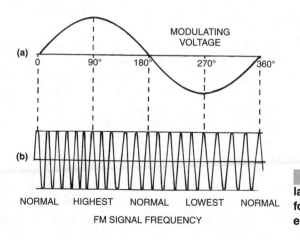

**Figure 18-6** Frequency modulation (FM): (*a*) modulating waveform, and (*b*) modulated envelope.

5 kHz, and the highest audio frequency is 3 kHz. Thus, the value of the deviation ratio is only 5 kHz/3 kHz = 1.67.

# Phase Modulation

In *phase modulation,* the instantaneous phase of the RF carrier is modulated by the audio modulating signal while the amplitude of the RF carrier is kept constant. The instantaneous amount of phase shift (away from its unmodulated value) is linearly related to the instantaneous voltage magnitude of the modulating signal. The rate of change in phase is equivalent to a shift in frequency, so the phase-modulated waveform is similar to the FM waveform. They can only be distinguished by reference to the modulating signal. Phase modulation is used primarily as an indirect method of creating FM by some FM transmitters.

# FM Transmitters

An *FM transmitter* produces a constant-amplitude sinusoidal output signal whose frequency varies in accordance with the audio signals being transmitted. There are two basic types of FM transmitters: *direct* and *indirect.* The direct FM transmitter is modulated by varying the frequency of the oscillator in response to the amplitude of the modulating signal. By contrast, the indirect FM transmitter is modulated by phase-modulating the output of a crystal-controlled oscillator in response to the modulating signal. The phase-modulated signal is converted to an FM signal by shifting it 90° in phase.

FM transmitters include preemphasis networks that provide equal signal-to-noise ratios for all of the frequency components in the FM signal being transmitted. These networks do this by delivering an output whose amplitude is proportional to the input frequency above a specified threshold frequency. Consequently, the output for the high-frequency components of the input signal is higher than that for the low-frequency components.

## DIRECT FM TRANSMITTERS

Figure 18-7 is a block diagram for a *direct FM transmitter.* The input audio signal is amplified by the *audio amplifier* and sent to the *preemphasis circuit* which accentuates the audio frequencies. This audio modulating signal is then sent to the modulator (JFET reactance or varactor diode) that applies either inductive or capacitive reactance to control the frequency of the *variable-frequency oscillator* (VFO). (The amount and type of reactance is determined by the amplitude of the audio modulating signal.) The VFO produces a sinusoidal RF output, which with no modulation is the center frequency of the transmitter. The modulated signal is sent through a *buffer stage* and on to *intermediate amplifiers* and two *frequency multipliers* that increase the effective frequency deviation of the FM signal. An RF driver and final RF power amplifier boost the RF signal to the power level for transmission.

An automatic frequency control (AFC) loop consisting of the *crystal oscillator, mixer and amplifier, discriminator,* and *DC amplifier* produces a correction signal that is propor-

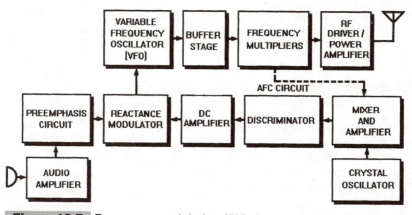

**Figure 18-7** Frequency modulation (FM) direct transmitter block diagram.

tional to any drift which occurs. When this signal is applied to the FM modulator, it returns the oscillator output to the correct frequency. The main drawback of the direct FM transmitter is the complexity of the AFC closed loop required to achieve the necessary frequency stability.

## INDIRECT FM TRANSMITTERS

A typical *indirect FM transmitter* contains a crystal oscillator that generates the basic transmitter frequency. The audio input is passed through a preemphasis network, an audio amplifier, and an audio correction network that processes the audio signal before sending it to the pulse-modulation modulator, which varies the oscillator frequency in response to the modulating signal. A buffer amplifier isolates the oscillator, and frequency-multiplier stages boost the signal and deviation frequencies. An RF driver and final RF power amplifier raise the signal to the desired power level. This transmitter does not include an AFC circuit because its crystal oscillator provides stable high frequency.

## FM STEREO TRANSMITTERS

Stereo audio signals are transmitted by FM transmitters in a process known as *FM stereo multiplexing.* Audio outputs from two microphones, designated the *left* and *right,* are applied to an adder and subtractor *matrix* circuit. The matrix has two outputs: the *sum* of the two instantaneous signal amplitudes (left + right) and the *difference* between them (left − right). The monaural and stereophonic frequency response is 50 to 15,000 Hz.

The sum signal is used to modulate the frequency of the main carrier. A monaural receiver will demodulate this main carrier into an audio signal containing both left and right signals, so no information is lost. The difference signal is used to modulate a double-sideband stereo subcarrier operating at 38 kHz. This subcarrier, which must be suppressed to a level of less than 1 percent of the main carrier, is a phase-locked second harmonic of the 19-kHz stereo *pilot* signal. The subcarrier then modulates the main carrier. The stereo demodulating circuits of stereophonic receivers are controlled by this pilot signal. The stereo demodulator provides a difference signal that can be combined with the sum signal to regenerate the original left and right audio signals at the receiver.

# AM Receivers

The *superheterodyne receiver* is a popular AM receiver circuit that overcame the short-comings of the earlier *tuned radio frequency* (TRF) *receiver* by converting all incoming frequencies to a single intermediate frequency (IF). It is the most widely used receiver circuit for personal radio receivers, and modifications of the circuit are also used for amateur and professional radio communications.

## SUPERHETERODYNE RECEIVERS

Figure 18-8 is a simplified block diagram of a *superheterodyne receiver* to receive signals from the standard 550- to 1500-kHz AM broadcast band. The RF signal received at the antenna is applied to an *RF amplifier* before being sent to the *mixer,* which simultaneously receives the output from a *local oscillator.* The mixer accepts only the signal frequency to which it is tuned, and it rejects all others. The local oscillator frequency always differs from the desired reception frequency by a value equal to the receiver's IF, typically 456 kHz.

In most superheterodyne receivers, local oscillator frequency is higher than the signal frequency. If the IF is 456 kHz and the mixer is tuned to 1010 kHz, the local oscillator is simultaneously tuned to 1466 kHz (1466 − 1010 = 456 kHz). If the mixer is then retuned to an 880-kHz signal, the oscillator frequency becomes 1336 kHz.

The mixer *heterodynes* the oscillator and input frequencies to produce the difference frequency or IF of 456 kHz which carries the same AM intelligence as the original RF signal. The IF output from the mixer is amplified by one or more *IF amplifiers* that will only amplify IF. The amplified output is then sent to the *amplitude-modulation detector* where the audio frequency is removed from the IF signal, amplified, and then sent to a speaker or headphones.

Higher-performance superheterodyne receivers typically include *noise limiter* and *automatic volume control* (AVC) circuits. The noise limiter prevents unwanted noise spikes riding on the signal from reaching the receiver by clipping the signal amplitude at a preset level. The AVC circuit holds the audio output from the receiver constant despite changes in

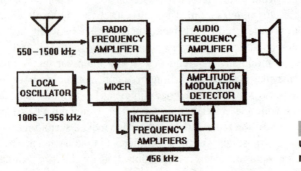

**Figure 18-8** Amplitude modulation (AM) superheterodyne receiver block diagram.

RF signal strength. It does this by developing a DC voltage whose amplitude is proportional to the signal strength of the audio delivered by the detector. This variable voltage is fed to the RF and IF amplifiers to control their amplification.

Conventional superheterodyne receivers recover the audio components from the IF signal, but they cannot receive continuous wave (CW) signals that do not include audio signals. A beat frequency oscillator (BFO) is added to the receiver to provide the missing audio if it is to receive CW.

# FM Receivers

## STANDARD FM RECEIVERS

A *standard FM receiver* receives and reproduces FM signals. Figure 18-9 is a block diagram of an FM receiver for receiving signals in the standard 88- to 108-MHz FM broadcast band. This block diagram is similar to the one for a standard AM broadcast receiver shown in Fig. 18-8. The FM signal selected from the antenna is amplified by the *radio-frequency amplifier*. It is then sent to the *mixer,* where it is mixed with a signal from the *local oscillator* which can generate frequencies of from 77.3 to 97.3 MHz. The output of the mixer is an intermediate frequency (IF) that retains the intelligence of the transmitted FM signal. This IF signal has a constant amplitude, but its frequency varies above and below 10.7 MHz to track transmitter modulation when tuned to the standard FM broadcast band. In this way the center frequency of the FM signal is converted to the receiver's IF.

After amplification by the *intermediate-frequency amplifier,* the IF is applied to a *limiter stage* which removes all amplitude variations carried on the IF. The 10.7-MHz limited signal is then applied to the *FM discriminator,* which provides an instantaneous audio output voltage that is directly proportional to the amount of instantaneous frequency deviation from the value of IF. This signal is then fed to a *deemphasis network* which reverses the preemphasis process performed in the FM transmitter. It restores the tonal balance of the FM signal in the receiver to the state it was in before preemphasis. This is done by circuitry that decreases the output to the audio amplifiers if the audio frequency increases and increases it if the audio frequency decreases. (The combination of discriminator and deemphasis circuitry replaces the detector in the AM receiver.) The audio signal is then amplified and sent to a speaker.

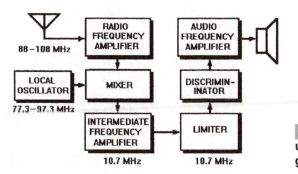

**Figure 18-9** Frequency modulation (FM) receiver block diagram.

## FM STEREO RECEIVERS

An *FM stereo receiver* can receive and reproduce FM stereo multiplex signals. The front ends of these receivers are the same as those of conventional FM receivers. Additional circuits are added after the FM detector to separate the individual components of the complex signal. These components are an L + R signal with a bandwidth of 15 kHz, and an L − R signal with a bandwidth of 23 to 53 kHz, and a 19-kHz pilot carrier.

These components are separated from the composite signal by filtering. A matrix circuit combines these signals so that the original left- and right-channel signals are recovered. Each of these recovered signals is then processed in separate audio channels containing deemphasis networks and audio amplifiers before being sent to the speakers. See also "FM Stereo Transmitters" preceding in this section.

# 19

# TELEVISION BROADCASTING AND RECEIVING TECHNOLOGY

# Overview

Television became one of the technical marvels of the twentieth century because it can produce moving images, voice, and music all on the same receiver. Crude television broadcasting based on mechanical scanning had been demonstrated before the turn of the century, but it was not until 1928 that the first television programs were broadcast in the United States. Those early systems depended on cumbersome and unreliable mechanical spinning disks to produce a picture. Modern television, however, depends on the compatible combination of a camera and picture tubes that are based on the synchronized scanning of electron beams.

The first cathode-ray tube, an evacuated glass envelope containing an electron-emitting cathode and a positively charged anode, was developed as a laboratory tool for investigating electrical phenomena in gases and in vacuum. An early version of the tube made possible the discovery of X-rays. Years later, researchers found that the electron beam could be scanned over a fluorescent screen by switching the polarities of internal electrodes to form images that remained long enough to be viewed. This discovery led to the invention of the oscilloscope.

But it was not until the iconoscope was developed that modern television became a reality. Instead of having the electron beam paint an image on a phosphor screen, the polarities of the internal electrodes were switched so that the beam scanned an optical image focused on an internal plate. The electrical signal taken from the plate was modulated by the electron beam so that it conveyed the image in the form of a picture signal (called *video* today). This signal, when transmitted with a timing pulse, could reproduce the image on the receiving tube. The pulse synchronized the picture tube beam with the iconoscope beam, thus creating the first electronically scanned TV system.

Television with electronic scanning was first publicly demonstrated in 1933, but commercial black-and-white TV broadcasting did not begin until the 1940s. Although color TV broadcasting began a few years later, more than 20 years would pass before the number of color sets in American homes would equal the number of black-and-white sets. By that time mass production and the economies of scale had made color TV sets far more affordable for most people.

The range of reliable reception for all broadcast TV signals is still limited to a radius of about 30 miles from the transmitter, placing a limitation on television viewers that is not imposed on radio listeners. Moreover, broadcast TV signals cannot reach many places surrounded by hills or mountains, imposing yet another restriction. This led to the development of community television, now called *cable television* (CATV). The cable service providers distributed signals to homes from receivers located high enough to receive line-of-sight transmissions. Later, the service providers distributed programming received by satellite relay, removing direct-broadcast limitations and giving subscribers access to many more TV channels. More recently, home satellite TV receivers have made possible direct reception of more than 150 channels.

The competition between cable and satellite service providers has been heating up, but the chances are that both will coexist for some time to come. Meanwhile, the giant telephone operating companies have been buying up cable companies with an eye toward making cable the universal medium for telephone service, interactive TV, E-mail, pay-per-view movies, accessing the Internet, and obtaining other services not yet defined.

With the development of digital TV and HDTV, the future technical direction of TV broadcasting remains a mystery, and many questions have yet to be answered. Is the average citizen willing to throw out a perfectly good analog color TV set and replace it with an HDTV-ready set? Will surfing the Internet be more fun on a TV set than on a computer? Do people really want to watch movies on a computer? With the prices of HDTV receivers now still above $3000, it might take years before installed HDTV sets catch up with analog sets, and the interim purchase of cheaper digital TV sets that are incapable of receiving HDTV is not expected to be popular.

# U.S. National Television Systems Committee (NTSC) Standard

The *National Television Systems Committee* (NTSC) *standard,* developed by RCA, was approved as the national standard in the United States in 1953. It is now in use in the United States, Canada, Japan, Mexico, and some South American and Asian countries. The other two worldwide standards, PAL and SECAM, although similar to NTSC, are mutually incompatible.

The NTSC standard for TV transmission is diagramed in Fig. 19-1. It allows a channel width of 6 MHz. The picture carrier is located 1.25 MHz above the lower boundary of the channel, and the audio center frequency is 4.5 MHz above the picture carrier. NTSC transmission is horizontally polarized. The composite picture is amplitude modulated (AM) and the audio signal is frequency modulated (FM). In each frame 525 lines are interlaced 2 to 1, and the scanning sequence is horizontal from left to right and vertical from top to bottom. It scans the 525 lines per frame at a rate of 30 frames (60 fields) per second.

The formation of a color TV picture requires that *luminance* (brightness), *hue* (color), and *saturation* (pastel quality versus vividness) be defined by a standard. Luminance is transmitted on the black-and-white signal, while hue and saturation values are carried by the *chrominance signal.* Changes in the amplitude of the chrominance signal are related to changes in saturation, and changes in its phase angle are related to changes in hue.

The chrominance and luminance signals are transmitted within overlapping parts of the same band by a process called *frequency interlacing.* A frequency band of at least 6 MHz is required to yield about 4 MHz for the transmission of luminance and color information and another 2 MHz is required for the transmission of sound and to provide a guard band against interference. The picture requires 3 to 4 MHz because the standard television screen can display up to 337,920 pixels. These pixels must be replenished about 30 times per second, the vertical scanning rate.

The standard NTSC television picture in the United States has an aspect ratio of 4:3, that is, 4 units wide and 3 units high. The scanning sequence when looking at the picture tube is shown in Fig. 19-2. The scanning spot starts at the top center of the screen (indicated by the

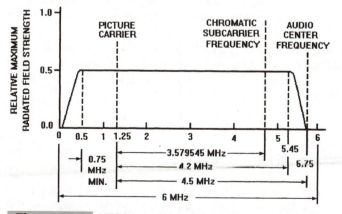

**Figure 19-1**  NTSC television transmission standards.

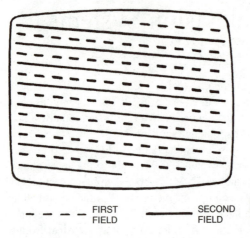

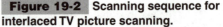

**Figure 19-2** Scanning sequence for interlaced TV picture scanning.

dashed line) and travels at a uniform rate from left to right along lines that lie at a constant distance below each other, as shown in the figure. When the end of the line is reached, the scanning spot quickly returns to the left edge to start a new line. During this return interval the spot is blanked out, so it is not shown in the figure. As the scanning spot travels back and forth across the screen, the spot moves downward at a constant rate. The lines in the figure are slightly sloped, and each line begins at a level that is a little below the end of the previous line.

The first *field* (shown as dashed lines) starts with the half line 1 and then jumps in sequence to line 3 and then 5, continuing on until all of the odd-numbered lines have been scanned. It is not complete until the spot is returned to the upper left corner of the screen. It follows a circuitous zigzag path in returning because it continues its back and forth horizontal line motion, but it is not seen because it is blanked out. It takes a time equal to that of 21 line scans or 1/60 s for the beam to return to the starting point for the next field. During this time, 262.5 lines are transmitted.

The spot then starts the second field (indicated by the solid line) by scanning line 2, and the sequence continues until all even-numbered lines have been scanned and the spot is again returned to the top left corner. At that time both fields are interlaced. The complete picture, called a *frame,* consists of 525 lines, and is transmitted in 1/30 s. The complete pattern shown in Fig. 19-2 appears on the TV screen in the absence of a picture, and it is called the *raster.*

The U.S. NTSC standard requires that 15,750 horizontal lines be scanned per second and that the vertical scanning rate be 30 lines per second. Each line has a duration of 63.49 µs. The color TV signal must be compatible with the black-and-white or *monochrome* TV signal in this standard.

# Color TV Transmitters

Transmitters of color TV have AM sections to process the video signal and FM sections to process the audio signal. The FM sections of color TV transmitters are identical to those in black-and-white transmitters. The video section also contains the same circuits as the black-and-white transmitter, as shown in the block diagram Fig. 19-3, but it also has additional circuits for processing the color signal.

Figure 19-4 is a simplified block diagram of those circuits in a color TV transmitter that process the red, green, and blue signals from the color camera signal to produce the video signal for transmission. These three signals are applied to a *matrix circuit,* which adds the signals in various proportions to produce three separate output signals: $Y$, $Q$, and $I$. The $Y$ signal carries the brightness, or *luminance,* variations of the picture. The $Q$ signal (for quadrature) corresponds to the green or purple information in the picture, and the $I$ signal (for in-phase) corresponds to the orange or cyan information. The $Q$ and $I$ signals combine to form the *chrominance* or $C$ signal, which contains all of the color information of the picture. The positions of the $Q$ and $I$ channels in the NTSC color TV frequency diagram are shown in Fig. 19-5.

The $Y$ signal is sent directly to the *adder* circuit from the matrix, but the $I$ signal is applied to a balanced $Q$ *modulator,* which also receives an input from the 3.58-MHz subcarrier oscillator. The $Q$ signal is also sent to a balanced $I$ *modulator,* and it also receives an

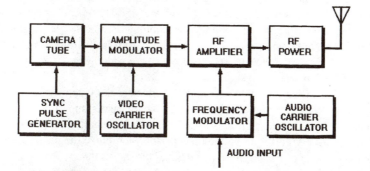

**Figure 19-3** Monochrome TV transmitter block diagram.

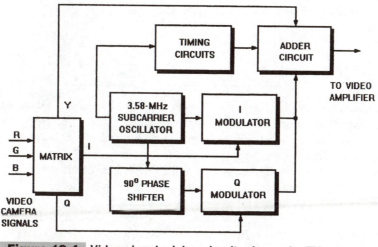

**Figure 19-4** Video signal mixing circuitry in a color TV transmitter.

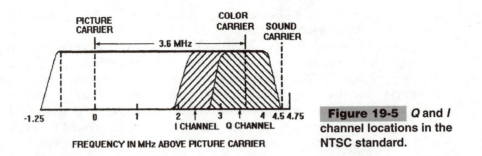

**Figure 19-5** *Q* and *I* channel locations in the NTSC standard.

input from the 3.58-kHz subcarrier oscillator, but it has been shifted in phase 90° by a *phase shifter.* Both the *Q* and *I* signals modulate the same subcarrier frequency and combine to form the chrominance signal, but they do not lose their separate identities because of the 90° phase difference of the subcarriers in the receiver. The *Q* and *I* modulator outputs are combined vectorially to form upper and lower side bands that are then applied to the adder circuit. The sidebands produced by the *Q* and *I* modulation are added vectorially to form the *C* signal. When the sidebands are combined, the 3.58-Hz subcarrier is suppressed.

The luminance *Y* and chrominance *C* signals are combined into a single video signal in the adder circuit. Horizontal and vertical sync pulses, as well as blanking and equalizing pulses, are generated by *timing circuits* and also sent to the adder, where they are inserted at the proper intervals into the video signal.

Another sync pulse, called the *color sync burst,* is added to the video signal. Each burst consists of a few cycles of the unmodulated 3.58-MHz subcarrier superimposed on each horizontal blanking pulse. Because the subcarrier was suppressed after the *Q* and *I* signals were generated, it must be reinserted by the receiver to synchronize the phase of the reinserted subcarrier with that of the original subcarrier at the transmitter.

The output of the adder circuit is a composite signal called the *colorplexed video signal.* It contains all of the picture and timing information. This composite video signal amplitude is amplified and sent to the video modulator. The other circuits in the color transmitter are the same as those in a black-and-white transmitter.

## TV Receivers

A *TV receiver* contains a speaker, a cathode-ray or picture tube, and associated circuits as shown in Fig. 19-6, a block diagram for a monochrome TV receiver. Deflection circuits provide current to drive a set of coils mounted around the neck of the CRT. Changing magnetic fields deflect the electron beams produced in the tube so that they trace closely spaced horizontal lines on the phosphor coating inside the faceplate, as described in "U.S. National Television Systems Committee (NTSC) Standards," earlier in this section.

A color TV receiver contains all of the circuits that are found in a black-and-white TV receiver as well as additional circuitry for recovering the original red, green, and blue signals from the modulated carrier. Both color and black-and-white signals consist of an FM sound carrier and an AM video carrier within the 6-MHz band. The video portion of the color signal, like that of the black-and-white signal, consists of horizontal lines of picture information, with the individual lines followed by sync and blanking pulses.

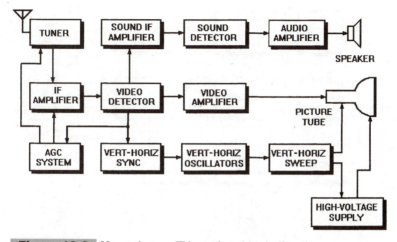

**Figure 19-6** Monochrome TV receiver block diagram.

The composite signal is processed by receiver circuits. Both the video and sound signals are amplified by an RF amplifier and converted to the receiver IF frequency by a mixer and local oscillator. The IF signal is then amplified by IF amplifiers and sent to the input of the detector. The detector demodulates the AM video signal and heterodynes the 4.5-MHz IF video carrier frequency with the FM sound signal.

The difference frequency produced by this heterodyning converts the audio signal to a new signal with a 4.5-MHz center frequency, as shown in the diagram Fig. 19-1. This is called *intercarrier detection.* The 4.5-MHz sound signal is amplified by an IF amplifier and sent to an FM audio detector. After demodulation by the detector, the audio is amplified to drive the speaker. The audio carrier of all TV signals is 4.5 MHz higher than the video carrier.

The video signal is demodulated when the 4.5-MHz audio signal is produced. The total video signal is then amplified by a video preamplifier stage. Then the luminance *Y,* chrominance *C,* and 3.58-MHz color sync bursts are separated by three filter networks, as shown in the simplified block diagram Fig. 19-7 for the color TV receiver circuits that reproduce the colors. The *Y* signal is then passed through the *Y-signal delay circuit* and amplified by the *Y luminance amplifier* before being sent to a matrix circuit. The delay circuit slows down the *Y* signal so that it reaches the matrix at the same time as the *C* signal, which requires additional processing.

The *C* signal is also separated from the video signal by a filter, amplified by the *C chrominance amplifier,* and then applied to both *I* and *Q demodulators* that extract the color information from the *C* signal. In addition to the *I* and *Q* signals, the *C* signal contains a 3.58-MHz *suppressed* carrier. This carrier must be reinserted to demodulate the signals. The 3.58-MHz carrier is generated by the *3.58-MHz oscillator,* and is kept in phase with the 3.58-MHz oscillator at the transmitter (Fig. 19-4) by the color sync burst signal, which is filtered out from the total video signal.

One output from the 3.58-MHz oscillator is sent directly to the *I* demodulator to demodulate the *I* signal. The carrier modulated with the *Q* signal at the transmitter was shifted 90° in phase before being sent to the modulator. Consequently, another output from the oscilla-

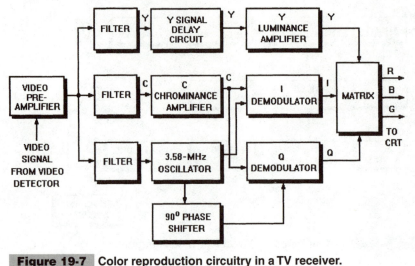

**Figure 19-7**  Color reproduction circuitry in a TV receiver.

tor is shifted 90° in phase by a 90° phase shifter before being applied to the $Q$ demodulator for demodulating the $Q$ signal. The demodulated $I$ and $Q$ signals are applied to the matrix along with the $Y$ signal. These three signals are combined in the proper proportions to reproduce the original red $R$, blue $B$, and green $G$ signals that are sent to the cathode-ray tube. The signals vary the intensity of the electron beams as they sweep across the CRT, and variations in brightness and color produce the picture.

# TV Camera Tubes

Most TV camera tubes have three principal parts: (1) *image section,* (2) *storage target,* and (3) *scan section.* A photoemissive surface in the image section and electron optics convert an optical image into an electron image, which is focused on the surface of the storage target to form a corresponding electric-charge image. The storage target stores the focused electric charge before readout and erasure by an electron beam generated in the scan section. The low-velocity electron beam repetitively scans the back surface of the target to generate a time-varying signal which is proportional to the magnitude of the spatial charge distribution. Simpler vidicon tubes, however, use a photoconductive target to perform the conversion function.

## VIDICON TUBES

The *vidicon tube,* as shown in section view Fig. 19-8, is a small, relatively simple TV camera tube with a photoconductive target. The image is optically focused by a lens on the faceplate which is coated on the inside with a transparent conductive film that forms the signal electrode. A thin layer of a photoconductive semiconductor film such as antimony trisulfide is deposited on the transparent electrode as the *target.* The electron beam from the electron gun is focused by a uniform magnetic field and swept by external deflecting coils so that it raster-scans the target.

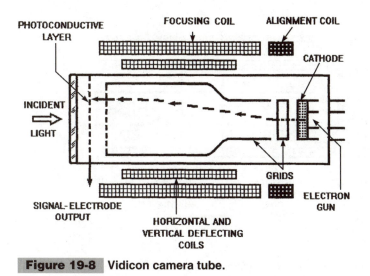

**Figure 19-8** Vidicon camera tube.

With about 20 VDC applied to the signal electrode or plate relative to the cathode, the scanned electron beam deposits electrons on the surface of the target, charging it to the cathode potential. When an image is focused on the target, its conductivity is increased in the illuminated areas, causing charge to flow. In these areas the scanned surface gradually becomes charged with respect to the cathode between successive scans. Electrons from the beam neutralize the charge accumulated, and generate a video signal in the signal electrode lead that is amplified. The charge-discharge cycle is completed when photoconduction through the target causes a positive-charge effect. The vidicon produces very little noise because it operates at a low current level. While its sensitivity is high, it exhibits some lag because of capacitive effects. It is a preferred camera tube where a small, light, low-power TV tube is required for remote, unmanned operation such as in closed-circuit surveillance systems. Increasingly, however, the CCD camera is taking over these duties.

## IMAGE ORTHICON TUBES

The *image orthicon TV camera tube,* shown in the section view Fig. 19-9, produces an electron image with its photoemitting surface. The image is focused on one side of a separate

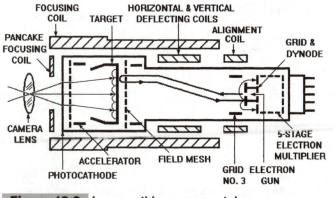

**Figure 19-9** Image orthicon camera tube.

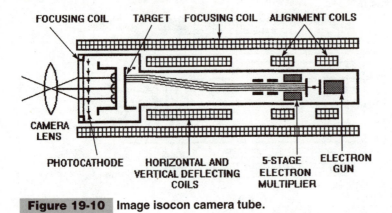

**Figure 19-10** Image isocon camera tube.

storage target that is scanned on its opposite side by an electron beam, usually consisting of low-velocity electrons. The output signal is generated by modulating the beam of electrons returned from the storage target. This return beam consists of scattered and reflected electrons, both influenced by their passage through electrostatic and magnetic fields. The image orthicon collects these electrons and both kinds are amplified by an electron multiplier. The sensitivity of this tube is high enough to pick up images in semidarkness. There are both color and monochrome image orthicons.

## IMAGE ISOCON TUBES

The *image isocon TV camera tube,* shown in Fig. 19-10, is an improved version of the image orthicon tube. It has special steering plates that separate the scattered and reflected electrons, permitting only scattered electrons to be amplified by the five-stage electron multiplier for signal generation. This improves the image isocons's signal-to-noise (SN) ratio. The scattered part of the return beam is proportional to the charges on the target. Several electrons are scattered for each electron that impacts the target to neutralize the positive charges. The higher the secondary emission, the greater the scatter gain.

# Charge-Coupled Device (CCD) Cameras

A *charge-coupled device* (CCD) *camera* is a TV camera that has a *CCD imager* as its image gathering component. The imager, as shown in Fig. 19-11, is an array of discrete photosensors on a silicon substrate that converts light energy into an electronic signal. A photocharge is generated on each sensor in the array that is proportional to the illumination incident on it. To produce a serial video signal from an array of thousands of sensors, their separate analog charge "packets" must be read out.

The CCD is a shift register that can shift groups of charge packets simultaneously in lockstep under clock control from one sensor to the next. This creates a two-dimensional

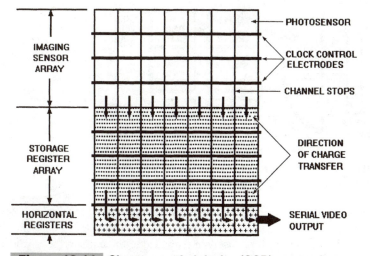

**Figure 19-11** Charge-coupled device (CCD) camera image transfer structure.

pattern which is transferred out of the imaging sensor array into a separate storage register array. A third array, the *horizontal register,* organizes a single serial train of analog charge packets. The process of moving charges out of the sensors is called the *transfer mechanism.* There are three transfer mechanisms: (1) *frame,* (2) *interline,* and (3) *frame-interline.*

Two separate arrays participate in the frame transfer, the *imaging sensor array* and the *storage register array.* The image is focused on the sensor array during the imaging period. The individual photosensors in the frame transfer structure accumulate photocharges that are proportional to the illumination of the optical image focused on them by a lens. At the end of the field-scan exposure, during the vertical blanking period, the photosensor array becomes a register. Under the control of a digital clock, the entire charge pattern is transferred rapidly in parallel through vertical CCD columns down to the field storage register, which is carefully shielded from the light.

The discharged image register is free to begin recharging for the next field. During this second field-scan time, the charge packets are transferred one row at a time out of the storage register into the horizontal serial register. The output of that register is the serial video output. When all rows of the storage register have been read out, the register accepts the next parallel transfer from the image register, and the operating cycle repeats. Both the interline transfer and the frame interline transfer mechanisms were invented to overcome image smearing under certain conditions. However, the explanation of the frame transfer mechanism is representative of all CCD imagers.

CCD TV cameras have replaced camera tubes in electronic news-gathering camcorders, and they have largely replaced camera tubes in TV broadcast studios. They offer higher dynamic range, elimination of lag, and complete removal of all highlight artifacts.

Solid-state imagers with from 300 to more than 700 horizontal elements and, with interlace, 500 vertical elements can be used in NTSC broadcasting. Color CCD TV cameras, as shown in the simplified block diagram Fig. 19-12, use three imagers; each has about 500 horizontal contiguous elements in a red-green-blue configuration. CCD cameras scan by

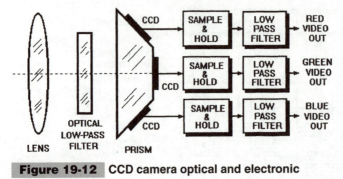

**Figure 19-12** CCD camera optical and electronic imaging elements.

the charge transfer process previously described. Transfer methods are used to assemble all of the array's discrete charges and format an output video signal.

The imager chips are mounted precisely and bonded to three prism faces. The interline transfer chip has gaps between sensor elements. The red and blue chips are each separately bonded to their respective optical ports so that their horizontal sensors lie precisely between those of the green horizontal sensors. The green CCD chip is spatially offset with respect to both the red and blue CCDs to ensure that each of the red, green, and blue signals are equally sampled by the imagers.

A CCD camera contains signal-processing and timing-control modules. The camera is connected by cable to a control unit consisting of a video output control, exposure control, video data rate control, and power supply. These cameras operate at voltages that are compatible with those needed by computers.

Less sophisticated CCD color cameras are in consumer-grade camcorders and digital cameras. They are also used in industrial computer vision systems that can automatically detect and classify nonconforming parts in process, measure the position, size, and shape of objects, and sort objects by size, shape, and color. Less expensive monochrome CCD cameras are used in some industrial vision systems as well as factory and office surveillance systems. See also "Computer Vision Systems" in Sec. 22, "Industrial Electronics Technology."

# Alternative TV Broadcast Standards

## PHASE-ALTERNATION LINE (PAL) SYSTEM

The *Phase-Alternation Line* (PAL) *system* is one of the two European standards that are nearly identical with NTSC except that improvements were made to produce better constancy of hue. It scans 625 lines per frame system at a rate of 25 frames (50 fields) per second. Phase information is reversed in the process of scanning, so that information on hue can be corrected in successive lines. Hue and saturation information are carried by quadrature modulation, but with one of the two modulations switched 180° from line to line at the transmitter. A delay line in the receiver restores the correct phase of the two modulations by delaying one modulation for the duration of the line. Its variations from the other standards can be traced to local differences in earlier black-and-white TV transmission standards.

PAL was developed in Germany, and many other European countries have adopted it, including Austria, Belgium, Brazil, Denmark, Finland, the Netherlands, Sweden, Switzerland, and Great Britain. It has also been adopted by Australia and some South American and African countries.

## SEQUENTIAL COLOR WITH MEMORY (SECAM) SYSTEM

The *Sequentiel Couleur Avec Memoire* (SECAM) *system* is one of the two European standards that is essentially identical with NTSC except for improvements made to produce better constancy of hue. This system scans 625 lines per frame at a rate of 25 frames (50 fields) per second. Alternate lines carry information on luminance plus blue; green is derived in the receiver by subtracting the red and blue information from the luminance signal.

SECAM has been adopted by France, Russia, and other Eastern European and African countries. It was also developed to make color TV fully compatible with monochrome (black-and-white) TV in those countries where it existed. The variations in SECAM are attributable to local differences in earlier monochrome TV transmission standards.

# Cable Television (CATV)

*Cable television* began as *community antenna television* (CATV) in the late 1940s to provide stronger signals from distant very high frequency (VHF) TV broadcast stations. The 12 TV channels at that time were assigned 6-MHz bands in the frequency spectrum between 54 and 216 MHz. A typical CATV installation consisted of an antenna aimed at distant transmitters from a tower located hundreds of feet above the outskirts of the community it served.

The distribution system consisted of a *head-end,* including the antenna and control station, and a coaxial-cable distribution system. A coaxial trunk cable carried the composite signal from the antenna to residential areas where customers' homes were located. Amplifiers along the trunk boosted the signal. A coaxial feeder cable from a trunk amplifier carried the signal through the neighborhoods, and customers' homes were connected by coaxial drop cables from a *multitap* on the feeder cable to the sides of their homes. Customer TV sets were then connected by coaxial cable within the home.

By the 1970s it had become economical for CATV providers to receive signals from satellites, so local head-ends were connected to distant master head-ends to form *supertrunks.* A simplified diagram of a typical system is shown in Fig. 19-13. The transmission of the TV composite signal through single-mode fiberoptic trunk cable with lasers began in the 1980s. Commercial laser transmitters and optical receivers can carry up to 110 channels in the 700-MHz bandwidth from 50 to 750 MHz on a single fiber. Coaxial feeder cable was connected to remote optical nodes on the fiberoptic cable containing receivers with PIN photodiode detectors. Up to four feeder cables can be fed by the amplifiers in the remote node. The distribution system remained the same beyond the node. CATV feeder cables now pass near more than 90 percent of all homes in the United States, and serve more than 60 percent of them.

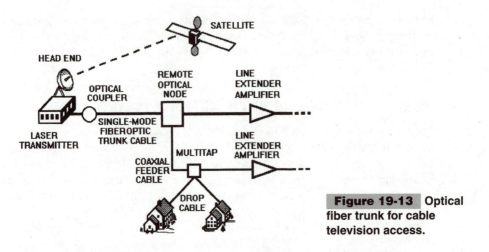

**Figure 19-13** Optical fiber trunk for cable television access.

In 1972, the FCC ordered all CATV systems serving more than 500 subscribers to make their distribution systems capable of two-way transmission. A band of frequencies from 5 to 35 MHz was set aside for this purpose, but few of the providers have made use of this band so far. Cable companies have completed the installation of digital cables that can serve 3.25 million cable subscribers nationwide. Cable lines now pass more than 90 million American homes, and in 1998, 65 million of them subscribed to cable services.

# High-Definition Television (HDTV)

*High definition television* (HDTV) is a new generation of color digital TV broadcasting offering higher definition or sharper pictures and better sound than is obtainable from any of the current three major worldwide television standards, NTSC, PAL and SECAM. It is based on the Motion Picture Experts Group (MPEG-2) video compression and transport protocol, multiple transmission formats, six-channel Dolby AC-3 digital audio surround sound, and vestigial sideband digital modulation.

The HDTV picture has a maximum of 2,073,600 picture elements (pixels), about 6 times the maximum of 337,920 pixels in the present NTSC image format. Progressive scanning will provide a temporal resolution of 60 frames per second, twice the 29.97 frames per second of standard NTSC images. The picture will be presented in a wider horizontal-to-vertical *aspect ratio* of 16:9 compared with the 4:3 aspect ratio of existing NTSC receivers. Because of its wider aspect ratio and extra features such as stereophonic sound, HDTV requires the transmission of at least 5 times as much information as existing worldwide television standards. Within each 6-MHz channel bandwidth, the HDTV system will deliver a digital channel of approximately 20 Mbps transmitted by digital vestigial sideband modulation.

The U.S. government has defined HDTV as a system that provides wide-screen pictures with twice the resolution of ordinary TV and with sound at least equal to the quality of that obtained from compact disks (CDs). An important criterion in the selection of an HDTV standard is its compatibility with the existing NTSC receivers. The U.S. Federal Communications Commission (FCC) had previously insisted that NSTC color TV broadcasting be compatible with black-and-white TV broadcasting.

Television stations are free to choose from 18 different formats for digital TV broadcasts, but four have emerged as the most likely digital formats to be broadcast. Every digital TV set will be able to receive any of the four.

1. *1080-I (Interlaced),* the most commonly used high-definition format, will have 1080 horizontal lines of resolution that will be interlaced with 1920 vertical lines. This compares with the 480 horizontal by 704 vertical lines for the NTSC format.
2. *1080-P (Progressive),* the highest-resolution format, is beyond the capability of existing TV channels. The 1080-horizontal-line signal is sent progressively without interlacing.
3. *720-P (Progressive),* the highest-level format that is compatible with all TV channels, will have 720 horizontal lines broadcast progressively. It is most competitive with 1080-I.
4. *480-P (Progressive),* considered to be standard definition, is a progressive 480-horizontal-line digital format. The NTSC 480-I format has the same number of lines, but they are interlaced. It offers slightly more than 304,000 pixels.

Digital TV receiver manufacturers plan to offer high-, medium-, and standard-definition receivers. Some sets will include circuitry to receive display formats that broadcasters prefer; others will receive formats favored by the computer industry; and still others will be able to receive advanced digital services that have yet to be defined.

# Set-Top Boxes

A *set-top box* or *converter* is a separate, stand-alone unit that acts as an intermediary between the cable and the cable subscriber's TV receiver or videocassette recorder (VCR) to provide special services on request. It can decode cable programs that were encoded by the cable service provider, and it can also compensate for inadequate tuning or extend the tuning range of the TV set. It might also permit interaction with the program or permit direct selection of pay-per-view cable programming with a keypad.

A *digital set-top converter* is a set-top box specifically intended for decompressing and decoding digital TV signals. Some versions are expected to include keypads that will permit the consumer to request pay-per-view movies, respond to questionnaires, or participate in live cable programs in progress. The new generations of set-top boxes will require more powerful microprocessors and will, in effect, be computers that depend on powerful computer operating system software. It is expected that they will be able to process high-speed digital signals moving to and from the home, making possible new and improved TV-based interactive services such as high-speed Internet access, E-mail, pay-per-view movies, dial-up electronic shopping, advertising directed to specific homes and interest groups, interactive games, and eventually high-definition video.

Some digital decoder set-top boxes are expected to be able to translate digital high- or standard-definition signals into analog signals for existing NTSC sets. HDTV provides an excellent picture, but its transmission consumes an entire channel, so many broadcasters do not want to use up their assigned channels for a single HDTV broadcast. Standard-definition digital offers a picture with the same resolution as NTSC TV but still leaves enough of the assigned channel for broadcasting other programs or providing other services. To receive an HDTV signal, at least 10 MB of digital memory will be needed at the receiver. A single decoder can decode both forms of transmission by sharing internal circuitry. The 8-bit data bytes that represent pixels can be reduced to 7, 6, or even 5 bits.

# Satellite Master Antenna Television (SMATV)

*Satellite master antenna television* (SMATV) is a satellite TV reception service restricted to subscribers living in a single building or a closely spaced group of buildings such as a college campus, industrial park, or resort. The subscriber leases a set-top box from the provider, who installs a common parabolic antenna, typically with a diameter of 6 ft (1.83 m), that provides reception for all subscribers in the cable system. The antenna is located high up on a building in a city or urban location that might be surrounded by other tall buildings. Received programs are sent by cable to customers in adjacent buildings.

# Digital Satellite Systems

A *digital satellite system* (DSS) is an all-digital direct satellite broadcast system that transmits high-quality TV signals to privately owned receivers, each equipped with an 18-in-(46-cm-) diameter paraboloidal antenna. Circuitry in the subscriber's receiver decodes the scrambled signals. Up to 175 channels can be accessed with this system. The antenna must be mounted outdoors or in a protective RF-transparent shelter and pointed at the satellite. There are several different systems now in operation that transmit on different frequencies.

# 20

# TELECOMMUNICATIONS
# TECHNOLOGY

## CONTENTS AT A GLANCE

# Overview

The term *telecommunications* generally refers to two-way transmission and reception of intelligent signals by wire and cable and through the air, as contrasted with the one-way transmission of radio and television broadcasting. It encompasses the transmission of voice, data, video, and facsimile (fax) signals from point to point over a frequency span from audio to microwaves. However, its primary focus is still on telephone communications that are carried by wire, coaxial, fiberoptic, and undersea cable, as well as terrestrial microwave and satellite relays, and cellular mobile telephone links.

The merging of computers and telephone networks has dissolved the boundary between the two technologies. The computer and telephone-line modem are vital for such services as fax, accessing the Internet, and for sending and receiving E-mail, but they are dependent on wire or cable links. Both the cordless and cellular mobile telephones depend on radio-frequency links, as do the pagers and handheld computers called *personal communications devices* (PCDs).

New concepts in telecommunications promise to keep the whole industry in a state of flux for many years to come. For example, fleets of commercial communications satellites have been lofted into earth orbit so that telephone calls can be made between any two locations on earth. Where alternative land lines exist, the lower-cost terrestrial links can be selected. The new low-altitude Iridium and Globalstar satellites are now competing for global telephone service business with the geosynchronous Inmarsat satellites.

The purchase of cable television companies by prominent long-distance multinational telephone service providers companies is seen as an indication of dramatic future changes in telecommunications. In the short run, long-distance telephone companies will gain access to local telephone business, and local telephone will gain access to long-distance service. However, the future of conventional telephone by wire seems to be in doubt, as the merged telephone and cable companies focus on the coaxial and fiberoptic cable that is expected to provide such services as high-speed Internet access, interactive TV, E-mail, and both local and long-distance telephone service.

The morphing of home computers into TV receivers with video card add-ons and the development of set-top boxes that will convert TV receivers into interactive and Internet terminals are clear evidence of a crossover between the two technologies. Another trend is the emerging market for low-priced home computers intended primarily for entertainment and surfing the Internet. That market has split off from the mainstream personal computer market with its seemingly insatiable demand for faster and more powerful microprocessors to carry out the traditional computer applications of data processing, accounting, word processing, and desktop publication. But even these higher-level computers will be linked by high-speed cables to the Internet. As if the situation in contemporary telecommunications was not confusing enough, some public electric utilities have announced plans for sending telephone signals over their power lines and grids.

# Communications

*Communications* is the transfer of information from point to point in the form of voice, digital data, video, and facsimile. *Service networks* are communications facilities that include *terminals* for the transmission and reception of information to and from the network. *Transmission facilities* transfer information from place to place, and *switching facilities* connect transmission facilities so that the information can be received at the intended location.

All communications can be classed as either analog or digital. In digital communication, binary digits or bits (ones and zeros) represent alphanumeric characters and the data stream of bits is generated and interpreted by a compatible terminal. The number of bits per character and the relation between the bit sequence and the character is the *code*.

The *American Standard Code for Information Interchange* (ASCII) is the most commonly used code for data transmission. It is based on 7 bits per character, plus a parity bit. (An 8-bit ASCII code is also available.)

*Protocols* are standard communications procedures that make both transmission facilities and receiving equipment compatible; they are analogous to spoken languages.

The *interface* is the boundary of the service network where the user interacts with the network using some kind of terminal. It is often needed to make the binary serial data compatible with the communications channel.

The functions of common data communications terminals and equipment are as follows:

- *Cluster controllers* interface computers or peripherals to communications lines.
- *Front-end processors* control communications lines for a host computer.
- *Modems* are bimodal circuits for converting digital data to a modulated analog form for transmission over telephone lines and reconversion of those signals back to digital data at the receiver.
- *Personal computers* can act as intelligent data terminals when connected in private networks or to the public switched-telephone network (PSTN).
- *Remote data concentrators* (RCDs) buffer messages from slow-speed lines and multiplex them over high-speed lines.
- *Time division multiplexers* (TDMs) combine two or more slow-speed lines into one high-speed line by time-sharing.
- *Video display terminals* (VDTs) provide access to a central computer for the exchange of information in such places as airline reservations desks, travel agencies, and securities offices.
- *Voice data private branch exchanges* (PBXs) manage large numbers of telephone and other communications equipment in office buildings and factories.

# International Transmission Standards

International transmission standards for data communications are identified by alphanumeric codes beginning with the letter V.

- *V.34* is the worldwide communications standard for modems operating at 28.8 kb/s or faster. Under this standard the modem can operate as slowly as 2400 b/s, and adjustment of speed for different line conditions is permitted.
- *V.42 bis* is the international standard for error correcting capabilities in modems.
- *V.90* is the new standard for 56-kb/s modem communication.

# Data Communications Channels

A *data channel* or *communications link* is a path for data transmission between two or more stations or terminals. The path could be a single wire, a twisted pair of wires, a parallel group of wires, a coaxial cable, or a fiberoptic cable. All channels have limitations on information-transmission capabilities imposed by their electrical and physical characteristics. The data-carrying capacity of the channel is called its *bandwidth.*

There are three basic types of data channels: (1) *simplex,* (2) *half-duplex,* and (3) *full-duplex.* These are similar to the transmission modes available in mobile, marine, and citizens band radio and telephony. The *simplex channel* permits the transmission of information or data in one direction only—from the sender to the receiver.

Time-shared transmission between two sender/receivers occurs over a *half-duplex channel.* Simultaneous transmission is not permitted. If a two-wire circuit is used, the line must be turned around to reverse the direction of transmission. The public telephone network is fundamentally half-duplex because the local loop is a two-wire circuit.

Simultaneous transmission between two sender/receivers can take place on a *full-duplex channel.* Both sender/receivers can "converse." A two-wire circuit permits full-duplex communications if different frequencies are assigned for the send and receive channels. However, four-wire circuits are most frequently used. In the United States, communications companies or common carriers offer both two- and four-wire channels.

The two-wire channel can be used in a simplex mode if terminals restrict the direction of transmission. In other words, a *transmit-only terminal* is connected to a *receive-only terminal.* A two-wire line can also be used in a half-duplex mode if line turnaround is initiated by the modem. Modems divide the two-wire public telephone network bandwidth into send and receive subchannels to create a full-duplex channel.

A four-wire channel can be used for full-duplex communications because it has separate wires for transmission and reception. It can also serve for half-duplex communication without line turnaround.

When messages are to be transmitted between a remote terminal and a computer, a series of signals must be exchanged to prepare the message. *Protocols* are predetermined signals that control the flow of messages and synchronize their transmission. The exchange of protocols is termed *handshaking.*

A data communication channel can be specified in terms of: (1) bandwidth, (2) private versus switched access, (3) propagation delay, (4) line configuration, (5) use of protocols, (6) availability, and (7) installation time and cost.

A line configuration can be *point to point, multipoint,* or *loop.*

# Channel Bandwidth

*Bandwidth* is a measure of the ability of a channel to carry information. The wider the bandwidth of the assigned channel, the higher the speed of transmission can be. Speed is measured in *bauds* or the number of line signal elements or symbols per second. If the signal element represents one of two binary states, the bauds are equal to the *bit rate.* When more than two states are represented, as in multilevel or multiphase modulation, the bit rate exceeds the bauds.

The bandwidth of a voice-frequency channel is 4 kHz, but only the frequencies from about 200 to 3500 Hz are usable for analog transmission. The actual data or bit rate depends on how many bits are encoded in each signal element. The signal element depends on the transmission used (analog or digital) and the coding scheme or modulation method.

# Analog Transmission

In *analog transmission* over a telephone line, a continuous range of frequencies and amplitudes is sent over a communications channel that accepts both voice and data. Linear amplifiers, attenuators, filters, and transformers are needed to maintain the signal quality. However, amplifiers increase the noise as well as the information content of a signal, so error rates are higher in analog transmission than in digital transmission.

The capacitance, inductance, and resistance characteristics of the telephone line delay and attenuate signals with different frequencies. These alterations of information content are called *envelope delay* and *attenuation distortion*.

Analog transmission systems require linear amplifiers and filters at fixed distances to boost the signal and filter out noise. The spacing of this equipment depends on the end-to-end media. For example, amplifiers are spaced about every 6000 ft (1830 m) for the twisted-wire pairs used in telephony. *Modems* are required for the transmission of digital data over the existing analog telephone lines.

# Digital Transmission

*Digital transmission* systems send data in the form of pulses over a communications channel at rates that are dependent on the digital carrier. Digitized data, voice, and video can be transmitted in digital format if sufficient bandwidth is available. Digital systems use regenerative repeaters to retime and reshape the digital pulses. This equipment recreates the original waveform more reliably than the linear amplifiers in analog systems and, as a result, there are fewer transmission errors per message. Error rates in digital transmission are typically 1 or 2 percent of those in analog transmission.

Different kinds of digitized information such as voice, text, computer files, E-mail, data, and fax can be mixed in the same digital system. In addition, digital data can be more effectively encrypted for security, and it can be compressed to save bandwidth.

The world's three major digital wireless standards are:

1. Time division multiple access (TDMA)
2. Code division multiple access (CDMA)
3. Global System for Mobile Communications (GSM)

TDMA permits transmission system capacity to be increased and makes it easier to transmit data over fiberoptic cables and microwave links. CDMA permits each transmission to be spread over the entire available band of frequency. CDMA messages are encoded with prearranged binary pulse trains that are decoded at the receiving end. The GSM standard combines both FDMA and CDMA technologies. GSM is available in 110 countries as GSM 900 or GSM 1900; both operate at higher frequencies than TDMA and CDMA.

The *Nyquist theorem* is a rule applied in signal processing for sampling an analog signal frequency to minimize the effect of high-frequency noise on its digital conversion. The theorem states that an analog frequency (called the *Nyquist frequency*) should be sampled at a rate (called the *Nyquist rate*) that is at least twice the highest frequency found in the signal. For example, a voice-grade channel with a 4-kHz bandwidth must be sampled at a minimum of 8 kHz to obtain an accurate digital representation of the signal's information content.

If unexpectedly high frequencies occur within the signal or noise accompanying the signal on the input channel, an effect called *aliasing* occurs. This shows up as spurious low-frequency signals caused by an intermodulation of high-frequency signal components (and input noise) with harmonics of the sampling frequency. In communications, aliasing is unwanted, but aliasing is intentionally used in cathode-ray tube displays to blend the jagged edges between regions of contrasting color or tone.

Because the intended transmission frequencies are known, the Nyquist rate can be set, but it might not be feasible to sample the data fast enough to prevent aliasing in all situations. Low-pass filters called *antialiasing filters* can attenuate these unwanted high frequencies.

# Channel Configuration

Data channels can be configured as single point-to-point links, a point-to-point network, a multipoint line, or a loop or ring. Two stations on a point-to-point line can exchange data after the connection has been made. A point-to-point network includes many point-to-point links between communications controllers and remote terminals.

A *multipoint line* requires either a *poll-select protocol* or dedicated frequencies for remote stations to regulate access to the shared inbound channel to the central computer. The loop or ring is assembled on the customer's premises with private wiring. This could be twisted-pair, coaxial, or fiberoptic cable. The loop has a master control station with a poll-select protocol that permits it to communicate with all of the secondary or slave stations. All stations in the ring can communicate with each other.

*Voice-grade* lines are telephone lines available by dialing the public switched telephone network (PSTN), and both conditioned and unconditioned private lines can be leased from providers. The usable bandwidth of all three lines is about 3.2 kHz, but the effective data rates differ. Each has a different specification for signal noise, amplitude attenuation, and envelope delay distortion.

*Dial-up lines* are two-wire pairs available from the common carriers on the PSTN. They permit one telephone to reach any point on the worldwide telephone network. The modem determines if these lines are used for half-duplex or full-duplex operation. They are organized as point-to-point links. Calls can be placed manually, by an operator, or with modems organized for autocall and autoanswer, permitting unattended operation.

There are many advantages to the PSTN but there are also disadvantages. A user can interpret weak voices over a noisy telephone line, request that a message be repeated, or request an alternate line if information is lost or not understood. However, computers or terminals cannot make decisions about the quality of digital information being transmitted, so they can easily lose or misinterpret data because of noise. A second disadvantage is *delay distortion* caused by the differing transmission speeds of the various frequency components of a signal being transmitted. This too can result in errors in the received data. A third disadvantage is time lost in the PSTN due to connect, disconnect, and turnaround times that limit the amount of data transmitted.

Private leased lines have advantages over the PSTN, including ready availability and freedom from busy signals. They also provide point-to-point or multipoint operation and can be conditioned for better data quality and higher transmission rates. Leased lines are generally four-wire circuits usable for half- or full-duplex operation. There is no line turnaround on a four-wire circuit, and simultaneous transmission and receiving is possible. Data integrity on unconditioned leased lines can be 10 times better than that on the PSTN. Microprocessor-based modems with automatic adaptive equalizers compensate for line impairments and greatly decrease errors at higher speeds.

The drawbacks to leased lines are their expense and location limitations, but they are cost-effective for users with demands for high-volume, high-quality telecommunications.

# Data Communications Line Sharing

Data communications lines are shared to reduce costs, improve reliability, and simplify maintenance. Communications lines and modems can be shared with equipment called *multiplexers, cluster controllers, remote data concentrators, modem sharing units, port sharing units, port selectors,* and *lineplexers.*

A multiplexer or MUX is a circuit that can transmit two or more messages on a single communications channel by dividing the bandwidth into frequency or time slots. Typically, there is one MUX connected to the serial ports of a central computer and another MUX at the remote site connected to each of the terminals connected to the computer. The MUX should have no effect on data sent between the computer and the remote stations, and it should have minimal effect on response time. The basic types of multiplexers are as follows:

- A *frequency-division multiplexer* (FDM) divides the available frequency bandwidth into narrower bands, each of which is used as a separate channel. FDM is used for low-speed synchronous full-duplex leased-line transmission.
- A *time-division multiplexer* (TDM) connects terminals, one at a time, at regular intervals, to the entire communications bandwidth. A TDM is usually used on T1-carrier systems at 1.544 Mb/s, although it can be used for synchronous or asynchronous full-duplex leased line transmission.

*Time-division multiple access* (TDMA) is a U.S. standard scheme for time sharing digital cellular telephone and satellite communication channels among multiple users assigned to specific transmission *time slots.* Digitized voice and data signals are compressed, stored, and multiplexed into as many time slots as there are voice channels. The time slots are then transmitted in a period of time called a *frame.* Each time slot is synchronized to the frame to avoid collisions. This is the basis for the term *synchronous transfer mode* (STM). Receivers extract the user's data bursts from the time slots and demodulate them in milliseconds.

*Code-division multiple access* (CDMA) is a broadcast system in which each transmitter spreads its modulated signal over the entire bandwidth assigned to all common carriers. The calls are encoded with prearranged binary pulse sequences programmed into both transmitter and receiver. Only the intended recipient can separate out the voice message from the background noise. CDMA increases communications capability 10 to 15 times the capacity of analog transmissions. (For more on CDMA, see "Code-Division Multiple Access (CDMA)," following in this section.)

*Statistical multiplexers* dynamically allocate time slots to attached devices based on their activity. By making use of idle time, more data streams can share a common communications line. Typically, four to eight terminals or computer ports are connected to a statistical multiplexer, which uses synchronous full-duplex leased-line transmission.

A *cluster controller* manages and directs messages to and from the connected remote devices when it receives poll and select commands from the central computer or front-end processor. The system devices include video display terminals (VDTs), personal computers (PCs), and printers. They are polled independently by the cluster controller for data to be sent to the central computer.

A *remote data concentrator* (RDC) is a communications processor such as a front-end processor (FEP) that is positioned at a remote site. The RDC functions are similar to those of the statistical multiplexer, but it is not transparent to the data flow. Multiplexers can be used in any computer system, but RDCs are made by computer manufacturers to be compatible with their product lines.

A *modem-sharing unit* (MSU) is a device that permits two to six synchronous terminals to share a synchronous modem. The terminals must be polled by the central computer so that the polled terminal recognizes its address and activates its request to send data. The MSU releases the modem to the first terminal that initiates a *request-to-send signal;* all other terminals are locked out.

A *port-sharing unit* (PSU) is similar to an MSU, except that it has its own timing source for transmit and receive clocks. A PSU connects from two to six polled synchronous terminals to a single computer port. The terminals can be local and directly connected or remote and connected through a modem. The PSU reduces the number of computer ports in a polled network.

A *port selector* (PS) or data PBX, also called a *port contention unit,* reduces the number of computer ports in a nonpolled network. The PS is used where terminals must contend for access to ports on the host computer. The PS allocates computer ports to inbound devices or communications channels on a first-come, first-served basis, where some devices or channels have priority over others.

A *lineplexer* splits a computer port into two or more communications channels. For example, it can split 19.2 kb/s, 16.8 kb/s, or 14.4 kb/s data streams into two 4-kHz bandwidth channels for leased lines or digital data services.

# Data Serial Transmission

*Data serial transmission* is the transmission of data single-file over a single conductor to minimize wiring and connector cost and space while simplifying data timing. *Parallel transmission* is the transmission of data in parallel files over parallel conductors where distances are short (typically less than 100 ft [30.5 m]) and high-speed operation is required.

The two basic data serial transmission data formats are *asynchronous* and *synchronous*. *Isochronous* transmission combines both asynchronous and synchronous methods.

*Asynchronous data transmission* is a method in which the time intervals between each transmitted character can be unequal in length. Each character is framed by 1 start bit and 1 to 2 stop bits. The transmission of a character begins when the line makes a transition from the "1" state to the "0" state. This transition is the reference that determines the character's bit cell boundaries. The receiver must detect this transition with a fast clock so that it can sample each bit in the center of the bit cell. Because characters can be sent at irregular intervals, this method is accepted for low-speed terminals where data entry is intermittent.

*Synchronous data transmission* is a method in which characters are transmitted contiguously without start and stop bits. The sending and receiving devices are synchronized by exchanging a predetermined set of synchronization signals, either periodically or before the transmission of each message. The sending device transmits a long stream of characters without start and stop bits. The receiver counts off the first 8 bits (ASCII code), assumes this to be the character, and passes it on to the computer. It then counts off the next characters until the message is completed.

Asynchronous transmission is favored for slow-speed, manually operated keyboard terminals because it permits characters to be transmitted at irregular intervals. Its disadvantage is that transmission is less efficient because each character must have start and stop bits. For example, with ASCII code, each 8-bit character of information requires a total of 11 bits for transmission. As a result, efficiency is 8/11 or about 73 percent.

*Isochronous data transmission* combines the elements of both asynchronous and synchronous techniques, although it really is a special case of synchronous transmission. Each character has start and stop bits so that characters can be transmitted at irregular intervals. Synchronous modems can be used, but there can be gaps in transmission because of the start and stop bits. The transmitter and receiver are synchronized during data transmission. This mode is faster than the asynchronous mode. Isochronous transmission permits speeds up to 9.6 kb/s without the need for large memory buffering of data, although synchronous transmission can even be faster. However, asynchronous transmission is limited to about 1.8 kb/s.

# Packet Networks

*Packet networks* were developed to reduce idle time in the transmission of data and make more efficient use of the transmission facilities. They also provide an equitable basis for charging users only for the data they actually send. A message sent from one terminal to another in a packet network is divided into *packets* of some definite byte length, typically from 53 to 128 bytes. Each packet contains a *header* that provides the network with a destination address. Thus packets can be sent individually from the source to their destinations. Packets from more than one terminal can be interspersed, each with its own address. The messages are reassembled and formatted at their destination switching centers before they are delivered. See also "Broadband ISDN (B-ISDN) and Asynchronous Transfer Mode (ATM)," following in this section.

# Synchronous/Isochronous Encoding

Synchronous data is often encoded to ensure that enough transitions exist in the data stream for the *phase-locked loop* (PLL) circuit in the modem or terminal to extract the receive clock from the received data. Encoding embeds the transmit clock in the data, while decoding extracts the receive clock from the data. The transmission encoding methods include:

- *Nonreturn-to-zero* (NRZ) has 1 representing a high-voltage level, 0 representing a low-voltage level, and bit value does not return to 0 voltage in the middle of the bit cell.
- *Nonreturn-to-zero inverted* (NRZI) inverts the binary signal state on a 0 of the message data and leaves it unchanged on a 1 of the message data.
- *Manchester Code* is a method for encoding clock and data bit information into bit symbols. Each bit symbol is divided into two halves, and the polarity of the second half is the inverse of the first half. A 0 bit is represented as a low polarity during the first half of the symbol, followed by a high polarity during the second half. A 1 bit is represented as a high polarity during the first half of the symbol and a low polarity during the second half. This encoding method is dependent on polarity.

*Data transmission rate* is usually stated in bits per second (b/s). This rate can be translated into characters per second by dividing it by the number of bits per character. For example, if a telephone line has a rated capacity of 4.8 bps, it is equivalent to 600 characters per second (4800/8) for an ASCII code with synchronous transmission.

A *baud* (Bd) is a unit of signaling speed or modulation rate that is given in signal elements per second or symbols per second. It is named for the nineteenth-century French communications engineer J. Baudot. The term *baud rate* is redundant because baud is a rate. Signaling speed in bauds is equal to the reciprocal of signal element length in seconds. A pulse and space are separate elements, so a printer producing 25 pulses per second is operating at 50 Bd. Baud typically has a value that is less than bits per second.

# T1-Carrier

The *standard T1-carrier* is a telephone service introduced in 1962 that permits digital transmission at 1.544 Mb/s to carry 24 voice channels with time-division multiplexing (TDM). This multiplexing arrangement is called a *digital loop carrier.* Pulse-code modulation (PCM) is used to code 8 bits per signal element for each voice channel sampled in 256 discrete amplitudes. The 4-kHz voice channel is sampled at the Nyquist rate (8 kb/s) to produce a channel data rate of 64 kb/s. An analog signal that is to be digitized must be sampled at twice its frequency to reconstruct the original waveform with no distortion. CCITT has set a standard sampling rate of 8000 samples per second for the 300- to 3400-Hz voice channel. A sample is taken every 125 μs.

The pulses are transmitted over copper wire pairs or microwave links. Repeaters are placed at 6000-ft (1830-m) intervals to regenerate and retime the digital waveform. The T1-carrier was designed to transmit voice signals between central offices that are less than 50 mi (80 km) apart. The T1-carrier system is the multiplexing structure for PCM systems in North America.

# Narrowband ISDN

An *Integrated Services Digital Network* (ISDN) uses a single digital transmission network to provide a wide variety of services such as voice, text, fax, videotext, and video, both switched and nonswitched. It can transmit data in its 128-kb/s bandwidth over telephone lines with special communications hardware and software. ISDN is offered by telephone service providers for a fee. It was developed to transmit both voice and data over the same lines at higher speeds than are possible with the existing public switched-telephone network (PSTN). The concept is based on a set of international standards issued in the 1980s.

Two versions of ISDN are in use today: a *basic-rate interface* that employs a two-wire connection from the telephone company's central office with a bandwidth capacity of 144 kb/s in each direction, and a *primary-rate interface* that employs a four-wire connection. The four wires can deliver 1.544 to 2.048 Mb/s. Primary rate is intended for private branch exchanges (PBXs) or high-end networking equipment, while basic rate is available to the general public.

The 144 kb/s of the basic rate interface is divided into three channels: two bearer (B) channels at 64 kb/s each, to match the rate at which all analog voice signals are digitized, and one delta (D) channel at 16 kb/s. The connection is full duplex, with data being sent over each channel simultaneously. Each B channel can carry the equivalent of an analog telephone call while the D channel transports signaling (such as call-setup and call-progress information). For data transport, the full 64 kb/s of each B channel can be made available, separately or together, to attached computer equipment, while packet data (at up to 9600 b/s) can share the D channel with signaling information.

The basic rate interface for the B and D channels is time-division multiplexed at 192 kb/s. The primary rate interface has two formats: the U.S. T1 digital standard and the European E1 digital standard. The T1-based primary rate interface includes 23 B channels and one D channel, and the E1 standard has 30 B channels and one D channel. The bandwidth of each B or D channel is 64 kb/s.

A data link connection is established in ISDN between a terminal and a central office switch on the D channel, and call-control signals can be exchanged at any time without concern for B-channel activity. Call-control information includes the called and calling numbers, bandwidth requirements, B-channel assignment, service request (voice or data, circuit or packet), and end-to-end protocol. Other services can also be requested.

The original ISDN concept has been extended to broadband networks, such as Synchronous Optical Network (SONET), at a speed of 155 Mb/s or higher and was named broadband ISDN (B-ISDN) to contrast it with the original narrow bandwidth ISDN (N-ISDN).

N-ISDN uses synchronous time-division multiplexing (TDM) for the two 64 kb/s B channels, while faster B-ISDN uses an asynchronous fast packet switching technique called *asynchronous transfer mode* (ATM). N-ISDN made a more significant change in the public telephone network than many earlier innovations including automatic switching, tone dialing, and the supply of power from the central office. While provisions have been made for digital transmission and intelligent network design, improvements have favored voice transmission. N-ISDN offers complete sender-to-receiver digital transmission by providing digital access on the local subscriber's loop from the customer's location to the central office. It also provides an integrated service by carrying voice and data over a single line.

The public telephone network has gradually transformed itself from analog to digital operation. Most central offices now have digital switching, and trunk lines are organized for digital transmission. However, the local subscribers' loop remains analog.

N-ISDN permits multiple connections and services over the same line. In addition to providing high-speed connections at 64 kb/s, not possible with modems, it allows multiple connections to be synchronized. Thus telephone lines can be shared rather than being dedicated for specific purposes. ISDN users can transmit their traffic in circuit or packet modes.

*Circuit-mode data service* provides an unrestricted channel so that the bit stream is transmitted unchanged from source to destination. This service is controlled by user equipment at each end, and it can transmit voice or data with such protocols as X.25 or frame relay.

*Circuit-mode speech and voiceband services* carry voice signals in the same way as the public switched-telephone network. The bit stream consists of analog signals that have been digitized by appropriate encoding standards, and it can be converted when passing through networks that use different encoding standards. The 3.1-kHz audio voiceband service is the same as the one used by modems.

*Packet-mode data service* gives customers access to the X.25 packet network service offered by the network to which the terminal is directly connected. It is available only on the basic-rate B and D channels.

# Broadband ISDN (B-ISDN) and Asynchronous Transfer Mode (ATM)

*Broadband ISDN* (B-ISDN) is an extension of narrowband ISDN (N-ISDN). It is the first technology to integrate voice and data communications into a common format that is efficient for both. The major feature of B-ISDN is its use of the asynchronous transfer mode (ATM), so B-ISDN and ATM are synonymous. ATM is the first technology to provide a common format for bursts of high-speed data and the ebb and flow of voice transmissions. In addition, the ATM cell format is suitable for all local-area networks (LANs) and wide-area networks (WANs) and it can interconnect them with even larger networks.

ATM is one of a general class of digital packet-switching technologies that relay and route traffic with an address that is contained within the packet. It is based on asynchronous time-division multiplexing (TDM) for carrying information in very short, fixed-length packets called *cells*. This contrasts with other packet technologies, such as X.25 or frame relay, which use long, variable-length packets.

ATM cells are 53 bytes long, consisting of a 5-byte header (containing the address and codes for identification and error checking) and a 48-byte information field. This contrasts with frame relay, which uses a 2- or 4-byte header and a variable-length information field. It is called *frame relay* because it transfers, or *relays,* frames of user data. These can range in length from 64 to more than 1500 bytes. ATM data channels are identified by *virtual path/virtual channel identifier codes* in each packet, which provide for different routing techniques. Cells are assigned to user channels on demand.

Packet technologies make more efficient use of communications channels than do the synchronous transfer mode (STM) technologies commonly used to transmit digitized voice.

The T1-carrier system and other STM services are routed over dedicated lines set up either by dialing or connection to a private line rather than by address. For example, the T1-carrier, a TDM system, is based on a frame which is divided into 24 voice channels. The time slots are multiplexed together with a frame bit to form the T1-signal. Because each slot is synchronized to the frame bit, it is an STM system (see "T1-Carrier"). Each time slot represents a voice call, so the digitized voice traffic is guaranteed access to the assigned slot for the duration of the call. The identity of the call is determined from its position within the frame rather than an address. Consequently, time slots within an STM frame cannot be shared among calls. This is acceptable for voice service but is wasteful for data transmission.

Packet and cell techniques are more efficient for data transfer because users have access to the entire communication channel when they need it. Demand can be at random intervals for random lengths of time. In ATM, the header containing the destination address compensates for the loss of the ability to identify data from its time slot within a frame. The short length of the ATM cell makes it suitable for data as well as voice, video, and other real-time traffic that would suffer from randomly varying transmission intervals and delays.

Consider the analogy of individual vehicles given access to multilane highways as opposed to truck convoys of varying lengths moving along single-lane roads. Thus ATM ensures that voice and video traffic, for example, can be given priority and never need wait more than one 53-byte cell time (3 µs at a 155-Mb/s data rate) before it can get on a communications channel. With STM technologies that wait could be several milliseconds.

ATM provides user communications at the T1-rate (1.5 Mb/s), and it provides integrated services for voice and nonvoice data, but at a wider range of bandwidth. B-ISDN permits interactive computer communications to run at higher speeds and allows distributed applications that make use of the computing and storage resources distributed over the network. It also permits such new applications as the distribution of digital video data.

Because B-ISDN was developed as a public telecommunications network, SONET, the standards for synchronous optical networks and synchronous digital hierarchy will be used. But where fiberoptic cable is not available, satellite communications can be served. Data rates of 155 Mb/s are possible with the satellites and modems today, and higher rates are possible.

# Pulse-Amplitude Modulation (PAM)

*Pulse-amplitude modulation* (PAM), as diagrammed in Fig. 20-1, is the process of amplitude modulating a pulse carrier waveform. Three different voice signals, Fig. 20-1*a, b,* and *c,* are sampled at different times so that they can be combined and transmitted in the correct time sequence over one transmission line, Fig. 20-1*d.*

# Pulse-Code Modulation (PCM)

*Pulse-code modulation* (PCM), as diagrammed in Fig. 20-2, is a modulation process in which the signal is sampled periodically by pulse-amplitude modulation. The sampling is performed by an electronic switch that acts like a stepping switch. Then each PAM sample is quantized

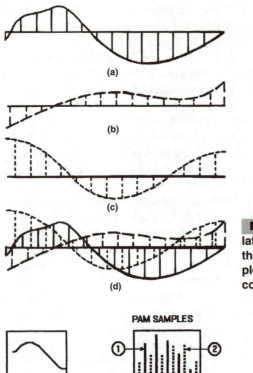

**Figure 20-1** Pulse-amplitude modulation (PAM) principles: (*a, b,* and *c*) three different voice signals are sampled at different times, and (*d*) they are combined and transmitted over one line.

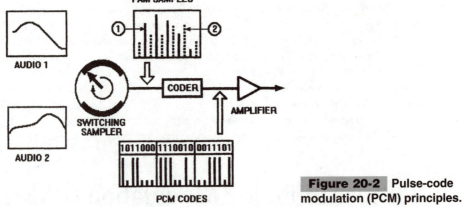

**Figure 20-2** Pulse-code modulation (PCM) principles.

or coded. The coder measures the height or amplitude of each sample and converts it into a binary number, such as 1011000 or 1110010. Thus the transmitted pulses (1s) all have the same value. At the receiving end, the binary code is converted back into PAM samples.

# Code-Division Multiple Access (CDMA)

In *code-division multiple access* (CDMA) both the transmitter and intended receiver are programmed to generate the same prearranged pseudorandom binary pulse train of +1s and −1s. This sequence is multiplied bit by bit by the binary +1s and −1s present in the digitized voice signal. The product sequence is then broadcast to the receiver. By comparing the sequence of incoming binary digits with the sequence the receiver is generating, it can *decode* or separate out the voice bits that were multiplied by the original pseudorandom binary pulse train.

Each transmitter in a network uses a different prearranged pseudorandom pulse train when it modulates the signal that it transmits. Thus all the transmissions in the network are distinguishable by their pseudorandom codes. Every transmitted message is spread over the entire allowable bandwidth, resulting in a *spread-spectrum signal,* which occupies far more bandwidth than would be necessary just to transmit its message content.

The pseudorandom pulse trains for CDMA spread-spectrum systems are selected to avoid mutual interference. This makes it unlikely that any receiver will lock onto a transmission not intended for it, even momentarily. The CDMA receiver "anticipates" what the next voice pulse will be (either a binary +1 or a −1). If it guesses wrong, it will encounter an adjacent sequence of pulses that has been inverted (+1s where −1s should be, and the inverse).

Spread-spectrum CDMA modulation is used in military communication systems, commercial communications satellites, and in the satellite-based Global Positioning System (GPS) navigation operated by the U.S. Department of Defense. The spread-spectrum concept has the following advantages:

- Messages can be directed to specific receivers in a communications network.
- High-resolution ranging is possible for fixing the position of a ship, aircraft, or vehicle and providing navigational aid.
- Natural noise and deliberate jamming do not interfere with transmissions.
- Higher-level transmission security is possible.

# Telephone Systems

The telephone set is still the most commonly used wired telephone communications terminal. Available in many styles such as desk, wall-mounted, cordless, and toll-calling models, it has two main parts: the *handset* and the *transceiver module.* Modern wired handsets contain the transmitter (microphone) and receiver (headphone) in a separate assembly connected by a flexible conductive cord to the transceiver module, which contains an active integrated circuit network, a tone ringer, a Touch-Tone keypad for entering telephone numbers, and a cradle or *hook switch* for activating the telephone. The transformers, bell ringers, and rotary dials of earlier telephones are no longer used.

Sounds, including the human voice, are alternating condensations and rarefactions of the surrounding air. Called *longitudinal vibrations,* they are caused by various sources of mechanical energy such as drums or vibrating strings (e.g., vocal cords or violin strings). Sound waves created by the human voice exert enough acoustic energy to cause a microphone diaphragm to vibrate, and this vibration converts the acoustic energy into an electrical voice frequency. The 4-kHz telephone channel bandwidth is wide enough to permit the transmission of intelligible speech over the range of 200 to 3500 Hz as well as tone dialing and other telephone control signals.

When a caller lifts the handset from its cradle or *off-hook,* connections are made to copper-wire pairs in the telephone line called the *loop* that connects the phone to the local central office for activating the phone. Direct current flows from the 48-V central office battery to the calling telephone, thus signaling the central office *electronic switching system* (ESS) that the caller wants service. The ESS responds with a dial tone. The caller then enters the sequence of digits in the number to be called on the Touch-Tone keypad that is connected to a dual-frequency signal generator. A different pair of two frequencies is sent to the central office for each digit keyed on the keypad.

The modem central office ESS is under the stored program control (SPC) of a digital computer that converts the address tones to digital signals and stores them temporarily in memory registers. The ESS then finds the shortest and least loaded transmission path through a *trunk line* to the far-end or called central office. An ESS in that central office sends a ring signal to the called telephone while simultaneously sending a ringing signal back to the caller. If the line is busy, a busy signal is sent back to the caller. But if the called party picks up the handset, DC current is drawn, signaling the central office to interrupt the ringing and complete the talking circuit. Local and regional calls can be completed over voice-grade trunks, but long-distance calls typically are digitized, multiplexed, and sent by other transmission media such as fiberoptic cables or terrestrial or satellite microwave links.

The two copper wires of the local loop are called the the *tip wire* and the *ring wire,* terms that originated with the early operator-switched telephone keyboards. When the handset is lifted from the cradle (off-hook), both ring and tip contacts close, completing the circuit between the handset microphone and headphone and the local central office ESS. However, the ring wire of the telephone is permanently connected to the central office on the central office side of the on-hook contacts, so that a ring signal can be sent to the called telephone when its handset is on hook.

Modern electronic telephones have lighter and more efficient microphones and headphones than the carbon microphones and permanent-magnet headphones typically found on the earlier mechanical telephones. See "Microphones" and "Earphones" in Sec. 17, "Electronic Sensors and Transducers." Some of the extra telephone services now offered on wired telephones include voice mail, caller ID, and call forwarding.

## OTHER TELEPHONE EXCHANGES

Many business offices have their own switching equipment called *private branch exchanges* (PBXs). PBXs allow calls to be connected within the organization without using the public telephone lines, but they also allow incoming calls to be directed to the proper party within the company. However, calls to parties outside the company must be switched through the PBX to the public network. A *private automatic branch exchange* (PABX) is an automatic electronic version of a PBX. *Centrex* switches perform the same functions as PBXs, but they are leased telephone company equipment located within central offices.

Some corporations have complete private switching networks that connect calls both within a single building or complex as well as to other parts of the organization located in different cities or even on different continents. These permit the companies to bypass local or interchange service providers and act as their own internal phone-service providers.

## TRANSMISSION METHODS

Telephone calls can be transmitted at voice frequency over twisted-wire pairs, or many voice frequency channels can be multiplexed together using frequency-division modulation (FDM) for analog carriers or time-division modulation (TDM) for digital carriers. The multiplexed signal can then be transmitted over paired-wire cable, coaxial cable, fiberoptic cable (including undersea cable), or RF relays. Terrestrial and satellite microwave links operate at frequencies of between 2 and 30 GHz. See "Satellite Communications" and "Global Telephone Satellites," following in this section.

## CELLULAR MOBILE TELEPHONE SYSTEM

The *cellular mobile telephone system,* as shown in block diagram Fig. 20-3, is a telephone system that depends on RF links to make connections with the PSTN. Voice and data can be transmitted and received from a *mobile unit* (cellular telephone) to a *cell site* for relay back to a *mobile telephone switching office* (MTSO) which makes the connection to the telephone network.

The cell site (also called a *base station*) provides radio links with the mobile units and voice and data links to the MTSO. It has antennas and transmitting and receiving equipment. The cell phone transmits at 824 to 849 MHz and the cell site transmits at 869 to 894 MHz. The MTSO (also called the *mobile switching station*) coordinates all cell sites and contains the cellular processor and switch. It has voice and data links to the telephone company offices, controls call processing, and handles billing.

The cellular mobile system increases the number of available radio channels by dividing the geographical area into *cells.* Frequencies are allocated so that adjacent cells do not broadcast or receive on the same frequencies. A geographical region is divided into generally hexagonal cells, each of which has a cell site with an assigned frequency. The MTSO circulates a *paging signal* from one cell site to another until it finds the mobile telephone whose number has been dialed. The located cellular phone responds by transmitting an acknowledgment to its local cell site. In this way the MTSO knows that both parties are ready to complete the call between the parties' respective cells.

Both cell phones and the MTSO use special *setup channels* to communicate the digital data for the call initiation. These setup channels are shared by all the users of a given cell. The MTSO assigns a channel pair to each of the mobile cellular phones, and the phones use these channels or voice links as long as the mobile units stay in their original cells. When a mobile unit moves from one cell to another during a call, the present cell site requests a *handoff.* The

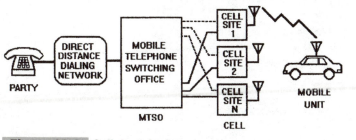

**Figure 20-3**  Cellular telephone system.

system switches the call to a new frequency channel in a new cell site without either interrupting the call or alerting the user. The call continues for as long as the user is talking.

A *cellular telephone* is a small, lightweight (typically less than 10 oz [284 g]) handheld, battery-powered transceiver, as shown in Fig. 20-4. It includes a lighted keypad and liquid-crystal or light-emitting diode display that indicates the number being called and provides other information about the status of the battery, and the local cellular network. These telephones can be powered by nickel–metal hydride, lithium-ion, or nickel-cadmium batteries.

Analog cellular systems predominate in the United States, but digital cellular systems are being introduced although complete U.S. geographical coverage is not now available. Many digital cellular and cordless technologies have been developed. Analog cellular systems are limited to frequency-division multiple access (FDMA) schemes, but the digital cellular systems can use FDMA, time-division multiple access (TDMA) and code-division multiple access (CDMA). When a multiple access scheme is selected for a particular system, all the functions, protocols, and networks must be compatible with that scheme.

North American TDMA is based on the IS-54/IS-136 standard. The North American FDMA and CDMA share the same 850-MHz band allocated for analog systems. Thus both work on phone-to-base frequencies of 824 to 849 MHz and base-to-phone frequencies of 869 to 894 MHz. CDMA, developed after TDMA, is based on the IS-95 standard. However, these digital cell phone technologies are not compatible.

The Global System for Mobile (GSM) standard, developed in Germany, combines both FDMA and TDMA technologies, and it is also compatible with ISDN. GSM 900, the European digital cellular standard, operates at phone-to-base frequencies of 935 to 960 MHz and base-to-phone frequencies of 890 to 915 MHz in 72 countries worldwide. GSM 1900, operating at the higher frequencies of 1710 to 1785 MHz and 1805 to 1880 MHz, is available in the United States, Canada, and 36 other countries.

Digital personal communications services (PCS) frequencies in the 1.9-GHz or 1900-MHz band as well as the 850 MHz-band have been allocated in North America. In the 1900-MHz band they are 1850 to 1910 MHz phone-to-base and 1930 to 1990 MHz base-to-phone. A specialized mobile radio (SMR) carrier service operating at 900 MHz is also being offered in North America. Some cellular phone service providers are offering special radio links from personal communications devices (PCDs) to cell phones for transmitting wireless data from the PCDs.

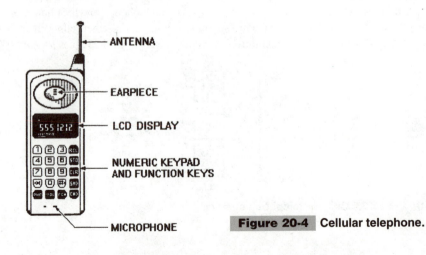

**Figure 20-4**  Cellular telephone.

The organization for a digital cellular system is the same as that shown in Fig. 20-3. Because of these different systems, subscribers have the option of purchasing analog service or one of the digital services. Digital services offer features such as sending and receiving E-mail and other text, voice mail, caller ID, call forwarding, and Internet access, as well as greater security against eavesdropping and theft of services. Both analog and digital cellular phones can be purchased, but dual-mode (analog and digital) models are also available. If a dual-mode phone is preferred, the model will be determined by the availability of digital service where it is to be used and by its format, CDMA or TDMA.

All cellular telephones have the following features:

- Authentication
- Ringer and earpiece volume controls
- Battery-strength indicators
- Audible keypad feedback
- Dual-tone multiple frequency (DTMF) key-tone signaling

Some cellular telephones also feature:

- Ability to store names as well as numbers
- Any-key answering
- Audible elapsed-talk timers

## CORDLESS TELEPHONES

A *cordless telephone* has a separate handset that transfers voice signals to its base module by a short-range radio link, as shown in the block diagram Fig. 20-5. The handset is a radio *transceiver* with a keypad for dialing. Battery powered, it has a range of about 200 ft

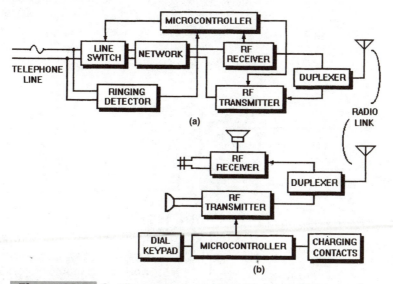

**Figure 20-5** Cordless telephone block diagram: (*a*) base unit, and (*b*) handset unit.

(60 m). The base module, also a transceiver, is powered by an AC-to-DC converter from the 120-VAC power line. The module is connected by wire to the PSTN. Most analog cordless phones operate at 49 MHz, but 900-MHz analog phones are available. There are also digital cordless phones that operate at 900 MHz and 2.4 GHz.

# Computer Networks

Computers can be linked together in networks to permit the easy exchange of information between persons in the same organization. One benefit is the *network server,* a computer able to store many different applications programs and large databases and produce them on demand. This conserves available hard-drive memory in the individual computers. Another benefit is the ability of all computers in the network to share printers or plotters. The three most common computer networks are the *local-area network* (LAN), the *metropolitan-area network* (MAN), and the *wide-area network* (WAN), designations that relate to the size and coverage of the system and their operational characteristics.

## LOCAL-AREA NETWORKS (LANS)

A *local-area network* (LAN) is a data communications network that interconnects computers, printers, computer servers, and terminals on the same floor of a building or distributed over a cluster of buildings, such as a university or corporation campus. LANs can transmit at the moderate to high data rates of 100 kb/s to 50 Mb/s. Higher-speed LANs in the 50 to 150 Mb/s range are becoming available. They comply with the fiber-distributed data interface (FDDI) and distributed-queue dual-bus (DQDB) standards. LANs can use their own switching equipment and do not depend on public utility carrier circuits. But they can have *gateways, routers,* or *bridges* to other private networks and the PSTN. The three basic LAN topologies or configurations are, as shown in Fig. 20-6, *star, bus,* and *ring*. Wireless LANS are available that will operate in either the infrared or radio-frequency bands.

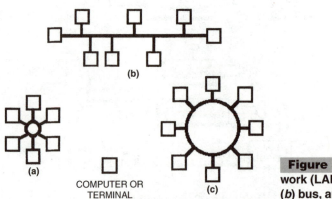

**Figure 20-6** Local-area network (LAN) topology: (*a*) star, (*b*) bus, and (*c*) ring.

## METROPOLITAN-AREA NETWORKS (MANs)

A *metropolitan-area network* (MAN) can cover a geographical area ranging from a few buildings to an entire city. There are several MAN standards. One has a dual-bus topology with coaxial or fiberoptic cable operating at rates of 44.736 MB/s and higher. The other is the *fiber-distributed data interface* (FDDI). MANs can be jointly owned by more than one company, and the responsibility for maintenance can be contracted out to specialized firms.

## WIDE-AREA NETWORKS (WANs)

A *wide-area network* (WAN) is a more extensive network than a MAN that can cover several cities or connect existing LANs across the country. WANs typically are based on X.25 packet switching. A national or multinational corporation can link its LANs in offices and factories around the country with a WAN. Typically large and complex, WANs are typically owned and maintained by large independent telecommunications providers able to offer packet-switched public data networks (PSPDNs).

# Satellite Communications

Telecommunications satellites are commercial spacecraft that are positioned so that they remain in essentially fixed *geosynchronous* or *geostationary orbits* above the equator, as shown in Fig. 20-7. Most civilian communications satellites use C-band frequencies: 6 GHz up and 4 GHz down. Some satellites are also capable of providing service in the Kuband frequencies: 13 GHz up and 11 GHz down. From the geostationary altitude, satellites can relay or repeat radio signals sent from a ground radio transmitter to one or more receiving stations elsewhere on the earth. They provide radio coverage far beyond what could be obtained with any terrestrial radio relay station and they are long-haul alternatives to microwave links and undersea cables.

Low-altitude orbiting balloons proved the feasibility of telecommunications satellites, but their orbits were too low for effective coverage, and as passive reflectors their returned signals were too weak for practical telecommunications. Active satellites with amplifiers

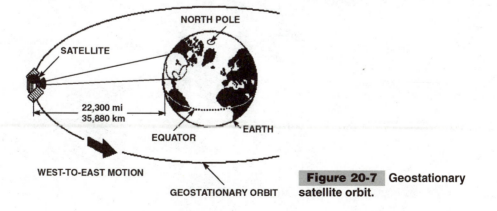

**Figure 20-7** Geostationary satellite orbit.

and transponders were developed to amplify the received signals, convert them in frequency, and retransmit them to earth receiving stations. But the first active satellites could only be lofted into low-altitude orbits because the existing booster rockets were not powerful enough to get them into geosynchronous orbits. Consequently, geographical coverage was poor and they were in range of a receiving station only for short periods each day.

With the development of more powerful boosters, the satellites could be sent into geosynchronous orbits at an altitude of 22,300 mi (35,880 km) above the equator, where their positions could be synchronized with the rotation of the earth. For all practical purposes they became fixed platforms in space, and their antennas could be directed for 24-h reception and transmission. A geosynchronous satellite can provide coverage over a wide belt around the earth extending from 60° north to 60° south latitude.

A geosynchronous communication satellite, such as is shown in Fig. 20-8, can have beam widths that cover up to 1000 mi (1610 km) on earth, so a cluster of satellites with partially overlapping coverage can reach about 40 percent of the earth's surface. But the disadvantages of satellites in geosynchronous orbit are their high signal losses and propagation delays resulting from the long distance the signals must travel. A one-way transit of a signal from the satellite to earth takes about 0.12 s, so the round-trip delay is about 0.24 s.

The transponders in telecommunications satellites are powered by arrays of *solar cells* that convert the sun's radiation directly into electric power. The large arrays are directed to face the sun continuously to generate maximum power. Batteries provide power during periods of eclipse.

*Spin-stabilized satellites* with solar cell panels positioned around their cylindrical bodies spin at about 60 rpm, but their antennas must counterrotate so that they remain pointed at the earth. *Three-axis-stabilized satellites* are equipped with solar-cell panels that unfold after they are in orbit. They are gyrostabilized by internal wheels that spin at high speed.

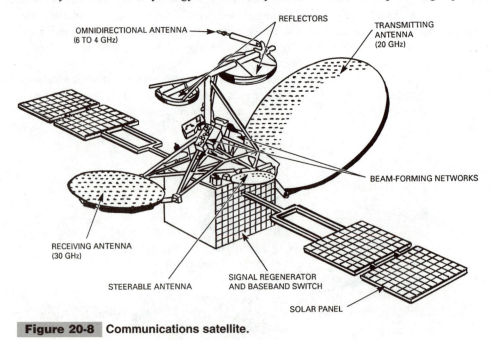

**Figure 20-8** Communications satellite.

Both types of satellite have separate antennas for receiving signals and transmitting them back to earth at other frequencies.

Most commercial communications satellites are linked to earth stations that transmit signals to them on uplink frequencies of 5.9 to 6.4 MHz and relay the amplified signals back to earth on downlink frequencies of 3.7 to 4.2 GHz. However, direct TV broadcast satellites use uplinks of 14 GHz and downlinks of 11 GHz, and U.S. military communication satellites use uplink frequencies of 8 GHz and downlink frequencies of 7 GHz.

The transponders in communications satellites have many different organizations. Some are designed for single carrier signals that can either be frequency- or time-division multiplexed. Others are designed to operate in either of two multiple access modes so that one transponder simultaneously carries signals from several different earth stations. The Intelsat V satellite, for example, is equipped with 24 transponders and has 12,000 telephone and 2 color TV circuits.

In FDMA, the frequency band of each transponder is subdivided and parts are assigned to different earth stations. This permits each station to transmit continuously in its assigned frequency band without interfering with the other signals. The earth stations receive all of the signals, but they demodulate only those signals assigned to them. In TDMA, each earth station uses the entire bandwidth of a transponder for a short period of time. The uplink signals are transmitted in bursts, and the satellite transponder interleaves signals that are downlinked to all stations.

Most satellites contain instruments that measure such variables as temperature, radiation, and magnetic field in and around them for transmission back to their earth supervisory stations to warn of potentially threatening environmental conditions such as solar storms. Some also carry receivers for signals that control onboard jets to make any orbital corrections needed.

Ground tracking stations for communications satellites have highly sensitive antennas with typical gains of 60 dB (1 million) for receiving weak signals. Tracking antennas can be rotated with a directional accuracy of about 0.001° (0.6 s of arc).

# Global Telephone Satellites

There are now three satellite systems or constellations dedicated to global telephone communications. The oldest of these systems is *Inmarsat,* which has been available since 1979. The four Inmarsat satellites are stationed at the geostationary altitude of about 22,300 mi (35,880 km). The *Iridium* and *Globalstar* are both in low-altitude orbits. The 66 Iridium satellites are in about 440-mi (710-km) orbits and the 48 Globalstar satellites are in about 880-mi (1420-km) orbits. Low-altitude satellites have shorter round-trip signal delays than geosynchronous satellites.

## INMARSAT

The *Inmarsat* constellation of four geosynchronous communications satellites provides personal communications services, primarily to terminals permanently installed on large oceangoing vessels. Two-way voice messages and computer data can be relayed to and from remote receivers. After the user selects the proper satellite, the telephone number can be dialed in the same way as an ordinary international telephone call. Inmarsat satellite communication frequencies are 1.6 GHz in the L band. Inmarsat satellites are positioned along the equator over Brazil, over the mid-Atlantic, over the Indian Ocean, and over the Pacific east of New Guinea. The constellation provides continuous global coverage except for the polar regions.

## IRIDIUM

*Iridium* is a voice-messaging system consisting of a constellation of 66 satellites, the world's largest satellite network, in low-altitude circular orbits of about 440 mi (710 km). Eleven satellites are symmetrically arranged in six orbital rings in essentially polar orbits (90° inclinations with respect to the equator). The satellite constellation can provide digital voice and data coverage of the entire surface of the globe. Each three-axis-stabilized satellite weighs 1600 lb (726 kg) and will include onboard switching and satellite-to-satellite messaging crosslinks to hand off messages to minimize delay in reaching their destinations. The uplink, downlink, and crosslink frequencies are 20 GHz and the subscriber frequencies are 1.6 GHz.

Iridium's personal communicators are about the same size as cellular telephones. They use TDMA to send and receive voice messages and digital data. Some communicators will be pagers capable of receiving digital data only, while others will relay all messages through Iridium satellites. A third class of communicators will be dual-mode, capable of automatically selecting cellular telephone links (if available) as an alternative to the more expensive satellite links. The system was named *Iridium* because it was originally planned to have 77 satellites, the same as the number of electrons orbiting the iridium nucleus. However, improved technology and cost considerations led to the reduction.

## GLOBALSTAR

*Globalstar* is another satellite voice-messaging system that is competing with Iridium. However, it will be simpler because the satellites will not crosslink with each other. Instead, they will depend on existing ground-based facilities for message switching. Globalstar plans call for 48 three-axis-stabilized satellites (including 8 spares). They will orbit in 8 circular planes inclined 52° with respect to the equator at an altitude of about 880 mi (1420 km) above the earth. Each of the 490-lb (222-kg) satellites will use CDMA modulation.

The constellation will provide digital voice and data communications over an area extending to 70° north and south latitude. Handheld mobile communicators, about the size of digital cellular telephones, will provide access to the satellites. Most will be dual-mode, also capable of accessing terrestrial cellular networks. Approximately 125 ground stations will link the Globalstar satellites to conventional land-line telephone networks around the world.

# CONSUMER ELECTRONICS PRODUCTS

# Overview

The term *consumer electronics* refers to electronic products or equipment that is primarily purchased by ordinary citizens for their personal or home use. While entertainment products predominate, many items formerly considered to be office equipment have become accepted as consumer products. Computers are the most obvious, but many homes now have color printers, copying machines, scanners, and facsimile (fax) machines.

The earliest consumer electronics entertainment products were battery-powered vacuum-tube radio receivers, and these were followed by record players with electronic amplifiers and motor-driven turntables that replaced the spring-wound mechanical players. A startlingly long list of electronics products has joined the ranks of consumer products over the past half century. These products include TV receivers, hi-fi stereo systems, VCRs, camcorders, tape recorders, audio CD players, microwave ovens, portable battery-powered radio–tape and radio–CD players, digital cameras, cordless and cellular telephones, and telephone answering machines. Add to this list smoke and carbon monoxide detectors, home security systems (including closed-circuit TV), weather-alarm radios, and pagers. The latest additions to this group are the digital video disk (DVD) and DVD players.

Electronic circuitry has displaced the gear-and-spring mechanisms of most watches and clocks, and the mechanical calculator has been banished in favor of electronic models, some of which can solve mathematical problems and plot the solutions on their displays. Many traditional household appliances, not considered to be electronic products, have nevertheless been transformed by electronic controls that have replaced their former clockwork timers and simplified their use. Consider the washing machine, dishwasher, air conditioner, range, and dehumidifier.

The consumer electronics industry is shifting its emphasis from analog sound reproduction to digital sound reproduction and is encouraging the replacement of audio equipment that still performs very well. The introduction of high-definition television (HDTV) and the development of other digital TV broadcasting formats is driving this transition.

The DVD has been introduced as a replacement for the compact disk (CD). It will be available in seven different formats; some are intended for players embedded in high-end personal computers, and others are intended for playback on TV sets. The DVD-video, for example, is inserted in a stand-alone DVD player that must be purchased. It sends its formatted output signal to the TV set in the same way as video cable. DVD-video disks can be purchased or rented, but the available selection of movies might be limited.

A commercial twist on the DVD concept is *Digital Video Express* (Divx). It seeks to replace the rental videotape cassette with a lower-cost proprietary rental video disk. The customer must purchase a proprietary Divx player that is connected by telephone line to a central computer. Upon rental the disk can be played for the length of time allowed by the rental fee. Extensions of playing time can be requested over the phone line at a billable rate. However, there is still life left in the audio CD, and major manufacturers are developing improvements on this format that could still be played on conventional CD players.

Over the past 10 years the electronics content of automobiles has increased dramatically, and a parade of innovations has been introduced. These include many electronically controlled functions: fuel injection, engine operation, transmission, antilock braking, and traction control, as well as entertainment, interior climate control, and security systems. Most recently, infrared viewing systems for extending night vision, vehicle locators based on the Global Positioning System (GPS), and distance-measuring electronic aids for parking have been added.

Software companies and consumer electronics manufacturers are developing a new generation of "smart" entertainment products and appliances that can communicate with each other and humans in home networks linked to the Internet. Each product or appliance will include an embedded processor/controller that conforms to an accepted global communication standard. The network will include units such as a TV set, stereo system, DVD player, and various kitchen appliances. They will be controlled via the Internet from outside the home or via a network that uses the household's electrical wiring within the home.

Two opposing software "platforms" have been proposed: (1) a *distributed* concept in which each networked unit will contribute processing power to the network and participate in the distribution of information and instructions to all of the other attached units, and have no central control point, and (2) a *centralized* concept in which all networked units will be controlled by a personal computer, TV set-top box, or infrared remote-control module.

Also, over the past 40 years the introduction of monolithic circuitry and fallout from military technology has caused explosive growth in another industry, marine electronics, which is a shared commercial and consumer marketplace. Examples of these products include low-cost VHF transceivers and Global Positioning System (GPS) receivers, depth finders, fishfinders, and short-range solid-state radars.

Radio receivers are discussed in Sec. 18, "Radio Transmitters and Receivers"; TV receivers are discussed in Sec. 19, "Television Broadcasting and Receiving Technology"; and telephones are discussed in Sec. 20, "Telecommunications Technology." Marine electronics are discussed in Sec. 24, "Marine Electronics Technology" and the Global Positioning System (GPS) is discussed in both Secs. 24 and 25, "Military and Aerospace Electronic Systems."

# Videocassette Recorders (VCRs)

A *videocassette recorder* (VCR) records and plays back video and audio signals on magnetic videotape in cassettes. It is capable of recording TV programs for later playback and playing prerecorded commercial movie videocassettes. Three different formats have been developed, VHS, Beta, and an 8-mm system, but the VHS format predominates.

VCRs record and play back tapes with bandwidths of up to 3.58 MHz by a method known as *helical scanning*. The videotape is pulled out of its cassette by a VCR mechanism, as shown in Fig. 21-1. This M-shaped arrangement of capstans and guide rollers grabs the tape and tensions it against the head. The tape is helically wrapped 180° around the read/write head, as shown in greater detail in Fig. 21-2. The drum, which contains the recording heads, moves in a direction opposite to that of the tape, permitting lower tape speeds and shorter tape lengths. This arrangement permits the reading or writing of video data on the tape in the *slant-track scanning method*. Each diagonal track contains the information necessary to complete one full TV frame. The audio signal is recorded on one edge of the tape, and the control signal is recorded on the other. VCRs can record in standard play (SP) and extended play (EP). Recording occurs at a head-to-tape speed of 229 in/s (5.8 m/s) on the cobalt-alloy-coated tape.

The *video home system* (VHS) format was developed by the Victor Company of Japan (JVC). The rectangular plastic cassette measuring $7\frac{3}{8} \times 4 \times 1$ in ($19 \times 10 \times 2.5$ cm) contains 0.5-in (13-cm-) wide videotape that moves between two internal reels at speeds of 1.3 or 0.66 in/s (33 or 17 mm/s).

The front panel of a typical VCR contains a digital display of the local time and date as well as the cumulative time for recording a program, playing a prerecorded tape, and

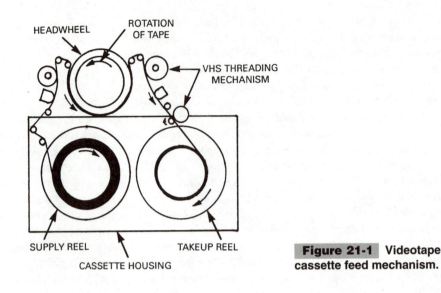

**Figure 21-1** Videotape cassette feed mechanism.

rewinding the tape. It also contains switches for selecting channels, setting the times, dates, and duration of programs to be recorded automatically, and rewinding and rapidly advancing the videotape. Battery-powered handheld remote controllers duplicate many of the VCR's front-panel controls.

Most VCRs are monophonic and are better suited for recording voice than music. However, hi-fi VCRs are now available. Standard VCR features now include four video heads and the ability to receive at least 125 cable channels. Most can be programmed for 365 days in advance for 8 events at a time (a show programmed to be recorded every day or every week counts as a single event). Some VCRs automatically switch from SP to EP to fit long recordings. VHS videocassettes can record up to 2 h of telecasts with SP and as many as 6 h with EP.

# Camcorders

The *camcorder,* a contraction of *video camera and recorder,* combines the functions of a charge-coupled device (CCD) video camera and a VCR. The basic parts of a camcorder, as

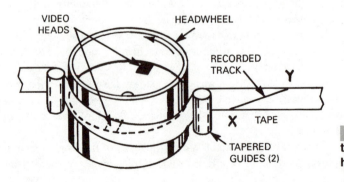

**Figure 21-2** Video-tape cassette read/write head.

shown in Fig. 21-3, are a *multicolor LCD viewfinder,* a *lens,* a *CCD detector,* a built-in *microphone,* a *conversion section,* and a *videotape recorder.* The audio and video signals are converted to electric charges by the conversion section. Each CCD detector uses about 250,000 picture elements (pixels).

Camcorders are available in many different sizes and formats: VHS, VHS-C, Hi8 mm, and digital video cassette (DVC). Full-size VHS camcorders are compatible with VHS VCRs. The VHS-C cassette, the most widely available, is smaller than the standard VHS cassette but it uses the same format. An adapter is required to play the VHS-C tape on a standard VHS VCR. VHS-C records for only 30 min in standard play (SP) and 90 min in lower-quality extended play (EP). The tape in the 8-mm cassette is not compatible with VHS VCRs. To view the recording made by an 8-mm camcorder, the unit must be plugged directly into a jack on the TV or VCR, or an 8-mm VCR must be used. However, sound quality is better with 8 mm than with VHS or VHS-C, and recording time is 2 h in EP and 4 h in EP.

*Digital video camcorders* (DVCs) offer higher picture quality than other formats, and the recordings can be dubbed and edited. These camcorders use tape cassettes, and playback is the same as with an 8-mm camcorder. Some models can feed video directly to personal computers.

The latest camcorders now include a multicolor LCD viewfinder, image stabilization, built-in video lights, a wide-range zoom lens, and stereo audio. Other features are a power zoom lens and a flying-erase head for making clean transitions between scenes. These units have controls for power, play, stop, fast forward, rewind, automatic and manual focus, fade, zoom, balance, sound, and playback through the viewfinder. They are powered by a rechargeable battery or alkaline cells. To edit the tape, the camcorder can be connected to a VCR so that it can be done while viewing the TV screen, or it can be done by looking through the viewfinder.

# Audio Compact Disks (CDs)

An *audio compact disk* (CD) is a 4¾-in (120-mm) optical disk that can store voice, music, or other sound as microscopic pits on a reflective surface. The audio is recorded on a mas-

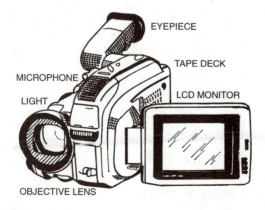

EYEPIECE

TAPE DECK

MICROPHONE

LIGHT

LCD MONITOR

OBJECTIVE LENS

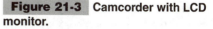

**Figure 21-3** Camcorder with LCD monitor.

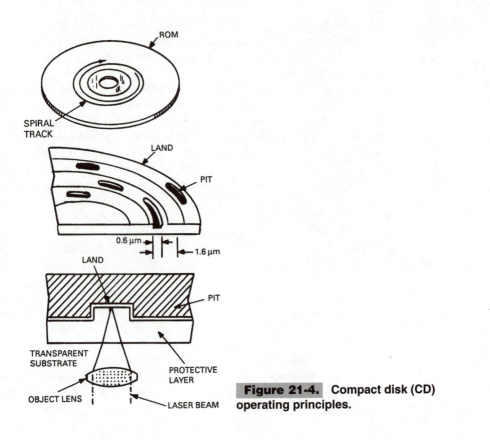

**Figure 21-4.** Compact disk (CD) operating principles.

ter disk by a laser in the form of a pitted spiral track. The pit pattern is permanently transferred to a plastic disk by stamping, much as is done in pressing vinyl long-playing (LP) records. The copy is played back by directing a low-power laser "stylus" on the moving pits, which modulate the reflected light to reproduce the original sound. The principles of CD recording and playback are shown in Fig. 21-4. The small slice of the CD illustrates how digitally encoded information is stored as a series of microscopic pits on the reflective surface.

The CD has replaced the vinyl LP and tape cassette as the preferred media for high-fidelity recording and playback. It can provide sonic detail and realism comparable to that of the best professional recordings. There are no annoying tick or pops between selections or during musical pauses.

# Audio CD Players

A *CD player*, as shown in the diagram Fig. 21-5, combines a motor-driven turntable with an electrooptical playback system. It converts laser light reflected from the moving pits on the disk to an audio signal which can then be amplified. The player can be a battery-

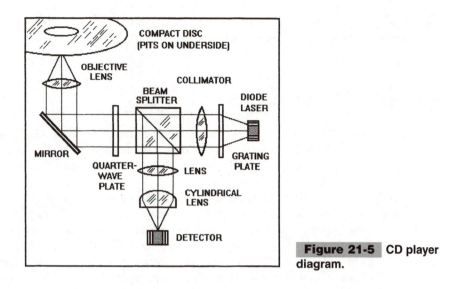

**Figure 21-5** CD player diagram.

powered single-disk portable unit for personal listening with headphones, as shown in Fig. 21-6, or a multidisk unit in a home. There are also single-disk automotive units.

Home CD changers typically hold five or six disks in a revolving carousel drawer; magazine-style changers are less common. Home CD jukeboxes store 25 to 200 disks.

Some players in stereo systems can hold up to 10 audio CDs in a rotary feeder to provide hours of uninterrupted listening. Controls permit random-play modes for modifying the music sequence. The disks are usually played back through a stereo system's speakers. Most CD changers provide instant access to any music on the disk. Many can be operated by handheld remote controls as well as by the panel controls on the player console. The user can skip the CD forward or backward to any selection or channel, or even repeat one track indefinitely. Most players also have digital displays to indicate playing time and the number of the selection on each disk. Some players can be programmed for 5 to as many as 32 selections in any sequence.

CD-ROM drives on personal computers can also play audio CDs. The computer monitor displays a virtual control panel that can be changed with the computer's mouse. This permits the listener to change volume, sequence, and order of play for speech or music and to keep track of such variables as the channel playing and the number of minutes that it has

**Figure 21-6** Portable CD player.

been playing. See also "Compact-Disk Read-Only Memory (CD-ROM)" in Sec. 16, "Computer Peripheral Devices and Equipment."

# Digital Video Disks (DVDs)

The *digital video disk* (DVD) is a higher-capacity modification of the CD that is also based on optical storage technology. There are seven different formats for DVDs, also called *digital versatile disks*. Some are intended for playing on personal computers and others are intended for playing through television sets.

The *DVD-video* has a DVD format that differs from that of the DVD-ROM intended for playback on a computer. It is a read-only disk with a storage capacity of 9 GB, permitting it to store two 2-h theater-quality full-length movies. DVD-video is a home entertainment medium that can store movies, television shows, and movie-type video presentations. It is the probable replacement for the videocassette. DVD-videos can be played on a separate DVD player connected to a TV receiver in the same way as a videocassette is played. Most DVD videos store a movie in both standard (4:3) and wide-screen (16:9) formats. This DVD is capable of 500 lines of horizontal resolution. Players for these DVDs will also play audio CDs.

The *Digital Video Express* (Divx) video disk is a proprietary form of the DVD-video for rental to compete against (and perhaps replace) rental videocassettes. It will contain theater-quality full-length movies and will be rented from local shops. A proprietary player must be purchased to play the Divx disk, and it must be connected by phone line to a central Divx computer. According to the plan, the discs can be rented at low cost and played for a period specified for the rental fee. If the renter wants to play the disk again, beyond the rental period, an order is sent to the Divx computer, which responds with a signal that "unlocks" the disk and then prepares billing for the additional playing time.

The Divx disk can store up to 17 GB. Plans call for a 48-h playing time for less than $5. Divx disks can also be purchased for $10 to $15 for unlimited play, and the players can also play standard DVDs. The advantages are said to be far lower cost per viewing than is afforded by the purchase of a DVD-video and the expectation that the concept will attract a wider program selection than will be available on standard DVDs.

*DVD-audio* is a DVD format for playing super-high-fidelity sound. These disks also can store up to 17 GB, equal to about 25 audio CDs. DVD-audio can also provide Dolby Digital (AC-3) surround sound signals for six channels, of which five are full-range. The DVD-audio is divided into about 40 sections, and a quick-search can be made among those sections by skipping forward and backward rapidly.

For further information on other DVD formats and players see "Digital Video Disks (DVDs) and Drives" in Sec. 16, "Computer Peripheral Devices and Equipment."

# DVD Players

The *DVD player* is an electromechanical system for playing DVD-videos. It is in a flat case about the same size as a VCR, with a digital clock and front panel controls. A cable connects it to the TV receiver. The player contains circuitry and mechanisms for reading the DVD-video and converting those signals to NTSC-compatible signals. Some models can also play CD-ROMs and audio CDs, but they cannot play Divx disks.

*Divx players,* similar to DVD-video players, are designed to play rental Divx disks. The player must be connected to a telephone line which forwards viewing data to and receives billing data from the Divx central computer for extensions beyond the paid rental period. The Divx players can also play standard DVDs.

# DVD-RAM Drives

The *DVD-RAM drive* permits recording, erasing and re-recording on the same digital disk an estimated 100,000 times. The DVD-RAM can store 2.6 GB of data on a single-sided disk and 5.2 GB on a double-sided disk. DVD-RAM drives can also read data from DVD-ROM, DVD-R, CD-ROM, CD-R, and CD-RW disks. Another competing format will offer a single-side 3-GB optical disk format.

# Home Theaters

A *home theater* is a room in a private home that is equipped as a multimedia entertainment center for listening to high-fidelity sound and watching television. It will typically include the following equipment:

- A large-screen or projection TV receiver
- A high-fidelity stereo receiver
- An audio CD player
- A DVD player
- A VCR
- Up to six high-fidelity speakers

A typical arrangement for the entertainment equipment is shown in Fig. 21-7. The room must be large enough to provide the necessary spacing between speakers to obtain the most effective response, and the walls, ceiling, and flooring must provide the necessary acoustical conditions for obtaining the full performance capabilities of the equipment. Viewers or listeners are seated at the back of the room (lower part of the diagram) in comfortable chairs. The two front speakers are positioned on the left and right sides of the TV screen, and a center speaker is positioned over the TV screen. Rear surround-sound speakers are located at the back of the room, and an optional bass speaker can be placed there.

A complete setup might include a stereo TV set with audio output jacks or a stereo high-fidelity VCR, a Dolby ProLogic or Dolby Digital Surround-Sound receiver capable of delivering 50 to 100 W of power per stereo channel, and the speakers previously mentioned. This equipment can all be operated by an infrared digital universal remote control.

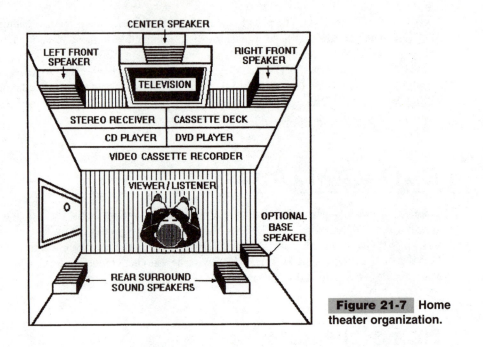

**Figure 21-7** Home theater organization.

Most CDs, tape cassettes, and FM radio broadcasts are produced for two channels, but additional center and surround-sound channels are encoded on the stereo tracks of many movies and TV broadcasts. The four channels can feed audio to five speakers: one channel each for the left and right front speakers, one for the center speaker, and one shared between the two rear speakers. The optional bass speaker is intended to accentuate bass sounds.

# Direct Satellite TV Receivers

A *satellite digital TV receiver* system consists of an 18-in- (45.7-cm-) diameter paraboloidal antenna connected by cable to the direct broadcast satellite TV receiver, as shown in Fig. 21-8. The receiver is connected to a standard TV receiver, much as a VCR or TV cable. The signal broadcast by the satellite can be decoded by the subscriber's receiver. Up to 175 channels are available with this system. The antenna must be mounted outdoors and it must be focused on the satellite, facing south in the Northern Hemisphere. Some TV receivers include built-in satellite receivers.

# Facsimile (FAX) Machines

A *facsimile (fax) machine* combines the dual functions of scanning documents with the sending, receiving, and printing of documents. The encoded documents are sent and received over the public dial-up telephone lines at speeds up to 14 kb/s.

The *subject copy* is inserted in the machine so it passes around a drum rotating at constant angular velocity. An optical carriage with a photocell is mounted on a traverse drive

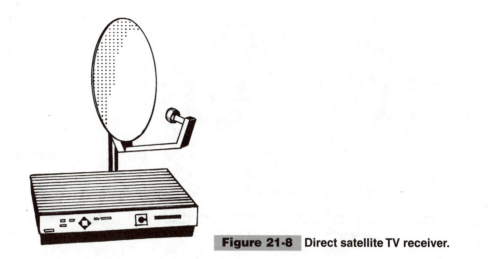

**Figure 21-8** Direct satellite TV receiver.

belt that advances a distance equal to the height of a specified rectilinear area or horizontal strip across the width of the subject copy. The photocell generates a stream of pixels encountered in a helical-scan track. The output of the photocell or cells is converted by a *modem* for transmission over the telephone line to a compatible fax machine capable of receiving it. The receiving fax machine must be in a mode for converting the data back to a signal that can be printed out on plain or thermal paper.

Fax machines are assigned exclusive telephone numbers so that they can be on standby status 24 hours a day without interfering with voice telephone service. Some fax machines also function as *scanners* or document digitizers for converting documents to a digital format for processing or storage by computer, and some can also serve as copying machines.

# Magnetic-Tape Recorders

A *magnetic-tape recorder* is a machine capable of recording and playing back audio-frequency signals on a magnetic tape. The principal parts are shown in Fig. 21-9. To record, it converts the signal to magnetic variations in the tape medium; to play back, it converts those magnetic variations back to audio-frequency signals. Typical recorders include an amplifier and speaker. Portable tape recorders are powered by batteries or AC-line transformers. Telephone-answering machines include magnetic-tape recorders.

# Microwave Ovens

A *microwave oven* is an appliance that can heat food rapidly with microwave energy. It is suitable for both thawing frozen food and cooking most homogeneous foods. A cutaway view of an oven is shown in Fig. 21-10. Food and other materials with high moisture content absorb microwave energy. In food, for example, molecules align themselves with the

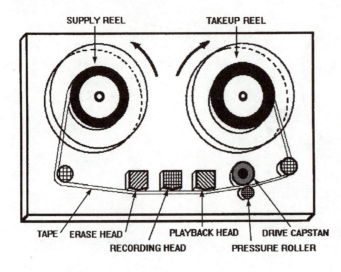

SUPPLY REEL    TAKEUP REEL

TAPE    ERASE HEAD    PLAYBACK HEAD    DRIVE CAPSTAN
RECORDING HEAD    PRESSURE ROLLER

**Figure 21-9**
Magnetic-tape recorder mechanism.

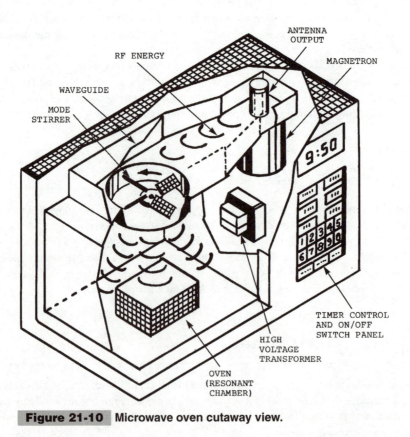

ANTENNA
OUTPUT

RF ENERGY

MAGNETRON

WAVEGUIDE

MODE
STIRRER

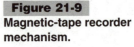

TIMER CONTROL
AND ON/OFF
SWITCH PANEL

HIGH
VOLTAGE
TRANSFORMER

OVEN
(RESONANT
CHAMBER)

**Figure 21-10**  Microwave oven cutaway view.

microwave energy, and rapidly changing radio-frequency polarity induces molecular friction, which heats the food. But, materials such as liquids, ceramics, and plastics can also be heated.

Microwave ovens are powered by microwave oscillator tubes called *magnetrons*. The magnetron, oscillating at 2.45 GHz, radiates microwave energy that is directed through a waveguide into the oven cavity. Food in the cavity is heated or cooked in time periods of from seconds to minutes, depending on its composition and bulk. The microwave energy is more evenly distributed within the oven cavity by a rotating fanlike reflector called a *mode stirrer.*

Typical small microwave ovens consume 400 to 500 W while large ovens consume 700 to 800 W. The microwave oven magnetron is an adaptation of a radar transmitter magnetron. Industrial-grade microwave ovens are used for drying wood and localized heating of other materials. See also "*Microwave Tubes*" in Sec. 7, "Microwave and UHF Technology."

# Pagers

A *pager* is a pocket-sized radio receiver that receives signals from a local paging service provider indicating that a specified telephone number (usually that of the pager owner) has been dialed. There are many different models with a wide choice of features. A simple low-cost model is illustrated in Fig. 21-11. The LCDs on pagers can display time, date, and length of message, and some can be scrolled to display short messages. Other pagers emit an audible signal or provide a tactile vibration for inaudible prompting. Pagers operate at 900 MHz. They are also called *beepers* (slang).

# Digital Cameras

A *digital camera* is a handheld still camera that takes photographs by recording images in a digital format on a digital storage medium rather than on photographic film. Digital cameras have optical lens systems, shutters, viewfinders, automatic built-in flash tubes, and other features found on conventional cameras. Some cameras have as many as five exposure settings, and some include 2× to 10× zoom lenses. The image is recorded by a built-in charge-coupled device (CCD) camera or CMOS devices for storage in its internal memory,

**Figure 21-11** Pager.

and it can be viewed immediately after it has been taken on the camera's small multicolor LCD monitor, which typically measures 1.8 to 2.5 in (4.6 to 6.4 cm) square.

The image can be recorded on semiconductor memory on a proprietary removable picture memory card, or on a 1.44 MB 3.5-in (90-mm) diskette. Cameras with diskette memory can store up to 40 pictures, while those with cards can store up to 192 pictures. The digitized images can be viewed on a computer monitor, TV set, or VCR with a video cable connection, and they can be reproduced on a computer-compatible printer. Pictures can be printed in black and white or color. Some digital cameras have resolutions as high as $1800 \times 1200$ dpi. Power for digital cameras is supplied by batteries, and imaging software is usually included in the camera's purchase price.

# Automotive Electronics

More than 60 years ago the first automotive radios became available. The first models had vacuum-tube circuitry and they required higher voltage than was available from the 6-V lead-acid batteries then being installed in cars. This higher voltage was supplied by a vibrating electromechanical DC-to-DC converter called a *chopper*. In the 1960s the circuitry was transistorized.

Within the last 10 years the electronics content of automobiles has mushroomed, and it is now of such importance that modern automobiles would no longer function without it. The critical ignition and engine-control functions as well as many other safety and comfort features are directed by microcontrollers, specialized microprocessors with many key computer peripherals such as memory and I/O ports integrated on the same chip. These microcontrollers are programmed at the factory for automatic or simple switch activation. The owner-operator need not be concerned with their programming and, for the most part, their control functions are invisible. Electronic circuitry now performs the following functions in automobiles:

- Throttle control to regulate engine power
- Gear-change sensitivity control for the automatic transmission
- Management of the electronic fuel-injection system, including the air-fuel mixture
- Antilock braking control
- Traction control
- Radio, CD, DVD, and audiotape player control
- Vehicle location and electronic map guidance based on the Global Positioning System (GPS)
- Interior climate control
- Automotive diagnostic functions
- Air bag, seat-belt tensioner, and inflatable curtain control
- Power window and door-lock control
- Instrument panel information control
- Driver's seat heating and automatic adjustment control
- Night-vision enhancement with a "look-ahead" infrared viewing system
- Security and alarm control

- Parking guidance with a rear-end distance-measuring system
- Miscellaneous control functions: map light, rear-view mirror, and defroster

The vehicle location and guidance system can include provision for transmitting an emergency position message to the local highway patrol in the event of an accident or breakdown.

However, some electronics that were introduced to much fanfare years ago failed to catch on and were eliminated because of lack of customer acceptance. Examples include citizens band radios, synthesized speech warnings (the talking dashboard), and instrument panels providing only digital readouts and pushbuttons for the entertainment and climate controls. Consequently, analog speedometers and rotary knobs returned to favor.

Carburetors were replaced by electronic fuel-injection systems that meter fuel precisely to each cylinder under all operating conditions, including engine idle. Electronic ignitions have replaced distributors and eliminated points and condensers. They provide the consistently hot spark needed for the leaner air-fuel mixtures required by clean air laws.

Most recently, route planning and guidance systems based on the reception of GPS satellite position signals have become available. Maps and text are presented on dashboard displays. Some cars now contain more than a hundred separate microcontrollers. See "*Global Positioning System* (GPS)" and "*Night-Vision System*" in Sec. 23, "Military and Aerospace Electronic Systems," and "Global Positioning System (GPS) Receivers" in Sec. 24, "Marine Electronics Technology."

## ANTILOCK BRAKING SYSTEMS (ABSs)

An *antilock braking system* (ABS) is an automotive hydraulic braking system controlled by electronics that minimize the driver's loss of control due to skidding when the brakes are applied and lock up on slippery road surfaces. The ABS has the following benefits:

- Helps the driver steer effectively during braking
- Maintains directional stability of the vehicle
- Can stop the car safely in a shorter distance than unassisted brakes

The ABS automatically pumps the brakes rapidly in short-duration pulses until the wheels regain traction. This pumping action can occur many times until the vehicle stops safely. The rapid pumping of automotive brakes at high frequency under electronic control applies the brakes more effectively on wet or icy road surfaces than a human operator can. ABS variations have been installed in aircraft and long-haul trucks for more than 25 years.

There is no standard ABS, and the variations between systems are based on different operating principles. There are two-wheel, four-wheel, and diagonally split systems. Each has different performance characteristics. A typical ABS includes wheel-speed sensors, an electronic control unit (ECU), and a hydraulic unit that can be modulated.

Sensors in the wheels (or transmission) generate voltage pulses at a rate proportional to wheel speed. These pulses are sent over wires to a microcontroller in the ECU which evaluates the signals and calculates wheel slip to determine if wheel lockup is imminent. If it is, the ECU sends modulating signals to the hydraulic unit. The modulated hydraulic pulses are then applied to the brake pads, releasing and reapplying them until the wheels regain traction and the vehicle is brought to a controlled stop. In some systems electric motor modulation has replaced hydraulic modulation for applying the brakes. Plans have been announced by some vehicle manufacturers to replace the interconnecting wires with fiberoptic cables.

## TRACTION-CONTROL SYSTEMS

A *traction-control system* includes sensors to detect when wheel traction is insufficient and the vehicle is beginning to slip on the road surface. A microcontroller begins to restrict the flow of fuel to the engine, slowing the vehicle, while applying progressive brake pressure to help the driver maintain control. It is now an optional feature on most automobiles whose manufacturers offer it.

# Automatic Teller Machines (ATMs)

The *automatic teller machine* (ATM) is a convenient banking terminal that permits customers to withdraw cash at all hours of the night or day. It is accessible to customers with the required banking card. The ATM is actually a terminal linked to a central computer by the public telephone system. First introduced in the United States 30 years ago, ATMs are now commonplace at banks, shopping malls, and airports. In 1998 there were more than 165,000 ATMs in the United States. They performed more than 11 billion transactions annually, or about 1.2 million per hour.

The banking card, similar in size to a credit card, has a magnetic strip containing the bank's electronic address and the customer's account number and personal identification number (PIN) in a digital code. The customer inserts the card in the reader slot, as shown in Fig. 21-12, and directions appear on the ATM's CRT screen. The customer enters his or her PIN and other information such as the amount requested at the keyboard. The ATM

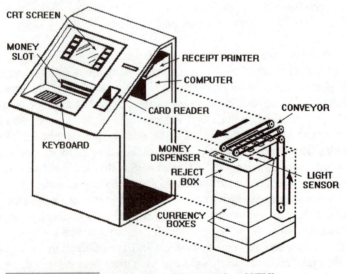

**Figure 21-12** Automatic teller machine (ATM).

then sets up a connection to the bank's computer, which verifies the customer's PIN and checks his or her balance to see if a sufficient amount is in the account. If there is, the computer authorizes the transaction.

The ATM then responds by delivering the requested amount to the customer after carrying out a sequence of security checks. Rotating rubber conveyor belts feed out one banknote at a time from the currency boxes below the conveyor. As the money passes an infrared sensor, IR energy is sent through the bill to verify that only a single bill of the proper denomination has been picked up. If a mistake is detected, the money is sent to a reject box.

Each bill is inspected until the correct number of bills has arrived at the holding area at the end of the conveyor near the dispensing slot. A set of rollers dispenses the money to the customer. When the money has been dispensed, the ATM sends a completion message to the bank, its computer updates the customer's account, and the card is returned to the customer.

ATM manufacturers are investigating various ways to increase ATM security. These include the addition of sensors that can recognize the customer's voice and/or face, or scanners that will match the customer's iris or fingerprints to his or her PIN.

# Bar-Code Readers

A *bar code* is a series of parallel black and white lines of different widths used to encode information about a product for sale in retail stores. Either printed on the package or on labels applied to the package, bar codes speed up the sale of items and provide the store with valuable data for inventory management. Primarily used in supermarkets and other large consumer goods outlets, they can be read by various kinds of readers or scanners which translate the light reflected by the white spaces in the bar-code pattern into digital codes that are sent to the store's computer.

The most widely used code in the United States is the *Universal Product Code* (UPC). It is a 12-digit code: the first digit identifies the general category of the product; the next 5 digits identify the product's manufacturer; the next 5 identify the individual item; and finally a *check digit* verifies that the code is scanned correctly. The UPC can obtain other information about the product such as price, inventory count, and taxes.

One type of bar-code reader is the handheld *pen scanner* or *wand*. It uses a light-emitting diode (LED) light source at the tip of the pen to sweep across the coded label, and it has a photodetector in its barrel. However, the tip must be dragged across the whole code label to read it, especially time consuming if the product has to be hand positioned for convenient scanning. This is most often necessary if the product has a cylindrical or irregular shape.

Stationary laser scanners or bar-code readers were invented to overcome this problem because they are able to read a bar code that is several feet away. The bar code-reader is located below the checkout counter, and a window in the countertop allows the laser beam to pass through to the product being scanned. The product is held so that the code label is directed toward the window, and then the code label is read as the product is swept over the window.

A functional diagram of a stationary laser bar-code reader is shown in Fig. 21-13. It includes a low-power visible-light laser diode and a motor-driven rotating pyramidal mirror. The laser beam passes through a parabolic mirror and strikes one facet of the rapidly

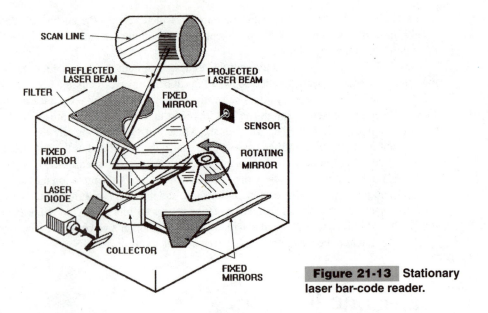

**Figure 21-13** Stationary laser bar-code reader.

rotating mirror, which focuses the beam on an array of tilted mirrors arranged around it. The rotating mirrors cause a continuous succession of beams to be reflected at different angles from the surrounding mirrors. All are directed upward through a filter toward the product with the bar-code label that is being moved past the window. This label is bombarded by beams striking it from different angles so that its orientation is not critical.

Within seconds enough energy will have been reflected from the label to obtain a valid reading. The light reflected from the label retraces the paths of the outgoing laser beams and is again reflected from the rotating mirror. But this time the reflected energy is focused by the parabolic mirror and strikes the photosensor on the opposite wall.

The photosensor converts the returned light into an electrical signal and the electronic circuitry converts the electrical signal into the digital code that can be interpreted by the store's computer. The computer then compares the digital code with a database of product names, sizes, and prices stored in memory. A customer receipt is printed out while the computer updates the store's inventory. The computer can also order stock replenishment if it has fallen below a quantity threshold level.

# INDUSTRIAL ELECTRONICS TECHNOLOGY

# Overview

*Industrial electronics* is the term for the specialized electronic devices, circuits, and systems developed to support the manufacturing, chemical, food processing, and other industries. It includes measurement instruments, data-acquisition and display systems, and process controls. These are typically heavy-duty components and systems because they are subjected to more abuse, a more demanding environment, and exposure to corrosive chemicals and vapors. Moreover, they must meet standards of reliability equal to if not greater than military electronics equipment to minimize downtime and increase productivity. Consequently, many components are packaged in heavier, more rugged enclosures than are comparable consumer or commercial electronics.

Industrial electronics make use of a wide range of hardware, from sensors and transducers to measuring and recording instruments for those variables associated with machine and process control such as temperature, pressure, speed, flow rate, acidity, current, voltage, and power. Industrial robots and manually controlled manipulators are examples of the marriage between mechanisms and industrial-grade electronics.

The subjects of closed- and open-loop control and servo- and synchrosystems discussed here could also have been placed in the section on military and aerospace electronics because of developments and advancements made in those areas for the control of military aircraft, ships, and weapons during and after World War II.

Computers have also had an impact on industrial electronics, primarily because of their ability to control machines and processes with software rather than their ability to process data and perform extensive computations. Computers for machine and process control eliminated banks of relays and extensive hardwired circuitry. They also made possible animated displays of processes in real time on video monitors. Machine operators and factory managers could see images of tanks being filled or emptied, valves being opened or closed, and chemical reactions taking place in real time in piping diagrams or electrical circuits. Computers made it easier to automate many processes, reduce production time, improve factory working conditions, and contribute to worker safety.

Today, as a result of price reductions in software, it is possible for desktop computers to act as virtual control consoles for monitoring and controlling many different kinds of industrial processes. In addition to presenting animated displays of system components, virtual instruments can present waveforms and digital readouts all on the same monitor.

Sensors and transducers are discussed in Sec. 17, "Electronic Sensors and Transducers."

# Data-Acquisition Systems

A *data-acquisition system* monitors *transducers* or *sensors* located in critical parts of machines or process lines, converts the output of the sensors or transducers to a common format for logging or display, and returns control signals to the machines or process. A block diagram for a typical system is presented in Fig. 22-1.

Sensors or transducers monitor the process. Their output signals are amplified and sent to an *analog multiplexer,* which acts as a kind of rotary switch to sample the outputs in repetitive order. A *sample-and-hold circuit* obtains a representative value of each output, which it sends to an *analog-to-digital converter.* In this system a digital computer is the controller and it can be programmed to record the data for record purposes or perform calculations to obtain other data in real time.

Sensor values can be compared with preset values, and if they are outside those limits the computer can send corrective signals. The digital control signals are converted to an analog format by a *digital-to-analog converter* to return to the source, thus completing a control loop.

Among the temperature sensors or transducers commonly used in data acquisition systems are *thermoswitches, thermocouples, resistance-temperature detectors* (RTDs), and *thermistors.* The force and pressure transducers include *strain gages* and *piezoelectric devices. Flowmeters* are commonly used to measure liquid or gas flow.

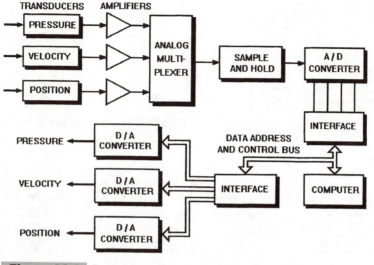

**Figure 22-1** Data-acquisition (DAQ) system.

# Virtual Instrumentation

Data-acquisition (DAQ) software now makes it possible to control and monitor industrial processes with virtual control panel displays on computer monitors. The monitor can display various meters and control switches in color that simulate such instruments as oscilloscopes and multimeters. It can also produce animated diagrams indicating the status of a process that includes warning alarms.

Analog signals provide the input for voltage, current, and power measurement, signal and transient analysis, and data logging, and analog output signals can be provided for machine and process control, waveform generation, and variable voltage sources.

The software interfaces either with plug-in DAQ boards or with external DAQ boxes. Included in the software is the binary code needed to configure the registers of the box or board for analog, digital, and timing input/output. The software controls the gain per channel, sampling rate, sampling order, digital levels, and counter and timer functions. It also controls the transfer of data to and from the board or box by polling, interrupts, or direct memory access. Some DAQ software is designed to control one specific application very well, but no other. Other general-purpose software can be dedicated to specific tasks by skilled programmers with a knowledge of the process to be monitored.

# Digital Panel Meters

The *digital panel meter* (DPM), as shown in Fig. 22-2, is a compact electronic system capable of measuring analog variables and converting them into an accurate digital reading on

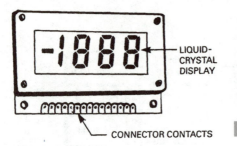

**Figure 22-2**  Digital panel meter (DPM).

the instrument's display. The principal circuits in a DPM are an *analog-to-digital converter* (ADC) and decoding and driving electronics for the digital display. The DPM is typically packaged in a case designed for mounting in a panel cutout. Battery- or line-powered DPMs can be modified to display a wide range of physical variables such as voltage, current, power, rpm, temperature, or flow rate. DPM displays are typically liquid-crystal displays (LCDs) or light-emitting diodes (LEDs).

DPMs can achieve accuracies of 0.1 to 0.005 percent of full scale, generally better than those obtainable from the analog panel meter (APM). The input voltages are obtained from sensors for temperature, pressure, force, velocity, weight, or distance. Other variables that can be measured include current, voltage, frequency, power, and acidity.

Originally intended to replace analog moving-coil meters, the DPM displays a numerical readout that can be read accurately 6 to 10 ft (1.8 to 3.0 m) from the DPM. For many applications this permits faster, more accurate measurements with less fatigue.

DPMs, initially only readout instruments, have evolved into single-channel data-acquisition systems. When matched to a specific transducer or sensor with the proper interfacing circuitry, the DPM is a stand-alone, self-powered measuring system capable of powering the sensor, generating and transmitting data, and providing a local readout. Some models also contain circuit cards for amplifying low-level signals from thermocouples, flowmeters, RTDs, and other sensors. Some also include circuitry for converting the variable being measured to ASCII code for transmission to a computer.

Factory-made DPMs are available in a wide variety of sizes and form factors with choices of character heights, number of digits per display, and power supplies. The standardized case size for industrial DPMs is $4 \times 2 \times 4$ in ($10 \times 5 \times 10$ cm), and most are 120- or 220-VAC line powered. Read-only DPMs intended as APM replacements are typically housed in smaller cases and are battery powered.

Typical DPMs have 3-, 3½- and 4½-digit displays. The ½ in the specification of the DPM's display refers to the use of the digit 1 in the most-significant-digit position. The full scale on a 3½-digit display, for example, is 999, but the additional digit 1 permits the display to show a value that is 100 percent higher (1999) in what is known as the 100 percent *overrange* condition. *Overload* occurs when the input voltage exceeds the 100 percent overrange condition. This is usually indicated by the repeated flashing of some element of the display. The terms *accuracy* and *resolution* with respect to DPMs are frequently confused. Resolution depends on the number of digits in the display. For example, a 3½-digit DPM can resolve 1 part in 2000 or 0.05 percent.

# Closed-Loop Control Systems

A *closed-loop control system,* as shown in the block diagram Fig. 22-3, has one or more *feedback control loops* that continuously compare system response with an input command that dictates a desired speed, position of the load, or other response. A *sensing device* in the feedback loop, such as an encoder, tachometer, or thermostat, senses any difference between the input command and the system response and generates an *error signal,* which is sent to a *controller and amplifier.* The signal from the amplifier alters the input to the *motor* and *load* so its output will cancel the error signal. The sensor's performance is directly proportional to the activity being controlled, such as the speed or position of the load or the ambient temperature. It is also called a *feedback control system.* Closed-loop systems are classified according to the variable being controlled. The most common control variables in electromechanical systems are *velocity, position, torque,* or various combinations of these.

## VELOCITY-CONTROL SYSTEMS

A closed-loop *velocity-control system* must contain a sensor in its feedback loop that can sense changes in velocity and produce an error signal proportional to deviations from the desired velocity setting. A *tachometer,* as shown in the block diagram Fig. 22-4, performs this function. It produces an electrical output that is proportional to the velocity of the motor. The error signal, proportional to a change in velocity, is sent back to the *controller and amplifier* to alter motor speed to maintain the desired velocity despite load changes.

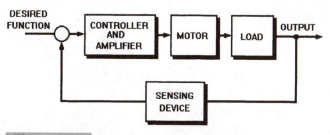

**Figure 22-3**  Closed-loop control system.

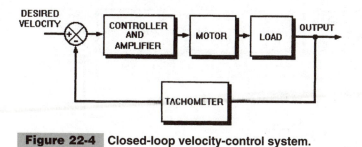

**Figure 22-4**  Closed-loop velocity-control system.

## POSITION-CONTROL SYSTEMS

A closed-loop *position-control system,* as shown in the block diagram Fig. 22-5, contains a sensor in its feedback loop that can determine the position of a moving mechanical element with respect to a preset reference. Examples of sensors capable of measuring position are *optical encoders* and *resolvers*. They can determine when a load, such as a cutting tool, shaft, or lever, has reached the desired position by counting pulses proportional to incremental movements. The pulse count is compared with an input setting, and the controller stops the movement when the counts are equal. Most position-control systems also include a separate velocity-control loop to stabilize the system.

## MOTOR TORQUE-CONTROL CIRCUITS

A *torque-control circuit* compares the motor's output current with its input current and amplifies that difference for use as an error signal to achieve loop closure. A constant current must be applied to the motor if it is to maintain the desired torque control because torque is proportional to motor current.

## INCREMENTAL MOTION-CONTROL SYSTEMS

An *incremental motion-control system* is a closed-loop control system that combines two or more control modes serially to accomplish an objective. A system with both velocity and position feedback loops can be programmed to follow a velocity profile before the desired position is reached. It could first ramp up to a constant velocity and then ramp down to zero before the system is switched so that the position-control loop stops the shaft or other moving element at a precise position.

# Open-Loop Control Systems

An *open-loop control system,* as shown in the block diagram Fig. 22-6, is one that has no means for comparing its output with its input for control purposes. In industry a machine or tool that depends on a *stepping motor* to position an object or load is an open-loop sys-

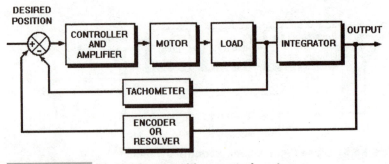

**Figure 22-5**  Closed-loop position-control system.

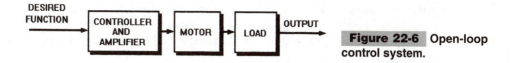

**Figure 22-6** Open-loop control system.

tem because the stepping motor responds to programmed input pulses that cause the motor's rotor to move in increments until the desired position is reached. A precise stepping motor with small increments of motion can receive a precise number of pulses that will position it at the desired location with negligible error. The pulses can be supplied by a variable pulse-generating circuit or by a computer.

# Servosystem Control

A *servosystem* is a specific type of electromechanical *closed-loop control system* as shown in Figs. 22-3, 22-4, and 22-5. In a servosystem, the output variable is measured, fed back, and compared to the desired input function at the summing point (symbolized as an X in a circle). Any difference between the two is a deviation or error, which is amplified as part of the correction process. The response of the system depends on how the loop is closed. A closed-loop *transfer function* is the relationship between the output and the input.

A servosystem will continue to operate as designed despite changes in load conditions, amplifier gain, wear on mechanical components, and even changes in the ambient temperature because the feedback loops can compensate for these changes. By contrast, the open-loop system shown in Fig. 22-6 will perform as designed as long as the system variables remain constant. Any change in load, amplifier gain, or wear in some mechanical component will result in increases in uncorrected errors.

Because servosystems are inherently unstable, care must be taken to prevent the system from responding to transient conditions which could cause it to go into uncontrolled oscillation. This generally calls for some form of damping or time delay that will allow the system to remain sensitive yet will avoid constant corrections as the transients pass around the loop. On the other hand, if the delay in error correction is too long due to poor system response, the error will increase until the system becomes unstable. An unstable control system is unable to cancel its error signal and can go into oscillation, a result that might cause damage to system components.

A servosystem with high amplification or gain must be able to correct its errors rapidly, but response time is less critical in low-gain systems. After any corrective action is initiated in the system, steps must be taken to prevent *overshooting*. This might be accomplished by compensating for any inherent system time delay. The objective of good servosystem design is to achieve a balance between load positioning accuracy and load stability.

The stability of a servosystem can be determined by subjecting the system to a large input error called a *step function* or *step command* and observing system response. This is done by introducing a large constant voltage. A stable servosystem will always return to a stable operating state unless there has been a component failure. However, a frictionless system will respond to a step function by going into oscillation, known as *hunting*. This will continue indefinitely unless the energy is somehow removed from the load.

The large positive error is amplified to start the motor and drive the shaft in the positive direction. With no friction or braking in the system, the load continues on past the input/output alignment or stability reference line because of energy stored in the load or inertia. This condition is termed *overshoot.*

When overshoot occurs, the error signal will be reversed, but it takes a finite amount of time for the reverse torque of the motor to increase sufficiently to stop the shaft and load. Because the amplified negative error persists, motor torque accelerates the shaft and load from the zero reference in the reverse direction. Once again, inertia causes the shaft to continue past the stopping position. This condition is termed *undershoot.* Hunting continues until the energy is removed from the load. One effective means for removing energy to damp out hunting is to apply friction brakes.

The three possible responses to a step command in a servosystem with friction or friction braking with respect to time are shown in Fig. 22-7: (1) *underdamped,* (2) *critically damped,* and (3) *overdamped.* The upper underdamped response appears as a damped oscillation curve. As friction increases, the number and amplitude of overshoots and undershoots decreases to zero over time. However, if excessive friction is applied to the system, it becomes overdamped, as shown in the lower response curve. But when the system is *critically damped* (just enough braking is applied to prevent minimal overshoot), the response is as shown by the middle curve. Most servosystems are designed for slight underdamping because an underdamped system is more responsive than either a critically damped or an overdamped system.

## SERVOSYSTEM BANDWIDTH

*Servosystem bandwidth* is the frequency band over which amplifier gain is substantially constant. It is defined as the difference in frequency between the half-power points $F_1$ and $F_2$, as shown in Fig. 22-8. The half-power points are determined by the frequencies at which the gain is 0.707 times the midband output, or about 70 percent of that value as shown on the plot of gain versus frequency.

## DIGITAL SERVOSYSTEMS

A *digital servosystem* is one that contains at least one digital component. For example, if the analog components are the motor, tachometer, resolver, and amplifier, the digital com-

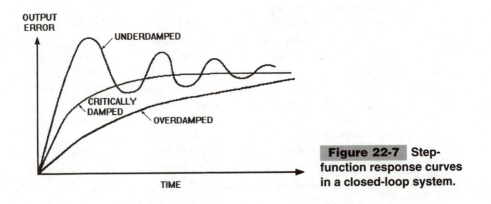

**Figure 22-7** Step-function response curves in a closed-loop system.

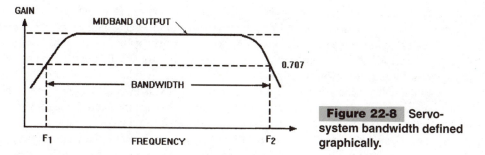

**Figure 22-8** Servo-system bandwidth defined graphically.

ponent could be a microcontroller or an optical encoder. The system might also include another digital component such as a resolver-to-digital converter. Digital components improve servosystem performance by allowing more design flexibility.

# Synchrosystems

A *synchrosystem* connects two shafts electrically so that the angular position or rotation of one shaft will always be synchronized with the angular position of the other shaft. Synchros, also called *selsyns,* are small, self-synchronous alternating-current machines. A synchrosystem acts as if two trains of gears were coupled together with a flexible shaft. However, electrical transmission permits the transmitter and receiver to be much farther apart than would be practical with mechanical coupling. Synchros are classified as angle-sensing transducers. Shaft angle is used in the measurement and control of position, velocity, and acceleration.

## CONTROL TRANSMITTERS AND RECEIVERS

The simplest synchrosystem consists of two units, a *control transmitter* and a *control receiver,* each with mechanically and electrically identical rotors and stators, as shown in the simplified schematic Fig. 22-9. Both transmitter and receiver rotors have single windings which are

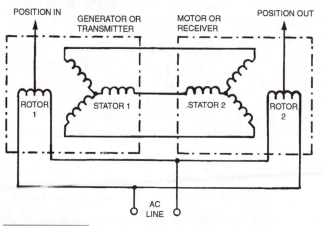

**Figure 22-9** Synchrosystem schematic diagram.

connected across the single-phase AC excitation source. Both stators have three wye-connected field windings spaced 120° apart. The ends of each winding on the transmitter stator are connected to the ends of windings on the receiver stator to form a three-wire circuit.

When the AC exciting circuit is energized, each rotor winding acts as the primary of a transformer, and a voltage will be induced in the three windings of its stator. The magnitude of the voltage in each stator winding depends on the position of the rotor. When the transmitter and receiver rotors are in corresponding positions, the stator voltages across corresponding stator terminals in the two units are equal in magnitude and opposite in phase. Thus there is no current flowing in the connecting wires between the stators of the two units.

However, if the rotor positions do not correspond, the voltages produced by transformer action in the transmitter stator windings will differ from those similarly induced in the receiver stator windings. The resulting currents that flow between the two stators provide the torque that turns the receiver rotor to a position which corresponds to the transmitter rotor's position. Signals representing an angular difference are transmitted by turning the transmitter rotor through an angle. The receiver rotor responds to those signals and immediately moves through the same angle.

## CONTROL TRANSFORMERS

A *control transformer* is similar to a control transmitter. However, in acting as a transformer, it accepts, at its three-wire stator terminals, the same kinds of signals that are produced by a synchro control transmitter which correspond electrically to some shaft angle. Thus it produces at its three-wire rotor terminals a three-wire set of carrier-frequency signals that are proportional to the sine of the angular difference between the electrical input angle and the mechanical angular position of its shaft.

## CONTROL DIFFERENTIAL TRANSFORMERS

A *control differential transformer* is similar to a control transmitter except that its rotor has a three-wire AC output. It accepts a set of signals at the carrier frequency (the kind produced by a control transmitter) at its three-wire stator terminals. The line-to-line amplitude ratios of this signal set correspond to a remote shaft angle. The differential transformer then produces a three-wire set of carrier-frequency signals at its three-wire rotor terminals whose line-to-line amplitude ratios are proportional to the difference between the input angle and the mechanical angular position of its shaft.

In addition to synchros, there are other angle-sensing transducers. These include the *encoder*, the *potentiometer*, and the *resolver*. Resolvers are related to synchros in that they also present information about the angular position of a shaft in the form of relative amplitudes of the excitation frequency. All signal and resolver signals, rotor and stator, input and output, are sine waves at the same frequency and in time-phase synchronization.

# Robotics

A *robot* is a reprogrammable, multifunction manipulator that can move material, parts, tools, or specialized devices through various motions for the performance of tasks under programmed control. The key words that distinguish a robot from either a specialized machine tool or a manually operated manipulator are *programmed control,* which today generally means *computer control.* A typical industrial robot system is shown in Fig. 22-10.

Most robots today are stationary industrial machines assigned to perform work in factories. Their work includes heavy-duty materials handling, welding, and painting. This work is often done in locations where there are hazards for human operators such as flying sparks from welding torches, fumes from paint or burning metal, extreme heat from proximity to furnaces, or intense persistent noise. Much of the work done by industrial robots calls for heavy lifting or contact with caustic or toxic chemicals. This working environment is stressful for workers, and can contribute to accidents because of constant distractions or discomfort.

However, many light-duty assembly and inspection robots are located in comfortable environments in close proximity to workers where they perform monotonous, repetitive tasks faster and more precisely than human operators. They can work 24 hours a day without rest breaks. Typical examples of work performed by these robots includes the picking and placing of electronic components on circuit boards for soldering, and the inspection and testing of finished circuit boards for quality.

Not all true robots are readily recognizable because they do not have the typical robotic arm. They are specialized machines or instruments designed to conduct scientific experiments in deep space, on planets, or at the bottom of the ocean under software control. A spacecraft or planetary lander can be computer programmed to carry out robotic functions, but human operators might intervene periodically to assist in specific tasks or redirect the robot. Less sophisticated hospital food tray and office mail delivery robots on wheels have been performing routine delivery tasks. Many different kinds of mobile home-housekeeping robots have been invented, but none has ever proved cost-effective and none has ever been produced in quantity.

Industrial robots have three major subassemblies:

1. A *manipulator* or *arm* that performs the required tasks
2. A *controller* that stores information, instructions, and programs to direct the movement of the manipulator
3. A *power supply* that drives the manipulator

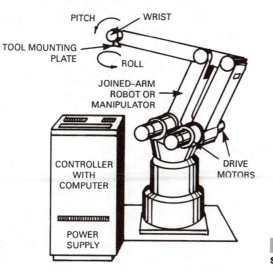

**Figure 22-10** Industrial robot system.

## ROBOT MANIPULATORS OR ARMS

The *robot manipulator* or *arm* defines the robot's capabilities. Typically a series of mechanical linkages and joints, the arm is able to move in different directions to perform work. It can be powered directly by electric, hydraulic, or pneumatic actuators coupled to the mechanical joints or links, or it can be driven indirectly through gears, chains, or ball screws.

The articulated sections of the arm are moved to position the robot's *end effector* or *tool holder.* The most widely used are grippers powered by electric solenoids or pneumatic cylinders which can grasp and release loads. Others are paint-spraying heads, gas torches, arc welding rod holders, glue applicators, and suction cups, to mention but a few.

Many robots are equipped with sensors that determine the positions of their various links and joints, and transmit this information continuously back to the controller. Feedback sensors can be as simple as limit switches actuated by the robot's arm, or they can be as complex as *encoders, potentiometers,* or *resolvers* for measuring position or *tachometers* for measuring velocity. The output from these sensors can be either digital or analog signals.

## ROBOT CONTROLLERS

The *robot controller* performs three functions:

1. Initiates and terminates motions of the manipulator
2. Stores position and sequence data in memory
3. Interfaces with external data-acquisition systems

The performance of the robot depends on the capability of the controller. A controller for a simple open-loop nonservoed robot can be a mechanical step sequencer with an electric counter, diode matrix, or a series of potentiometers as its memory. However, the controller for a closed-loop servoed robot could be a programmable controller or a computer with hard and diskette drives. Some robot programs can be stored in read-only memory (ROM), and others require either hard drives or some removable storage media. The controller can be part of the manipulator or it can be housed in a separate cabinet.

The controller initiates and terminates the motions of the manipulator through interfaces with the manipulator's control valves or electric feedback sensors. It might also be required to perform the calculations needed to control the path, speed, and position of the *end effector* which performs the work. The motion and position control loops of the servoed robot are closed through the controller.

## ROBOT POWER SUPPLIES

The *robot power supply* provides the energy to drive the manipulator's actuators. In electrically driven robots the power supply regulates the incoming AC line power and provides the required AC and DC voltages to drive the joint motors and other actuators. However, the power supply of a hydraulic robot includes a hydraulic reservoir and pump. If the robot is powered by compressed air, the air is usually supplied from a shop air compressor.

# Robot Classifications

Robots can be *nonservoed* (open loop) or *servoed* (closed loop). Nonservoed robots depend on the accuracy of the actuators and the precision of their gears and links to position the end effector correctly at or on the work (which is usually prepositioned). Most nonservoed robots are *limited-sequence* machines. By contrast, servoed robots include appropriate sensors in their closed loops to report the position of the end effector to the controller at all times.

A second classification relates to the motion of the end effector or tool in carrying out a programmed task: *point to point* or *continuous path.* The end effector of a nonservoed or limited-sequence robot is usually limited to point-to-point motion. These tools can be powered by hydraulic or pneumatic cylinders, vane motors, or electric stepping motors.

By contrast, servoed robots are capable of continuous-path motion. Their end effectors are moved under computer control in continuous sweeping motions that reach the work by the shortest paths, thus avoiding contact with nearby machines or structures. These robots can be programmed to control both speed and path contour.

*Degrees of freedom* refers to the motion that can be carried out by the manipulator or arm. The number of degrees of freedom determines the robot's possible applications and flexibility. In general, the number of degrees of freedom equals the number of the robot's articulated joints. Most limited-sequence robots have only two or three primary degrees of freedom, and their tasks are usually limited to opening and closing grippers.

However, servocontrolled robot arms have at least three primary degrees of freedom for the placement of the wrist or the end of the forearm in the most favorable position to perform useful work. The wrist can have two or three additional degrees of freedom so that a tool or gripper can be positioned at the optimum angle of attack to perform the required task.

The *robot envelope* is the three-dimensional shape described by the motion of the end effector or wrist as it moves completely through its outer limits of motion.

# Computer Vision Systems

*Computer vision,* also called *machine vision,* depends on a computer to acquire, interpret, and process visual information. Most industrial computer vision systems today include charge-coupled device (CCD) TV cameras as their primary sensors. The camera forms a video image of an object, and image-processing techniques are used to give the system the ability to recognize, classify, position, or orient objects in the camera's field of view.

Industrial vision systems are used for the inspection, positioning, counting, measuring, and classification of objects. However, the most common application for computer vision inspection systems is parts inspection, as shown in Fig. 22-11, permitting them to sort good from bad parts. Vision system inspection procedures are developed with conventional computer programming techniques. These call on previously written subroutine packages to perform the visual processing steps.

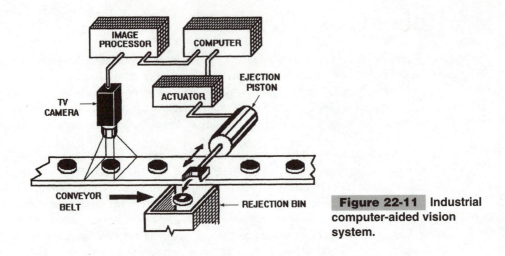

**Figure 22-11** Industrial computer-aided vision system.

The host computer contains the visual characteristics of different prototype objects in its memory in the form of templates so that it can classify an unknown object in its field of view by comparing it with those different templates. The templates are a collection of feature values that represent numerical quantities or measurements. They are independent of the part's position and orientation so they can be used as a basis for comparison between two images.

Vision systems can also provide position and orientation information about an object in its field of view. Typically, the system must be given the zero position information about each object that is imaged so it can report displacement from that position. Vision systems can also be used in noncontact measurement. The measurements are made when the system is directed to find one or more dimensions of the object. The accuracy of visual measurement depends on the resolution of the image, the placement of the camera, the lighting, and other factors such as the object's reflectivity and composition.

CCD cameras can scan at a rate of 30 complete frames per second, but this rate can be varied with external clock signals. The output voltage of the camera is proportional to the time integral of light intensity during the frame scan. The lens in the camera determines the field of view. A short focal length lens gives a wide-angle view of the object and a long focal length lens gives a telephoto view. The camera is connected to the computer that performs the signal processing, and the interface hardware performs the timing, digitizing, and circuit isolation or buffering. The digitizing circuit for the display can be a *thresholding circuit* for a binary system or an analog-to-digital converter (ADC) for a *gray-scale system*.

A thresholding circuit represents all gray levels in the image above a certain threshold as 1 (white) and all those below the threshold as 0 (black). A gray scale is a series of achromatic tones that covers a full range of grays between white and black, usually divided into 10 steps.

# Programmable Controllers

A *programmable controller* (PC) is a digital controller that provides the logic required for decision making in industrial control applications. It includes solid-state logic that can be altered without wiring changes, making it a direct replacement for relays and hardwired solid-state electronics for the control of motors, solenoids and other actuators. Designed for operation in hostile industrial environments, the controllers do not need fans, air conditioning, or electrical filtering. They are programmed with simple relay ladder diagram language.

Direct electrical connections are made between the controller and the actuators. The desired logic sequence for the actuators is programmed into the controller with front panel switches or an ancillary programming device and stored in the controller's memory. Changes in logic sequence can be made rapidly by reprogramming, avoiding the need to change any wiring. Consequently, programmable controllers are reusable if the equipment being controlled is no longer needed. Indicator lights are provided at its major diagnostic points to simplify troubleshooting. Maintenance procedures are simple to assure minimum downtime and maximum equipment productivity. In most cases only input/output (I/O) modules need be replaced.

Programmable controllers were introduced in the late 1960s in the automobile industry to avoid the costly and time-consuming rewiring of machine control systems at times of model changeover. They have evolved with the availability of advanced solid-state circuitry and now include many of the features of computers. Because they are compatible with computers, the activities of many programmable controllers in a factory can be monitored and controlled by a cable-connected computer.

The main functional blocks of the PC are its central processing unit (CPU), memory, I/O modules, and power supply. The control logic program stored in the memory directs the CPU's selection of events and their logical sequence. For example, if switch $A$ is closed, then solenoid $B$ will be energized. The CPU scans all of the I/Os in a fraction of a second, and records the status of each input and output in memory.

Necessary control actions are initiated after the instructions are executed and the results are scanned. The logic determining new output states is solved after the I/Os are scanned. Scanning is repeated frequently so that any changes in inputs and outputs can be detected by the CPU. Inputs are signals from sensors and switches in the control loop, and outputs are signals that actuate the machine or industrial process being controlled.

Some controllers can perform data acquisition and store data. They can also report on the execution of commands, perform complex mathematical algorithms, and control both servo motors and stepping motors. Many controllers include provision for self-diagnosis and system troubleshooting. Controllers have as few as 32 I/Os and as many as 8000.

The controller is organized to execute a series of sequential scanning tasks and emulate, as nearly as possible, the operation of a relay panel. These tasks include checking the status of all I/Os, solving logic problems in accordance with the logic program in memory, performing self-diagnostics, and communicating with its programming panel.

Programming languages for controllers manage many I/O points that are part of the control loop. Four major languages are used by controllers: (1) *boolean equations,* (2) *mnemonic programming,* (3) *logic diagrams,* and (4) *ladder diagrams.* The most popular language is ladder diagram logic, which is related to relay ladder logic.

I/O modules for PCs reduce input voltages to logic signal levels that can be handled by the CPU and also convert logic signal levels from the CPU to voltage levels for the control of system actuators. In addition, they isolate the CPU components from electrical noise. I/O modules accept inputs from limit switches, pushbuttons, strain gages, thermocouples, and other sensors. They also provide ASCII serial interfacing. Some I/O modules now include microcontrollers to preprocess information before it reaches the controller's CPU, increasing its speed of signal processing.

Controllers ordinarily give only on-off commands that cannot directly control machine speed or position. However, if their scan rate is fast enough for motion control, the controllers can be equipped with digital position sensors, actuators, and control panels. With this equipment they can perform motion control in either closed-loop feedback systems or open-loop positioning systems. In a closed-loop feedback system, the controller can control the movement of a sliding part of a machine. For example, command signals to direct the part to move a specified distance can be entered digitally by a keyboard or preset program. The analog output of a digital-to-analog converter (DAC) is sent to an electric gearmotor drive which turns a leadscrew to move the machine element. When the part reaches the desired distance, a resolver, also geared to the leadscrew, sends a feedback signal to stop the motor at the position specified.

Similarly, open-loop control can be achieved with a module that controls the motion of a stepping motor by producing drive pulses. The controller sends data and motion commands to the module and compares the motor's actual and programmed positions. If the positions are not the same, a new command is sent to the motor to adjust the actual position of the load.

# MILITARY AND AEROSPACE ELECTRONIC SYSTEMS

# Overview

The armed forces require electronic systems that are substantially more rugged and reliable than commercial and industrial electronic systems, especially those that will be installed on combat aircraft, ships, and vehicles. These requirements are spelled out in military specifications and standards that cover all phases of procurement from design of components to manufacturing and testing. The qualification of basic components from resistors and capacitors to ICs is more thorough, and some semiconductor devices are subjected to 100 percent testing, and those tests must be documented. Military electronics equipment is subjected to more severe environmental stresses than comparable commercial, industrial, and consumer electronics. Emphasis is on the survivability of equipment operating at temperatures from the extremes of those encountered in the Arctic to those encountered in the desert, as well as exposure to shock, vibration, salt spray, and radiation from nuclear weapons. The lives of soldiers, sailors, and airmen depend on the reliability of that equipment.

Among the specialized electronics purchased only for military applications are long-range surveillance radars and systems for secure communications, missile guidance, active night-vision, and fire control for artillery, tanks, and antiaircraft guns. Nevertheless, many important parts of the military electronics inventory such as radio transceivers, telephone equipment, aircraft instruments, navigation aids, and ground-controlled landing systems are more rugged versions of comparable civilian equipment.

Electronics developed for military systems during World War II and for many years after, because of the cold war, were at the cutting edge of electronics technology, and much of this knowledge has filtered back to the civilian electronics industry. Many common consumer electronics products owe their origins to military equipment. The military-aerospace industry was at one time the largest employer of skilled scientists, engineers, and technicians, and the Department of Defense paid out billions of dollars for basic research and development and manufacture of military hardware. Funds were provided for various projects ranging from the advancement of electronics component manufacture to the development of lasers and satellite navigation, communication, and surveillance systems.

But the technical leadership in electronics seems to have reverted back to civilian hands with the introduction of digital computers. Computers, software, large-scale ICs and memory, and microprocessors were initially developed to meet growing commercial and industrial needs for faster and better data processing and computation capabilities. It was soon apparent that those items had a place in military hardware but because of the mandatory rigorous qualification processes, military electronics had trouble keeping up with the rapid advancements being made in civilian electronics. Their introduction into military systems occurred after they were modified to meet military standards and specifications for reliability and durability.

Most military electronics systems are custom made, with relatively few ever produced in quantities comparable to consumer or commercial products. Many are little more than advanced development models that are subject to frequent fixes and modifications in the field. Avionics for fixed-wing aircraft, helicopters, and missiles accounts for the largest share of the military electronics budget simply because they are produced in far larger numbers than combat ships, combat vehicles, and permanent communications and radar installations.

Military electronic systems are usually procured directly by the branch of military service that ordered them as government-furnished equipment (GFE), and they are sent to the system contractor for installation and integration with other equipment on a specific platform. Combat ships and large aircraft such as strategic bombers are likely to contain an intermix of older, reliable, and proven equipment along with newly developed or updated systems that reflect the latest changes in the technology of warfare.

Military electronics systems can be generally classified into six categories:

**1.** Communications equipment
**2.** Navigation and guidance aids

**3.** Weapons aiming and fire control
**4.** Target detection
**5.** Surveillance from land bases, ships, aircraft, and spy satellites
**6.** Electronic countermeasures (ECM) and counter-countermeasures (ECCM)

Military electronics equipment has benefited from the same advancements made in commercial, industrial, and consumer electronics: increasing transistor density per IC with decreasing IC line widths, faster microprocessors, and improved power sources. These advancements have reduced the weight, size, and power requirements for equipment while improving their reliability, but in some instances space and weight savings have been offset by the addition of more circuitry to improve equipment sensitivity, range, response, and versatility.

# Military Communications

Radio-frequency communications accounts for a large share of military communications equipment. Man-pack and mobile vehicular RF systems combine the transmitter and the receiver in a single transceiver case. This equipment now includes microprocessors for channel switching and other functions. Each branch of military service has its own ground microwave communications links, satellite links, and field telephones interconnected by wire or fiberoptic cable. Military command and control networks duplicate many of the functions of commercial networks.

In addition very low frequency (VLF) radio systems communicate with submerged submarines, acoustic transceivers permit underwater communication, and infrared lasers permit covert, highly directional communications day or night.

# Navigation and Position Finding

Military aircraft and ships use radar for navigation and weapons fire control. Short-range navigational radar on ships provides an image of the shoreline and other ships in the vicinity. Some have the ability to distinguish moving objects from fixed objects. Airborne radar screens show geographic features such as rivers, lakes, and shorelines. Accurate determination of a ship or aircraft's position in latitude and longitude can also be determined with parabolic navigation systems such as Loran C and satellite navigation systems such as the Global Positioning System (GPS). Aircraft altitude can be determined precisely with radar altimeters and GPS receivers. Water depth under a ship can be determined precisely with acoustic depth sounders.

Automatic microwave beacons in the identification, friend or foe (IFF) system on ships and aircraft respond to encoded interrogation signals from friendly ship or aircraft radars. The IFF system is intended to protect ships, aircraft, and vehicles from accidental attack by friendly forces. Inertial guidance systems are installed on submarines and long-range aircraft as navigational aids. This equipment might now include laser gyroscopes in place of conventional mechanical gyros and accelerometers. In general, military navigation and position-finding electronics are similar to the civilian versions installed in ships and aircraft.

# Weapon Aiming and Fire Control

Specialized fire-control radars have replaced earlier optical range finders and electro-mechanical computers for determining target range and bearing for aiming ground-based artillery or shipboard guns. Specialized radars have also been developed that can determine the locations of hidden enemy gun positions by extrapolating data obtained by tracking incoming projectiles.

Other radar fire-control systems aim and guide ground-to-ground, ground-to-air, air-to-air and air-to-ground missiles. Radar systems that support ground and shipboard missile launchers provide instantaneous target position and relative motion data as well as the beams for guiding "beam-riding" missiles. Airborne radars provide comparable capability for aircraft-launched air-to-air and air-to-ground missiles.

Antimissile missiles are designed to home in on active radar or jamming transmitters operating either as pulsed- or continuous-wave systems. The only defense for a radar installation against attack by a radio-frequency-sensitive missile is infrequent and random transmission of RF and the immediate shutdown of its transmitter when it has been warned of possible attack. Independent guidance systems on air- or ground-launched missiles can make the necessary course corrections to reach the target area, and radar fuses can detonate warheads at some distance from the target without actually striking it. Cruise missiles with a range of about 500 mi (800 km) carry onboard navigation systems based on GPS guidance or computerized map reading to make pinpoint strikes on designated targets.

Active laser range finders can determine the precise range of targets for conventional artillery or guns mounted on mobile platforms such as tanks. Lasers can also be used for terminal guidance of ground-to-ground, ground-to-air, and air-to-ground missiles equipped with sensors that permit them to automatically track infrared energy reflected from a target. Weapons with these sensors are called "smart" weapons. This capability permits highly selective targeting to reduce collateral destruction of nonmilitary targets, minimize civilian casualties, and safeguard friendly forces in the vicinity.

Both short-range antitank missiles fired from launchers and naval torpedoes launched from submarines feed out thin wires that are used to transmit midcourse correction signals to the weapons to assure that they strike the intended target. Some missiles are equipped with passive infrared-seeking circuits that permit them to home in on heat sources such as an enemy tank or aircraft's engine exhaust. But for this weapon to be effective, the target must be isolated from background heat sources and positively identified.

# Target Detection

Large, powerful, long-range over-the-horizon radars can detect the approach of ballistic missiles and alert defensive systems thousands of miles away. Moreover, powerful phased-array search radars on large ships such as aircraft carriers and cruisers as well as airborne radars on early warning aircraft (AWACS) can detect aircraft and tactical missiles hundreds of miles away. Shorter-range fire-control radars can detect immediate local threats less than 50 mi away and dispatch intercept missiles.

# Enemy Surveillance

Surveillance or long-term monitoring of the enemy can be carried out with high-definition spy satellites, long-range early warning and communications (AWACS) air-craft, or dedicated high-altitude U-2 type reconnaissance aircraft. Enemy activity can be tracked with film or video cameras sensitive to natural light or infrared. Radar mapping can be carried out at night or through fog and haze that would defeat conventional pho-tography.

# Electronic Countermeasures (ECM)

*Electronic countermeasure* (ECM) radio-frequency jamming equipment is an important military weapon for depriving the enemy of effective radio communications and radar return signals. It can be accomplished by powerful broadband transmitters having the necessary agility to change transmission modes and frequency rapidly in response to enemy reaction. The transmitters can be on ships, aircraft, or vehicles. However, any active jammer, like any active radar, invites retaliatory strikes from enemy missiles capable of homing in on the transmitter's RF emissions and destroying ground-based antennas or air-, sea-, or land-based platforms containing the transmitter.

Passive monitoring by satellites and high-altitude aircraft seeks to locate and identify possible electronic threats to ships or aircraft over or near enemy territory. In addition, most combat aircraft and helicopters carry receivers that permit them to detect such electronic threats as missile guidance beams and determine the bearings of their sources so they can take evasive action.

Specialized electronic countermeasure (ECM) aircraft called *weasels* are equipped with more extensive broadband radar and radio receivers that can also pinpoint sources of radi-ation, either defensive countermeasures or offensive guidance beams, as they fly low over enemy territory. These aircraft are fitted with antiradiation missiles that can home in on enemy antennas and transmitters to destroy them. If the targets are electronic countermea-sure (ECM) transmitters, the aircraft are performing an *electronic counter-countermeasure* (ECCM) function.

Radar-guided missiles can be misdirected by transmitting false echoes with the same modulation patterns and frequencies as the missiles' guidance systems so that enemy radar operators misjudge the location of the targets. *Chaff* is also widely used as a defensive mea-sure. It consists of thin metal wires precisely cut to dimensions that are even multiples or fractions of the wavelength of the radar frequencies to be jammed. They form dipoles that reradiate incoming microwave energy. The wires are packed into rockets that are fired into the air in the direction of an enemy threat from ships and aircraft so that they form reflective clouds to confuse radar-guided missiles. Chaff produces *volume clutter,* and its effects are more pronounced at the higher X-band frequencies. To be effective the chaff wires must be cut to the lengths that will resonate in the band of frequencies to which the radar to be jammed is tuned. The shell explodes at a preset height and scatters the miniature dipoles into clouds that drift with the wind as they fall.

Infrared- or heat-seeking missiles can be deceived if the military ship or aircraft trying to defend itself against attack fires flares in the general direction from which the threat is expected. As an alternative to flares, some military aircraft and helicopters are equipped with defensive lasers that can be focused on approaching heat-seeking missiles to confuse their guidance systems.

# Avionics Systems

*Avionics* are electronic equipment designed specifically for mounting in aircraft. In military and commercial aircraft they typically are packaged in modular metal boxes that can easily be disconnected and removed for replacement or maintenance. Avionics equipment includes radio and radar transmitters and receivers and navigational equipment such as GPS receivers, radar beacons, and inertial guidance systems. Aircraft require special antennas that conform to their fuselages or are streamlined to minimize air resistance or drag. Radar antennas typically are mounted behind plastic cones at the nose of the aircraft or in streamlined domes under or on top of the fuselage.

## DISTANCE-MEASURING EQUIPMENT (DME)

*Distance-measuring equipment* (DME) is a two-way aircraft ranging system that includes an airborne interrogator and a ground-based transponder. The airborne interrogator transmits 3.5-$\mu$s, 1-kW pulses at the rate of 30 per second on one of 126 channels which are 1 MHz apart in the 1025- to 1150-MHz band. The transponder replies with similar pulses on another channel 63 MHz above or below the interrogating channel. The signal received by the aircraft is compared with its transmitted signal, their time difference is derived, and the distance is computed and displayed. There are about 2000 ground stations worldwide. Tacan is an enhanced military version of DME.

## TRAFFIC COLLISION AND AVOIDANCE SYSTEM (TCAS)

The *traffic collision and avoidance system* (TCAS) for aircraft is based on *secondary surveillance radar* (SSR). It consists of microwave-frequency transponders, computers, display screens, and voice alarms on each aircraft that protect it from possible midair collision. The system alerts all pilots to the potential for midair collision in enough time to allow them to take the necessary evasive action such as changing course, climbing, or descending to avoid the other aircraft.

When a TCAS-equipped aircraft interrogates another nearby TCAS-equipped aircraft, it receives a reply from the other aircraft's SSR transponder. The reply permits the interrogating aircraft to determine and display, the relative altitude and position of the other aircraft, and compute its closing velocity to determine if a collision threat exists. Some versions of TCAS include cockpit displays of complementary escape maneuvers for each converging TCAS-equipped aircraft.

Most commercial passenger aircraft flown by U.S. carriers are now TCAS equipped. Each transponder transmits a signal identifying itself and giving the altitude of its host aircraft. All TCAS-equipped aircraft in a given area are, in effect, simultaneously transmitting the same information to all other computers, giving their altitude, speed, bearing, and whether they are climbing or descending.

The onboard computers process all of the information being received from other aircraft and present it on a plan-position indicator (PPI) display that can show all other TCAS-equipped aircraft present within a 40-mi (64-km) radius. Although the screen normally is set for the 40-mi range when each aircraft is clear of immediate traffic, it can be set to zoom in to a 20-, 10- or 5-mi (32-, 16-, or 8-km) radius as the aircraft approaches its landing field or encounters higher traffic density. Aircraft within the displayed area appear on the screen as white, open diamonds.

The computer is programmed to concentrate on an envelope that defines the aircraft's protected airspace. Oval-shaped, the envelope encloses an area 1200 ft (366 m) above and below the TCAS-equipped aircraft and a variable distance ahead of and behind it. When another aircraft nears the envelope of a protected aircraft its computer alerts the pilot that his airplane's protected space will be penetrated in 40 s unless evasive action is taken. The message "Traffic, traffic" is loudly announced, and simultaneously the intruding white diamond becomes an amber circle. This change is accompanied by the appearance on-screen of a numerical readout of the approaching airplane's relative altitude and an indication of whether it is climbing or descending.

Should the aircraft get even closer—within 25 s of an intrusion into the protective envelope—the computer gives another warning. The amber circle representing the threatening aircraft becomes a bright red square, and a voice announcement orders the pilot to climb, descend, increase or reduce the rate of climb or descent, or take one of a number of other possible actions to avoid collision.

At the same time, the pilot of the airplane making the incursion receives instructions directing him or her to take actions correlated with those given to the first aircraft. If the pilot of the other aircraft follows these instructions the separation distance will be assured, and the risk of collision between the two aircraft will be minimized.

## VERY HIGH FREQUENCY OMNIDIRECTIONAL RANGE (VOR)

*Very high frequency omnidirectional range* (VOR) is a ground-based aircraft navigational system that operates in the VHF band. It is useful for high-flying aircraft out to distances of about 230 mi (370 km) and low-flying aircraft out to about 30 mi (48 km) line-of-sight distance from a VOR station. There are more than 1000 VOR stations in the United States and more than 1000 in the rest of the world. The VOR can be conventional or Doppler. Both operate on 160 channels between 108 and 118 MHz.

The Federal Aviation Administration (FAA) plans to retain parts of the VOR system for backup after it adopts the Wide-Area Augmentation System (WAAS) for commercial aircraft guidance. See "Wide-Area Augmentation System (WAAS)" in this section.

# Radar Systems

*Radar* (for *radio detection and ranging*) is an active radio transmitting and receiving system capable of detecting and determining the range and bearing of distant targets such as ships or aircraft. A radar system illuminates distant objects with RF energy and then receives, detects, and displays the reflected energy. Radars are installed in ground stations, aircraft, ships, spacecraft, and vehicles.

A typical pulsed-radar system is shown in the simplified block diagram Fig. 23-1. The transmitter emits high-powered RF in short pulses through a directive antenna that illuminates the object or target. The returned echo is received, usually by the same antenna, passed by a *transmit-receive* (TR) *switch* and amplified by a high-gain, wideband receiver. The output of the receiver can then be displayed on a cathode-ray tube in any of many different formats or on a raster-scanned liquid-crystal display.

Target bearing or azimuth is determined from the direction of the antenna's axis when the echo is received. Range is measured in units of time (typically microseconds) for direct conversion into distance (nautical miles, yards, meters, or feet). RF energy travels at the speed of light, 300 m/µs, so it takes about 7 µs for RF energy to make a 1-km round trip.

Radars transmit high frequencies, typically in the microwave band, because short wavelengths permit higher antenna directivity, more accurate range readings, and better resolution of targets. Analysis of the display yields information on target bearing, course, speed, and closest point of approach (CPA).

The duration and repetition rate of the transmitted pulses can be altered to obtain the most favorable echoes. The emission of short pulses at a rapid rate permits the reception of returns from nearby targets such as ships in a harbor and navigational aids such as marker buoys.

Some radars operate with continuous-wave (CW) RF transmissions that determine the range of an object by measuring the Doppler shift in the returned echo. This form of radar

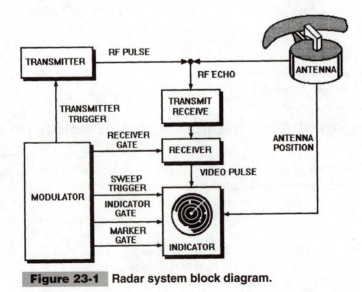

**Figure 23-1**  Radar system block diagram.

can determine a target's velocity. The transmitted RF can also be frequency modulated (FM) so that the target's range can be determined by comparing phase or modulation characteristics of the outgoing and returned signals.

Radar has been adapted to function in many different locations, and has been modified to perform many different tasks. The largest and most powerful ground-based military radars are capable of searching for targets over the horizon or identifying and distinguishing differences between spacecraft and discarded objects or trash in orbit. On the other end of the size scale, small packaged radars available for less than $2000 permit small ships and private yachts to navigate safely at night or in fog. Specialized airborne radars provide navigational aid in commercial aircraft and help them to avoid dangerous weather conditions such as thunderstorms or hurricane cloud formations.

Government radars have been designed specifically for air traffic control, ship traffic control, meteorology, and geological surveys. However, the widest variety of radars are found in the military services where, in addition to performing all of the functions required of commercial systems, they perform such duties as ground surveillance, aiming and firing weapons and missiles, tracking enemy shells and tactical missiles, and providing long-range early warning of the approach of ballistic missiles or long-range enemy aircraft. The largest military radars are ground-based phased-array radars with electronically steerable antennas for tracking ballistic missiles, spacecraft, and space-orbiting trash. Other very large and powerful radars are located on ships and long-range early-warning aircraft.

Self-contained air-transportable radar systems, as shown in Fig. 23-2, have power generators, radar screens, and radio communications equipment all in a single shelter with the antenna mounted on top. This system is part of a ground control system for guiding aircraft to advanced airfields. It can be disassembled and its antenna can be stowed in the shelter to fit into a cargo aircraft. Some of the smallest radars are altimeters for aircraft and portable man-pack battlefield surveillance systems.

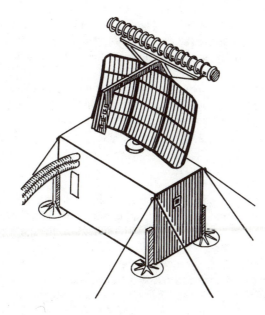

**Figure 23-2**  Air-transportable radar system.

All radars, regardless of size, have four principal functional groups: (1) *transmitters*, (2) *antennas*, (3) *receivers*, and (4) *displays*.

## RADAR TRANSMITTERS

In a pulsed radar system the *transmitter* provides the RF pulsed output. Its power must be sufficient for the intended application such as long-range search or navigation within a 20-mi (32-km) radius. The important transmitter characteristics for pulsed radar are: *pulse length, pulse rate* or *repetition frequency, duty cycle, peak power, average power,* and *carrier frequency.*

*Pulse length* is measured in microseconds. Longer pulse lengths are necessary to detect distant objects with search radars. But for short-range resolution (measured in feet or meters) pulses must be relatively short. *Pulse repetition frequency* (PRF) must be slow enough to permit the pulse to reach the outer limit of its range and return before the next pulse is transmitted. Most pulsed radars are designed so the PRF, antenna beamwidth, and rotation rate permit 20 to 40 pulses to be transmitted while the antenna is still focused on the target, thus allowing a repetition of returns.

*Peak power* and *average power* in watts are determined by the intended application. To improve range resolution when performance is limited by transmitter peak power, pulse-compression techniques are used. These waveforms permit range resolutions that are shorter than those corresponding to the radiated pulse width. The choice of carrier frequency is also determined by the radar's application, and it is influenced by the available power supply, permitted or practical antenna size, and space constraints on or in the platform.

The RF power source in radar transmitters can be a power tube such as a *magnetron* or *power klystron,* or a *master oscillator–power amplifier chain.* The oscillator might be a solid-state device, but the power amplifier for the chain is usually a *traveling-wave tube* (TWT).

The power tube oscillator or final power amplifier stage converts most of the electrical input power to RF pulse power in response to the *pulse-forming modulator.* It supplies a faithful replica of the pulse to the modulating grid of the amplifier or the cathode of the magnetron.

The *duplexer* or *transmit-receive* (TR) *switch* allows one antenna to be used for both transmission and reception. It is a protective device that shorts out and blocks strong transmission signals from entering the sensitive receiver and damaging it. See Sec. 7, "Microwave and UHF Technology."

## RADAR ANTENNAS

*Radar antennas* are made in many different shapes and sizes, but the most common for military applications are parabolic sections made of sheet metal that look like barrel staves or open-frame structures. Both types are mounted on motor drives that turn them at variable speeds through 360°.

A shipboard antenna for surface search or navigation is typically a paraboloidal section with its long axis horizontal. It produces a vertical fan-shaped beam as much as 30° high but only 2 to 4° wide to permit targets to be detected even if the ship is pitching or rolling.

An antenna for determining the altitude of approaching aircraft or missiles is usually a parabolic section with its long axis vertical so it will form a horizontal fan or beaver-tail beam. This beam might be 30° wide but only 2 to 4° high. In an aircraft ground-controlled approach (GCA) system the antenna might be mounted so that the wide part of the beam can be moved through a 90° angle with respect to the ground or sea to follow the changing altitude of the approaching aircraft. In some GCA systems the height-finding antenna is independently mounted so that it can rotate through 360° and in others it is mounted on the same 360° drive as the log-axis parabolic antenna for determining the aircraft's range and bearing.

The size and shape of the antenna reflector is determined by the transmitted frequency (multiples of half wavelength). Long-range radars that operate in the UHF band or at low microwave frequencies (500 MHz to 2 GHz) have large reflectors, but short-range radars that operate at the higher microwave frequencies (3 to 10 GHz) have smaller reflectors. Some are covered by RF-transparent *radomes* for protection against the weather. See Sec. 6, "Antennas and Horns."

## RADAR RECEIVERS

The received signals from the TR switch are mixed with local-oscillator signals to produce an intermediate frequency (IF) signal that is easier to amplify and process. Typical radar intermediate frequencies are 30 or 60 MHz. Most local oscillators today are solid-state devices, which have replaced reflex klystrons. Mixing occurs in a crystal cavity, and the resulting IF signal is amplified by the IF amplifier and then fed to the *detector,* which produces a video signal. The radar might have several parallel receiver channels to process more than one return simultaneously. The voltage of the video signal is proportional to the strength of the received signal, but that video signal must be amplified to drive the cathode-ray tube (CRT) or other form of radar display.

## RADAR DISPLAYS

The range and bearing of target can be indicated on many different kinds of CRT presentations. Of these, the *plan-position indicator* (PPI), shown in Fig. 23-3, is the most popular because it is the easiest to interpret, a desirable quality in emergency situations. It is a polar-coordinate presentation with the radar antenna represented as a spot in the center of the screen. Targets (ships or aircraft) show up as illuminated spots or pips, and land masses show up as bands or stripes on the periphery of the screen as the radial beam sweeps past. All images are refreshed on each pass of the beam, which is synchronized to antenna rotation. The positions of all targets and land masses are oriented with respect to compass or true magnetic direction, whichever is selected. Target movements on the screen are relative to the movement of the radar, which might or might not be moving.

The range and bearing of all targets can be determined approximately by observation of the screen and a knowledge of the range setting and the compass graduations. However, most modern radar systems now provide a numerical readout on the face of the screen when a visible *cursor* is placed on the target image. Systems with microprocessors can present the target's range, bearing, course, and speed directly on the screen in tabular form.

An airborne radar PPI presentation is similar to a shipboard presentation but it can be more difficult to interpret because the antenna is looking down on a large land mass. For

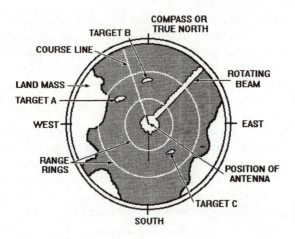
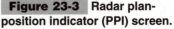

**Figure 23-3** Radar plan-position indicator (PPI) screen.

this reason, the presence of a shoreline, a river, lakes, a mountain range, or a distinctive landmark is helpful for navigation and targeting.

There are other more specialized radar displays in use today, primarily for such military applications as determining the altitude of aircraft. Fire-control radars have B displays, as shown in Fig. 23-4, because they are used to display the range and bearing of a specific target or targets in a narrow sector of the antenna's sweep. The H display, as shown in Fig. 23-5, is another special-purpose presentation of the target's range, bearing, and elevation on the screen at the same time.

Many modern military radars systems digitize the analog radar information obtained from the antenna and present it on a TV-like raster-scanned video display in which only moving targets are shown, and they have been converted to symbols. These computer-generated displays have removed all extraneous information from the screen that is not of immediate interest to the radar operator, weapons officer, or the ship's captain. These displays trade off the higher accuracy of a conventional analog PPI display to obtain a more simplified and easier to interpret symbolic presentation. These displays are similar to those used in aircraft traffic-control centers. See Sec. 13, "Display Devices and Systems."

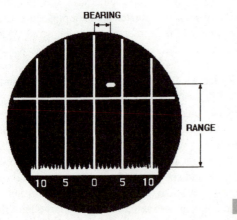

**Figure 23-4** B display for radar.

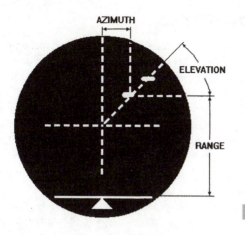

**Figure 23-5** H display for radar.

## RADAR FREQUENCY BANDS

The radar carrier frequencies are given in the microwave frequency designation chart Table 23-1. The spectrum is divided into bands.

■ Long-range air-search radars operate in the *L band*—1 to 2 GHz (15 to 30 cm wavelength but called the *20-cm band*).

■ Long-range marine radars operate in the *S band*—2 to 4 GHz (7.5 to 15 cm wavelength, but called the *10-cm band*).

■ Weather radars operate in the *C band*—4 to 8 GHz (3.75 to 7.5 cm wavelength).

■ Short-range marine radars operate in the *X band*—8 to 12 GHz (2.50 to 3.75 cm wavelength, but called the *3-cm band*).

■ Airport ground-control radars operate in the *K bands*—18.0 to 27 GHz (1.1 to 1.7 cm wavelengths, but called the *millimeter bands*). They provide the highest target definition, but their range is limited in rain or fog by water vapor that attenuates those frequencies.

| Frequency (GHz) | 0.1 | | 0.15 | 0.2 | 0.3 | 0.4 | 0.5 | 0.6 | 0.75 | 1 | 1.5 | 2 | 3 | 4 | 5 | 6 | 8.0 | 10 | 15 | 20 | 30 | 40 | 50 | 60 | 75 | 100 |
|---|---|---|---|---|---|---|---|---|---|---|---|---|---|---|---|---|---|---|---|---|---|---|---|---|---|---|
| Wavelength (cm) | | 300 | 200 | 150 | 100 | 75 | 60 | 50 | 40 | 30 | 20 | 15 | 10 | 7.5 | 6 | 5 | 3.75 | 3 | 2 | 1.5 | 1 | 0.75 | 0.6 | 0.5 | 0.4 | 0.3 |
| Military bands | | VHF | | | | UHF | | | | | L | | S | | C | | | X | $K_u$ | K | $K_a$ | | MILLIMETER | | | |

IEEE
STD 521-1976

**TABLE 23-1** Military frequency band designation chart.

## RADAR TARGET CHARACTERISTICS

The strength of RF signals reflected by a target is a function of its size, shape, and composition. The return will change if the target shifts its orientation with respect to the receiving antenna or if the radar switches to another frequency. Both aircraft and ships become nearly invisible if they are viewed head-on. Objects or targets made of fiberglass, wood, or other nonmetallic substances might be invisible to the radar. The concept of stealth ships or aircraft is based on the knowledge that some materials will absorb radar energy and that both absorbent and reflective surfaces sloped backwards with respect to incoming radar beam will deflect the energy away from the radar rather than toward it. Stealth aircraft take advantage of both of these effects by absorbing and deflecting radar energy to minimize or eliminate any echoes, thus permitting the aircraft to elude radar surveillance.

By contrast, small sheets of metal or metallized plastic joined so that they form corners are highly reflective of radar signals. They can return more energy to the radar receiver than far larger vertical metal sheets. Corner reflectors on channel-marking buoys and corner reflectors hung on ship's masts enhance their radar visibility, especially important at night or in fog.

## RADAR NOISE

A radar echo can be detected only if its strength is high with respect to various competing RF signals or *noise.* Interference from the RF and IF signals originating within the receiver must be isolated and minimized. External radio noise from natural and humanmade sources as well as deliberate jamming signals by an enemy can interfere with reception.

## RADAR CLUTTER

*Clutter* is any interference with radar reception that masks valid returns from targets of interest such as ships or aircraft. This nuisance return can be surface clutter caused by radar sidebands scattering RF energy from the sea near the antenna. It increases with wave height or turbulence in sea-state conditions. *Volume clutter* is caused by fog, clouds, rain, snow, and birds.

Deliberate volume clutter for jamming military radars can be produced with *chaff* or fine metal wires that act as resonant dipoles when they are scattered in the air. Chaff is most effective at the higher microwave frequencies. See "Electronic Countermeasures (ECM)" previously in this section.

## HEIGHT-FINDING RADARS

A *height-finding radar* has an antenna capable of searching in three dimensions to determine the height of multiple targets such as aircraft or missiles. One example of an antenna capable of height finding has a vertically mounted parabolic cylindrical reflector with multiple feed horns. The RF energy reflected from the horns produces a stack of horizontal *fan beams.* Another type of three-dimensional antenna has a motor-driven elevation scanning feed on a parabolic dish reflector that provides a rapidly oscillating horizontal fan beam. Both of these antennas can be rotated through 360° to obtain accurate bearings.

# TRACKING RADARS

A *tracking radar* is one that is capable of tracking one target very accurately. It usually has a *Cassegrain dish antenna* that is steered in the direction of a target that has been identified. After the tracking radar has acquired the target, it receives a continuous stream of data. The feed horn, mounted at the focal point of parabolic dish, produces a narrow *pencil beam*. By nutating the feed horn, the beam is scanned conically around the reflector axis in the direction of the target. The amplitude and phasing of the reflections generate error signals used as feedback so that the antenna-positioning motor drives the antenna into alignment with the target. When locked on the target, the antenna provides a very accurate target angle.

# MONOPULSE RADARS

A *monopulse radar* is a radar whose operation depends on a tracking technique called *monopulse angle measurement*. It is the alternative to the earlier conical-scanning method, and is less likely to become a target for radar-seeking missiles. A pair of vertical feed horns and a pair of horizontal feed horns are offset from the antenna reflector's axis so they view the target from different angles. The radar transmits a single pulse (monopulse) and, because the horns are not aligned, four different reflected signals are returned. The differences between the returned signals create error signals that permit the antenna's axis to be positioned exactly on the target. Monopulse antennas are used for weapons fire control and missile guidance. They are difficult radars to jam because of their low repetition rate, meaning that the time between pulse transmission when the transmitter is off far exceeds the pulse width. This infrequent pulse pattern can confuse radar-seeking missiles and cause them to go astray. However, monopulse radars can be jammed by oppositely polarized signals.

# PHASED-ARRAY RADARS

A *phased-array radar* is a radar with an antenna that can be electronically scanned in three dimensions to transmit and receive RF energy, eliminating the need for moving the antenna in either bearing or elevation. A simplified block diagram of the command and control functions of a phased-array radar is shown in Fig. 23-6. The antenna consists of an array of transmit-receive elements stacked to form a planar surface. Phased arrays are steered by selectively delaying and coordinating the transmission of energy from each of the elements so that a narrow cone or pencil beam is formed. Its apex is at the center of the array, as shown in Fig. 23-6.

Arrays are classified as either active or passive. *Active arrays* have duplexers and amplifiers behind each element or group of elements. As a result, they are capable of higher power than conventional antennas. By contrast, *passive arrays* are driven from a single radio-frequency source. Both passive and active arrays must divide the signal from a single transmission line among all elements of the array. This is usually done with an optical or a corporate feed.

An *optical feed* is a single source, typically a horn, that illuminates the back of the array with a spherical phase front. Power collected by the rear elements of the array is transmitted through components called *phase shifters* that produce a planar front and steer the array. The energy then is radiated from the other side of the array, as if it were a lens.

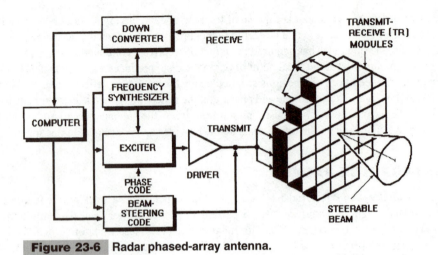

**Figure 23-6**   Radar phased-array antenna.

There are two kinds of *corporate feed networks:* series and parallel. The *series networks* are fed from one end and the hierarchical *parallel networks,* which branch like trees, are fed from the source, analogous to a trunk. Both kinds use transmission-line components to divide the signal among the elements. Phase shifters can be located at the elements or within the dividing network.

Phase shifters produce controllable phase shifts over the operating band of the array. Both digital and analog phase shifters are made with ferrites or PIN diodes. An example of a ferrite phase shifter is the *Reggia-Spencer phase shifter.* It is a waveguide section with a bar of ferromagnetic material located axially inside and a solenoid wrapped around the outside. The phase shifter delays the RF signal passing through, thus shifting its phase. The amount of phase shift can be controlled by the current through the solenoid. The longitudinal magnetic field produced when the solenoid is energized changes the permeability of the ferrite. This produces a variation in the RF energy propagation constant, permitting phase shift to be controlled by drive current. The Reggia-Spencer is a reciprocal phase shifter so it provides the same phase shift for signals passing in either direction. See also "Reggia-Spencer Phase Shifters" in Sec. 7, "Microwave and UHF Technology."

In large systems, such as those intended for space surveillance or long-range search, the output signals of hundreds of individually fed TR elements made as replaceable modules form the pencil beam. Large phased-array radars have gradual failure modes, because the failure of a small number of individual TR modules and amplifier tubes will not significantly degrade system performance.

Airborne phased-array radars typically have optical feeds. These radars can be scanned electronically far faster than is possible with a mechanically oscillating paraboloidal antenna. Moreover, the complete radar is lighter than conventional airborne radars because scanning motors are eliminated and the complete system occupies less space.

Some U.S. Navy cruisers are equipped with the AEGIS system, a fully automatic and integrated system for air defense. It is based on the AN/SPY-1A/B multifunctional phased-array radar system that includes four fixed planar antenna arrays mounted in near vertical positions on deck houses so that they face 90° apart. Each 10-ft (3-m) diameter array contains over 4400 radiating elements. Computers rapidly adjust the phase relationships between groups of elements to shift each array's beam direction electronically to track targets in about a 100° sector. It can supply tracking data for aiming defensive weapons on more than 250 airborne and surface targets while continuing its hemispherical scanning.

Massive ground-based phased-array antennas can detect and track multiple satellites, missiles, and other space vehicles. Mobile phased-array radar systems are included in ground-based antimissile batteries for tracking ballistic missiles. Other phased-array radars have been designed to track mortar or artillery fire. Computers compute the trajectories of the shells and from this information calculate the geographic coordinates of the weapons that fired them so that they can be destroyed.

## CONTINUOUS-WAVE (CW) DOPPLER RADARS

A *continuous-wave* (CW) *Doppler radar* permits its operator to discriminate between fixed and moving targets because it has a CW transmitter and the returned RF energy is detected by mixing it with some of the transmitter's RF power. Fixed targets produce a constant voltage, but moving targets produce an alternating voltage at the Doppler frequency difference between the transmitted and received signals. This radar is best suited for measuring the radial velocities of targets and for detecting the presence of moving targets rather than determining their positions accurately. It cannot determine range accurately.

## PULSED DOPPLER RADARS

A *pulsed Doppler radar* is a modification of a CW Doppler radar that can obtain range information from a CW Doppler radar beam. The received pulses are small segments of the CW returns. A fixed target produces uniform pulses, but pulses from a moving target vary in amplitude periodically because of phase coherence. Each time a fixed target echo returns, it is mixed with a voltage that has undergone the same difference in phase since the instant of its transmission.

## MOVING-TARGET INDICATION (MTI) RADARS

A *moving-target indication* (MTI) *radar* is a CW radar that can obtain precise range information if the CW carrier is pulse modulated. The received pulses are small samples of the CW return. A fixed target produces uniform pulse returns, but returns from moving targets vary in amplitude periodically. MTI radars differentiate between stationary and moving targets by subtracting echo pulses from exact replicas of their transmitted pulses. This process produces constant amplitude pulses for stationary targets and pulses of varying amplitude for moving targets. It is then possible to display only the moving targets by canceling out the constant-amplitude returns.

## FIRE-CONTROL RADARS

A *fire-control radar* is a specialized radar for tracking identified targets to determine their range and bearing. This precise information is used to direct artillery, antiaircraft, or antimissile fire and to guide missile launchings. This type of radar obtains its contacts in a handoff from long-range *search radars,* generally after the targets have been identified as aggressive or threatening.

## LIMITED-SCAN ARRAY RADARS

A *limited-scan array radar* is a specialized radar whose antenna beam scanning is limited to about $\pm 10°$ for such applications as controlling the approach of aircraft from the ground and locating the trajectories of enemy artillery or mortar shells. Its multifunction array antenna is simpler than that of a wide-angle, beam-scanning radar because it does not requires as many phase shifters. It is also known as a *limited field-of-view* (LFOV) *radar.*

## MULTIFUNCTION-ARRAY RADARS

A *multifunction-array radar* combines both search and track functions. It typically includes a phased-array antenna with high-power RF phase shifters under the control of digital signal processors. Agile antenna-beam scanning permits it to perform dual functions if it transmits at a frequency that is a compromise between the optimum frequency for searching and the optimum frequency for tracking.

## SEARCH RADARS

*Search radar* is a general term that applies to any radar system capable of examining a hemispherical volume of space around the antenna for targets on land, sea, or in the air. It can have a continuously rotating mechanical antenna or phased-array antenna whose electronically scanned beam is in a continuous-scan mode. These radars typically operate in the S or L frequency bands and hand off information to higher-frequency, higher-definition, shorter-range radars for target tracking or weapons fire control. It is also called a *surveillance radar* when long-range targets are being monitored, as in the civilian air traffic control system.

## TRACKING RADARS

A *tracking radar* is a radar system whose mission is to keep one or more targets under continuous surveillance so that a more accurate determination can be made of the target's location. Targets of interest detected by a search radar are often handed off to and acquired by the tracking radar. Some radars combine both search and track functions by time-sharing the agile beam of a phased-array antenna.

## SECONDARY SURVEILLANCE RADARS (SSRs)

A *secondary surveillance radar* (SSR) is a unit of air traffic control radar that identifies and tracks civil and military aircraft worldwide. It is a component in the FAA's *Air Traffic Control Radar Beacon System* (ATCRBS). An SSR radar is typically combined with an air traffic control radar. The SSR antenna is mounted on top of the ATC radar antenna. The ground station interrogates the aircraft's transponder with a narrow fan-shaped beam at

1030 MHz, and it receives replies at 1090 MHz for determining the aircraft's range and bearing. The airborne transponder returns a train of pulses that identify the aircraft and reports its altitude. It was developed from the military *identification, friend or foe* (IFF) system. See "Identification Friend or Foe (IFF)" below.

# Radio-Frequency Transponders

A *transponder* is a beacon that replies with a unique identification code after being interrogated on a specific frequency. There are RF transponders on military and civil aircraft and on military ships. The *interrogation signal* is usually sent from a transmitter that could be on the ground, in an aircraft, or on shipboard. The transponder's response is called a *reply*. It replies only when interrogated with the proper signal for security reasons or to conserve power.

## AIRCRAFT TRAFFIC CONTROL TRANSPONDERS

Radio-frequency transponders are important equipment in the commercial air traffic control (ATC) system. A transponder on a commercial aircraft permits ground-based ATC personnel to monitor the position of the aircraft by interrogating its transponder when the aircraft has entered a designated airspace, typically near an airport. The ATC radar transmits an interrogation code, and the transponder on the aircraft replies in its own unique code that gives its airline, flight number, and altitude. This reply is converted to an alphanumeric message that appears on the ATC radar screen in the correct relative position so that the range and bearing can be determined. These transponder replies provide a more distinctive return than could be obtained by radar echoes from the aircraft, and they also permit the ATC controllers to rapidly distinguish between aircraft in the immediate airspace, as shown on a crowded ATC radar screen.

## SATELLITE TRANSPONDERS

Military and commercial communications satellites are equipped with electronic circuitry, also called *transponders,* that receive RF transmissions from a ground station on one frequency (uplink) and retransmit them back to earth at another frequency (downlink). The reception area covered by the downlinked signal is determined by the satellite's downlink antenna, transmission frequency, and power. These transponders are used in the relay of television programs and other commercial and military voice and data transmissions.

## IDENTIFICATION, FRIEND OR FOE (IFF)

*Identification, friend or foe* (IFF) is a military identification system based on microwave-frequency transponders installed on friendly aircraft that are interrogated by search or fire-control radars on land, on shipboard, or in aircraft. It is intended to prevent aircraft from being shot down accidentally by friendly aircraft, antiaircraft guns, or missiles. The IFF transponders are installed on aircraft, and their reply codes are changed daily to avoid their use by enemy aircraft seeking to slip past radar screens without detection. The IFF transponder is interrogated by radar pulses, and the reply shows up on the interrogating radar's PPI screen as a coded pattern indicating the aircraft's range and bearing.

# Sonar Systems

*Sonar* (an acronym for *sound navigation and ranging*), is a system that projects acoustic energy into the water and measures the speed of its return from a reflective surface as an echo to detect an underwater object and determine its range and bearing. A simplified block diagram of an *active sonar system* is shown in Fig. 23-7. It is useful for detecting submarines, sunken ships, obstructions such as rocks, and naval mines. A sonar system typically includes a video display that provides a visual indication of the target and an audio speaker that provides audible information about the identity of the target or contact. The change in pitch of the audible signal indicates relative motion of the target with respect to the sonar system in accordance with the Doppler principle.

An active sonar system is analogous to an active radar system except that the transmission medium is sound rather than radio energy. A sonar transducer or transducer array is analogous to an antenna. Some transducers can be trained through 360°, while others are trainable through a more limited sector such as 90°. From knowledge of a target's change in range and bearing over time, its course, speed, and closest point of approach can be determined either by manual plotting or by computer.

The transmitter formats the signals that drive the transducer or transducers which project the sound into the water. Sonar transducers are made from materials that demonstrate either the magnetostrictive or piezoelectric effect.

The receiver section contains the electronic circuitry that detects and amplifies the echoes and converts them into video signals to drive a display for monitoring targets. The

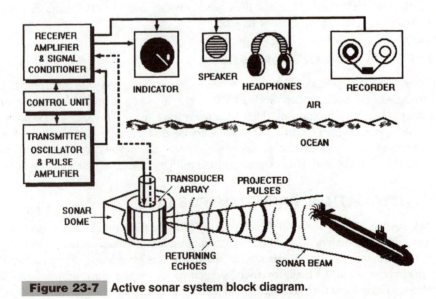

**Figure 23-7** Active sonar system block diagram.

sounds can be obtained from the reciprocal action of the transducers or from *hydrophones,* essentially speakers modified for use underwater. It also provides the audible sounds for a speaker, and headphones give the sonar operator important information about the target and its motion.

Sound pulses or *pings* from an active pulsed sonar strike the target, and are returned as attenuated signals. Pulse transmission is delayed in time to permit weak echoes from the limit of the sonar's range to be received. Range is determined by measuring the elapsed time between pulse transmission and the receipt of the echo. Bearing is determined by encoders or other electromechanical sensors that track the position of the centerline of the transducer or array as it moves.

Active sonar systems on naval surface ships are used for antisubmarine warfare (ASW) and mine detection. By contrast, submarines depend primarily on passive sonar for tracking and attacking surface vessels and other submarines to avoid detection that would result from the use of active transducers. But they also have active sonars for navigation when close to shore or for detecting floating mines.

While sonar systems are primarily military detection and surveillance systems, active commercial or modified military sonar installed on scientific research ships and submarines has proven invaluable for underwater navigation and exploration. Depth sounders, which determine the depth of water under a ship's hull, and *fishfinders,* which can track and locate schools of fish, both operate on the same principles as sonar, and are widely used on commercial and pleasure boats. See also "Depth Sounders" and "Fishfinders" in Sec. 24, "Marine Electronics Technology."

## SOUND TRANSMISSION IN WATER

The velocity of sound in saltwater is approximately 4800 ft/s (1463 m/s). It increases with temperature from about 4700 ft/s (1430 m/s) at 32°F (0°C) to 5300 ft/s (1615 m/s) at 85°F (29°C). (This is more than 4 times the speed of sound in air at sea level.) A 20-kHz sound in seawater has a wavelength of about 3 in (75 mm). The frequency selected is based on the intended application. Low frequencies of 10 to 50 kHz provide better depth penetration, but the higher frequencies of about 200 kHz provide better target definition.

Water temperature and salinity are extremely important factors in tracking targets with sonar. The detection of a submarine by a surface vessel, for example, can be difficult where there are pronounced differences in temperature and salinity between the ocean's surface and its floor. Sound beams entering water that is warmer on the surface than below the surface are bent downward by a refractive effect, but beams that enter water that is colder at the surface than below it are bent upward. In long passages through the water the beam might be bent many times into a serpentine shape that will cause it to miss some targets completely.

The presence of a *thermocline* or layer of water at a transition temperature between warm and cold water can shield a nearby submarine from detection although it might be within the range of an active sonar. The active sound beam is bent so that it does not reach the submarine. Similarly, the sounds made by a submarine traveling beneath a thermocline can be deflected so that they do not reach the surface and be detected by an enemy ASW ship.

Background noise sources can seriously impair the effectiveness of both active and passive sonar. One important background noise source is the machinery and propeller noise from the host ship. Some ASW ships tow hydrophone arrays on long cables to avoid the interference of their own machinery and propellers. Moreover, the towed array can sink deeper into the ocean to penetrate subsurface thermoclines to detect submarines below them.

Common sources of hydrodynamic noise are air bubbles, turbulence, and cavitation. Another is reverberation, which results from multiple sound reflections between the surface and the ocean floor caused by storms and other natural phenomena. These noises increase in shallow water. Shrimps, whales, and some kinds of fish also cause background clicking or singing noise.

## SUBMARINE SONARS

Passive sonar is the most important source of information for submarines because their success depends on stealth. Their presence would be easily detected if they were to use active transducers when near enemy ships. As a result, submarines limit the use of their active sonar to such functions as mine detection and navigation. However, active sonar has been used by submarines to determine ice thickness when traveling under the polar ice cap.

Modern U.S. Navy attack submarines are equipped with spherical arrays of nearly 1000 hydrophones mounted within a fiberglass nose cone at the bow of the ship. The sound passes through the cone to the hydrophones, which can detect noise emitted by sources miles away. The receiving beam of the hydrophone array can be electronically steered in three dimensions in a way that is analogous to the receiving beam of a phased-array radar.

The sonar sphere listens for machinery and propeller noise from surface ships and submarines. Sounds are gathered, recorded, and compared with noise profiles or *signatures* of thousands of known ships stored in a computer database. The profiles can discriminate between natural noise and ship machinery and propeller noise, and can be used to identify ships and submarines by type and even country of origin. Further computer analysis of these signals can also provide range information. A transducer is mounted below the spherical hydrophone array for use when it is safe to use an active system without revealing the presence of the submarine.

Because the forward sonar sphere is limited to a coverage of about 220° around the bow of the submarine, it is supplemented by other towed active and passive sonars. Housed in the stern planes, the sonar heads are paid out on long cables. A passive towed array can be deployed out to 2500 ft (760 m), and an active sonar array can be deployed out to 1500 ft (460 m). There are also towed front-deployed passive and active arrays.

Modern submarine torpedoes with a range of more than 10 mi (16 km) are both wire and sonar guided. Target information is entered by computer into the torpedo at the time it is fired. After launching, updated guidance information can be sent to it through a thin wire which remains connected to the submarine. The torpedo homes in on its target with active or passive sonar, and as it nears the target the wire is cut. If the torpedo misses, it can turn around and find the target with its own active sonar.

## MAGNETOSTRICTIVE TRANSDUCERS

*Magnetostrictive transducers* for projecting sonar signals underwater are made from bundles of nickel-alloy tubes because each tube exhibits the *magnetostrictive effect* of elongating in the presence of a strong magnetic field. This effect is comparable to the piezoelectric effect in crystals. The bundles of tubes are enclosed by coils which receive electrical pulses from the sonar transmitter. The resulting electromagnetic flux elongates the array of tubes, causing them to exert a mechanical force against a diaphragm which is in contact with the water. Acoustic waves are propagated into the surrounding water at frequencies related to the frequency of the current applied to the coils.

Between pulses, the echo returns to the diaphragm and applies pressure to it, shortening the lengths of the tubes and inducing a voltage in the surrounding coils. This voltage is amplified and converted into a video signal for display and audible frequencies for the headphones and speaker.

## PIEZOELECTRIC TRANSDUCERS

A piezoelectric transducer depends on the *piezoelectric effect* exhibited by some crystals and ceramics. These materials change their physical dimensions when they are placed in an electrostatic field. Lead-zirconite-titanate and barium-titanate ceramics are used to make sonar, depth sounder, and fishfinder transducers. They are formed into disks or bars that are subjected to electrical pulses from the sonar transmitter. The pulses cause the transducer to expand, exerting a force on the surrounding water at the frequency of the transmitter pulses. The shape of the transducers determines the shape of the sound beam.

The echoes returning between pulses apply pressure to the transducer, deforming it and causing it to produce a voltage. That voltage is amplified and converted into video and audio signals. Unlike magnetostrictive transducers, piezoelectric transducers provide more efficient reciprocal energy transfer, and they produce sound energy at the same frequency as the exciting voltage. Piezoelectric ceramic transducers can be exposed directly to the sea.

Similar ceramic piezoelectric transducers cause vibrations at ultrasonic frequencies in ultrasonic cleaning tanks which can clean intricate objects when filled with appropriate cleaning solutions. The crystal microphone and the surface acoustic wave (SAW) device are other examples of piezoelectric transducers.

## VARIABLE-DEPTH SONARS (VDS)

A *variable-depth sonar* (VDS) is a sonar system whose transducer array and supporting electronics is located in a torpedo-shaped case that is towed behind an antisubmarine warfare (ASW) ship on a long umbilical cable. The depth of the housing can be set by the towing ship to permit it to sink to regions where reception is improved because it is far enough below the surface to avoid reverberations from the ocean's surface and far enough behind the towing ship to avoid the interference from its engine and propeller noise. It can also be set to travel below thermoclines. A further advantage is that if the towed VDS array is active and it becomes a target for a submarine-launched acoustic-homing torpedo, its destruction will not endanger the towing ship.

## CONTINUOUS-WAVE (CW) SONARS

A *continuous-wave* (CW) *sonar* is a short-range sonar developed for guiding miniature scientific submarines engaged in undersea archeological, geological, or biological exploration. CW sonar gives the crew of the submarine a clearer display of the surroundings than a pulsed sonar under conditions of reduced visibility. It can be used to assist the deep-diving submersible in navigating safely over or around undersea obstructions such as cliffs, trenches, and rock formations that could damage or trap it.

## SIDE-LOOKING SONARS (SLS)

A *side-looking sonar* (SLS) is a towed sonar system similar to variable-depth sonar that makes continuous scans of the sea floor on both sides of the ship's heading. These scans can be plotted to chart the contours of the sea floor and provide indications of sunken objects of interest. The torpedo-shaped housing that is towed with an umbilical cable behind a ship is shown in Fig. 23-8. The transducers within the housing are scanned at right angles to the course of the housing and its tow ship. The signals from the transducers can be displayed on a monitor or printed out on chart paper. SLS was developed for strip mapping the sea floor, and it has proven to be useful in oil exploration and the recovery of sunken submarines and treasure ships. Moreover, the plotted returns provide useful data for topographical maps of wide strips of the ocean floor.

## DIPPING SONAR

A *dipping sonar* is an airborne sonar system that includes a display, control, transmitter, and receiver section in an ASW helicopter and a transducer in a streamlined housing that is attached to it by a combination electrical/electronic and lifting cable. A winch lowers the transducer unit into the sea below the hovering or slow-moving helicopter during search operations and reels it back in when the mission is completed. The transducer housing is streamlined and equipped with fins so that it can be towed steadily in the same direction behind the helicopter. The housing can be submerged to different depths so that the helicopter crew can listen for and pinpoint the location of the submerged submarine. The transducer is towed over long distances in the most likely location of a submarine initially detected by the array of *sonobuoys* dropped from an ASW aircraft.

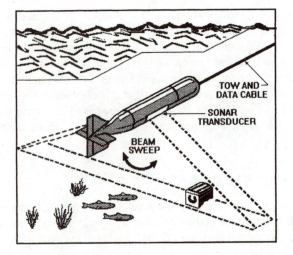

**Figure 23-8** Side-looking sonar (SLS).

A *sonobuoy* is a floating buoy containing a hydrophone to listen for submarine noises and a transmitter to send those sounds back to an ASW aircraft. Sonobuoys are dropped in large numbers in patterns that bracket the area where the submarine is believed to be located.

The dipping sonar can be maneuvered more precisely and rapidly than sonars on surface ships, making it difficult for the submarine being tracked to escape once it has been detected. The ASW ships then converge on that location.

# Underwater Communicators

An *underwater communicator* is essentially an acoustic megaphone for transmitting voice or code signals through the water from surface ships to divers or sunken submarines. They are useful for giving instructions or warnings to divers or persons trapped underwater. They are based on piezoelectric transducers that are modulated by code generators or the human voice.

# Acoustic Transponders

An *acoustic transponder* is an acoustic beacon that is actuated upon receipt of a coded interrogation message. The electronic circuitry is packaged in a waterproof pressure vessel able to withstand long-term immersion hundreds of feet deep in the ocean. It will respond with a coded identification signal when interrogated by an acoustic transducer. An acoustic transponder is analogous to an RF transponder, and it can be useful for staking out specific positions on or near the ocean floor. In a typical application three or more transponders are anchored in a pattern around a sunken object so that a surface ship, such as a research vessel, can leave the area and return to the same precise location. They are useful for establishing the exact positions of sunken ships, oil or gas fields, or archeological sites. Transponders conserve power by responding only when interrogated, and are designed to remain in a standby mode for years.

# Underwater Laser Systems

An *underwater laser scanning system,* as shown in Fig. 23-9, consists of an argon laser, motor-driven rotating mirrors, and light-to-digital converter circuitry that is mounted in a sealed, torpedo-shaped case which is towed behind a ship. The laser beam is scanned over the ocean floor in regions under surveillance, and reflected laser light is returned to the case for conversion into a raster scan signal that can be displayed on a CRT monitor or converted into still pictures. It provides useful information for ocean engineering, salvage operations, archaeology, and military surveillance. It can provide clearer pictures of objects on the ocean floor than side-looking sonar.

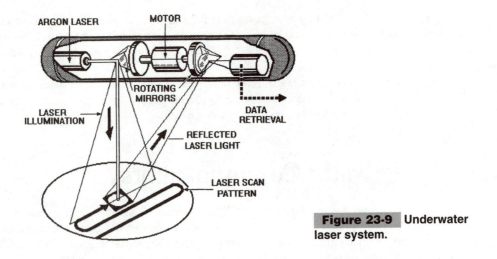

**Figure 23-9** Underwater laser system.

# Military Night-Vision Systems

*Military night-vision systems* permit the viewing of objects and scenes at night without illuminating them with searchlights that would reveal the position of the observer and invite attack. They are based on *image-intensifier* and *image-converter tubes.* Image-intensifier tubes, also called *night-vision light-multiplier tubes,* permit observations to be made at night by the weak natural night light from the moon, stars, or cloud reflection. By contrast, image-converter tubes require that the scene or object be illuminated by invisible infrared (IR) radiation, typically from a laser. They convert the IR illumination of objects or scenes to visible light.

Passive night-vision systems based on image-intensifier tubes are especially valuable for soldiers on night patrols or aircrews attempting to land aircraft on blacked-out landing strips. By contrast, active night-vision systems based on an image-converter tube and IR illumination allow the viewing of scenes for aiming weapons on enemy targets in total darkness. The tube is capable of converting the infrared-illuminated scene to a visible image that is bright enough for precise targeting. These systems are installed on aircraft with missiles or "smart" bombs, tanks, and other military platforms.

During a military operation any vision system that requires IR illumination must be used sparingly to prevent detection by an enemy with a night-vision scope. Active night-vision systems are typically used during surprise attacks where aggressive actions would negate any requirement for concealment.

## NIGHT-VISION SCOPES

A *night-vision scope* is a passive monocular *night-vision telescope* based on an image-intensifier tube. A commercial version, as shown in Fig. 23-10, includes an image-intensifier tube as shown in Fig. 12-13. It is in a housing with an objective lens for enlarging the image and ocular lenses for focusing the image. The lower compartment is occupied by a high-voltage power supply, typically a voltage-multiplier circuit, and batteries. The complete optical instrument weighs less than 1 lb (454 g) but is easy to hold in one hand.

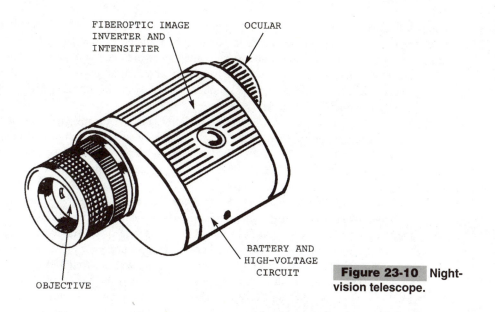

FIBEROPTIC IMAGE
INVERTER AND
INTENSIFIER

OCULAR

BATTERY AND
HIGH-VOLTAGE
CIRCUIT

OBJECTIVE

**Figure 23-10** Night-vision telescope.

The night-vision scope functions at night in low natural light from such sources as the moon, stars, and cloud reflection. Available light can be magnified up to 30,000 times. This instrument typically is used for night scouting and patrolling, although there are versions that can be mounted on sniper rifles, machine guns, and other light weapons. The passive night-vision scope is also called a *nightscope, night-vision telescope, sniperscope, snooperscope,* and *starlight scope.* See "Image-Intensifier Tubes" in Sec. 12, "Optoelectronic Components and Communication."

## NIGHT-VISION BINOCULARS

A *night-vision binocular* or *night-vision goggles* is a pair of image-intensifier scopes with the necessary optics and power supply arranged for binocular viewing to provide a better sense of depth. Each of the two viewing scopes is identical to a monocular night-vision scope, but the power supply is shared. Helmet-mounted military night-vision goggles make flying aircraft (particularly helicopters) or driving vehicles at night safer, particularly in blacked-out combat zones. They leave the pilot or driver's hands free for other duties.

Some night-vision systems are sensitive to all radiation from visible light at 350 nm through the near-IR region at 900 nm. However, the output response of night-vision goggles suitable for aircraft use is selected to be insensitive to the blue and green in aircraft cockpits. These instruments respond to light from visible red light at 600 nm to the near-IR wavelength of 900 nm. The goggles are designated as ANVIS for *airborne night-vision imaging system.* See "Image-Intensifier Tubes" in Sec. 12, "Optoelectronic Components and Communication."

# Global Positioning System (GPS)

The *Global Positioning System* (GPS) is a satellite-based radio-navigation system that uses passive triangulation to permit receivers on land, sea, and in the air to pinpoint their positions with greater accuracy than existing radio-navigation systems. It is officially called the *Navstar Global Positioning System* because the system includes Navstar satellites. Accurate three-dimensional position, velocity, and time can be obtained with passive receivers. The GPS system has been accepted by both military and civilian users. The military services use GPS receivers in aircraft, ships, and land vehicles, and handheld receivers can be carried by foot soldiers.

GPS was developed by the U.S. Department of Defense to provide military ships and aircraft with a more accurate method for determining position anywhere in the world than an inertial guidance system. It was also intended to permit accurate targeting of cities and military installations in wartime. The satellites and their ground-support equipment are still maintained by the Department of Defense.

Two classes of signals are broadcast from the satellites: the *Precise Positioning Service* (PPS) for authorized military users and the *Standard Positioning Service* (SPS) or GPS-C (degraded nonscrambled commercial), a signal available free of charge to anyone, anywhere with a commercial receiver. Horizontal position for accuracy for PPS is within about 50 ft (15 m) but for the uncorrected GPS-C signal fixes are within about 300 ft (90 m).

The GPS signals are transmitted in a spread-spectrum coded division multiple-access (CDMA) format, but the PPS signal is spread over a spectrum 10 times as wide as the GPS-C signal (20 MHz versus 2 MHz) to make it more difficult to jam.

The GPS system consists of 24 Navstar satellites, including 3 operational spares. The satellites are in 12-h orbits with 4 satellites in 6 orbit planes inclined at 55°, all at an orbital altitude of 10,900 nmi (15,250 km). The GPS system configuration is illustrated in Fig. 23-11.

Each Navstar GPS satellite weighs about 1 ton (910 kg) and is made from 65,000 parts. Solar-array panels on the sides of each satellite generate 710 W of electrical power to drive onboard transmitters. Each satellite is equipped with four mutually redundant atomic clocks: two cesium clocks and two rubidium clocks. If one clock fails, another is switched on to take its place.

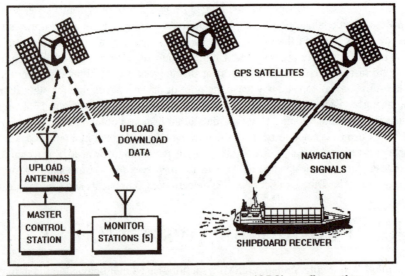

**Figure 23-11** Global Positioning System (GPS) configuration.

The satellites broadcast precisely timed, coded signals on two frequencies at 1.575 and 1.228 GHz. These signals include data on their position and clock errors. Radio waves travel 1 ft (30 cm) in a billionth of a second, so every billionth of a second timing error creates at least a 1-ft (30-cm) error in the receiver's position. To obtain the military 50-ft (15-m) accuracy on a global basis, the satellite clocks must be mutually synchronized within 13 billionths of a second.

A ground-based master control station (MCS) and five monitor stations track the satellites and precisely calculate each satellite's position and orbital track. Data is then transmitted back to the satellites from three uplink stations to correct errors and resynchronize their clocks.

The prearranged sequence of timing pulses from the satellites, which can be received anywhere on earth, permit the receiver to determine the exact time of its departure. When a receiver picks up a sequence, it registers the signal arrival time and compares it with the time the signal was sent from the satellite. The difference between these times is measured in microseconds. The receiver's microprocessor then multiplies the measured signal's travel time by the speed of light (186,300 mi/s [300,000 km/s]) to obtain the distance to the first satellite.

This procedure is repeated four or more times to triangulate the locations of other satellites within visual line of sight above the horizon. As many as eight will be above the radio horizon simultaneously within the range of each receiver. The microprocessor then solves a set of simultaneous equations to determine the receiver's exact location. The accuracy of a fix is determined by the precision of the transit-time measurements of the signals from the satellites to the receiver. The microprocessor can also be programmed to calculate bearings and ranges to other places on earth based on the location information. Positions can be displayed in three mutually orthogonal coordinates such as longitude, latitude, and altitude.

Three range measurements will define the receiver's current location, but in practice information from a fourth satellite is needed to compensate for any clock errors in the receiver. Commercial receivers currently available are equipped with quartz-crystal oscillators that lose or gain a second in about three years. Low-cost commercial receivers now include 12 separate channels for receiving signals from as many as 12 satellites simultaneously. The atomic clocks on each satellite are 10,000 times more stable than the quartz-crystal oscillators in the receivers, and provide an accuracy of 1 s in 300,000 y.

For many civilian users, particularly small boat and private aircraft owners, accuracy within about 300 ft (90 m) is satisfactory, but higher precision is needed for commercial ships that must navigate in narrow channels, commercial aircraft attempting instrument landings, and for effective guidance of automobiles within cities. To correct the GPS-C signal so that more accurate navigational fixes can be obtained, two systems have been developed: the *Differential Global Positioning System* (DGPS) and the *Wide Area Augmentation System*. For more information on the DGPS, see "Differential Global Positioning System (DGPS)" in Sec. 24, "Marine Electronics Technology."

## WIDE-AREA AUGMENTATION SYSTEM (WAAS)

The *Wide-Area Augmentation System* (WAAS) is another system for improving the accuracy of location fixes obtained from GPS-C primarily for commercial aircraft. The GPS-C signals are corrected in real time with a supplementary signal that is generated at ground stations and transmitted to aircraft by geosynchronous satellite. This system is expected to be mandatory for commercial aircraft flying over the United States, but the disposition of the existing network of more than 1000 radio beacons and radio aids as well as about 100 long-distance radars for use by ground controllers has yet to be decided. All or some of these could remain to supplement the WAAS.

The FAA has built ground reference stations around the country that will compare their locations with the positions estimated by the GPS. They will then calculate the error in positioning and send this data to two central stations for collection and broadcast to two satellites in geosynchronous orbit. The satellites will then broadcast the error-correction data over the continental United States. When these signals are received by the aircraft they will be used to correct the onboard GPS receiver, giving it more precise three-dimensional position information. WAAS will be able to pinpoint the location of an aircraft's GPS-C receiver to within about 20 ft (6 m) anywhere in the continental United States. It will warn pilots within 6 s of a system malfunction—enough time to abort a landing.

The WAAS will perform essentially the same function as the Differential Position Global System (DPGS), except that satellites will relay the correction signals to obtain wider coverage for aircraft in flight than the ground-based radio beacon of DPGS. See "Differential Position Global System (GPGS) in Sec. 24, "Marine Electronics Technology."

# Laser Range Finders

A *laser range finder* is a military laser system for determining the range of a target and is the light wave analogy of a microwave radar. The laser range finder, like the *laser geodimeter,* determines distance by measuring the return time of reflected laser energy. The military range finder typically operates in the pulsed mode in the infrared (IR) region. By contrast, civilian geodimeters for surveyors operate in continuous-wave (CW) mode in the visible red region.

Laser range finders measure the time for a coherent pulse of IR energy to travel to the target and for part of that energy to be reflected back to a detector. Because the emission is not visible to the eye, suitable detectors are used to indicate that the system is functioning. IR laser range finders are impaired by smoke, fog, and dust. Most operate in the 3- to 5-$\mu$m or 8- to 14-mm bands.

The design of a laser range finder for a military mission is determined by the following factors:

- Characteristics of the target (the larger the better)
- Laser output wavelength
- Atmospheric conditions
- Detector to be used
- Size and weight restrictions imposed by its mission

Presentday field-portable military laser range finders and related targeting systems typically employ neodymium:YAG (Nd:YAG) lasers. Peak laser power varies from several hundred kilowatts to several megawatts per pulse, and pulse repetition rates can be up to 100 pps. In addition to the laser, a system includes a battery, power supply, transmitter, range receiver, range counter, Pockels cell driver, and flash-lamp trigger modules. These systems typically use either silicon PIN or avalanche photodiode detectors. Avalanche detectors require less pulse energy for a given maximum range under most atmospheric conditions.

# Laser Guidance

*Laser guidance* is a semipassive technique for guiding flying objects equipped with sensors that respond to laser illumination of targets. This form of terminal guidance permits the precise aiming of missiles or the dropping of "smart" bombs on intended targets. An active laser guidance system includes a laser target *designator* whose beam illuminates the target. The missile must be able to stay within the IR beam so that it can home in on the reflected laser energy. A simplified block diagram of an active laser guidance system is shown in Fig. 23-12.

Military laser guidance systems use IR-emitting lasers as designators because their beams are not visible to the unaided human eye, and they would not reveal the location of the laser to the enemy. Weapons capable of seeking targets illuminated by IR energy from a laser require sensitive electronics and closed-loop feedback flight controls. When the missile's imaging IR seeker circuitry "locks on" to the illuminated target, it generates signals that move its flight wings or fins as necessary to keep the missile within the cone of reflected energy. The vertex of that cone is the target.

The laser target designator could be on the same attacking aircraft as the laser-seeking missile or bomb or it could be on separate aircraft. (For certain missions the laser designator is more effective if it is on the ground.) The most likely targets for IR-guided missiles are buildings of tactical or strategic importance such as command centers, weapons factories, or power stations located where precision bombing will minimize the ancillary destruction of civilian homes and buildings and the resulting civilian casualties.

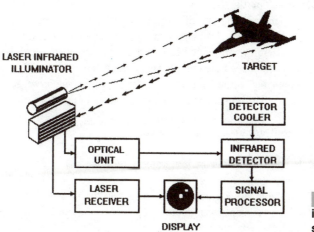

**Figure 23-12** Active infrared laser guidance system.

Both the airborne laser and the missile's seeker circuitry are mounted on gimbals so they can acquire and track targets in three mutually perpendicular directions. Some laser guidance systems are capable of identifying many targets with multiple illuminators. The targets are distinguished from each other by lasers that are pulse or frequency modulated with different codes.

Most IR guidance systems have silicon detectors in the seeker missile because they are sensitive to the 1.06-$\mu$m emissions of the neodymium:YAG lasers most widely used as target illuminators. Longer wavelength detectors are required for $CO_2$ laser illuminators.

An example of a missile specifically designed for laser-beam guidance is the air-to-ground Hellfire missile. Conventional bombs can be made "smart" if they are equipped with nose-mounted assemblies that include both laser guidance sensors and flight-control fins.

A *forward-looking infrared* (FLIR) *unit* is a U.S. military active infrared detection system that includes a laser transmitter and receiver based on an image-converter tube. A FLIR can detect, identify, and determine the range and bearing of a target by sensing its radiation. FLIR systems include discrete detector arrays that scan for objects in space and generate video signals when they are encountered. A signal processor amplifies the electrical signal and sends coded target information to the display. A detector made from a material such as mercury-doped germanium (Ge:Hg) must be cooled cryogenically to a temperature of about 30 K to obtain its highest performance. The returned signals are then scanned into a video display.

# Laser Gyroscopes

A *laser gyroscope,* as shown in the simplified diagram Fig. 23-13, is an assembly of lasers arranged in a ring so that they sense the rate of turn about an axis. The area between the cathode and the anodes is the active region where a voltage discharge provides the excited neon atoms for lasing. Two opposing laser beams combine to form interference fringes. A rate-of-turn signal is obtained when laser beams of the same wavelength are constrained to move in opposite directions in the ring-shaped cavity. The difference in frequencies between the two beams depends on the rate of rotation of the gyroscope mount. Each laser oscillates at a frequency which depends on its apparent path length. The laser beam in one direction sees a shorter path length, while the beam in the other direction sees a longer path length. The difference in frequencies between the two beams is a function of rotation rate of the ring about the axis normal to the ring. This frequency difference is sensed and read out as a rate-of-turn signal. Position information is obtained by integrating the rate output signal.

There are two general configurations of laser gyros:

1. *Active ring,* in which the laser beams are generated within the cavity or ring filled with helium and neon (He:Ne) gas
2. *Passive ring,* in which the lasers are outside the ring, and the laser beams are directed into the ring formed from fiberoptic conductors

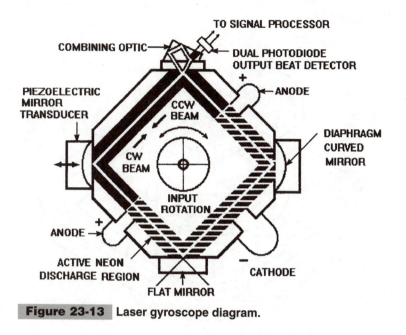

**Figure 23-13**  Laser gyroscope diagram.

Ring laser gyros have demonstrated drift rates of 0.005° per hour, and their performance is comparable to that of navigation-grade mechanical gyroscopes. However, mechanical gyroscopes produce an output in the form of an angular displacement rather than a rate-of-turn signal. Solid-state ring laser gyros that use gallium-arsenide (GaAs) injection lasers have been developed.

# Inertial Guidance Systems

An *inertial guidance system* is a self-contained navigational system for aircraft and ships, particularly submarines, based on the continuous output of accelerometers mounted on a gyroscopically stabilized platform. At the start of a trip precise coordinates and local time are entered into the system's computer, which is then able to *dead reckon position* automatically by measuring the amount and direction of acceleration in three mutually perpendicular directions. The computer calculates the speed at a particular instant and the distance traveled by a process of integration of deviations from the reference coordinates. As shown in Fig. 23-14, the gimbaled stable platform contains three gyroscopes with mutually perpendicular axes and three accelerometers that sense acceleration along three mutually perpendicular axes. Where possible, midcourse corrections of estimated position can be made automatically by making observations of the celestial navigation stars.

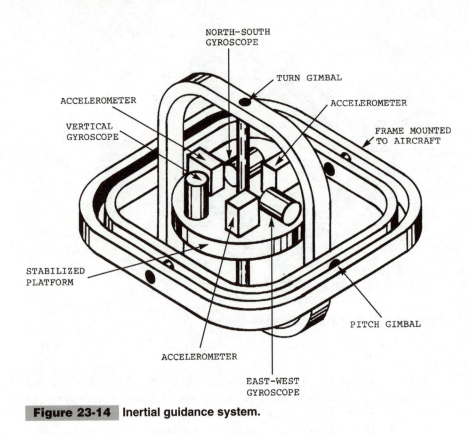

**Figure 23-14** Inertial guidance system.

# MARINE ELECTRONICS TECHNOLOGY

# Overview

The first electronics on passenger and merchant ships were radio transmitters and receivers for routine communications and calling for help in distress situations—fire, collision, sinking, or the presence of seriously ill or injured persons on board. Then came radio direction finders as navigational aids. But after World War II commercial versions of military radar and loran were added, permitting accurate navigation at night and in inclement weather, significantly improving safety at sea. Fishing vessels took advantage of commercial depth sounders and spin-offs from sonar called *fishfinders* to increase the efficiency of fishing and the size of their catches. They were also able to make safer passages to and from the fishing grounds, guided by the information gained from radar and loran.

But it took the development of solid-state circuitry and ICs to reduce the weight, size, and cost of all of this marine electronics equipment, making it affordable and available to smaller commercial ships and private boat owners. The miniaturization of shipboard electronics that formerly required large and heavy cabinets and consumed significant amounts of power, to small cases and even handheld units led to explosive expansion in the commercial and consumer marine electronics industry.

The first miniaturized transistorized equipment was UHF and VHF transceivers. These were followed by solid-state depth finders, radars, loran receivers, and radio direction finders. Further reductions in power requirements led to raster-scanned video displays, fishfinders, chart plotters, and handheld battery-powered Loran C and Global Positioning System (GPS) receivers.

Among the latest marine electronic products are electronic charting systems that can integrate data from radar and GPS receivers on images of local navigational charts inserted in the form of cartridges. There are also CD-ROMs available that will provide the display of regional navigational charts for all coastal areas of the United States on a notebook computer which can be carried on and off the ship. Standardized bus connections permit the GPS receiver to supply the data for moving markers designating the ship's position on the CD-ROM-generated charts.

Among the other innovations based on miniaturized electronics are emergency position-indicating radio beacons (EPIRBs) for use beyond the range of VHF transceivers. They can send distress signals to orbiting satellites that identify the specific ship in distress for forwarding to the nearest air-sea rescue station.

Radio transmission and reception are covered in Sec. 18, "Radio Transmitters and Receivers." The principles of radar, sonar, and GPS are covered in Sec. 23, "Military and Aerospace Electronic Systems."

# Depth Sounders

A *depth sounder* is a marine electronic instrument that can determine the depth of water beneath a ship's keel by measuring the time taken for sound energy from a transducer to travel down to the sea floor, be reflected, and return to the transducer. Figure 24-1 shows the principles of depth-sounder operation.

The primary function of the depth finder is distance measurement, but the presence of fish or other underwater objects will affect depth readings. Depth sounders and sonar are based on the same principles, so their components are similar.

The principal components of a depth sounder are a *display/control head* and a *transducer*. The display/control head includes a digital display and a *transmitter-receiver* in the same case, and it is connected to the transducer by coaxial cable. The transducer, acting as an antenna, is a solid disk of piezoelectric ceramic that is exposed directly to the water. It oscillates in the ultrasonic range, typically at about 200 kHz. The ultrasonic energy is directed vertically in a narrow beam to the sea floor. The output pulse triggers a counter in the display/control head, and the transducer receives weak echo signals that

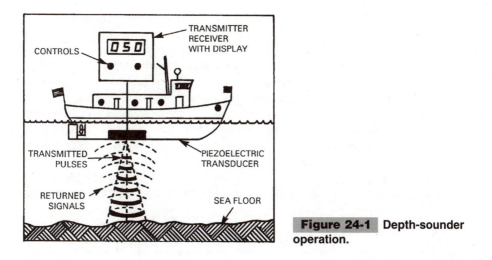

**Figure 24-1** Depth-sounder operation.

stop the countdown. The return signals are processed to provide a digital readout, which is translated by electronic circuitry into a depth reading in feet, meters, or fathoms. Water absorbs high frequencies more readily than low frequencies, so a 50-kHz transducer is more effective in water deeper than about 400 ft (120 m), but 200 kHz provides better resolution.

The narrow sound beam can provide accurate depth readings only if the host ship remains within $\pm 10°$ of vertical. The time required for a signal to travel from the transducer to the sea floor is the product of water depth (in feet) times the speed of sound in water, typically about 4800 ft/s (1460 m/s) in seawater. In 100 ft (30 m) of water the pulse makes a round trip from the bottom in about 1/25 s. Ocean and lake bottom surfaces are more likely to be rough than smooth, so the reflected signal is widely scattered, and only a small fraction of it returns to the transducer.

A depth sounder is a useful navigational aid for piloting at night or in the fog when nearby landmarks are not visible. If the pilot knows the ship's approximate position and the state of the tide, a more reliable fix can be obtained by comparing depth readings with the values shown on the local area nautical chart.

Some multipurpose depth sounders combine depth measurement with fish finding and course plotting. A profile of the ocean floor with depth information is displayed on one part of a split LCD screen along with symbols representing passing schools of fish. The other part of the screen can simultaneously display a digitally produced chart showing the ship's position, obtained from a GPS receiver, and the range and bearing to selected waypoints of interest as well as the time required to get there at the ship's present speed. Depth sounders are also called *depth finders* and *fathometers*.

# Fishfinders

A *fishfinder* is a specialized form of *depth sounder* for finding schools of fish. The system includes a *display/control unit* with a graphic display and a *transmitter-receiver,* typically mounted in a single case which is connected to a hull-mounted *transducer* by coaxial cable. The display can be either an LCD panel or a CRT monitor that presents a graphic profile of the ocean floor under the host ship. It also provides symbols on the display indicating probable schools of fish as well as underwater vegetation and structures. The latest fishfinders include microprocessors to control the display. Some can provide a three-dimensional image of the ocean floor by splitting the transducer beam into sections.

Steerable scanning transducers on some models can be adjusted to look straight down into the water or be moved to look in front of, behind, or on either side of the moving ship, making them true *sonar* systems. More sophisticated fishfinders have sensors that permit them to display water temperature and cable connections to other instruments that provide a readout of the ship's speed and position. Commercial and scientific fishfinders are equipped with printers for printing out the display as a continuous strip on paper for study and record purposes.

# Smallcraft Radars

Affordable radar systems for small ships and pleasure craft that offer many of the features of large radars on passenger and merchant ships have been made possible by the development of raster-scanned monochrome color LCD displays and solid-state circuitry. The useful range of these radars is about 20 mi (32 km), adequate for slow-moving vessels, but the radar's power consumption has been reduced to less than 30 W and weight has been significantly reduced. Some smallcraft radars with longer ranges have raster-scanned monochrome CRT displays, but they draw more power. While these displays cannot provide the accuracy and resolution of the larger CRT PPI displays, they can still provide navigational information that is accurate enough for small craft.

The radars with LCD displays offer 320- × 240-pixel resolution on a 7-in (18-cm) VIS backlit display mounted in a book-sized case with a control keypad located next to it. They include variable range markers and electronic bearing lines. Some models offer digital displays of key navigation data obtained by cable from other onboard instruments. The most common connection is from a GPS receiver that gives the radar information about navigating to the next waypoint. A chart reader can also be interfaced on some radars so that they can display either image with the push of a single button.

The radomes for these systems have diameters of 12 to 18 in (30 to 46 cm) and weigh 10 to 12 lb (4.5 to 5.4 kg). Typical characteristics of these radars include:

Maximum range  16 to 36 mi (26 to 52 km)

Radiated output  2 to 4 kW

Vertical beam width  25 to 30°

Horizontal beam width  4 to 7°

Power consumption  2 to 3 A

Bearing accuracy  ±1°

Range accuracy  1 to 1.5° or about 20 m

See also "Radar Systems" in Sec. 23, "Military and Aerospace Electronic Systems."

# Electronic Chart Plotters

An *electronic chart plotter* can display nautical charts on a backlit black-and-white LCD screen. The 6-in (15-cm) VIS screen is in a pocketbook-sized case that can be mounted on a ball-jointed bracket. Memory cartridges that store charts for various coastal regions around the world can be inserted in the unit to provide information about local waters. The charts provide latitude and longitude, depth contours, and coastal features. An omnidirectional trackpad permits the user to move an on-screen cursor to any point on the screen for gaining further information or plotting routes.

Routes can be planned by guiding the cursor over the chart and touching a single key to enter *waypoints* (locations specified by latitude and longitude over the ocean floor) where desired. Waypoint selection and connecting course lines, along with a constantly updated position relative to them, is automatically displayed on the chart. The position can be obtained from a GPS receiver built into the chart system or from external GPS or loran receivers. A cable can provide on-screen depth readout or send an output signal to an autopilot.

Events and marks can be entered anywhere, and the range and bearing between any two points will be automatically computed. Zoom-in and zoom-out controls permit the magnification of any desired area in the center of the screen for a close-up view.

Navigation software on CD-ROM now permits a battery-powered laptop computer to function as a portable navigational display. Multicolored reproductions of official nautical charts can be displayed on the laptop computer's screen. Position data from an external compatible GPS satellite receiver or loran can be introduced by cable connection into the computer to establish a continuously updated position plot of the user's vessel.

This form of electronic navigation system provides continuous navigation updates. Present location, progress, and route status are available. Cross-track error, range and bearing to the next waypoint, and estimated time of arrival can be displayed on the screen. In addition, the software permits panning and zooming in on sections of the chart of special interest, and the chart scale can be changed. Moreover, the user can print out sections of the chart as well as prepare *floatplans,* or nautical travel itineraries.

# Loran C

*Loran* is an electronic navigation system for ships and aircraft based on radio signals. Its name is derived from *long-range navigation.* It is a passive or receive-only hyperbolic system that can be used to obtain an accurate navigational fix from the latitude and longitude coordinates determined by the receiver. The original Loran A, also called *standard Loran,* has been phased out and replaced by Loran C.

Loran C transmitters operate on 100 kHz in "chains" of a master and two to four slave stations. The *master station* is common to each set of *slave stations.* It transmits its pulses first, and these propagate outward over the region covered by the system while simultaneously traveling toward the slave stations. When the master pulses are received at its slaves, they are triggered to transmit their own pulses. Each station actually transmits groups of pulses on the same frequency. The chains are identified by their pulse group repetition intervals. The loran transmitter signals and frequencies are controlled by atomic clocks.

The master and slave stations are a fixed distance apart, so there is a constant difference in time between the transmission of pulses from each of the stations. This results in a *time difference* (TD) in the arrival of pulses from each pair of stations that is measured by the loran receiver. That time difference is proportional to the distance between the receiver and the two transmitters.

The master station's pulses are received first by the loran receiver, which then measures the time for the arrival of the slave pulses. This time difference in microseconds is the basis for further calculations performed by the receiver. Because the plot of constant time difference from each pair of stations is a parabola, complete families of parabolas from many station pairs are printed on standard navigation charts.

By locating the parabola on the chart representing the time difference between a known pair of loran stations, the navigator can establish a *line of position* (LOP). The intersection of a second time-difference hyperbola or LOP with the first LOP provides an accurate navigational *fix*. Additional LOPs further increase the accuracy of the fix. The latitude and longitude coordinates on the chart permit a navigator to establish an accurate position. Loran accuracy depends on the constant speed of radio signals (186,000 mi/s or 300,000 km/s). The time of RF travel is accurately converted into distance by the loran receiver.

The availability of low-cost microcontroller-based circuitry made it possible to produce lightweight, compact, and affordable loran receivers for private and commercial vessels and aircraft. Loran receivers can provide a direct readout in latitude and longitude after local coordinates are entered, and they can also display other navigational information on LCD displays such as range and bearing to a waypoint (intended destination), speed along the track, and estimated time to reach a waypoint at the ship's present speed.

Loran C has a ground-wave surface range up to about 1200 nmi (2200 km), with sky-wave reception out to about 3000 nmi (5500 km). Receivers with microcontrollers can fix their positions within 500 ft (150 m) of their true positions within about 500 nmi (930 km) of a loran transmitter under favorable weather conditions. The receivers can be connected with radars, chart plotters, and depth finders to share information.

A Global Positioning System receiver can establish a ship's position far more accurately than Loran C, so it has displaced loran as the navigational instrument of choice. Because of the rapid worldwide acceptance of GPS, the U.S. Coast Guard may shut down Loran C after the year 2000. However, it has been estimated that there are still more than a million loran users in the United States.

# Global Positioning System (GPS) Receivers

The *Global Positioning System* (GPS) is described in Sec. 23, "Military and Aerospace Electronic Systems." It has a constellation of 21 active Navstar satellites and 3 active spares that broadcast two sets of signals on different frequencies: one set is coded for reception by military receivers with the necessary decoding circuitry, and the other is an open GPS-C (degraded commercial) signal available free of charge to anyone, anywhere in the world with a commercial receiver. The GPS system is officially called the *Navstar Global Posi-*

*tioning System.* Military GPS receivers can establish a position fix that is within about 50 ft (15 m) of true location, but the GPS-C receivers can provide position accuracy within 300 ft (90 m) of true location.

This level of position accuracy is adequate for most pleasure boat and commercial ship navigators, but the real value of GPS is its ability to direct navigators to their destinations. After a *waypoint* or location determined by latitude and longitude coordinates is entered into the receiver, it will give the distance and direction to it, off-course errors, and the time to reach that point at its present speed.

Many different models and grades of handheld GPS receivers are available, some for less than $200. A typical handheld GPS receiver is shown in Fig. 24-2. It can be powered by alkaline cells, and it features an internal flat-plate antenna. Models intended for fixed mounting typically cost more, but they offer larger displays and keypads as well as external antennas for better reception. A basic handheld GPS receiver, priced under $200, will provide the following information:

- Latitude and longitude of position within about 300 ft (90 m) of the exact terrestrial position
- Speed and course over the sea floor or actual ground
- Probable time of arrival at a selected waypoint based on present course and speed
- Direction and exact distance along the great circle route of any of up to 500 waypoints and 20 reversible routes stored in the receiver's memory

The latest GPS receivers have as many as 12 parallel channels, each of which can track a different satellite simultaneously. Most GPS receivers have graphic LCDs capable of a selection of graphic displays showing steering information, a plot of where the ship has been, and chart information. The time required to obtain the first fix has been reduced from over 2 min with 1- to 3-channel receivers to less than 45 s with 12-channel models.

Some GPS receivers are combination units that integrate GPS with VHF radios, depth sounders, and fishfinders. Signals from receivers can show a ship's position superimposed

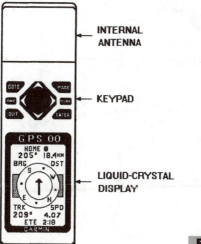

**Figure 24-2** Handheld GPS receiver.

on an electronic chart plotter generated from plug-in cartridges or CD-ROMs. Some of these GPS–chart plotter's screens can display important functions such as range and bearing, heading error, and time to go, all on the same display.

The *Differential Global Positioning System* (DGPS) provides more precise position information for navigating in narrow channels at night or in fog. It does this by providing a correction for the GPS-C signal. Most GPS receivers can be interfaced with DGPS. Another positioning system available for commercial aircraft navigation is the *Wide Area Augmentation System* (WAAS). Both of these systems require special beacon-receiving circuitry and an additional antenna to receive this supplemental data.

GPS receivers are no longer limited to maritime use. Some models have been designed for over-the-road and off-road travel. They contain road-map data, including landmarks such as lakes, rivers, and interstate highways, and they can store 500 waypoints and 20 reversible routes.

GPS receivers built into digital wristwatches are now available as consumer products. In addition to telling time, each watch can show latitude and longitude and a simple map indicating the watch's position relative to an established reference point on its multifunction LCD display.

## DIFFERENTIAL GLOBAL POSITIONING SYSTEM (DGPS)

The *Differential Global Positioning System* (DGPS) compensates for the inherent inaccuracies of a position when it is determined from GPS-C signals. DGPS uses existing transmitters from the Radio Beacon System (RBS). Figure 24-3 is a block diagram of the system. Active ground-based transmitters now provide coverage to the East, West, and Gulf Coasts, as well as the Mississippi River and Great Lakes regions. It is operated by the U.S. Coast Guard.

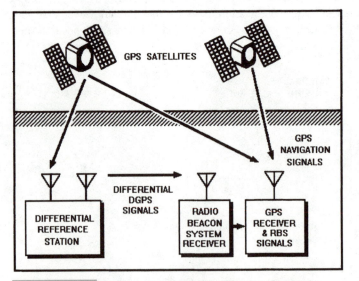

**Figure 24-3** Differential Global Positioning System (DGPS) configuration.

A *differential reference station* calculates a correction signal based on the error between its known position and its position as given by the GPS-C receiver. That signal is transmitted at 100 to 200 b/s to a 4-ft (1.2-m) loran whip antenna connected to an RBS which, in turn, is connected to the GPS receiver. (Some GPS receivers now include RBS receiving circuitry.) The signal can improve the accuracy of an uncorrected GPS receiver to within 15 to 30 ft (5 to 9 m) and its speed accuracy to ±0.1 knot. The RBS signal can be received within a 500-mi (804.5-km) radius. Some DGPS RBS receivers are about the size of a handheld GPS receiver.

# Emergency Position-Indicating Radio Beacons (EPIRBs)

An *emergency position-indicating radio beacon* (EPIRB) is a battery-powered emergency radio beacon for use by ships in distress that require rescue. It is intended for use beyond the range of VHF transceivers only for Mayday emergencies. One model of a 406-MHz EPIRB with an automatic turn-on switch is shown in Fig. 24-4. EPIRB transmissions reveal its position to both aircraft and satellite receivers which, in turn, relay that information to the nearest shore station to alert search and rescue services.

EPIRBs can be triggered legally only when a vessel is on fire, in danger of sinking, or has a person on board with a life-threatening injury or disease. The more than 20-year-old 121.5-MHz (civilian) and 243-MHz (military) beacons are being superseded by the newer, more effective 406-MHz beacons. The 121.5- and 243-MHz EPIRBs will continue to serve in parts of the globe where a satellite is within line of sight of both the beacon and a ground station so that real-time links can be made.

The 406-MHz EPIRB can operate outside of those limits because its signals are stored in satellite memory until a link can be made to transfer the data to a ground station. Its signals provide more accurate vessel location and time data than the earlier 121.5- and 243-MHz systems. In addition, the 406-MHz EPIRB continually transmits a digitally encoded identification signal that permits the ground station to identify the ship in distress by name, type, size, and home port. Orbiting satellites relay requests to the National Oceanic and Atmospheric Administration [NOAA]. The U.S. Coast Guard is then alerted by NOAA on 406 and 121.5 MHz.

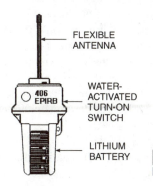

FLEXIBLE
ANTENNA

406
EPIRB

WATER-
ACTIVATED
TURN-ON
SWITCH

LITHIUM
BATTERY

**Figure 24-4** Emergency position-indicating radio beacon (EPIRB).

Category I 406-MHz EPIRBs are designed and mounted so that they deploy automatically, float free of their shipboard mounting, and activate. Category II EPIRBs must be manually deployed to be activated. The earlier 121.5- and 243.0-MHz EPIRBS were referred to as Class A if they were automatically activated and Class B if they were manually deployed and activated.

EPIRB transmitters and batteries are packaged in waterproof buoyant plastic cases with the antenna and activation switch on top. The latest models are powered by lithium batteries with shelf lives of up to 10 years which permit signals to be transmitted for up to 5 days. Earlier models were powered with water-activated batteries.

Category I and II EPIRB signals can be received by aircraft, and both U.S. search and rescue satellite-aided tracking (SARSAT) and Soviet COSPAS satellites in low-level (500- to 600-mi [800 to 970 km]) polar orbits. At least one geostationary satellite (22,300 mi [35,880 km] altitude) can also receive the signals. But only the polar-orbiting satellites can relay the distress signals to a series of ground stations or local user terminals (LUTs) in the worldwide system.

Electronics on the satellites compute the location of the beacon by measuring its Doppler *inflection point*—the time of closest approach when the received signal pitch switches from increasing to decreasing. Decisions for starting search and rescue missions are made when the information from a LUT is received at a mission control center (MCC). The MCC then alerts the appropriate rescue coordination center (RCC). COSPAS-SARSAT is an international cooperative program with many different countries participating at all levels.

# Radio Direction Finders (RDFs)

A *radio direction finder* (RDF) is a radio receiver with a directional antenna and a visual null indicator. When visibility is poor, an RDF can help fix a ship's position, and the navigator can home in on a transmitter near his destination. The directional antenna on most RDFs can be rotated so the receiver is fastened in a fixed position. An RDF is basically an improved version of a portable radio receiver with a rotating antenna on top. The RDF antenna can be either a loop or a bar. The loop is about 1 ft (30 cm) in diameter and the plastic-covered bar measures about $1 \times 6$ in ($2.5 \times 15$ cm).

As the antenna is rotated through 360°, the output signal strength will pass through two positions of maximum signal strength and two positions of minimum sensitivity called *nulls*. Theoretically the peaks will be 180° apart while the nulls are located 90° on either side of the peaks. However, because the nulls are sharp and precise while the peaks are broad and poorly defined, the nulls are used for direction finding. On most sets the antenna is turned by mechanical gearing to keep the user's hand from interfering with the reception.

The operator can judge by ear the position of the antenna at a null or peak, but most RDFs have an analog meter to give a visual indication. RDFs normally cover three frequency bands: (1) a low frequency (LF) beacon band, (2) the standard AM radio broadcast band, and (3) the 2- to 3-MHz communications band. The marine radio beacons along or just off the U.S. coast and near the Great Lakes are operated by the U.S. Coast Guard on frequencies between 285 and 325 kHz. They provide the most accurate bearings for shipping. However, aeronautical beacons that operate at lower and higher frequencies can also be used by ships for direction finding.

# SCIENTIFIC AND MEDICAL
# INSTRUMENTATION

# Overview

The first major development in medical electronics was the discovery of X-rays by the German physicist, Wilhelm Roentgen, in 1895. He was experimenting with a rather crude cathode-ray tube when he discovered that mysterious rays had penetrated soft materials and left shadow profiles of the bones of his hand on a fluorescent screen. He called them *X-rays* because, at that time, they were not known to be short-wave electromagnetic radiation. This discovery made possible, for the first time, the nonintrusive examination of the internal organs of living persons, so it had far-reaching implications for scientific and medical research.

Subsequent developments in X-ray machines improved the resolution of X-ray radiograms while minimizing patient exposure to the dangerous rays. More recently closed-circuit television, multicolor CRTs, and computers have made it possible to enhance the basic radiogram with digital signal processors and software algorithms that produce video "slices" or section views of living organs. The technique, known as *computed tomography* or *CT scanning,* provides color-coded images to highlight specific internal body functions.

The images are formed from data taken from a circular array of sensors surrounding a rotating X-ray scanner. Further data processing can yield three-dimensional views of human organs such as the brain or heart for further study or analysis.

Another noteworthy advance in medical instrumentation was made when it was found that atoms in living tissue will emit resonant RF signals in the presence of strong magnetic fields and radio waves. This led to the invention of *magnetic resonance imaging* (MRI), another nonintrusive process that can produce three-dimensional multicolor images of organs within living persons to complement the findings of CT scanning. Other diagnostic processes that depend on computer data processing are *positron-emission tomography* (PET) and *sonography.*

This section discusses only a few examples of advanced medical instrumentation made possible by electronics and computers, but many more mundane physicians' tools, invented before the era of solid-state electronics, have been updated with microcontrollers and electronic circuitry to provide faster and more accurate readings. Examples are the digital thermometer, digital blood-pressure meter or sphygmomanometer, and the electrocardiograph.

Some of the new medical instruments are byproducts of research in other areas of science, and the same instrumentation that is valuable for diagnostics and treatment of patients is useful in medical, pharmacological, and biological research. For example, considerable knowledge of brain function has been gained by observing changes in activity levels of certain regions of the brain when comparing the results from normal and abnormal subjects.

The *electron microscope,* a direct descendant of the early cathode-ray tubes, has made possible important discoveries in chemical, biological, medical, metallurgical, and even electronic research. There are now different kinds of electron microscopes for examining different subjects in different ways.

A full discussion of all of the ways in which electronics has contributed to scientific research is beyond the scope of this handbook, but electronics and computer science have effectively revolutionized the sciences of astronomy, meteorology, and oceanography, to mention but a few. The field of radio astronomy, for example, developed out of research to find the source of mysterious background radio noise that was interfering with communications.

# Computer-Aided Medical Imaging

*Computer-aided medical imaging* refers to many different computer and video-based techniques that permit physicians and surgeons to view organs and tissues within the human body noninvasively. Five different computerized body-scanning techniques are based on the ability of the computer to enhance or construct more detailed images than could be derived directly from the basic sensors or recording methods:

1. Computed tomography (CT) or computer-aided tomography (CAT)
2. Magnetic resonance imaging (MRI)
3. Digital subtraction angiography (DSA)
4. Positron-emission tomography (PET)
5. Sonography

## COMPUTED TOMOGRAPHY (CT)

*Computed tomography* (CT) is computerized body scanning based on X-rays. Scanners convert a sequence of X-ray pictures into digital code that, with the aid of specialized computer software, is interpreted by the computer to form high-resolution video images on a CRT monitor. The principles of CT scanning are illustrated in Fig. 25-1. The X-ray tube in the CT scanner is mounted on a circular frame that revolves through 360° in a plane that passes through the patient. The scanner produces a thin, fan-shaped X-ray beam as it rotates around the patient's body, exposing all sides within the frame to X-rays. Sensitive detectors mounted inside the fame opposite the tube convert the changing patterns of radiation into signals that are processed by computer.

Specialized software digitizes the detector data and integrates it into a signal suitable for video scanning. The TV monitor can display cross sections or thin "slices" of the patient's body at any point along its length, and colors can be assigned by tissue density or other variables. Multicolor two- and three-dimensional images can also be formed.

X-rays are absorbed by dense body structures such as bones or metallic foreign objects such as bullets, but they pass through soft tissues. Dark shadows with sharp edges indicate

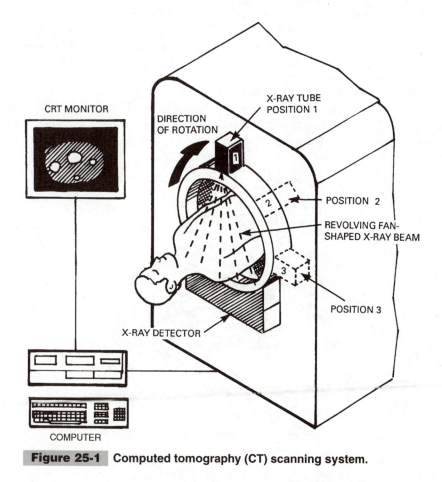

**Figure 25-1** Computed tomography (CT) scanning system.

dense structures on conventional X-ray film. Softer tissue such as muscles, organs, and skin are more easily penetrated by X-rays and show up only as vague shadows. Consequently, even the best of conventional X-ray radiograms, which view the body from only one angle, require interpretation by a specialist in radiology, and those interpretations are open to differences of opinion. Analysis can be difficult because the shadows of bones, muscles, and organs are superimposed on one another, and they can partially or completely mask some clinically significant anomalies that lie behind them.

Recent improvements in X-ray machines have resulted in clearer radiograms with lower patient radiation exposure to harmful X-rays, and now images from conventional X-ray machines can be digitized and computer enhanced to improve their contrast.

## MAGNETIC RESONANCE IMAGING (MRI)

*Magnetic resonance imaging* (MRI), formerly called *nuclear magnetic resonance imaging* (NMRI), is a method for observing internal organs and tissues of the human body based on the action of radio waves on living tissue in a strong magnetic field. The data obtained can be displayed on a computer monitor as two- and three-dimensional images in color for real-time analysis and diagnosis and later playback. The principle parts of an MRI system are illustrated in the cutaway drawing Fig. 25-2.

MRI is based on a phenomenon known as *atomic resonance*. The patient enters a cylinder surrounded by electromagnets and radio-frequency coils. Because the human body is 70 percent water, it has a high hydrogen content. Protons within the hydrogen atoms normally spin like tops and point in random directions. However, within the strong electromagnetic field inside an MRI scanner, the protons align themselves in the direction of the electromagnet's poles. The field does not hold them in rigid alignment, so they wobble or *precess* at a known frequency or rate. This frequency is proportional to the strength of the magnetic field.

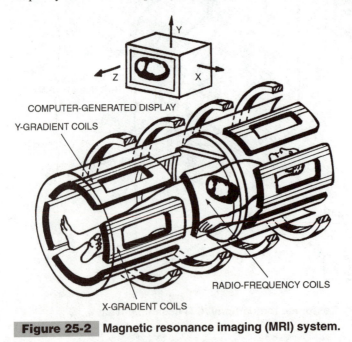

**Figure 25-2**  Magnetic resonance imaging (MRI) system.

The MRI scanner excites those protons with radio-frequency pulses synchronized with the proton precession frequency, destabilizing them and forcing them out of alignment. Within milliseconds they resonate and realign themselves. However, before they regain their initial orientation they emit faint characteristic RF signals. The computer converts these faint signals into an image of the area scanned.

MRI is capable of showing the densities of hydrogen atoms and their interaction with surrounding tissues in a cross section of the body. Tissues and organs have differences in water content that are proportional to hydrogen density, so the system can distinguish between them.

The controllable cylindrical electromagnet in the MDI scanner is supercooled by liquid helium. It has an axial bore with a diameter of about a meter, large enough to admit a prone patient. In a typical MRI system, this electromagnet weighs more than 20 tons (18,100 kg) and is capable of producing a uniform magnetic field of 1.5 tesla (T) within its axial bore. This is approximately 30,000 times the strength of the earth's natural magnetic field.

The RF is generated with coils. Signals are produced either by surface coils, a cylindrical head, or larger "whole-body" coils housed within the bore of the magnet. A computer and color monitor provide the video images. The complete MRI system must be shielded from external radio frequencies with an appropriate metal shield.

The computer establishes a grid of boxes in three dimensions, *x, y,* and *z,* called *voxels* (volume elements). First the magnetic field is varied in the *z* direction to define a plane of interest where the body will be scanned. RF coils within the magnet emit a pulse at the frequency necessary to cause the nuclei of the selected atoms to resonate and produce faint radio signals. An MRI scan translates the decay rates of the radio signals into levels of brightness, and then yields a computer-generated image of the organs.

Before the protons realign themselves and the resonance is damped out, other coils are varied in magnetic strength in the *y* direction. This causes the protons to precess at different rates from the top of the plane to the bottom. The computer is then able to locate voxels in the *y* direction after detecting hundreds of resonance cycles.

Coils then vary the magnetic field in the *x* direction, causing protons to resonate at different frequencies as they stabilize themselves. After each voxel is located in the *x, y,* and *z* directions, the computer displays the voxel on the screen of the CRT monitor as a *pixel.* The brightness of the pixel is determined by the number of hydrogen protons within the voxel and the magnetic properties of the tissue. The pixels form a readable image when raster scanned on the video monitor screen.

The parts of the body containing high percentages of water show up more clearly in the MRI images than those parts containing little water, such as teeth and bones. The MRI image differs from the X-ray CT scanned image because of its ability to show tissue and bone marrow clearly, even when it is surrounded by bone. MRI, for example, can distinguish between the brain's white matter and water-rich gray matter better than an X-ray CT scan. It has also been found that the radio signals emitted by cancerous tissue take longer to decay than signals from healthy tissue. Thus MRI images complement rather than compete with CT scans.

MRI can pinpoint otherwise invisible tumors, allow a tumor's reaction to chemotherapy to be monitored, and permit dosages to be applied directly to the tumor site. It can also be used to diagnose joint and bone problems and isolate ligaments and soft tissues that cannot be distinguished by X-rays.

Hydrogen is used as the basis for medical MRI scanning, but other elements such as iron, sodium, and phosphorous will resonate, and their presence or absence can provide early warning signs of strokes or heart attacks. MRI scanners with magnetic fields of 4 T—

80,000 times the strength of the earth's magnetic field—have been built. Higher magnetic fields significantly enhance image quality on high-resolution monitors.

## DIGITAL SUBTRACTION ANGIOGRAPHY (DSA)

*Digital subtraction angiography* (DSA) is a specialized form of computed X-ray imaging for the diagnosis of heart problems. It employs opaque dyes injected in the arteries in conjunction with X-rays. A computer converts images into digital codes, and these codes can be used to compare a sequence of X-ray pictures made from different angles. DSA deletes everything from the image except the specific veins or arteries under examination.

An X-ray picture of the heart is first made by the digital X-ray scanner to provide a reference. Then a contrast agent or dye is injected through a catheter into the coronary arteries. A second X-ray image is made showing the agent moving through the heart's vessels. The computer subtracts the first image from the second, leaving only the difference image—blood vessels containing the agent. By highlighting dynamic aspects of the human body, such as the passage of blood through the heart, DSA is effective in the study of heart disorders and the prediction of possible disease or heart attack.

## POSITRON-EMISSION TOMOGRAPHY (PET)

*Positron-emission tomography* (PET) is one of two forms of imaging trace amounts of radioisotopes. Radioactive tracers are well suited for studying epilepsy, schizophrenia, Parkinson's disease, and stroke. A PET scanner, as shown in the cutaway diagram Fig. 25-3, consists essentially of a ring of radiation-detection sensors mounted around a supporting ring with a diameter large enough to permit the passage of the patient's body.

A small, low-energy cyclotron, not part of the equipment, prepares *isotopes* with short half-lives for the PET scan. These substances can lose half of their radioactivity within minutes or hours of creation. When injected into the body, the radioactive solution emits positrons that can be detected wherever they flow by the ring of sensors around the patient.

When in the human body, the positrons collide with electrons, and the two annihilate each other, releasing two gamma rays. The emitted rays move in opposite directions, leave the body, and strike the crystal radiation detectors arrayed in the ring configuration inside the PET scanner. The crystals respond to the incident gamma rays by emitting a flash of energy that is then converted electronically into digital signals.

A computer records the location of each energy flash and plots the source of radiation within the patient's body. It then translates that data into a PET-scan image on a color CRT monitor. The concentration of isotopes can be displayed on the monitor in colors that indicate differing levels of biological activity.

## SINGLE-PHOTON-EMISSION COMPUTED TOMOGRAPHY (SPECT)

*Single-photon-emission computed tomography* (SPECT) is another method of imaging that depends on trace amounts of radioisotopes. SPECT differs from PET because it can use commercially available radioisotopes, greatly reducing the cost of the medical diagnosis.

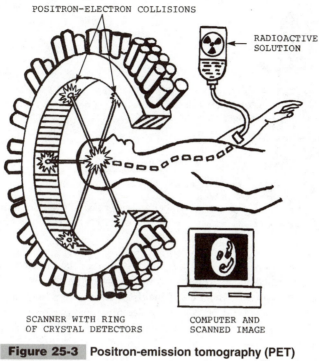

POSITRON-ELECTRON COLLISIONS

RADIOACTIVE
SOLUTION

SCANNER WITH RING
OF CRYSTAL DETECTORS

COMPUTER AND
SCANNED IMAGE

**Figure 25-3** Positron-emission tomography (PET) system.

## SONOGRAPHY

*Sonography* is a noninvasive scanning technique in which high-frequency sound waves are projected into the human body and their echoes are used to create an image. It is actually a specialized short-range sonar system designed for viewing internal organs and tissues. Short-range, high-frequency sound waves from a piezoelectric transducer are beamed into the human body, and the reflected sound waves are converted into electrical signals. Those signals are processed and displayed on a computer monitor. It can show the differences in densities and reflective properties of the organs scanned in black and white or color. The latest sonography equipment is able to provide three-dimensional views of the objects being examined.

Sonography is the only computer-aided scanning method recommended for the examination of pregnant women. It can provide a detailed picture of a fetus and reveal any abnormalities in its position; it is also suitable for examining other body organs such as heart, liver, and gall bladder.

# Electron Microscopes

## SCANNING ELECTRON MICROSCOPES (SEMs)

The *scanning electron microscope* (SEM) is based on the principal of scanning the specimen to be observed with a fine electron beam and collecting electrons scattered from it to

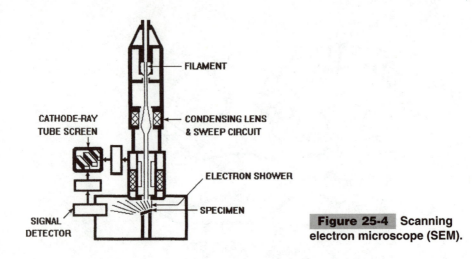

**Figure 25-4** Scanning electron microscope (SEM).

form an image on a cathode-ray tube monitor. A simplified cross section view of an SEM is shown in Fig. 25-4. The microscope is contained within a vacuum chamber, and its electron beam is formed by a heated filament coated with an electron-emitting material. The electron beam, in a high-voltage field, is formed by electromagnetic coils called *condensing lenses* and swept by a sweep circuit in a raster scan, as is done with an electromagnetic CRT in a TV monitor.

The electron beam strikes the specimen and electrons spray out in all directions like light rays after striking an object. The scattered electrons are collected and detected by a signal-detection circuit and sent to a video monitor. Specimens viewed on the monitor appear in three dimensions, and can be magnified up to 10,000 times.

Some specimens are first flash coated with a thin film of precious metal in a vacuum deposition chamber to enhance their ability to scatter electrons. SEMs are used to view sample integrated circuit chips, and this examination is a requirement for the qualification of some high-reliability semiconductor devices. SEMs are also used to view microminia-

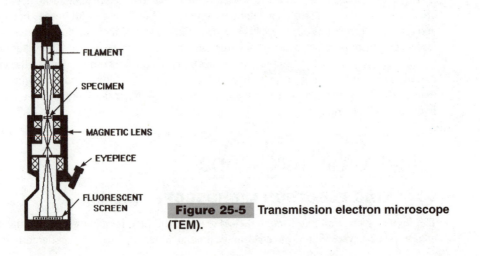

**Figure 25-5** Transmission electron microscope (TEM).

ture transducers and machines formed by semiconductor processing methods, biological specimens such as bacteria and viruses, and minute features of plants, insects, and animals.

Advanced SEMs called *cold-field emission microscopes* eliminate the need for flash coding metal on specimens that can alter or destroy them. These microscopes operate in a cool partial vacuum, and can magnify specimens up to 200,000×.

## TRANSMISSION ELECTRON MICROSCOPES (TEMs)

A *transmission electron microscope* (TEM) is an electron-beam analogy of a conventional visible-light microscope, and it is also closely related to a cathode-ray tube, as shown in the simplified cross section Fig. 25-5. The microscope is housed in a vacuum chamber. The electron beam is formed by a heated filament coated with an electron-emitting material and it is shaped by electromagnetic coils that act as a glass lens in transmitting light. The beam, in a high-voltage field, first passes through the specimen and then through another series of condensing "lenses" that focus it on a fluorescent screen. The image formed on the screen can be viewed through an optical eyepiece. However, unlike images seen in SEMs, the images of specimens in TEMs bear little resemblance to the specimen's actual appearance. TEMs are used primarily for viewing specimens in medical and genetic research.

# SEMICONDUCTOR DEVICE MANUFACTURING

# Overview

The invention of the transistor, closely followed by the invention of the monolithic integrated circuit, has revolutionized the electronics industry. Since the 1960s whole new industries have been created to manufacture semiconductor devices that range from pinhead-size diodes to microprocessor chips the size of postage stamps. While much of the technology for manufacturing ICs was adapted from earlier work on transistors and other discrete semiconductor devices, the fabrication of devices with microscopic features demanded new and improved lithographic methods and equipment. Billions of dollars have been spent on tools, equipment, and plant facilities, and millions of new jobs have been created around the world in fabricating and packaging these devices. Simultaneous advancements in computer technology and applications software merged with this improved process technology to make possible quantum leaps in device performance over the past 30 years.

In 1964, Gordon Moore, a founder of Intel Corporation, predicted that the number of transistors in an integrated circuit would double every 18 months, and that prediction has turned out to be accurate. Increases in circuit density have been accompanied by reductions in feature size. Design rule dimensions have decreased from tens of micrometers to a quarter of a micrometer. These dramatic size reductions have resulted in significant improvements in device speed and reliability as well as reductions in power consumption. Work is progressing on the next generation of ICs with dimensions smaller than one-fifth micrometer (0.18-$\mu$m). Integrated-circuit prices continue to fall, illustrating textbook examples of economies of scale driven by worldwide demand. The fortunes of the manufacturers have risen and fallen according to their ability to keep up with this ferocious pace.

Regardless of the type, size, or complexity of semiconductor device, their manufacturing processes exhibit more similarities than differences. All start life as polished and oxidized semiconductor wafers cut from large crystals before undergoing a dizzying series of layering, patterning, doping, and heat-treating processes. Surviving devices must pass rigorous testing procedures before they end up as packaged products suitable for sale.

Silicon remains the predominate material from which semiconductors are made. However, certain kinds of commercial and military analog, digital, and interface devices, both discrete and integrated circuit, have been made from gallium arsenide (GaAs) for more than 20 years because of certain advantages they offer over comparable silicon devices. These include speed, frequency, and radiation resistance. But GaAs technology is now being challenged by another technology called *silicon germanium* (SiGe) for radio-frequency and microwave applications. Germanium was used in the manufacture of early transistors but it was later superseded by silicon.

# Crystal Growth

A *crystal* is a solid material whose atomic skeleton forms a definite geometric pattern (lattice), like a jungle gym, scaffolding on a building, or a geodesic dome. Snowflakes, salt, quartz, diamonds, rubies, and emeralds are all crystals. Germanium, silicon, and gallium arsenide are important semiconductor crystals used in the manufacture of semiconductor devices. The *Czochralski (CZ) crystal growth method,* illustrated in the cross-section diagram Fig. 26-1, is the most popular process for growing large, single crystals for manufacturing semiconductor devices. Czochralski crystal growth is performed in a *crystal puller.*

Most silicon crystals are grown by the CZ method. Small chunks of polycrystalline silicon are placed in the quartz (silica) crucible with small amounts of the chemical dopant elements needed to produce either N- or P-type silicon. Examples of dopants that produce N-type silicon, called *donors,* are arsenic, phosphorous, and antimony, and an example of

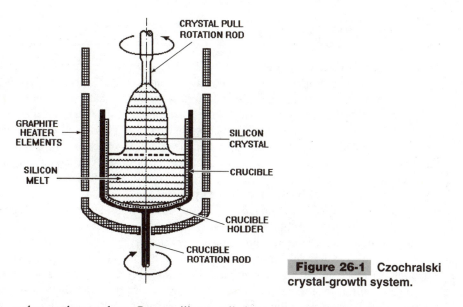

**Figure 26-1** Czochralski crystal-growth system.

a dopant that produces P-type silicon, called an *acceptor,* is boron. Donors have an extra electron and acceptors are deficient in an electron.

The silicon is first heated by the puller's heating elements until it melts down to form a liquid, called the *melt.* A *seed,* or small crystal of doped silicon taken from a previous melt, is required to start the process. It is clamped in the rotary chuck above the crucible and then lowered into the melt. The seed is rotated in one direction, and the crucible in its holder is rotated in the opposite direction.

Crystal growth begins as the seed is slowly raised above the melt, bringing with it a film of the molten silicon that has adhered to the seed and cooled. Crystal diameter and length depend on the temperature, the pulling rate, and the inside dimensions of the crucible. The temperature gradients must be minimized during the growth process to avoid dislocations in the crystal structure. While cooling, the silicon atoms in the melt orient themselves with the seed's crystalline lattice structure, and that structure will extend throughout the crystal as it grows in both diameter and length.

During growth the retraction or pull rate of the seed is controlled by a closed loop through a computer to assure that the crystal grows to the desired diameter. Counterrotation of the seed crystal and crucible reduces radial temperature gradients, and slow withdrawal of the rotating seed results in uniform crystal growth. The conditions for optimum growth vary widely, but pulling rates can be as rapid as a few inches an hour.

Silicon single crystals with diameters of up to 200 mm (8 in) and lengths up to several feet are now standard in the semiconductor industry, but equipment is now being built to grow 300 mm (12 in) wafers, expected to become the next industry-standard size. The larger wafer size will permit doubling the number of chips per wafer while decreasing production costs by as much as 40 percent. The growth process is observed by closed-circuit TV in a thick-walled control room as a safety measure.

A variation of the CZ method called the *liquid-encapsulated Czochralski* (LEC) *method* is used to grow Group III-V semiconductor materials such as gallium arsenide (GaAs) and gallium phosphide (GaP).

# Wafer Preparation

Wafer preparation starts with the removal of the finished and cooled silicon crystal from the puller. It is then trimmed, and both tapered ends are sawed off. The next step is grinding the entire diameter of the crystal in a centerless grinder so that wafers cut from it will fit the standardized diameter dimension of the process holders. The crystal is then tested by X-ray diffraction or collimated light refraction to determine its orientation and resistivity. An electrical conductivity test is performed on the crystal to verify that the crystal doping that took place during growth was uniform and correctly done. After connecting a probe from a polarity meter, the conductivity type (N or P) is displayed. The amount of dopant in the crystal is determined by a resistance measurement.

While the crystal is still mounted in a cutting block, a flat surface, called the *major flat,* is ground down the length of the crystal. This is done so that each wafer sliced from the crystal will have a reference edge positioned parallel to one of the crystal's natural planes, which were determined by the orientation test. The flat will provide a visual orientation reference during all further wafer processing. After the wafer is sliced, all pattern masks will be positioned with reference to this major flat. A smaller secondary flat is also ground on the crystal to designate wafer orientation and conductivity for the rest of the processing.

The wafers are sawed from the crystal with the inside edge of a thin, ring-shaped circular saw blade. The cutting surface is first coated with diamonds that do the actual cutting. Each sawed wafer is about 0.5 mm (0.02 in) thick. It then undergoes a two-step polishing process to assure that the surface is free of irregularities and saw damage and that it is absolutely flat. The first rough polishing is done by a lapping process with an abrasive slurry to remove any surface damage caused by the sawing process. The second and final polishing to a mirror finish is done by a combination of chemical etching and mechanical buffing. Finally, the edges of the wafer are rounded by grinding to minimize the chances that the edge will chip and cause the wafer to break.

After passing inspection, a protective layer of silicon dioxide is grown on the wafer by passing it through a furnace where it is exposed to the flow of hot gas at 1000 to 1200°C. The precise thickness of the oxide layer can be set by controlling the temperature, time, and gas flow rate. This layer shields the wafer's surfaces from scratches and contamination. Those wafers to be used in fabricating CMOS devices will receive a layer of epitaxially deposited silicon before they are oxidized.

# Mask Preparation

*Mask preparation* is the process of preparing the *photomasks* or *reticles* required for fabricating all semiconductor devices, discrete and IC. The manufacture of semiconductor devices is done in a series of steps that require from 5 to 20 or more masking operations, each calling for a unique mask. The masks are usually prepared by computer-aided design (CAD) and computer-aided manufacturing (CAM) techniques.

The complete design of a semiconductor device can be carried out on a computer workstation supported by design software. The end product might be an original product or an enhancement of an existing product. "Paperless" design now permits the preparation of functional block diagrams, complete circuit schematics, and masks on a workstation in many colors. A new design for an IC chip starts with the appropriate schematics for gates and cells (termed *macrocells*) called up from a database in computer memory and arranged on the workstation screen. By contrast, a revised design can start with the data from its predecessor.

The operation of the new or revised device can be simulated dynamically on the workstation screen. After the design has been verified in schematic form, the actual device elements can be laid out to scale on the screen, again by calling up patterns from a database stored in computer memory. The result will be a complete multilayered master drawing in color on screen, many times larger than the actual device, from which the actual production masks are derived. All of the individual layers must be precisely registered.

The digitized data defining each layer is then sent to an *x-y* plotter which produces master masks for the fabrication process. These masters are also many times larger than the actual device. They are then photoreduced to prepare the reticle, a pattern formed on glass or quartz from a thin, opaque film of chromium. The pattern can be either a negative or positive image. The reticle can then be reproduced many times to form a multiimage mask for step-and-repeat imaging that permits hundreds of images on a wafer to be exposed at one time.

# Wafer Fabrication

*Wafer fabrication* is a series of processes that are carried out in the manufacture of semiconductor devices. The procedure starts with the polished and oxidized blank wafer that has passed all of its quality tests. There are basically four different operations performed on a wafer during the fabrication process: (1) *layering,* (2) *patterning,* (3) *doping,* and (4) *heat treatment.* Some of these steps are performed many times during the process, depending on the design and complexity of the device.

## LAYERING TECHNIQUES

*Layering* is the operation in which layers of material are added to the wafer surface. The layers can be insulators, semiconductors, or conductors of different materials grown or deposited by different techniques. Oxidation is one method for growing silicon dioxide layers on the silicon wafer, as was done during wafer preparation. The most common deposition methods are chemical vapor deposition (CVD) performed in furnaces and evaporation and sputtering of metals performed in vacuum chambers. See "Vapor-Phase Epitaxy (VPE)," following in this section.

*Epitaxy* is a layering process for growing a perfect crystal layer on a wafer surface that has different dopants or even minor defects but a similar crystal lattice structure. The layer is grown by exposing the site to the molten or vaporized crystal epitaxial material. The atomic orientation of the crystalline growth layer is controlled by the structure of the substrate. See also "Epitaxial Processes," following in this section.

## PATTERNING TECHNOLOGY

*Patterning* is a series of steps in which well-defined patterns are formed on previously added layers by selective removal of unwanted material. The terms *photomasking, masking, photolithography,* and *microlithography* all refer to patterning processes. The pattern formed on the wafer surface defines the functional parts of the device in the exact dimensions required by the design in the precise location on the wafer. Patterning steps are the most critical operations in the fabrication process. Errors in the placement and dimensioning of patterns at any stage of the process can alter the electrical operation of the device or, worse yet, prevent it from functioning. This process has become more critical, particularly for IC memories and microprocessors, as line widths or design rules become smaller. The prevailing industry standard is 0.25 µm, with even smaller dimensions on advanced devices. This is complicated by the fact that some ICs require more than 20 patterning operations.

Patterning steps make use of *reticules* or photographic masks formed on transparent substrates, *photoresist masks* formed by dissolving unwanted areas of photoresist that cover oxide coating or metal film that has been deposited on the wafer, and *oxide masks* formed in the oxide coating on the semiconductor wafer by dry or wet chemical etching after photoresist masking.

*Photoresist* is a light-sensitive polymeric film that is spun onto the wafers before they are exposed to high-intensity UV radiation through a mask and the photoresist is developed. Negative photoresist cross-links and polymerizes in the transparent pattern of the mask where it is exposed to UV. This makes it resistant to solvent, while the regions protected by the opaque parts of the mask are dissolved by the solvent. The result is a negative image of the mask on the wafer. By contrast, positive photoresist cross-links and polymerizes under the opaque pattern of the mask that protected it from UV. This makes it resistant to the solvent, while the exposed areas are dissolved by the solvent. The result is a positive image of the mask. Etching takes place in those areas where the photoresist has been removed.

## DOPING METHODS

*Doping* is the process of introducing specified amounts of dopants into the surface of the wafer through "windows" patterned in surface layers. Doping of the wafer occurred when the crystal was doped during the growth process by adding donor or acceptor elements to the melt. However, doping during wafer fabrication is achieved by *thermal diffusion* or *ion implantation.* Doping creates the N- or P-type pockets in the wafer surface that form the diodes, transistors, resistors, and capacitors of the device.

*Thermal diffusion* is the process of doping a wafer by heating it in a furnace under controlled conditions to about 1000°C and then introducing a vapor of the dopant material that penetrates the wafer surface. The composition of both *intrinsic* (undoped) or *extrinsic* (previously doped) semiconducting materials can be altered by this method. Intrinsic material can be doped, and extrinsic material can be switched from P- to N-type or the converse.

*Ion implantation* is a selective doping process that drives ionized dopant atoms into the semiconductor wafer at room temperature. After stripping one or more electrons from the dopant atoms, the resulting ions are accelerated by a high voltage and formed into a narrow beam that is directed into windows patterned in the wafer surface, as shown in Fig. 26-2. The ions have enough energy to penetrate the surface, and then slow to rest within the wafer

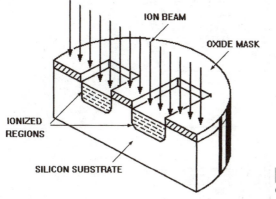

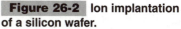

**Figure 26-2** Ion implantation of a silicon wafer.

at depths determined by their mass and energy. The accelerated ions cause damage to the crystal lattice, but this can be repaired by annealing the wafer in a furnace. Ion implantation permits the doping levels to be accurately controlled. It can also introduce certain dopants that are difficult to deposit by thermal diffusion.

## HEAT TREATMENT

*Heat treatment,* a process of annealing by heating and cooling a wafer, repairs lattice damage caused by ion bombardment and removes stresses in its crystal structure caused by thermal diffusion. The wafer is baked at about 1000°C and then cooled. In another heat-treating process, the wafer is heated to about 450°C and then cooled so that metal layers, now typically aluminum, are bonded more securely to the crystal substrate, improving the electrical conductivity of the terminals.

# MOSFET Transistor Fabrication

This description of the processes for fabricating a metal-oxide semiconductor field-effect transistor (MOSFET) has been simplified to avoid technical details that are not necessary in a general description of the manufacturing process. MOSFETs are no longer made in the way described, but the processes remain essentially the same. Optical lithographic processes using ultraviolet (UV) light are applied in the patterning steps. The process is illustrated by the six parts of the simplified drawing Fig. 26-3.

The process begins with a lightly doped P-type silicon wafer that has been oxidized in a furnace, as shown in Fig. 26-3a. A film of photoresist is applied over the oxide layer, and the wafer is exposed to UV through a photomask, changing the composition of selected parts of the photoresist layer. The unwanted areas of the photoresist are removed with a chemical solvent, exposing the underlying oxide layer. A wet or dry etching process is used to open source and drain "windows" down to the bare P-type silicon, as shown in Fig. 26-3b. In an ion implantation process, N-type dopant ions are shot through the open windows in the oxide layer to convert the extrinsic P-type region to N-type "wells" of controlled depth.

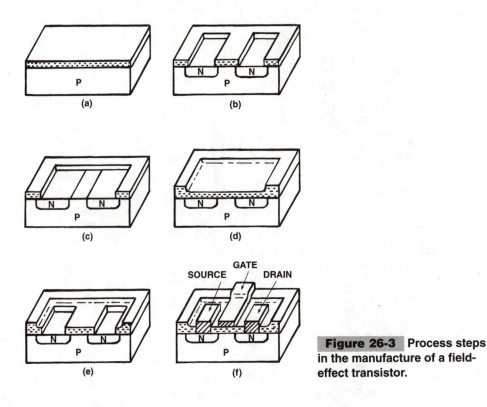

**Figure 26-3** **Process steps in the manufacture of a field-effect transistor.**

Next, the oxide layer between the source and drain windows is removed, exposing the P-type gate region, as shown in Fig. 26-3c. A new oxide layer is then grown over the remainder of the oxide layer and the exposed N- and P-type silicon regions, as shown in Fig. 26-3d. Another photoresist, masking, and etching step, Fig. 26-3e, opens new windows over the N-type source and drain regions. A metal layer, typically aluminum, is deposited in direct contact with the N-type regions and it also forms the metal gate. Another photoresist, mask, and etching process removes excess metal between the gate, source, and drain contacts and their bonding pads to provide electrical isolation, as shown in Fig. 26-3f.

The wafer is then passed through a nitrogen-atmosphere furnace to alloy the metal in the source and drain regions to the underlying silicon to improve the bond for enhanced conductivity and improve the mechanical bond between the metal gate contact and the underlying oxide. The final steps in this procedure (not shown) are the deposition of a protective passivation layer over the entire wafer to protect the surface contacts during testing, packaging, and the life of the device. This passivation layer is then etched away from the device's bonding pads to permit wire bonding.

Other devices such as bipolar junction transistors (BJTs) and silicon-gate MOSFETS are made with the same processes, but typically more masking and etching steps are required. Integrated-circuit fabrication calls for many more layering and patterning steps than discrete devices, but passive components such as resistors and capacitors are formed at the same time as the active diodes and transistors.

# Integrated-Circuit Manufacturing

The four basic operations described under "MOSFET Transistor Fabrication," layering, patterning, doping, and heat treatment, are also used in the manufacture of integrated circuits.

## MOS VERSUS BIPOLAR IC FABRICATION

The methods for fabricating bipolar junction transistors (BJTs) differ from those used in fabricating metal-oxide silicon (MOS) ICs because of their different geometries and operating characteristics. CMOS IC fabrication technology, for example, is simpler than bipolar IC fabrication technology because P-type wells can be diffused into N-type substrates in fabricating N-channel MOS transistors, as shown in Fig. 26-3. Thus fewer steps are required for making a CMOS IC than are needed for making an NPN bipolar transistor IC.

Typically, more than 10 masking and etching operations as well as dozens of other procedures are necessary for the manufacture of even the simplest integrated circuit. A CMOS memory IC, for example, might have 14 masking levels and more than 100 processing steps. Nevertheless, IC manufacturing technology is an extension of discrete-device manufacturing technology, and the similarities exceed the differences.

The CMOS cross section shown in Fig. 26-4 is based on silicon-gate technology, which differs from the metal-gate technology shown in Fig. 26-3. The figure is vastly simplified because a section drawing of an actual CMOS IC is far more complex and difficult to interpret. The silicon gate process has permitted higher density and finer line structures, and has made the next-generation 0.18-μm and smaller technology feasible.

A simplified cross-section view of an NPN transistor in an integrated circuit is shown in Fig. 26-5. The significant difference between this transistor structure and that used for discrete signal-level and power transistors is the formation of all metallized contacts on the top surface of the IC. This permits all wire bonding to be performed on the metallized top layer of the IC. For most discrete and power NPN transistors, the collector contact is the metallized underside of the die. It is typically the surface bonded to the leadframe so that only two wire bonds need to be made.

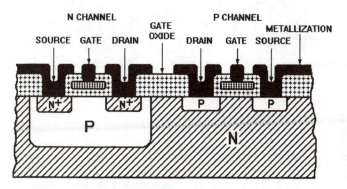

**Figure 26-4** CMOS IC combines NMOS and PMOS silicon-gate transistors.

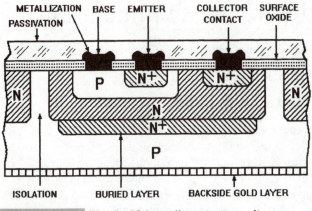

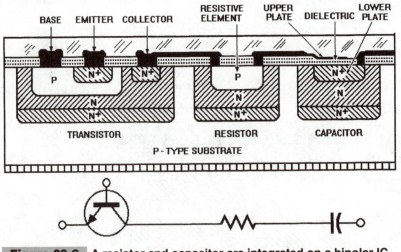

**Figure 26-5**  Bipolar IC has all contacts on its upper surface.

## ACTIVE COMPONENT FORMATION

Active circuit elements such as bipolar and MOS transistors and diodes are formed on the integrated circuit wafers simultaneously with the formation of passive components such as resistors and capacitors. The patterning steps are designed to include elements of both active and passive components that share the same processes. A simplified cross-section view of a small part of a bipolar integrated circuit, Fig. 26-6, shows an NPN transistor integrated with a resistor and a capacitor.

Because all components are formed on a wafer of conductive silicon, the integrated components must be electrically isolated from each other. This isolation is achieved by diffusing phosphor into the P-type substrate to form the heavily doped $N^+$ wells as barrier regions for all components.

**Figure 26-6**  A resistor and capacitor are integrated on a bipolar IC.

## MONOLITHIC-TRANSISTOR FORMATION

The N-type well for the NPN transistor's collector region is formed by diffusing a lighter N-type dopant into the $N^+$ barrier wells. Then boron is diffused into the N-type collector well to form the P-type base region of the transistor. This step is followed by another N-type diffusion to form the emitter region. As shown in Fig. 26-6, the emitter region is heavily doped $N^+$ and another well in the collector region is heavily doped $N^+$ for better conductivity with the metallized layer. All of the IC terminals are formed at the same time by patterning the final metal layer.

## MONOLITHIC-DIODE FORMATION

An *integrated diode* (not shown in Fig. 26-6) can be made by forming a PN junction at the same time as the NPN transistor's collector-to-base or base-to-emitter junctions are formed. Both diode terminals are brought to the top surface of the chip. If faster diode switching is required, the diodes can be made during transistor emitter-base diffusion; the P-type diode anode is formed during the base diffusion, and the N-type cathode is formed during emitter diffusion. As in fabricating the transistor, the diode terminals are formed by patterning the final metal layer. To avoid unwanted transistor action, the anode terminal short circuits the P-type anode region to the N-type transistor collector region. If reverse voltage is applied to the diode's emitter-base junction, it will act as a zener diode.

## MONOLITHIC-RESISTOR FORMATION

An *IC resistor* can be formed by diffusion into the IC chip at the same time as the NPN transistor base regions are formed. One method for forming the resistor is shown in the center of Fig. 26-6. The N layer below the P layer will be reverse biased for isolation from the substrate. Silicon dioxide is used as the insulation. The resistor terminals are also formed by patterning the final metal layer. The resistance values of the monolithic resistor can be tailored by adjusting the ratio of length to width of the resistive strip or by performing special ion implantation procedures in those regions. The tolerance of these monolithic resistors is typically only $\pm 30$ to 50 percent. As an alternative to monolithic resistor formation, resistive films can be deposited directly on the upper surface of the oxide layer.

## MONOLITHIC-CAPACITOR FORMATION

An *IC capacitor* can be formed as a junction or as a MOS component. Junction-type capacitors depend on the inherent capacitance of a reverse-biased PN junction. This method limits maximum capacitance values to about 100 pF, enough for bypassing or coupling functions. Capacitance value depends on the reverse voltage applied across the junction.

A section view of a MOS capacitor is shown on the right side of Fig. 26-6. A heavily doped N layer is diffused into the wafer at the same time as the transistor emitter and collector interface regions are formed. This layer becomes the low-resistance bottom capacitor plate. A thin layer of silicon dioxide forms the dielectric, and the upper plate is formed by the final metallization process. Contact with the bottom plate is made by an isolated metal-film terminal through a small window etched in the oxide. A monolithic capacitor formed this way can have a capacitive value of 3 to 30 pF, depending on the dielectric used and its plate area. Unlike the junction-type capacitor, it is not polarity or voltage sensitive.

# BiCMOS-IC Fabrication

More than 25 years ago a compromise technology called *bipolar-CMOS,* or more familiarly *BiCMOS,* was developed to take advantage of the best properties of both bipolar and MOS transistors. Bipolar ICs switch faster than CMOS and offer stronger current drive. By contrast, CMOS ICs consume less power, generate less heat, and permit higher component density than bipolar ICs. BiCMOS ICs are more complex than either bipolar or CMOS ICs because more masking steps and more precise control of epitaxial layer thickness are required.

A simplified cross-section view of a BiCMOS IC is shown in Fig. 26-7. Because of the integration of bipolar transistors, an epitaxial layer must be grown on the P-type substrate. This is why the BiCMOS process is more complex and expensive than a CMOS-only process. However, well structures can be diffused into the base layer to trap alpha particles which might cause errors in high-density CMOS memory. An example is the buried N+ layer under the N well of the PMOS transistor in Fig. 26-7.

By carefully designing ECL circuitry into a CMOS IC, it is possible to create a BiCMOS chip with circuit density that is nearly as great as its CMOS counterpart. Moreover, it will be faster and can be built to less demanding design rules. This applies to gate arrays as well as memories, especially static RAMS.

BiCMOS is especially suited for input/output- (I/O-) intensive applications. Bipolar and CMOS circuit elements can be combined in various ways to meet application requirements. It has been used successfully to make analog and digital ICs, memories, microprocessors, and semicustom ICs, primarily gate arrays.

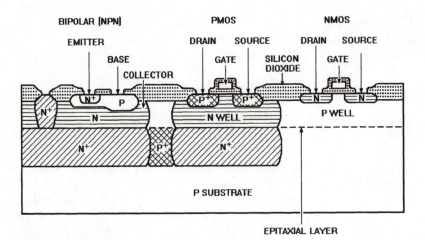

**Figure 26-7**  BiCMOS IC has an epitaxial layer for forming bipolar transistors.

# Wafer Sorting

*Wafer sorting* takes place after wafer fabrication or all semiconductor devices is complete. During this procedure each chip or die is tested for electrical performance. The wafer is mounted on a vacuum holder, and electrical probes that align with all of the device bonding pads are brought in contact with the wafer in automatic computerized test equipment. Power sources provide the required test voltages, and the equipment automatically records the results for all of the devices on the wafer. The number, sequence, and type of tests specified have been programmed into the equipment, and testing is carried out without operator assistance.

The equipment records the characteristics of all devices, passes as well as failures, by wafer and production lot. This complete record gives the process engineers valuable trend information about why and where devices are failing or approaching rejection limits. The records also indicate how to take corrective action to head off excessive reject rates.

The automatic test equipment determines the yield of all acceptable devices. Individual devices that pass all of the tests can still be classified for speed or other characteristics to determine their disposition, and perhaps even selling price. Rejects are marked with a visible ink spot for identification, and some equipment also generates maps of the wafers to record the locations of the rejects on the wafer.

# Microlithography

Optical lithography is still the most widely used method for transferring the complex patterns required for each layer to the wafer in the manufacture of microcircuits. The process is called *lithographic* because it has been derived from lithographic processes long used in printing and photography. It is also referred to as *microlithography* in recognition of the minute size of the device features being transferred. The following three alternative methods to optical lithography are still under development or being used experimentally:

**1.** Electron-beam lithography
**2.** Ion-beam lithography
**3.** X-ray lithography

## OPTICAL LITHOGRAPHY

The *optical lithography* method, previously described in "MOSFET Transistor Fabrication," makes use of UV radiation between 248 and 436 nm to transfer the features through masks onto the photoresist film on the oxidized or metallized wafer surface. This method is still favored over the alternatives because, in addition to being cost-effective and reliable, it has been continually updated with technical improvements that have permitted it to keep pace with other advances in semiconductor device manufacturing.

Within recent years the principal source of UV for optical lithography has been a mercury lamp filtered to emit the *g-line* at 436 nm. Stepper manufacturers have also introduced

*I-line* lens steppers that emit at 365 nm. Excimer lasers (see "Excimer Lasers" in Sec. 12, "Optoelectronic Components and Communication") and mercury lamps are being used to obtain image wavelengths that are shorter than 308 nm in the deep-UV region. Both permit the mask patterns to be transferred to the wafer with high accuracy. The next goal of production manufacturing is to move from 0.25- to 0.18-μm line widths, believed to be within the capability of optical lithography.

The photoresist is exposed through the mask by contact printing with a well-collimated and filtered UV light source. The mask and resist-coated wafer surface can be held parallel and in close contact by a vacuum clamping arrangement in a machine called a *stepper,* or the mask can be positioned above the wafer so that the image projected can be either larger or smaller than the mask pattern. If a larger mask image is used, it usually is reduced optically during the projection step-and-repeat process. The stepper must shift the photographic plate between exposures with great accuracy, typically ±1 μm or better.

Each mask must be registered precisely with all the others so that when the masks are superimposed, the registration between masks is kept within extremely narrow tolerances. Careful control of the movement of the mask permits accurate mask alignment. Use of the same stepper permits subsequent masks to be precisely registered over previous patterns.

The excimer laser can emit UV wavelengths as short as 248 nm, and it is considered to be the longest wavelength that can achieve 0.18-μm and smaller line widths. A process known as *phase shifting* is also seen as a way to improve the resolution of UV lithography. Special masks with an additional pattern exploit the interference effects of coherent light or higher-contrast images.

## ELECTRON-BEAM LITHOGRAPHY

In *electron-beam* (E-beam) *lithography,* electrons from an electron gun are focused to produce a small-diameter spot that traces out patterns in the same way as an electron beam traces out waveforms on an oscilloscope screen. It is now being used to produce high-quality masks and reticles and to do short-run production of devices such as application-specific ICs (ASICs). However, it might be used in the future for volume wafer production. The process takes place in a vacuum chamber and the beam can be steered and turned on and off by a computer. It uses an alignment and exposure technique called *direct writing.* The pattern is exposed in the resist by either raster or vector scanning. Large wafers are moved under the beam on *x-y* tables.

In an electron-beam system there are no distortions introduced from masks or optical effects such as diffraction. Existing machines are capable of 0.25-μm feature sizes, but future machines are expected to be able to resolve 0.18-μm feature sizes and smaller. The electron-beam process is slow because of the relatively long time required for the beam to trace out each pattern, and it takes time to exhaust the vacuum chamber. Moreover, electrons tend to scatter in the photoresist.

## ION-BEAM LITHOGRAPHY

In *ion-beam lithography,* ions trace out patterns in the photoresist. The mass of the ions (usually silicon) is larger than that of the electrons, reducing the scattering and minimizing any additional steps necessary to correct distortion. Its use is largely limited to research.

## X-RAY LITHOGRAPHY

In *X-ray lithography,* low-energy soft X-rays (0.4 to 5 nm) form patterns in the photoresist with resolution at 0.02 μm or lower. Soft X-rays are not readily absorbed by photoresist, and they offer a large depth of focus, permitting straight-walled profiles in thick films. But conventional X-ray tubes do not provide X-rays with enough density for use as production emission sources. Plasma tubes and synchrotrons are better sources of radiation.

A single synchrotron can provide X-rays for many lithography systems, but the cost and large size of a synchrotron system make it prohibitive in cost for all but the largest IC manufacturers.

By contrast, plasma-based X-ray sources of radiation are about the size of a conventional X-ray machine, and they are more cost-effective than synchrotrons, but they deliver less power and production quantities would be lower. Moreover, plasma systems require special lens elements that are far more refined than optical stepper lenses.

# Compound Semiconductors

*Compound semiconductors* have properties that are similar to those of silicon and germanium, which have four valence electrons. *Binary compounds* suitable for making semiconductor devices are formed from the Group III and V elements of the periodic table that have an average of four valence electrons. A common binary compound is *gallium arsenide* (GaAs). Gallium has three valence electrons (Group III) and arsenic has five valence electrons (Group V), giving GaAs an average of four valence electrons. Other examples are gallium phosphide (GaP) and indium phosphide (InP). Semiconductor compounds such as *cadmium sulfide* (CdS) are also formed from Group II and Group VI elements. The carrier mobility and energy gaps of these binary compounds differ from those of germanium and silicon.

Group V elements such as antimony can also be combined with Group II elements to form compound semiconductors. Examples of these are *indium antimonide* (InSb), *gallium antimonide* (GaSb), and *aluminum antimonide* (AlSb).

The preparation of N-type and P-type compound semiconductors is similar to that for elemental semiconductors. The doping of III-V compounds is accomplished by adding an element with six valence electrons (Group VI) such as *tellurium* as a donor impurity to produce N-type material, or by adding an element with two valence electrons (Group II) such as *zinc* to produce a P-type material.

A major advantage of III-V semiconductors is that junctions between dissimilar semiconductor materials can be grown on them. Called *heterojunctions,* they have defect-free interfaces without stress or dangling bonds. These heterostructures make possible new devices and ICs that can operate in the high end of the microwave band. In addition, III-V devices such as LEDs and laser diodes emit photons at useful wavelengths.

Adding a third chemical element, such as aluminum, to a crystalline binary alloy forms a *ternary alloy.* Its optical and electrical properties, such as electronic energy band structure, differ from those of binary compounds, but its lattice structure remains the same as that of the binary compound. An example is *aluminum gallium arsenide* (AlGaAs). The tailoring of materials to obtain desired properties for specific applications is called *bandgap engineering.*

## GALLIUM ARSENIDE (GaAs)

*Gallium arsenide* (GaAs) is an important binary compound used in the fabrication of both analog and digital integrated circuits, semiconductor lasers, diodes, and infrared-emitting diodes (IREDs). GaAs crystals with diameters of 100 mm (4 in) are grown by the liquid-encapsulated Czochralski (LEC) method. The GaAs melt is prevented from dissociating at the high temperatures with a protective layer of boric oxide ($B_2O_3$). Pressure in the crucible is held at about 75 atmospheres or 1100 psi. The crucibles are made from pyrolytically deposited boron nitride to prevent crystal contamination.

# Gallium-Arsenide Transistor Manufacturing

Transistors originally developed in silicon are being made from gallium arsenide to take advantage of the higher speed and higher-frequency operation possible with a GaAs substrate. Because GaAs is a compound, it does not form natural oxides like silicon. Consequently, silicon oxides are deposited on the GaAs substrates to fabricate GaAs bipolar and MOSFET transistors.

The most common GaAs transistors today are *metal semiconductor field-effect transistors* (MESFETs). Most discrete GaAs RF transistors are MESFETs and they are integrated into most GaAs ICs. The MESFET has a structure similar to a MOSFET, but its deposited metal-gate structure forms a Schottky-barrier diode, as shown in Fig. 2-12.

The length of the metallized gate is critical in both discrete transistors and ICs. Typically measuring 0.5 to 1.0 μm in most discrete transistors, they can be as small as 0.2 μm in ICs. However, the gate structure is much wider than its length—typically 900 to 1200 μm.

The active region of MESFETs is usually doped by ion implantation. A 0.1- to 0.2-μm-thick N-doped region is typical for depletion-mode or D-MESFETs. The enhancement-mode or E-MESFET and the enhancement-mode JFET or E-JFET are other GaAs transistors. Both E-MESFETs and D-MESFETs can be combined in an IC to form enhancement/depletion-mode (E/D) logic.

The *high-electron-mobility transistor* (HEMT), as shown in Fig. 2-13, was designed for integration into ICs. A layer of aluminum gallium arsenide (AlGaAs) grown on a GaAs substrate, known as a *heterojunction,* improves device performance and permits high levels of integration. Heterojunction E/D technology was developed as a method for making GaAs digital LSI and VLSI devices more economically. Another GaAs transistor developed on a heterojunction is the *heterojunction bipolar transistor* (HBT), as shown in Fig. 2-14. The HBT was designed to achieve higher levels of integration. Both HEMTs and HBTs require special processing to achieve precise, sharp heterojunctions. Experimental

HBTs have been fabricated by the silicon germanium (SiGe) technology process as an alternative to GaAs HBTs.

# Gallium-Arsenide IC Manufacturing

*Gallium-arsenide* (GaAs) *ICs* have been fabricated to perform a wide range of digital and analog functions at high speed and at frequencies up to 20 GHz. They are based on the integration of transistors on semi-insulating GaAs substrates, generally following techniques developed for the fabrication of silicon ICs. However, many new and exotic processes have been developed for fabricating larger and faster GaAs ICs.

GaAs IC technology continues to improve, but the future of GaAs ICs is seen in custom and application-specific ICs (ASICs) rather than in the families of standard products that are common among silicon ICs. Ongoing improvements in silicon ICs have eroded the performance margins that GaAs ICs once held over silicon ICs. GaAs ICs are now generally classed as *high-speed digital, high-speed analog/interface,* and *microwave.*

The digital and analog/interface devices are niche parts that supplement or complement existing silicon ICs in applications where higher speed and lower power are requirements. These devices can operate at higher temperatures with lower noise levels than comparable silicon ICs, and they can withstand nuclear radiation.

## GaAs DIGITAL ICs

GaAs ICs are faster and consume less power than equivalent silicon digital bipolar ICs. They are made to be compatible with other logic families, particularly emitter-coupled logic (ECL). But GaAs ICs lag behind bipolar ICs in integration density. GaAs small- and medium-scale digital logic, memories, and gate arrays have been made.

## GaAs ANALOG/INTERFACE ICs

GaAs analog/interface ICs supplement slower silicon bipolar and CMOS linear and interface ICs in systems that require higher data rates or faster conversion. They are also used to interface microwave and digital systems. GaAs operational amplifiers, comparators, and analog-to-digital and digital-to-analog converters have been made.

## GaAs MICROWAVE ICs

GaAs amplifiers and oscillators are able to amplify or oscillate efficiently at frequencies higher than 2 GHz, the limit for silicon devices. GaAs microwave monolithic ICs (MMICs), as shown in Fig. 7-16, have extended the capabilities of GaAs microwave transistors into higher integration levels to replace frequency-limited silicon MMICs and hybrid circuits containing either silicon or GaAs transistors. They are cost-effective in the microwave range of 500 MHz to 2 GHz, and are required for higher frequencies. GaAs MMICs are installed in phased-array radars and electronic warfare systems operating in the C and X bands. They are now made as amplifiers, oscillators, mixers, and switches.

# Epitaxial Processes

*Epitaxy* is the process of growing new layers of semiconductor crystals on wafers with matching crystalline structures. The word is derived from the Greek *epi* meaning "on" or "upon" and *taxis* meaning "arrangement" or "order." Epitaxy is analogous to spraying water on an ice-skating rink where it freezes to form a smooth surface. The three principal types of epitaxy are *vapor phase* (VPE), *liquid phase* (LPE) and *molecular beam* (MBE), but there are also modified versions of these technologies. An epitaxial layer has the same crystallographic structure as the substrate or wafer on which it is deposited.

## VAPOR-PHASE EPITAXY (VPE)

*Vapor-phase epitaxy* (VPE) is the process of adding a layer of semiconducting material to a wafer in a furnace by introducing the materials in vapor form to condense on the wafer. The heated wafers are exposed to a heated flowing stream of gaseous elements that flow across them inside the furnace and condense on them to form a new crystal layer. VPE is used to grow silicon layers on integrated circuits and to produce light-emitting diodes (LEDs). It is also called *chemical vapor deposition* (CVD) and the advanced technology is called *metal-organic chemical vapor deposition* (MOCVD).

## LIQUID-PHASE EPITAXY (LPE)

*Liquid-phase epitaxy* (LPE) is a process for adding a layer of semiconducting material to a wafer in a furnace by sliding the heated wafer over the surface of a molten solution of the material to be deposited. The solution temperature is controlled to permit the crystalline growth of a layer on the wafer. LPE is used in the production of light-emitting diodes (LEDs), semiconductor laser diodes, and photodetectors.

## MOLECULAR-BEAM EPITAXY (MBE)

*Molecular-beam epitaxy* (MBE) is a process for depositing semiconductor materials, one monolayer at a time, onto a substrate in an ultra-high-vacuum chamber. Each monolayer is less than 10 Å (1 nm) thick. The quality of the wafer can be determined by its diffraction pattern and monolayer growth can be observed and monitored by *reflection electron diffraction* (RED). The growth process is computer controlled.

The atoms or molecules line up in ultrathin layers that duplicate the crystal structure of the wafer. MBE is performed in an ultra-high-vacuum chamber with an array of small furnaces projecting into it. The axis of each furnace and its opening are directed toward the wafer mounted on a substrate manipulator, and each furnace has a crucible containing materials such as aluminum, arsenic, gallium, or silicon that will be deposited on the wafer.

As each oven is heated, the material inside evaporates and some of its molecules or atoms escape in a stream or *thermal beam* through the opening. Because of the high vacuum in the chamber, there are no atoms of air to impede the particles as they travel directly to the heated wafer. The thickness and chemical composition of each layer can be precisely controlled by varying the intensity of the beam or closing furnace shutters. The added layers can be of the same material (*homoepitaxy*) or different materials (*heteroepitaxy*). It takes 1 s to grow a monolayer and about 1 h to grow a layer 1 µm thick.

The precision of the MBE process makes it possible to grow the following structures:

- Complex crystal "sandwich" structures with millions of layers one atom thick
- Hybrid crystals with sharply defined boundaries between the two different materials
- Structures with the molecules of different materials melding into each other
- Windows of one material embedded in a layer of another material

Semiconductor devices have been fabricated by MBE for both research and development and commercial photonic and microwave systems. MBE was used to make the fastest transistor, a selectively doped heterostructure device made as a multilayered sandwich of GaAs and AlGaAs.

# SEMICONDUCTOR DEVICE PACKAGING

# Overview

Many different cases and packages have been developed to protect discrete and integrated circuit devices. The wide selection has been dictated by cost as well as the ability of the package to protect the device under the most severe environmental stresses it is likely to encounter in its intended application. The chips or dies inside must withstand shock, vibration, both operating and storage temperature excursions, chemical spills, and humidity swings. Circuit designers select package styles with one eye on cost and the other on the minimum amount of protection needed for a long useful service life.

The cases chosen for military and aerospace electronics will obviously differ from those chosen for the more tolerant industrial and automotive electronic circuits, and those, in turn, will differ from the cases typically specified for consumer products. In military, aerospace, commercial aircraft, and high-reliability applications, metal and ceramic packages are the first line of defense. Whether small-signal, IC or power devices, they are likely to be exposed to far wider temperature excursions than most industrial, commercial, or consumer-grade devices. The far higher prices charged for military-qualified devices are influenced more by the higher costs of screening, inspection, testing, and documentation mandated by military specifications than the cost of the more rugged, temperature-resistant cases.

Semiconductor devices that will be exposed to the more rigorous industrial and automotive environments are also likely to be packaged in ceramic or metal cases, but they will not undergo the same quality inspections as those conforming to MIL-S-19500, MIL-M-38510, and MIL-STD-883C. Rugged, higher-temperature-resistant cases are customer specified for power semiconductor and radio-frequency devices because, in addition to protecting them from wider ambient temperature swings, they must be able to dissipate internally generated heat nondestructively. By contrast, commodity devices for consumer entertainment products are usually packaged in plastic cases.

Semiconductor device packaging has lagged behind the advancements made in wafer fabrication because dramatic increases in transistor density have made it necessary to invent new forms of packages, especially those that lend themselves to automated packaging. The newer, higher-density chips have more bonding pads and I/O pins that require more electrical connections. In addition, the higher transistor and gate populations of the latest generation memories and microprocessors have complicated the problems of heat dissipation.

A long-term trend seen in the industry has been the upgrading of plastic packages to replace metal packages for power devices. The inclusion of heat-dissipating metal tabs within plastic packages has provided heat-dissipation capability to match the metal cases of earlier-generation devices. This trend toward plastic cases has led to weight, space, and cost reductions. It is most apparent in the packaging of rectifiers, transistors, and SCRs for industrial and automotive applications, so many flat plastic cases have replaced metal cases.

Another trend is toward increasing use of minimal, leadless plastic cases for ICs, making them suitable for surface mounting. The lead pitch on these cases has been reduced from the standard 0.10 in (2.5 mm) to 0.05 in (1.3 mm) or less to conserve circuit board "real estate." The time-honored dual-in-line package (DIP) is no longer being selected for new designs of advanced products.

Most cases are standardized and registered with the Joint Electron Device Engineering Council (JEDEC). There are a relatively small number of general styles, but so many variations in size that their identifying numbers can be confusing. It is now difficult even for veteran designers to recognize a package style by number at a glance. In addition, many semiconductor manufacturers have introduced their own versions of standard cases to adapt them to new or modified devices. However, some of these modifications are adopted later as standard industry packages.

The packaging of devices is considered to be the first level of interconnection in an electronic system. After the die or chip is bonded to a metal leadframe or ceramic case, wire bonds are made between the pads on the die or chip and the I/O leads. These interconnections are now made by highly automated methods and are not separable or repairable because they are enclosed by the device package or molded plastic case.

Technical problems have been solved permitting the use of copper as a replacement for aluminum bonding pads and wire bonds, and these techniques are expected to become more widely accepted in the coming years as existing production equipment is replaced.

# Die and Chip Preparation

After the acceptable chips or dies have been identified and the rejects have been marked, the wafer typically must undergo thinning, and some will receive a backside thin-film gold layer. Only after one or both of these steps have been completed are the dies or chips separated from the wafer. After separation, most will be packaged, but some will be sold as bare devices for mounting directly on the ceramic substrates of hybrid circuits or on the prepared circuit cards of multichip modules.

## DIE BACKSIDE PREPARATION

Wafers will be thinned to make the dies or chips easier to package, and to repair any damage that might have occurred to the reverse side of the crystal structure during processing. The thinning is done either by mechanical grinding or chemical etching. The original wafer thickness is typically reduced to 0.2 to 0.5 mm. If the chips are to be attached to packages by gold-silicon eutectic solder, a thin film of gold will be applied by evaporation or sputtering to improve the bond.

## DIE SEPARATION

Dies or chips are separated from the wafer by sawing or scribe-and-break methods. *Scribing* is done by passing a diamond-tipped scribe through the center of a thin border between chips, and separation is done by surface pressure, much as glass sheets are snapped apart after scribing. *Sawing* is the preferred and more precise method because the edges will be sharper, and the chips are less likely to be chipped or cracked during separation.

## DIE PICK AND PLACE

Acceptable chips identified during wafer sorting are picked from the wafer after they have been separated and placed in carriers. This can be done manually or with an automated vacuum wand under computer control that distinguishes the acceptable chips from the marked rejects.

## DIE INSPECTION

All acceptable dies are inspected optically for sharp edges and freedom from contaminants and defects. This inspection can be done manually with binocular microscopes or with a computer-aided vision system.

# Die or Chip Attach

*Die attach* or *die bonding* is the process of mounting a semiconductor die or chip on a lead-frame or ceramic substrate by soldering or adhesives. This bond, which must be strong and secure, acts as a medium for transferring heat from the chip to the package, and it provides either an electrically conductive or insulating interface with the package, depending on requirements. Many different bonding methods have been developed including the use of conductive metal alloy preforms, lead-tin solder preforms, or silver-filled powdered glass frit that is melted between the die and leadframe or package base. Organic adhesives such as epoxies and polyimides can form either electrically conductive or insulating bonds.

## EUTECTIC DIE ATTACH

The most common electrically conductive method for attaching dies or chips to ceramic cases or metal cans is bonding with gold-silicon (Au-Si) eutectic. Figure 27-1 shows how this eutectic bond is made in a TO-39-style metal case. Gold melts at 1063°C and silicon melts at 1415°C, but when mixed to form a eutectic they begin to liquefy or alloy at 380°C. The packages suitable for this process have gold-plated die-attach surfaces, and the chips or dies have backside gold deposition.

A thin gold-silicon preform is placed over the gold-plated attach surface, and the package is heated to about 425°C, until the eutectic preform melts. The chip is then placed in the die-attach area, pressed down into the liquid eutectic until a bond is formed, and then the package is allowed to cool. This method is favored for packaging high-reliability devices because of its strength and ability to dissipate heat. An alternative die-attach method uses soft lead-tin solder, also a eutectic. Alloys of 95 percent lead and 5 percent tin melt at about 300°C.

## ADHESIVE DIE ATTACH

The use of liquid adhesives such as epoxy or polyimide is an alternative die-attach process. These adhesives form an insulating barrier between the die and package, but it can be made electrically and thermally conductive if powdered gold or silver is added. Silver-filled epox-

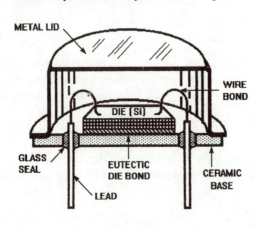

**Figure 27-1** Metal can package for military and high-reliability semiconductor devices.

ies are popular for bonding chips or dies to copper leadframes, and silver-filled polyimide is favored for leadframes made of Kovar, an iron-nickel-cobalt alloy. The adhesive is deposited in the die-attach area, and the chip is positioned over the area and forced into the viscous liquid to form a thin, uniform layer under the chip. The bond is cured by heating the assembly in an oven. Adhesive die attach is more economical than eutectic die attach because the process is simpler and the materials are cheaper, but the adhesive bond lacks the strength of the eutectic bond and is subject to decomposition at elevated temperatures.

# Wire Bonding

*Wire bonding* is the process of bonding thin (0.7- to 1.0-mil) wires from chip bonding pads to the inner leads of package leadframes, as shown in Fig. 27-2. Most wires are drawn from gold and aluminum, and both are good conductors and ductile enough to withstand deformation during bonding and still remain strong and reliable connections. Each metal has advantages and disadvantages, and different methods must be used to bond them. Some devices require dozens of wire connections, and they are made automatically at high speed.

## GOLD WIRE BONDING

Gold wire is an excellent conductor because it resists oxidation and can be melted to form strong bonds with aluminum pads. It is used in *thermocompression* (TC) or *ball bonding* and *thermosonic bonding.* Before a thermocompression bond is made, the chip or die and package are heated to 300 to 350°C. The gold wire is fed from a tube called a *capillary,* as shown in Fig. 27-3a. An electric spark or small hydrogen flame melts the tip of the wire into a ball that is positioned over the first bonding pad. The capillary then moves down and squeezes the molten ball against the pad to form a strong bond, as shown in Fig. 27-3b. The capillary then feeds out more wire to form a loop before traveling to the related pad or lead, as shown in Fig. 27-3c. The capillary moves down again, and with heat and pressure applied, melts the wire onto the gold-plated lead or pad and pinches it off, as shown in Fig. 27-3d. The spark or flame then severs the wire and forms a ball for the next pad bond, as shown in Fig. 27-3e. This procedure is repeated until all chip and package lead connections have been made.

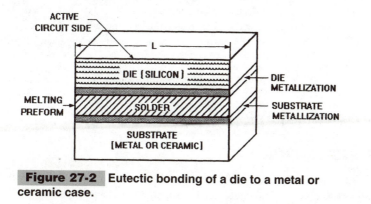

**Figure 27-2**  Eutectic bonding of a die to a metal or ceramic case.

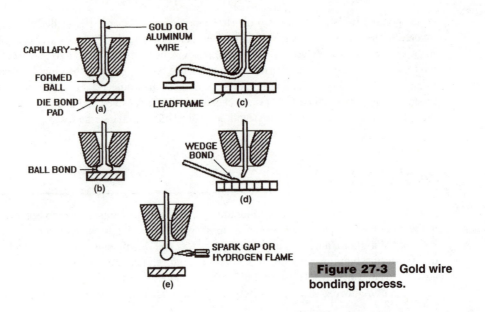

**Figure 27-3** Gold wire bonding process.

Thermosonic ball bonding is essentially the same as thermocompression bonding, but the bonds are made at a lower temperature. A pulse of ultrasonic energy sent through the capillary heats the wire to a temperature high enough to form a strong gold-aluminum alloy bond.

## ALUMINUM WIRE BONDING

Aluminum wire exhibits lower conductivity and is less resistant to corrosion than gold wire, but it costs less and forms effective aluminum-to-aluminum bonds with the pads. This wire can be bonded at lower temperatures than gold. The procedure for aluminum wire bonding is essentially the same as for gold wire bonding, as shown in Fig. 27-2, but it is performed by a process called ultrasonic or *wedge bonding,* similar to thermosonic bonding.

In ultrasonic aluminum wire bonding, the package is mounted in an *x-y* table and the table, rather than the capillary, is repositioned for each bond. The chip bonding pad is positioned under the wire in the capillary and clamped to it. A pulse of ultrasonic energy is sent through the capillary to form the bond. Then the table retracts, and a loop of wire is pulled out before the table is positioned for the next bond. Finally, the wire is pinched off at the package lead. These steps are repeated until all of the chip and package wire bonds have been made.

# Inspection, Marking, and Testing

## PRESEAL INSPECTION

The wire-bonded chip attached to leadframe or package is then inspected optically. This quality examination includes checking to verify proper chip placement, strong chip attachment (indicated by a bead of solder or epoxy showing around the periphery of the chip), completion of all required wire bonds from chip pads to leadframe leads or package pads, and freedom from any visible surface contaminants.

## MOLDING OR SEALING

The chip leadframe assemblies are then placed in molds and epoxy compound is injected to form the plastic DIP, SOT, and SOIC packages. The ceramic packages are covered and sealed.

## PLATING, TRIMMING, AND MARKING

The exposed leads of the plastic package extending beyond the body of the case are plated to improve their solderability. The frames that supported the outer ends of the leads on the leadframes are trimmed to separate them. The outsides of the packages are then marked with the manufacturer's logo, part number, country of origin, and other identification information.

## FINAL TEST

Each device is given the specified final electrical and environmental tests to ensure quality. Statistical sampling might be permitted for this test. Power is applied to some devices for a specified period, called a *burn-in test,* to identify possible early failures or "infant mortality." The most extensive tests are given to military specification and high-reliability space-qualified devices, and the verification documentation of the results is prepared.

# Alternative Chip Packaging

## BEAM-LEAD TECHNOLOGY

*Beam-lead technology* is a method for bonding chips or dies to cases with gold beams that have been formed as an inherent part of the device during wafer fabrication. The exposed beams extend over the edges of the chip after it has been separated from the wafer. The chip is then inverted or placed facedown, and several beam leads are attached simultaneously to the package substrate. The body of the chip remains slightly elevated above the substrate because of the thickness of the beam leads.

## FLIP-CHIP AND SOLDER-BUMP BONDING

Semiconductor device chips are now being made with soft solder bumps deposited on each bonding pad. Each solder bump corresponds to a matching inner lead on the package. The connections are made after the chip is inverted by applying heat and pressure to cause the solder bumps to reflow. This connection method eliminates wire bonds which can pose reliability problems, and it permits lower overall package height because extra space is not required to cover the wire loops. Consequently, flip-chip packages offer the benefits of a lower profile and shorter electrical paths by avoiding chip bonding pads to the I/O leads.

## TAPE AUTOMATED BONDING (TAB)

*Tape automated bonding* (TAB) is an alternative chip-mounting method in which chips are automatically mounted on a thin, flexible plastic substrate. The chip pads are bonded to thin-film metal leads that have been deposited by sputtering or evaporation on the surface of the plastic tape. The tape is positioned precisely over the chip so that the inner leads mate with the bonding pads of the chip. The bond is then made by applying heat and pressure with a tool called a *thermode.* This process has made possible the extremely thin electronic circuits in pocket radios, calculators, watches, cellular telephones and other small electronic products.

# Discrete Device Cases

A wide variety of ceramic, metal, and plastic cases have been developed specifically for packaging discrete devices such as diodes, transistors, or thyristors, while others are used to package diode arrays. Small-signal, low-power diodes are packaged in axial-leaded glass cases such as the DO-41 or radial-leaded molded plastic packages such as the DO-15 case. Many diodes and low-power rectifiers are also packaged in axial-leaded plastic cases.

Figure 27-4 shows some of the cases widely used for discrete devices. The plastic TO-226 style (formerly TO-92), shown in Fig. 27-4a, is used for packaging small-signal transistors and thyristors.

Molded plastic flatpack cases have been developed for such discrete devices as radio-frequency transistors, power transistors, and thyristors, rated to 15 A. An example is the three-terminal, transfer-molded TO-220 case, shown in Fig. 27-4b, that is similar in appearance to the TO-218 case. These cases are alternatives to the metal cans. The chip or die is mounted on a copper tab that can act either independently as a heatsink or as a metal-to-metal interface with a larger heatsink or heat-dissipating busbar. The tabs are pierced to permit the cases to be mounted on a larger heat-dissipating surface with nuts and bolts.

The plastic dual-in-line packages (DIPs), such as the TO-116 shown in Fig. 27-4c, widely used to package integrated circuit chips, are also used to package diode arrays.

Metal cases are widely used to protect small-signal and power transistors, as well as triacs and SCRs that must withstand the wide ambient temperature excursions encountered in industrial, military, and aerospace high-reliability applications. These cases are made in

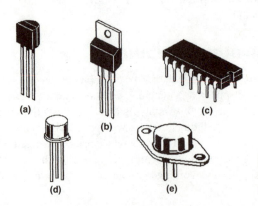

**Figure 27-4** Discrete device case styles: (*a*) TO-226, (*b*) TO-220, (*c*) TO-116 (DIP), (*d*) TO-205 and (*e*) TO-204.

two parts—the *cap* or lid and the *header* or base. The header is a flanged disk with three or more insulated pins projecting perpendicular to its bottom surface, and the die-attach area is on the inner surface. The flanged cup-shaped cap or lid is welded to the header flange to form a hermetic seal. Radio-frequency transistors are packaged in standard metal and plastic cases, but proprietary ceramic cases developed by the manufacturer predominate.

The three-pin TO-205 (formerly TO-39) metal case, shown in Fig. 27-4*d,* is suitable for packaging three-terminal high-reliability discrete devices. It is similar in appearance to the TO-18 and TO-52 cases. The larger two-pin TO-204 case (formerly TO-3), shown in Fig. 27-4*e,* is suitable for packaging low-power transistors. The pins are base and emitter terminals and the case is the collector terminal. It looks like the metal TO-213 (formerly TO-66) case.

The acceptance of surface mounting as an alternative to through-hole mounting of devices on circuit boards has increased the demand for flat packages with leads that can be soldered directly to the surface of circuit boards. Figure 27-5 shows a flat surface-mount package suitable for transistors and other three-terminal devices. Called a *small-outline transistor* (SOT) *package,* the case is designated the TO-236 (formerly SOT-23). A larger version, designated TO-261, was the SOT-223. The short stub pins of these cases are bent outward so that their ends lie flat on mating solder pads on the surface-mount circuit boards.

# Optoelectronic Device Cases

## LIGHT-EMITTING DIODE (LED) PACKAGES

The most popular case styles for packaging both visible-light-emitting diodes (LEDs) and infrared-emitting diodes (IREDs) are the T-1 and T-1¾ radial-leaded, bullet-shaped plastic

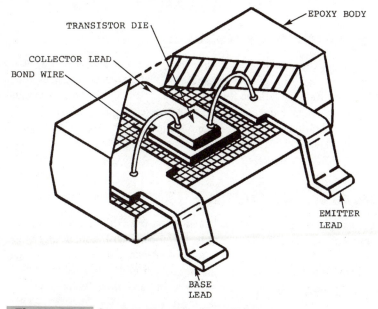

**Figure 27-5**  Small-outline transistor (SOT) case.

packages. The N-type layer of the LED die is metallized before the wafer is diced. After the die is attached to a radial leadframe, a fine gold or aluminum wire is bonded between the upper P layer and one of the radial leads, as shown in Fig. 13-5. The assembly is then molded in a bullet-shaped epoxy T-1 or T-1¾ lens package, as shown in Fig. 13-6. The plastic package can be clear, contain diffusing particles, or be colored red, amber, or green depending on the emission color of the die and its end-use application. The die's axial position with respect to the lens determines the emission angle of the IR beam. It can be from 30 to 110°.

Other LED packages include flattop and surface-mount styles as well as rectangular molded cases which emit a rectangular light pattern from their end surfaces. In military and high-reliability applications, LEDs and IREDs are packaged in hermetically sealed TO-39-style metal cases with glass lenses in the ends of their caps.

## INFRARED-EMITTING DIODE (IRED) PACKAGES

Infrared-emitting diodes (IREDs) are packaged in the same style packages as LED dies except that the plastic cases typically are transparent rather than translucent or dyed a color. Some IREDs are hermetically sealed in TO-39-style metal cases with a glass lens forming the end surface of the cap for use in extreme environments.

## LASER DIODE PACKAGES

Laser diodes typically are packaged in metal TO-39-style cases with a glass lens similar to those used to package LEDs and IREDs. The laser die is located at the center of the package flange so that its main beam is radiated through the lens. Some light is also emitted from the back facet of the die, but it is directed into the stem and does not add to the forward beam.

## LED NUMERIC DISPLAY MODULES

A seven-segment LED numeric module is made by bonding individual dies to a substrate and covering the dies with a molded plastic case that has slots to define the segments. When filled with translucent plastic (typically dyed a color), the slots act as light pipes, as shown in Fig. 13-8. Each character is formed by the segments illuminated by an LED die molded within the light pipe. All numbers and the letters from A through E can be formed. A 16-segment display can form 64 of the ASCII characters, and a $5 \times 7$ dot matrix of LEDs can form all 128 ASCII characters.

## OPTOCOUPLER PACKAGING

Standard optocouplers typically are packaged in six-pin DIP packages, as shown in Fig. 12-6. The industry standard for electrical isolation between the input and output devices within the device is a minimum of 5000 VAC peak. Package styles 4N25 through 4N38 are assigned to popular optocouplers with phototransistor output devices, and 4N29 through 4N33 are assigned to optocouplers with photodarlington output devices.

# Integrated-Circuit Packaging

For many years the *dual-in-line package* (DIP) has been the most popular case for packaging integrated circuits. DIP cases are available in many different sizes made from ceramic or molded from epoxy. However, the ever-expanding sizes of chips, particularly for microprocessors and semiconductor memories, combined with the trend toward surface mounting, has created a need for new package styles with emphasis on lower profiles and smaller dimensions between pin centerlines. It has spurred the development of new methods for soldering the package pins to surface-mount circuit boards or for mounting them in special high-density sockets.

## DUAL-IN-LINE PACKAGES (DIPs)

A molded plastic *dual-in-line package* (DIP), as shown in Fig. 27-4*c,* is still widely used to package IC chips. Its case length is determined by the number of pins required in the parallel rows on each side of the package. The standard DIP pitch is 0.1 in (2.5 mm) on centers. Pin rows are spaced 0.3, 0.4, 0.6, and 0.9 in apart, depending on the number of pins in the package. The flat pins, stamped from sheet metal alloy, have shoulders that provide a gap between the bottom of the DIP and the circuit board to permit post-solder cleaning for the removal of any contaminants trapped under the package.

DIP packaging begins with a *leadframe* stamped from thin metal in the form of a web that includes all of the package pins and a die-attach surface. The chip is attached to the leadframe and wires are bonded from the chip bonding pads to the inner surfaces of the pins, as previously described in the "Die or Chip Attach" and "Wire Bonding" paragraphs of this section.

The completed assembly is then placed in a mold, and epoxy is injected to form the rectangular epoxy block that encloses the chip and the inner ends of the leads or pins. The exposed leadframe receives a protective plating, and the ends of the pins are released by trimming off the frame before they are bent downward. Circuit boards designed to receive DIPs include uniform rows of plated-through holes that mate with corresponding DIP pins when they are inserted either manually or automatically. Alternatively, matching sockets can be used if it is expected that the devices will be removed and replaced before the end of the service life of the host circuit board. These sockets include formed or machined spring-loaded sockets for each pin that clamp the device securely in position.

Standard DIPs have from 4 to 64 pins. A 64-pin DIP is 3.2 in (8.1 cm) long with the parallel pin rows spaced 0.9 in (2.3 cm) apart. It has a "footprint" of about 3 in$^2$ (19.4 cm$^2$). It is also called a *DIL package.* A notch or mark on one end of the case indicates the position of pin 1.

*Skinny DIPs* are narrower than standard DIPs. They have from 20 to 28 pins with a pitch of 0.1 in (2.5 mm) on centers and row separation of 0.3 in. *Shrink DIPs* have a pitch of 0.07 in (1.8 mm) and row separation of 0.4, 0.6, and 0.75 in for 28, 42, and 64 pins, respectively. Other nonstandard DIPs have 0.5-in (12.7-mm) pitch.

Two different *ceramic DIP* case styles are available: CERDIP (ceramic DIP) and a ceramic DIP with side-brazed pins. These cases are primarily intended for military and high-reliability applications where they will be able to withstand wide temperature excursions. The CERDIP has two alumina ceramic parts: the *base* and the *lid*. Both are recessed to form a cavity for the chip. The leadframe is embedded in the base. Eutectic die attach and gold wire bonds are standard processes in assembling chips to these cases. The lid is then fused to the base with low-melting-temperature glass powder that forms a hermetic seal.

The side-brazed ceramic DIP is made of three layers of alumina ceramic with the leads previously brazed to the base. The die-attach and wire-bonding methods are the same as those for CERDIP cases. Metal lids are furnace brazed to the base section with gold-tin solder to form a hermetic seal, but ceramic lids are bonded to the base with low-melting-temperature glass powder in a furnace to form a hermetic seal.

## CHIP CARRIERS

A *chip carrier* is a low-profile, square IC package for surface mounting with metal pins projecting from all four edges. Its cavities or internal mounting areas occupy most of the package. Chip carriers can be leaded or leadless and made of either ceramic or plastic. *Ceramic leaded chip carriers* (CLCCs), as shown in Fig. 27-6a, are used for multipin ICs that are to be mounted in sockets either because they are subject to replacement or because they are too delicate to undergo any of the conventional soldering processes during circuit-board population. CLCCs are available with up to 84 pins for packaging analog, digital, and memory ICs for military and high-reliability applications. They can be hermetically sealed to protect the enclosed IC chip from environmental contamination.

So-called leadless chip carriers actually have short J-type leads that are bent under the case for soldering to mating pads on surface-mount circuit boards. The *plastic leadless chip carrier* (PLCC), shown in Fig. 27-6b, is the most popular and lowest-cost chip carrier. These cases are available with from 18 to 124 leads. PLCCs are being used to package analog and logic ICs, semiconductor memory, microprocessors, and microcontrollers. A 124-lead PLCC occupies the same circuit-board area as a 64-pin DIP. Both chip carrier styles provide shorter conductive paths from the chip to the external leads than DIPs, and they occupy less board space than DIPs with the same number of pins. Pin pitch is either 0.040 or 0.050 in (1.0 or 1.3 mm) on centers.

(a)

(b)

(c)

**Figure 27-6** Integrated-circuit case styles: (*a*) CLCC-32, (*b*) PLCC-44, and (*c*) SOIC-16.

## SMALL-OUTLINE INTEGRATED CIRCUIT (SOIC) PACKAGES

A *small-outline integrated circuit* (SOIC) *package,* as shown in Fig. 27-6c, is a plastic case intended for packaging ICs with 8 to 28 pins. It is called an SOIC-16 or 16-SOIC because it has 16 pins. The ends of the stub pins are bent outward. A variation of this package, the SOJ, has 20 to 28 pins bent under the case body in a J-bend style. Their chip-attach and wire-bonding processes are similar to those for plastic DIPs, case styles that they replace in surface-mount technology.

## BALL-GRID ARRAYS (BGAs)

A *ball-grid array* (BGA) is a flat plastic IC package with a profile less than 0.1 in (2.5 mm) high for surface mounting large-scale ICs with high pin counts on circuit boards. Similar in form to the pin-grid array (PGA), they have a series of solder bumps (balls) on the bottom of their case instead of pins. The solder balls are arranged to bond the package to mating contacts on the circuit board during solder reflow.

## FLATPACKS

A *flatpack* is a small, lightweight IC package with its pins projecting parallel to the base of the case. Some flatpacks have pins projecting from only two parallel sides, and others with higher pin counts have them projecting from all four sides. Plastic flatpacks that are 0.55 in (14 mm) wide have 44 to 100 pins spaced 0.026 to 0.039 in (0.7 to 1.0 mm) on centers. Ceramic flatpacks, also called CERpacks, have widths of 0.28 to 0.39 in (7.1 to 9.9 mm) and 16 to 24 pins spaced 0.050 in (1.3 mm) on centers. The die-attach and wire-bonding techniques are the same as those used with ceramic DIPs.

## LEAD-ON-CHIP (LOC) PACKAGES

A *lead-on-chip* (LOC) *package* is a molded epoxy DIP package for very large scale ICs that have their bonding pads arranged down the center of the chip.

## MICRO-BALL-GRID ARRAYS (MBGAs)

A *micro-ball-grid array* (MBGA) is a miniature BGA package for surface-mounting IC chips.

## MICRO SMT (MSMT) PACKAGES

*Micro SMT* (MSMT) packages are minimal plastic surface-mount cases. Only slightly larger than the bare dies or chips to be packaged, they permit very high density circuit-board populations. Micro SMTs are smaller than either the DIP and SMT cases for packaging chips with the same number of pins. Some are smaller and thinner versions of existing standard packages, and some are only 1 mm high. Among the recent introductions in this category are: (1) *thin small-outline packages* (TSOPs), (2) *ultrathin packages* (UTPs), and (3) *thin quad flatpacks* (TQFPs).

## PIN-GRID ARRAYS (PGAs)

A *pin-grid array* (PGA) is a square ceramic package with an array or grid of pins spaced 0.10 in (2.5 mm) apart in a "bed of nails" configuration. PGAs are available with from 64 to 256 leads. A PGA with 256 leads occupies the same circuit-board space as a chip carrier with 124 leads. Sockets with arrays of vertical holes are available for PGAs.

## QUAD FLATPACK (QFPs)

A *quad flatpack* (QFP) is a flat molded-epoxy case for large-scale IC chips with leads projecting from all four sides; hence, the term *quad* (for quadrant). A typical QPF will have 64 leads with 16 projecting from each side. There are also *thin quad flatpacks* (TQFPs).

# Power-IC Packaging

Power ICs are packaged in the same conventional IC packages as signal-level ICs except that more attention is given to keeping junction temperatures below critical values. Some manufacturers use copper rather than Kovar leadframes. As in other power devices, IC power-handling ability is improved by mounting them on heat sinks and cooling them with forced air. Power ICs are in dual-in-line packages (DIPs) with 8 to 28 pins and small-outline transistor (SOT) cases if the dissipated power is less than 2 W. More complex parts are in single-in-line package (SIP) cases with from 11 to 23 leads. Plastic TO-220-style packages are used for power ICs that dissipate from 5 to 10 W. Some power ICs are packaged in plastic leaded chip carriers (PLCCs).

# ELECTRONIC CIRCUIT PACKAGING
# AND ASSEMBLY

# Overview

The miniaturization of circuit boards and cards, made possible with higher-density integrated circuits and surface-mount technology (SMT), has been an ongoing trend in electronic packaging. The growing complexity of ICs as transistor densities increase has, to some extent, been offset by shrinking IC line widths or design rules. Semiconductor package sizes, profile heights, and pin spacing or pitch have been reduced to permit higher device populations per circuit board or card.

The use of standard size dual-in-line packages (DIPs) for ICs is declining as original equipment manufacturers (OEMs) turn increasingly to surface-mount packages that are reflow soldered to circuit boards without plated-through holes. These packages have permitted finer conductive traces on the circuit boards with narrower spacing between them. The elimination of most of the plated-through holes on circuit boards that must contain leaded components has also resulted in increased component population density. Some components such as transformers and power semiconductors do not lend themselves to SMT packaging. Lead-tin solder bonds might not reliably retain objects the size and weight of transformers, and they might not survive immersion in molten solder. Also, the heat dissipated by power semiconductors might cause the solder bonds to fail.

SMT construction is a requirement for the manufacture of many of the latest-generation electronics products such as notebook and palmtop computers, cellular telephones, pagers, CD players, and handheld GPS receivers. Moreover, miniaturized circuitry has made possible many plug-in computer peripherals that were, in earlier generations, separate products. Examples include modems and hard-disk, diskette, CD-ROM, and DVD drives.

However, smaller, high-density circuit boards have also provided benefits for many electronics products that are not enclosed in handheld or pocket-size cases, such as desktop computers, stereo components, VCRs, television sets, and test instruments. Miniaturized circuits have permitted higher speeds and greater reliability.

# Printed-Circuit Boards and Cards

*Printed-circuit boards* (PC boards or PCBs) and *cards* are thin, rigid insulating substrates that include conductive patterns and pads for the mounting and interconnection of electronic and electromechanical components. They are also called *printed-wiring boards* (PW boards or PWBs). Figure 28-1 is the artwork master for a simple, single-sided PC board. All black lines and circles and squares with holes in them represent conductive paths and pads; parallel rows of squares or circles represent mounting areas for DIPs. The objective of all PCB manufacture is to convert this pattern into hardware with the conductive lines and pads in the desired locations. This applies to single- and double-sided boards as well as multilayer PCBs. The most popular and economically significant boards are made from rigid phenolic-impregnated paper or epoxy-impregnated glass cloth.

It can be seen from Fig. 28-1 that both the terms *PCB* and *PWB* are misnomers because neither is an accurate description. The boards actually consist of disconnected metallic

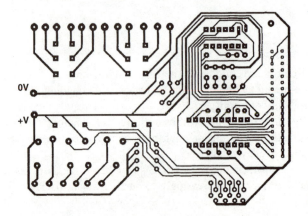

**Figure 28-1** Printed-circuit board master artwork.

traces on an insulating substrate rather than wires or complete circuits. They become circuits only when active and passive components have been bonded to them. Moreover, the traces typically are formed by the etching away of a layer of copper foil from the insulating substrate, or plating a thick film of conductive metal, usually copper, on a chemically prepared insulating substrate (additive method), or combinations of these methods. The only *printing* done is the application of resist masks through prepared fine-mesh screens in some processes. Nevertheless, both of the terms PCB and PWB have persisted for years, and it is unlikely that they will ever be replaced with a more accurate term in common usage.

Because of the economic and practical importance of rigid planar boards, the discussion of circuit boards in this section is confined to their fabrication. Rigid planar boards are insulated substrates suitable for mounting electronic components, and they are found in a wide range of products, from consumer appliances and toys to computers, scientific instruments, and military equipment. In addition to providing mechanical support for the components, they also afford protection for the components during handling. They might also include conductive surfaces or planes for radiating heat, distributing power, or providing radio-frequency shielding for sensitive components.

There are many variations of the rigid planar circuit board including the *multiple-wire board* made by the automatic placement and bonding of thin insulated wires on the surface of the rigid board to act as conductors. This technique has proven to be economical for small-scale production runs or prototype development, but it is uneconomical for large-scale production.

Yet another variation is the *wire-wrap board* that contains an array of vertically mounted rectangular posts. Insulated wires are wound on those posts by special hand tools or automated machines which tightly wind and bind the wires to the posts under tension. However, the wires can be removed to permit reconfiguring the circuit if and when changes are needed to modify or customize equipment in the field. They are practical for certain kinds of communications equipment and automated test equipment subject to customizing for the performance of specific tasks or to accommodate field modifications needed to reconfigure a system.

*Backplanes* or *backpanels,* similar in construction to conventional PC boards, are unifying electrical "spines" for plug-in *daughterboards.*

*Flexible circuits* or *flex-circuits* are variations of rigid circuit boards made with thin, flexible substrates of such insulating sheet plastic as polyester or Mylar. They are used in thin or narrow electronic products whose circuitry is divided between a lower case and a hinged cover, and are likely to include flexible parallel conductors that act as a ribbon cable.

# Classification of Rigid Circuit Boards

Conventional PC boards can be classified into three categories:

1. *Single-sided boards* (SSBs) have conductive traces and pads on only one side. They are typically made of rigid phenolic-reinforced paper or polyester-reinforced glass fiber with a lamination of copper foil.
2. *Double-sided boards* (DSBs) have conductive traces formed on both sides of the rigid board. They are more likely to be made from polyester- or epoxy-reinforced glass-fiber board than phenolic-reinforced paper. Some DSBs have plated-through holes (PTHs)

formed by metallizing the walls of holes through the board so that a conductive path interconnects both sides.

3. *Multilayer boards* (MLBs) typically have from 3 to 16 layers, but some might have as many as 60 layers. MLBs with 4 or 6 layers are the most common today. They are usually constructed by bonding together layers of uncured glass-fiber laminate with preformed copper traces. After lamination is completed the resulting boards are finished by processes similar to those used for DSBs with plated-through holes. Some layers might serve as power planes and others might serve as ground planes.

# Circuit-Board Manufacturing

## PHOTOMASK PREPARATION PROCESS

The first step in the manufacture of a PCB is to make a scaled drawing that includes the outline drawings or "footprints" of all components that are to be mounted on the board in their proper positions within the borders of that board. Figure 28-1 is an example of artwork for a single-sided board. The designer can evaluate the advantages or disadvantages of single-sided, double-sided, or multilayer boards for the end application.

The drawing includes the positions of all holes that must be formed through the board for leaded components and mounting hardware. The designer is expected to be familiar with all of these components, their seated height above the board, and their electrical isolation and heat-dissipation requirements. This artwork is the basis for reproducing the pattern to form a mask. The artwork for the simplest boards for hobbyists or design prototype circuits can be drawn by hand, formed from adhesive strips, or produced on a computer screen with the aid of circuit-board design software. If the artwork for a single-sided board is drawn to true dimensions, the pattern can be transferred directly to a copper-laminated blank by tracing, photographic methods, or xerographic copying methods. But more typically the artwork will be prepared oversize and reduced photographically to prepare photomasks.

For commercial purposes the designer might be required to resolve any conflicts between artwork for optimum performance or lowest production cost. In many cases, the form factor of the circuit board and its dimensions will be influenced by the size and shape of the housing or case of the host product or equipment. If the board is a modification or improvement of an earlier circuit, the design might be governed by the packaging for its host product. Other decisions to be made include the choice of board material and the selection of design rules for thickness and spacing of conductive trace widths. Simple, single-sided boards are the easiest to design and the least expensive to produce.

Commercial and industrial circuit design facilities typically have access to large databases of accumulated data, often in the form of previous designs. Thus new circuit designs can be extrapolations or revisions of existing designs where decisions about dimensions and materials and fabrication method have been made previously. Today many of these facilities do all of their circuit-board design work on computer workstations and, as a result, have large libraries of drawing elements and complete designs stored in computer memory for ready reference and reuse.

Commercial and industrial PCB artwork is usually drawn larger than true scale, typically 10 times true size or 10:1. This permits later reduction to actual size by photolithographic or computer techniques. Original artwork for complex PC photomasks drawn on the computer screen are typically downloaded to *X-Y* plotters for producing masks of each layer of the board. The master drawing has opaque and transparent areas outlining conductor and pad patterns. Depending on requirements, the actual production photomask can be prepared as either a positive or negative image of the master artwork.

## MASK AND ETCH PROCESS

To manufacture fine-line, high-definition PCBs, the photomask is placed directly over the copper-foil-laminated blank that has been coated with an ultraviolet- (UV-) sensitive photoresist, either in the form of a liquid or a dry film. There are two kinds of photoresists: *negative* and *positive.* The photomask is placed over the photoresist-covered blank and both mask and blank are exposed to UV radiation. After the photoresist is developed and placed in a chemical solvent, the desired photoresist pattern has been hardened and remains while that part which is soluble in the solvent is washed away. The result can be either a negative or positive image of the conductive traces, depending on the etching process selected, as shown in Fig. 28-2.

A *negative photoresist,* as shown in Fig. 28-2 (left) is a resist that cross-links and polymerizes if it is under the transparent parts of the photomask so it is exposed to UV, making it resistant to solvent after development, while the areas of resist protected from exposure to UV by the opaque parts of the mask are removed by the solvent. This results in a negative image of the mask. The copper foil can then be etched and removed from the areas of the pattern where resist has been removed.

By contrast, a *positive photoresist,* as shown in Fig. 28-2 (right) is a resist that cross-links and polymerizes if it is under the opaque parts of the photomask so it is protected from UV. This makes that part of the pattern resistant to a solvent after development, while the resist areas exposed to UV through the transparent parts of the mask are removed by the solvent. The result is a positive image of the mask. Again, the copper foil can be etched from the areas of the pattern where resist has been removed.

One advantage of using positive photoresist is that the board can be recoated with photoresist and exposed a second time to define additional conductive features. The choice of

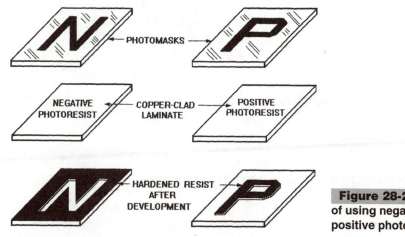

PHOTOMASKS

NEGATIVE PHOTORESIST — COPPER-CLAD LAMINATE — POSITIVE PHOTORESIST

HARDENED RESIST AFTER DEVELOPMENT

**Figure 28-2** Results of using negative and positive photoresists.

resist is based on such factors as desired conductive line width and spacing, conductor density, and the complexity of the mask.

## SUBTRACTIVE PATTERN-PLATING PROCESS

The most widely used commercial PC board manufacturing technique is a form of *subtractive* or *print-and-etch process* called the *pattern-plating process,* as shown in Fig. 28-3. It differs from the simple etching process widely practiced by hobbyists and lab technicians for producing one or a few prototype boards rapidly. In that process unwanted copper is simply etched from the laminated board, leaving a bare copper trace pattern. While this is acceptable for prototype projects, the copper pads must be tinned or lead-tin plated to accept lead-tin solder. Any copper left bare is subject to oxidation, so that overall tin or lead-tin plating is the only satisfactory way to protect the circuit board for long service life.

However, it has been found that a more economical way to produce simple, reliable boards is with the pattern-plating process. It turns out to be easier to perform the plating steps during the etching process rather than performing them after etching is complete.

Photoresist is applied to the copper-laminated blank and the photomask is placed over it prior to UV exposure, as shown in Fig. 28-3*a*. After the resist is developed, the part of the resist that defines the conductive pattern is dissolved, leaving that pattern in the exposed copper foil. The hardened part of the resist forms over the part of the foil that eventually will be removed, as shown in Fig. 28-3*b*. The exposed copper is then electroplated with either lead-tin solder, tin, tin-nickel, or nickel-gold. This plating enhances the solderability of copper during the subsequent solder-reflow process. The plated copper foil then becomes the mask for etching the unwanted part of the copper foil from the board, as shown in Fig. 28-3*c*.

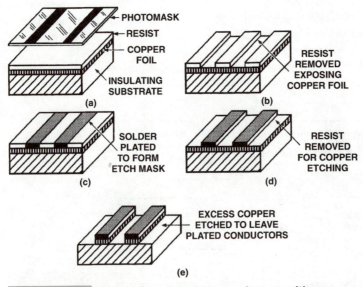

**Figure 28-3**  Printed-circuit board manufacture with an electroplated etch mask.

In the next step, the hardened resist is removed by another solvent, exposing the unwanted copper surfaces, as shown in Fig. 28-3*d*. The last step is to chemically etch away the unwanted copper, completing the board, as shown in Fig. 28-3*e*. Thus the electroplating forms the mask for the final etch process.

## PRINTED-RESIST PROCESS

The *printed-resist process* is used to make relatively simple one- and two-sided boards in commercial production. It is an alternative *subtractive process* in which production photomasks are used to prepare screens for the direct printing of resistive paints on a copper-foil-laminated board. Originally silk screens were used, but they have been replaced by reusable and more durable fine stainless-steel-wire mesh. By using photolithographic methods, part of the screen is filled with a viscous substance that leaves a pattern of open mesh for the penetration of photoresistive ink. A wiping or squeegee process applies the ink through the open mesh to the bare board.

## ADDITIVE PROCESS

The alternative to the subtractive process is the *additive process,* in which copper is deposited by the *electroless process* on an adhesive coating that has previously been applied to a blank board. The adhesive permits the required copper pattern to be selectively plated on the board. In later steps, the thickness of the initial copper deposition is augmented by copper plating. As in the subtractive process, the conductors can be tin-lead soldered or tin plated to improve their solderability.

The advantages claimed for the additive method include savings in the cost of etching copper and the later recovery of that copper from the etchant chemicals. It is also intended to reduce the amount of used chemicals that require proper special disposal equipment and procedures to meet the legal requirements for pollution control.

The additive process is combined with the subtractive process in forming *plated-through holes* (PTH) in double-sided and multilayer boards. This combined process is called the *multiple metal pattern-plating process.*

## PLATED-THROUGH HOLES (PTH) PROCESS

The development of the *plated-through-holes* (PTH) *process* was an important advancement in the manufacture of reliable double-sided and multilayer PC boards. The *additive process* eliminated the practice of inserting metal eyelets in drilled holes for conductive layer interconnection, a costly, time-consuming, and unreliable process. Plated-through holes also act as internal vias or conductors to connect embedded layers to one or both sides. The plated holes become sockets for component leads. During wave soldering, molten solder is drawn up around the leads by capillary action to form a secure bond.

Hole drilling can be done on the circuit-board blanks before or after the circuit conductors have been formed. Holes are drilled at those locations where interconnections are to be made between both sides of a double-layer board or between one or more internal layers and one or both sides of a multilayer board.

The inside walls of the holes are then coated with an adhesive and copper plated by *electroless deposition*. When this process is performed as part of the subtractive process, the initial electroless copper deposition is followed by copper plating and then lead-in solder or gold-nickel plating. By contrast, in the additive process only the additional reinforcing copper plating is added.

## MULTILAYER BOARD FABRICATION

*Multilayer boards* are made by forming pads and traces on partially cured laminate by any of the photolithographic and patterning methods previously described. The individual registered layers, alternating with insulating layers of the multilayer board, are then stacked and bonded under heat and pressure to cure the laminate to form a monolithic board.

The internal layers of multilayer boards, typically with four or more layers, can have internal conductors or internal ground or power-conducting planes. The standard thickness for commercial, industrial, and military PC boards, regardless of the number of layers, is about 0.0625 in (1.6 mm).

Board edge contacts are formed at the same time as other conductors and component mounting pads if *card-edge connectors* are to be used. These edge contacts act as the plugs for mating card-edge connectors. They are usually gold plated to permit mating and removal of the connectors without damage to the contacts. These contacts are not needed with two-piece printed-circuit board connectors.

# Surface-Mount Circuit Boards

*Surface-mount technology* (SMT) has eliminated the need for costly plated-through holes in boards or cards adapted to that technology. The "leadless" components are soldered directly to the surface conductors and pads of the "hole-less" surface-mount PC boards. The SMT process has extended the use of single-sided boards, but those suitable SMT boards must have thinner line widths and higher circuit density to contain the same components on one side that formerly occupied both sides of a double-sided board or even a multilayer board. This increased density is accomplished with SMT components whose stub or bent lead spacing has a finer pitch than the 0.10 in (2.5 mm) on-center spacing of conventional leaded components.

# Soft Soldering Processes

In modern mass-production manufacturing facilities there are five methods for soft soldering components to the boards: (1) *wave*, (2) *vapor-phase reflow*, (3) *infrared oven*, (4) *hot-plate conduction*, and (5) *laser*. These methods differ in the way heat is applied to remelt the solder. Each is based on one or more of the three methods for transferring heat: (1) conduction, (2) convection, and (3) radiation. The solder used in all of these processes is typically 63 percent lead (Pb) and 37 percent tin (Sn) or 60 percent tin and 40 percent lead. Solder is typically applied to components for later reflow soldering by screening or stenciling solder paste on metallized contact pads. The solder paste might be solidified or cured after the components have been placed to eliminate component movements by sliding on the paste.

## WAVE-SOLDERING PROCESS

Prior to *wave soldering,* both the circuit-board traces and the bonding pads of the components have been lead-tin plated so that the molten solder will successfully "wet" the surfaces to be bonded. Leadless or stub-leaded components must be temporarily glued to the circuit board with a drop of fast-drying adhesive to permit the boards to be handled and placed in the soldering machine without falling off. This is typically performed as part of the pick-and-place operation.

*Wave soldering* is an automated process for soldering leadless or leaded electronic components to circuit boards. The solder wave is formed by pumping molten solder through a nozzle to form the wave before it flows back to the bottom of a holding tank where it is reheated and recirculated continuously, as shown in Fig. 28-4, a simplified drawing of a dual-wave unit.

The first step in the process is to invert the populated circuit boards and position them on the inclined conveyor that can move at speeds in excess of 20 ft/min (6 m/min). The wave solder machine includes a *preheat stage,* an *active wave stage,* and an *exit.* The velocity profile of the wave can be controlled to optimize the soldering task.

The first wave, narrow and turbulent, deposits a continuous coating of solder onto the underside of the PC board, "washing" all of the leaded or temporarily bonded leadless components. This turbulent first wave is intended to complete all solder connections, eliminating any solder skips. It also provides positive solder delivery for soldering components temporarily glued to the underside of the PC board. The second *laminar-flow wave* smoothes out the solder deposited by the first wave to assure the proper flow of solder into any plated-through holes and attached connectors.

Some modern wave-solder machines have a hot-air jet called an *air knife,* which emits from a slotted outlet located just beyond the laminar-flow wave. It levels the solder on the backside of the PC board to eliminate any *solder bridging* before the solder solidifies. It also stresses the solder, lessening its adhesion while the solder is still molten, and blows

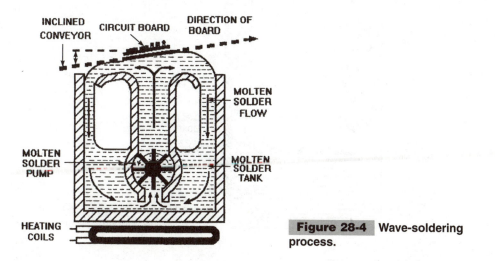

**Figure 28-4** Wave-soldering process.

molten solder out of any connection with improper wetting or if there are air holes below the surface. *Solder stressing* lessens adhesive forces that would prevent the air knife from removing solder from the connection.

## VAPOR-PHASE SOLDERING PROCESS

*Vapor-phase soldering* is a *reflow-soldering* method, shown in Fig. 28-5, for bonding electronic components to circuit boards, primarily in *surface-mount technology* (SMT). Components whose leads have been coated with lead-tin *eutectic solder* are positioned on the board with similarly prepared pads. The assembly is placed in a vapor-soldering chamber, and the 215°C (419°F) heat from the vapor of boiling fluorinated hydrocarbon liquid causes the solder to reflow and bond the components to the board. The melting temperature of the chamber is precisely controlled by the liquid's boiling point. Some systems use a secondary vapor "blanket" at a temperature of 48°C (117°F) to reduce the evaporation loss of the primary vapor. It is also called *vapor-phase reflow soldering*.

# Surface-Mount Technology (SMT)

*Surface-mount technology* (SMT) is a manufacturing technology in which leadless components are soldered to lands or pads on the surface of a circuit board without holes, as shown in Fig. 28-6. SMT covers all aspects of component assembly from the design and manufacture of electronic components adapted to that technology to the design and preparation of hole-less circuit boards and the tools and equipment needed to pick, place, and temporarily cement the SMT components to the board prior to soldering. In addition, it covers cleaning, testing, and quality control.

The benefits of SMT include increased population density of components, reduction in cost, complexity, and size of circuit boards or cards, and a reduction in assembly costs through automated assembly methods. The elimination of hole drilling and plated-through

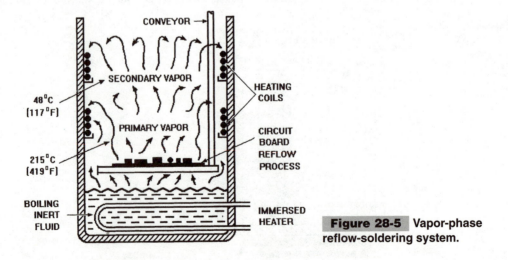

**Figure 28-5** Vapor-phase reflow-soldering system.

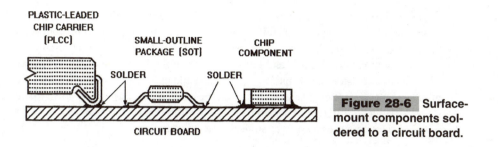

**Figure 28-6** Surface-mount components soldered to a circuit board.

holes contributes to the savings. Packaging density has increased because component manufacturers, particularly semiconductor device manufacturers, have reduced device package sizes and pin pitch specifically for SMT. The smaller, more densely populated boards can contain the same number of components as a conventional leaded component board while occupying less space because of their lower profiles. This permits reduction in the size of product cases or enclosures because populated boards or cards can be stacked closer in card cages, leading to improved volumetric efficiency and lower weight. Size and weight reduction is important for such products as cellular telephones, pagers, and certain kinds of personal communicators. Circuit-board or card area reductions in excess of 40 percent have been reported.

The pitch or spacing between the centers of pins on SMT components is typically half or less of the 0.10-in (2.5 mm) lead spacing of conventional components. In addition, SMT circuit boards can be smaller than conventional circuit boards of the same complexity because no space needs to be allotted for the alignment of the leads into through holes by automatic insertion machines.

Increased component density shortens the length of interconnections between components and enclosures or cases in smaller products. This minimizes radio-frequency and electromagnetic interference (RFI and EMI), and permits the circuitry to operate at higher frequencies.

Significant cost and time savings are achieved through the use of automated, computer-controlled component pick-and-place machines for high production runs such as are needed in the high-volume manufacture of consumer entertainment, telecommunications, and automotive electronics products.

The automated machines can be equipped with oscillating feed hoppers, dispensers of taped components from reels, and automated grippers that pick up components such as resistors, capacitors, diodes, transistors, and even integrated circuits that are packaged in uniform chips in standard case sizes. The grippers can place the components precisely at the locations on the boards or cards that have been specified by the placement program, and they can do this at extremely high speeds.

SMT is an outgrowth of *hybrid circuit technology* developed during the early days of solid-state circuitry for military, aerospace, and high-reliability microcircuitry before the introduction of very large scale integration (VLSI). Bare diode and transistor dies, as well as IC chips, can be bonded directly to ceramic substrates with thin-film metal conductors. Chip and die bonding can be done with precious-metal eutectic solders in furnaces with protective atmospheres to eliminate oxidation. Resistors, capacitors, and inductors can be formed by thick- or thin-film technology directly on the substrate and additional resistors or capacitors in chip form can be added as needed.

The design of SMT circuit boards must take into account the thermal coefficients of expansion of both components and boards. These must be matched to prevent solder-bond failure because the SMT component leads are soldered directly to the circuit board and receive any transmitted shock and vibration without the buffering effect of flexible leads. The shorter transistor and IC package stub leads and chip terminations preclude flexibility of movement. Unless care is taken to match the coefficients of temperature expansion, the stresses set up by different expansion and contraction rates can cause destructive fractures of the component cases as well as solder-bond failure.

# Solder Resists

A *solder resist* is a coating that masks off a PC board surface and prevents those areas from accepting any solder during vapor-phase or wave soldering. Its prime function is to restrict the molten solder pickup or flow in those areas of the PC board that are not covered by the solder resist. It also reduces the possibility of inadvertent *solder bridges* forming during wave soldering and protects the PC board conductors from contamination.

There are both permanent and temporary solder resists. Temporary resists are usually applied to selected areas of the PC board to protect certain features from accepting solder. These resists usually consist of a latex rubber material or adhesive tapes, some of which dissolve in the cleaning processes. Permanent resists are in the form of liquid photoresists, liquids for screen printing, and dry films.

# Multiple-Wire Circuit Boards

*Multiple-wire circuit boards* are an economical alternative to conventional circuit boards when small lot quantities are being ordered because they save the cost of artwork and photomask preparation. Thin insulated wires are placed and bonded in a point-to-point configuration on blank epoxy-reinforced glass-fiber board. The wires are placed by computer-controlled *x-y* motion machines that have been programmed to feed the wire from reels to make the interconnections continuously. The wires, which are permitted to cross, are then bonded to the surface of the laminate with heat-cured adhesives. These boards, which can achieve component densities equal to those of double-sided boards, are also about 0.0625 in (1.6 mm) thick.

# Wire-Wrap Boards

*Wire-wrap boards* are epoxy-reinforced glass-fiber boards with arrays of vertical square posts staked in a uniform pattern in the boards. The posts permit point-to-point interconnection by tightly wrapping the bare ends of insulated wire around them. Wire wrapping can be performed manually with hand tools or with automatic wire-wrapping machines. These boards are used in equipment whose functions typically are customized or subject to change after installation. Examples are automatic test equipment (ATE) or telecommunications equipment. All or part of the original or factory wiring can be altered in the field to accommodate changes in function or system configuration.

# Backplanes

A *backplane,* also known as a *backpanel,* is a circuit board that functions as an interconnecting "spine" between two or more boards or cards within an enclosure of an electronic product or system. It is a rigid circuit board on which printed-circuit-board connectors have been mounted in uniform rows, as shown in Fig. 28-7. Backplanes are often referred to as *motherboards,* and all of the other interconnected circuit boards are termed *daughterboards.* Backplanes usually are custom made by PCB fabrication techniques. A typical backplane is made of glass-fiber epoxy laminate FR-4 with a thickness from 0.060 to 0.250 in (1.5 to 6.4 mm). The circuitry can be formed on one or both sides of the backplane and it can be multilayered with up to 12 layers. Multilayer backplanes can provide better electrical properties than two-sided backplanes.

Capacitance, voltage drop, and ground loops are minimized. Heavy internal copper layers up to 0.030 in (0.8 mm) thick provide the necessary high current-carrying capability for backplanes that include high-speed logic and ICs.

Some military specification backplanes are fabricated as sandwiches of aluminum sheets enclosing a thin vinyl inner layer. Components are isolated from the aluminum surface laminations by insulated bushings that are pressed into a series of very accurately positioned holes in the panel. Backplanes laminated from sheet aluminum provide greater package strength than conventional backplanes, and they also provide radio-frequency shielding. In a typical military-style backplane, one aluminum lamination is the *ground*

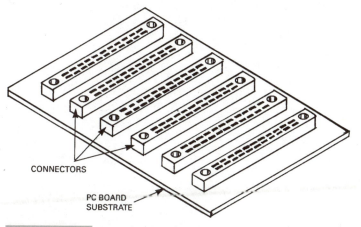

CONNECTORS

PC BOARD
SUBSTRATE

**Figure 28-7**   Backplane.

*plane* and the other is a *voltage plane.* These panels can be designed to have the desired capacitance and impedance characteristics while also providing a rigid support structure for card-edge connectors.

Backplanes can be substrates for either one-piece edge connectors or two-piece box-style PC connectors. Connector pins are terminated with solder eyelets or wiring posts. The spacing between connectors determines the spacing between adjacent circuit boards. Where backplanes are fitted with two-piece PC-board connectors, the connector half with the pins (male side) is mounted on the backpanel while the connector half on the daughter PC board contains the sockets (female side). The case or enclosure provides protection against accidental misalignment of the rows of pins.

Backplanes, like PC boards or cards, are fabricated to both commercial and military specifications. If the electronic system is subject to changes during its operating life, the wire connections can be *wire wrapped.* This technique is also employed if the equipment is dedicated to a specific application at the point of installation. Wire-wrapped backplanes can be made with the necessary impedance characteristics to permit efficient use of emitter-coupled logic (ECL).

Press-fit connectors are used where it is necessary to mount connectors in prepared holes in the backplane. Press-fitting involves driving connector contacts into precisely formed plated-through holes with inside diameters of 0.040 to 0.043 in (1.0 to 1.1 mm). The press-fit interface is both mechanically secure and electrically stable. Gas-tight electrical connections are made between the copper-plated side walls of the plated-through holes and the contact. The contacts have square cross sections that deform the sidewalls with enough pressure to cut into the copper plating to prevent the intrusion of oxides and contaminants. The shanks of some contacts are made so that they are compressed when they are inserted into the plated-through hole and then expand to provide more surface contact between the pin and the inner diameter of the hole than rigid shanks.

After the insertion of the pins, an insulated housing is seated over the contacts to form the connector receptacle. Some systems allow for individual contact replacement and accessibility to the printed circuit conductors for circuit changes or repair.

# ELECTRONIC HARDWARE:
# WIRE, CABLE, AND CONNECTORS

# Overview

Electronic hardware has evolved from earlier electrical hardware, and it still includes many of the same products such as wire, cable, connectors, insulating blocks, and fasteners. However, over the years new kinds of hardware have been developed specifically for electronics, such as specialized connectors, sockets, and adapters, and this is an ongoing process. The transition from electron-tube to solid-state circuitry reduced the operating voltages of circuits and permitted lightweight circuit boards to be substituted for metal chassis with their rows of tube sockets and heavy transformers. This, in turn, largely eliminated point-to-point wiring and allowed smaller connectors, cases, and enclosures.

Demand continues for conventional electronic hardware such as insulated hookup wire, telephone and flexible cords, coaxial and flat-ribbon cable, and knobs that have been available for years. But the imposition of international safety standards and requirements for component interchangeability has had an impact on the design and manufacture of all electronic products for sale in the expanding global market. Meeting the demand for reliable, low-loss, and low-cost parts has led to an astonishing variety of interconnection products. This part of the electronics industry continues to meet ever-changing and more challenging requirements for interconnection products and sockets with ever-finer pitch and lower profiles.

# Wire for Electronics

The most popular forms of wire are hookup wire, wire-wrap or backplane wire, and twisted pairs. The discussion in this section will be limited to those categories. However, other types of wire include antenna wire, high-voltage leads, magnet wire, and test-probe wire.

## CONDUCTIVE MATERIALS

Copper, copper-covered steel, high-strength copper alloys, and aluminum are commonly used as conductors in electronics, but pure nickel, pure silver, copper-covered aluminum, gold, and tungsten wire are in demand for components and instruments. Nevertheless, copper still remains the most widely used metal for wire because of its high electrical and thermal conductivity; ductility; malleability; solderability; high melting point; high resistance to corrosion, wear, and fatigue; and, perhaps of most importance, its low cost.

The copper grade most widely specified for electronics is *electrolytic tough pitch* (ETP). Although highly refined, it contains minute amounts of copper oxide which are objectionable in many applications. Additional refining produces *oxide-free, high-conductivity* (OFHC) *copper*. The stiffness of solid copper wires is reduced by stranding for greater flexibility. Nineteen-strand wire retains some stiffness but truly flexible wire has 26 or more strands. The temper or hardness of copper wire ranges from soft-drawn annealed (SD) and medium-hard drawn (MHD), to hard drawn (HD). The tensile strength of MHD and HD copper wire exceeds that of SD copper of the same gauge.

## SOLID WIRE GAUGE

The gauge of solid wire for electronic applications rarely exceeds No. 10 American Wire Gauge (AWG), the U.S. standard for identifying wire sizes up to and including No. 4/0. The next size larger than No. 4/0 AWG is measured in circular mils (CM) and has a value of 250,000 CM or 250 MCM.

## STRANDED WIRE

*Stranded wire* is specified where there is a requirement for more flexibility than is afforded by solid wire with equivalent cross-sectional area. Moreover, stranded wire can withstand more vibration and bending than solid wire before fracturing. Surface damage or even the accidental cutting of a few stranded conductors will have less effect on the wire's electrical resistance and performance than nicks or cuts to a solid wire because their effect on the current-carrying capability of the wire is negligible.

Conductors widely specified in electronics have 7, 10, 16, 9, 26 or more strands, with 7 and 19 strands the most widely accepted numbers. For any given wire size, increasing the number of strands (with corresponding decrease in strand diameter) makes the wire more flexible but increases its cost.

Conductor weight and resistance is proportional to circular mil area which, in stranded wire, depends on the number and diameter of component strands. But the circular mil area of a stranded wire only approximates that of its solid wire equivalent.

## WIRE PLATING

Tin plating is the protective coating most widely applied to copper wire to prevent oxidation and improve its solderability, but silver or nickel plating is also used for some applications. Tin-plated copper wire is acceptable for conductors that are not exposed to temperatures greater than 150°C. Because it has lower conductivity than copper, tin plating increases the resistance of copper wire. By contrast, silver plating soft-drawn copper wire can increase its maximum operating temperature from 140 to 200°C without increasing its resistance. It is useful in high-frequency circuits because silver's higher conductivity enhances surface conduction. Nickel plating also increases the operating temperature of copper wire from 140 to 260°C, and it is recommended for Teflon TFE insulated hookup wire capable of operating for prolonged periods over a temperature range of 200 to 260°C.

## PROTECTIVE INSULATION AND JACKETING

Various thermoplastic, thermoset, and elastomeric plastic resins are applied as primary insulation on solid and stranded conductors. They are also used as jackets over primary insulation, braids, shields, and cables. The extrudable primary insulation materials include polyvinyl chloride (vinyl or PVC), the polyolefins (polyethylene and polypropylene), and the fluorocarbons (principally Teflon TFE, FEP, and PFA).

## HOOKUP WIRE

*Hookup wire* is single-conductor solid or stranded insulated wire manufactured to meet UL, CSA, and military specifications. As with other commercial wire and cable products, this classification encompasses a wide range of products available with different wire gauges, insulation, and colors. Some is made as standard off-the-shelf or catalog wire, but much of it is made to customer specifications. Hookup wire gauge ranges from 32 to 12 AWG.

## WIRE-WRAP WIRE

*Wire-wrap wire,* also called *backplane wire,* is fine insulated hookup wire for making solderless point-to-point wiring connections on circuit boards or backplanes with square terminal posts. It is silver-plated solid or stranded wire, 32 to 24 AWG, typically with Tefzel and Kynar insulation because of their physical toughness and resistance to cut-through as well as ease of stripping, absence of set when spooled, nonflammability, and wide temperature range. Its small outside diameter allows the placement of many wires on wire-wrap boards.

Insulation is stripped a short distance at both ends, and each end is mechanically stretch wrapped around the square terminal posts under tension as high as 130,000 lb/in$^2$ by hand tools or automatic machines. See "Wire-Wrap Boards" in Sec. 28, "Electronic Circuit Packaging and Assembly."

## TWISTED-WIRE PAIRS

*Twisted pairs,* the least expensive conductors for signal transmission, are available in a wide range of wire sizes and insulation. They are still used in signal and data transmission despite the competition from coaxial cables and optical fibers. The conductors are typically stranded 30 to 20 AWG, and the most common insulations are PVC, irradiated PVC, and low-dielectric Teflon FEP or polyethylene.

# Cable for Electronics

Cable for electronics is made from two or more insulated conductors, typically solid or stranded copper wires in a common jacket. Cables can be classified as parallel insulated and bundled and twisted insulated conductors.

There are four kinds of parallel insulated conductors:

1. Uninsulated conductors within insulated 50- and 95-ohm transmission line, 300-ohm twin lead, and antenna rotator cable
2. Uninsulated conductors laminated between two layers of insulating materials (flat cable or flat flexible cable)
3. Individually insulated wires bonded or woven together (ribbon cable)
4. Individually insulated wires enclosed in an extruded tubular jacket

## COMMON-INSULATION CABLE

*Common-insulation cable* is produced by drawing the desired number of uninsulated wires simultaneously through the extruder to form an integral plastic covering over them. The size and shape of the extrusion die determines the wall thickness and the shape of the finished cross section. The conductors can be distinguished and polarity can be identified by the following features:

- Contrasting color stripe along one edge of the insulation
- Tinned and bare copper conductors
- Ridges or fins along one edge

## FLAT CABLE

*Flat cable,* also known as *planar cable,* is an assembly of parallel conductors bonded together by insulation. It is specified where space limitations make the installation of round cables difficult. It saves space and weight and is flexible enough to be bent around sharp turns and follow contours. Also, the large exposed surface dissipates heat easily. Its outstanding feature is uniform conductor-to-conductor spacing, making termination with mass-termination insulation-displacement connectors (MTIDCs) easy and economical.

The most common form of flat cable is made by simultaneously extruding insulation over as many as 64 parallel round solid or stranded conductors. Another type is *bonded ribbon* or *rainbow cable.* A third type, *flat laminated cable,* is made by sandwiching parallel flat conductors between layers of insulation and bonding them with adhesives or heat and pressure. The spacing between flat-cable conductors measured between centers (pitch) is typically 0.050 in (1.3 mm), but 0.100-in (2.5-mm) pitch cable is also available.

## FLAT-CABLE INSULATION

The insulation for flat cable includes polyvinyl chloride (PVC), polyester composites, cross-linked polyolefin, Teflon, and polyimide. PVC is used for commercial cable because of its low cost. *Polyester* composites exhibit good heat conductivity, shrink, stretch, and cold flow characteristics as well as dielectric strength. They also have high flex life, good tensile strength, and excellent abrasion resistance. *Polyolefin* has excellent electrical properties, a 150°C temperature rating, and good resistance to hot solder and chemicals. Both *Teflon TFE* and *FEP* have excellent electrical properties, and remain chemically stable when subjected to a wide range of environmental stresses. TFE is an effective insulation for temperatures up to 260°C, while FEP is useful to 200°C.

## SHIELDED FLAT CABLE

*Shielded flat cable* is flat cable with copper-braid shielding and a protective vinyl jacket that permits its use outside of a shielded enclosure, as between two separate components of a system. The center sections of the cable can be rolled for greater bend and twist flexibility, leaving the ends flat for termination with post-and-box or D-type connectors.

## EXTRUDED FLAT CABLE

*Extruded flat cable,* as shown in Fig. 29-1, consists of parallel 28 AWG stranded tinned copper conductors that have been both joined and insulated by extruding insulation resin over them, typically fire-retardant, flexible PVC. The cable can have from 9 to 64 conductors. This construction permits easy separation of the individual insulated conductors. Manufacturing processes permit insulation thickness and wire spacing to be closely controlled to optimize the impedance, capacitance, and inductance properties of the cable. The conductors for standard cable are spaced 0.050 in (1.3 mm) on centers (pitch).

## RIBBON CABLE

*Ribbon cable* or *rainbow cable* is flat cable formed by heat sealing or bonding parallel lengths of colored wire that have previously been insulated with PVC. It saves more weight and space than any other type of flat cable. A principal benefit of this cable is ease of tracing its color-coded wires. Each insulated conductor can be separated easily without damaging its insulation. In appearance it is similar to the flat cable shown in Fig. 29-1. The conductors for standard cable are spaced 0.050 in (1.3 mm) on centers (pitch).

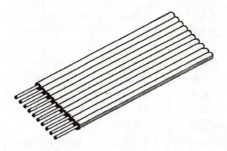

**Figure 29-1** Extruded ribbon cable.

## WOVEN RIBBON CABLE

*Woven ribbon cable* is flat cable made by weaving colored PVC-insulated wire into a planar configuration. It is more flexible than other kinds of flat cable. The weaving process gives the cable inherent strain relief, useful in preventing conductor flexing fatigue.

## COAXIAL CABLE

*Coaxial cable* has a single inner conductor, either solid or braided, and an outer shield of copper braid over a dielectric core. It is jacketed with PVC for protection, as shown in Fig. 29-2. This cable is effective for efficient transmission of RF signals from 1 kHz and 4.0 GHz, a range that includes both the FM and TV bands. Coaxial cable with two conductors is called *twinax* and with three conductors is called *triax*. Semirigid coaxial cable with an external metal sheathing is also specified for microwave- and near-microwave-frequency transmission. It minimizes RF losses because the bending radius is larger than that of conventional coaxial cable. See also "Coaxial Cable Transmission" in Sec. 7, "Microwave and UHF Technology."

## FIBEROPTIC CABLE

*Fiberoptic cable* consists of single or multiple strands of glass fiber enclosed in a protective jacket, permitting it to be handled and installed without damaging the glass fibers. The fibers are transparent monofilaments of glass or plastic capable of transmitting modulated analog or digital light signals in the visible and infrared regions. A simplex fiberoptic cable is shown in Fig. 29-3.

Glass fiber is the preferred medium for the transmission of optical signals over distances greater than 0.6 mi (1 km) because it introduces lower losses or attenuation than does plastic fiber. A single fiber in a protective sheath can be an optical cable. There are both *single-mode* and *multimode* fibers. Single-mode fibers are specified for long-haul transmission, and multimode fibers are used for short-haul transmission of less than 0.6 mi (1 km). The

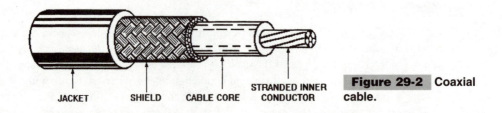

JACKET    SHIELD    CABLE CORE    STRANDED INNER CONDUCTOR

**Figure 29-2** Coaxial cable.

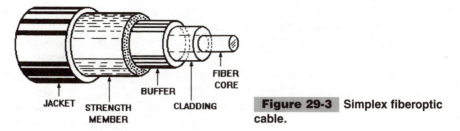

**Figure 29-3** Simplex fiberoptic cable.

insulating and jacketing materials applied to fiberoptic cable are the same as those used on copper wires and cables.

# Electronic Cable Manufacturing

## COMMON-INSULATION CABLE ASSEMBLY

In *common-insulation cable assembly,* individual conductors, either wires or smaller cables, are bundled for insertion in cable jackets. Color coding on individual wires and cables makes cable installation and hookup easier and is helpful in circuit tracing and problem diagnosis. Large bundles of small-diameter insulated conductors form a generally round cross section, but small numbers of large-diameter conductors result in a fluted cross section.

*Fillers* are included in cables assembled from a small number of large-diameter conductors to round out the depressions that would otherwise appear under the outer jacket. Fillers improve the appearance of the cable and also provide proper conductor separation in low-loss transmission lines. They also act as cushioning in heavy-duty cables subject to frequent flexing and impact. The most commonly used fillers are lightweight, nonconducting cotton, jute, vinyl, polyethylene, and twisted-polyethylene monofilaments.

*Binders* are threads wound spirally over individual groups of insulated conductors to hold them together for jacketing. Colored binders separate and identify identical groups of conductors. Nylon is widely used as a binder for electronic cables, while polyester or polypropylene are more commonly used in telephone cables.

*Tape* is often wound around insulated conductor bundles as added protection against mechanical abuse and to prevent damage to the conductor insulation. For example, tape can be wound between shields and adjacent conductors. Tape makes it easier to strip a jacket, and its presence ensures a smooth jacket surface. Polyester is the most common tape material, but paper tape is also popular. Tapes can be applied either spirally or longitudinally.

*Jackets,* also called *sheaths,* cover and protect the enclosed cable conductors against abrasion, mechanical damage, spilled chemicals, and fire. Jackets can cover either single conductors or an entire cable. Nylon inner jackets, where used, are typically 0.002 to 0.006 in (0.05 to 0.15 mm) thick, but outer jackets, most frequently vinyl, polyethylene, neoprene, or polyurethane, typically account for 10 percent of the cable's core diameter. Jackets are usually pulled on over the core so they adhere rather loosely and are easy to strip. However, some neoprene jackets are pressure extruded to fill all the voids and convolutions in the cable core. While providing a smooth, firm surface, they are more difficult to strip unless the underlying conductors are insulated with a different type of plastic or covered with a separator or barrier.

Vinyl (PVC) jackets, inherently flame and abrasion resistant, are specified for indoor and general-purpose cables. Neoprene is specified where the cable will be exposed to abuse and severe handling. It will not stiffen in subzero temperatures and is oil, ozone, and weather resistant. Cables with neoprene jackets can be buried in the ground or placed in conduits, trays, racks, or ducts. *Hypalon* offers most of neoprene's properties, but it also exhibits superior resistance to ozone, oxidation, and heat.

Many electronic product manufacturers prefer premanufactured cable assemblies for external interconnections. In applications where the cable is subject to physical or environmental abuse, molded assemblies are specified. Molding cable with an appropriate sealant keeps moisture and dust out of the critical spaces under the cable jacket. The sealant enters the terminating connectors because the molding bonds the jacket to the connector. Other jacketing materials include ethylene propylene rubber (EPR), polyethylene, Kynar, polyurethane, thermoplastic elastomer (TPE), Teflon FEP, and Tefzel.

## COAXIAL CABLE MANUFACTURING

Inner conductors for coaxial cable are typically made from solid or stranded annealed copper wire that can be bare, tinned, or silver coated. Cable attenuation is lower with solid conductors, but cable flexibility is greater with stranded conductors. Annealed copper wire is generally preferred because of its excellent electrical properties. However, for mechanical strength, copper-covered steel or copper alloy conductors are used.

Five different core dielectric materials are now used in the manufacture of coaxial cable. Four of these are based on polyethylene and the fifth is Teflon FEP. Extruded polyethylene is the most widely used core dielectric because it is inexpensive and offers very good electrical properties. But it is not generally suitable for applications where the ambient temperature will exceed 80°C. Polyethylene is flammable, but the shield and PVC cable jacket reduce this hazard. It has the highest dielectric constant of the five materials: 2.27.

*Cross-linked (irradiated) polyethylene* is used where ambient temperatures can reach 125°C. It has about the same electrical characteristics as conventional polyethylene, but its additional toughness is obtained from an irradiation process that transforms it into a thermosetting material. This dielectric has a greater resistance to abrasion, ozone, solvents, heat from soldering, and cracking due to stress. Its dielectric constant is 2.45.

*Cellular (foamed) polyethylene* is used where lower dielectric constants are required. Dielectric constants can be as low as 1.5. Cellular polyethylene can also be cross-linked for greater strength and toughness. Flame-retardant polyethylene contains additives which improve its ability to withstand flame, but this feature increases its dielectric constant to 2.5.

*Teflon FEP* has excellent electrical properties, and it is used where ambient temperatures will reach 200°C. Its low dielectric constant of 2.15 makes it suitable for use in miniature coaxial cables.

The dielectric constant $K$ of coaxial cable core material determines the outside diameter and weight of the finished cable. Where attenuation and impedance are to be kept constant, the separation between the inner and outer conductors can be reduced if the value of $K$ decreases. This permits the manufacture of thinner and lighter cable. Core materials with the lowest dielectric constants are recommended for airborne applications where space and weight reduction is critical. However, the dielectric constant can be lowered with air-spaced (semisolid) cores. A low-loss filament is wrapped around the inner conductor in place of the solid cable core. A jacket made of the same material is then extruded over the spiral, leaving an air space. The effective *K* will depend on the air-to-dielectric ratio.

*Shielding* (or the outer conductor) of coaxial cable is braid that is woven from small-diameter bare, tinned, or silver-coated copper wires. Double braids provide even more effective shielding. Where crosstalk and noise must be minimized, a double-shielded, double-jacketed triaxial cable is specified. U.S. Military Standard MIL-C-17 mandates braided shields for RG-style coaxial cable. But lower-cost, less effective shields can be made from spiral-wound wire, aluminized sheet Mylar, and conductive plastic. These alternative shields are acceptable only for commercial use in coaxial cable suitable for data transmission.

*Jackets* protect coaxial cable against environmental stresses. Both polyethylene and PVC are effective jacket material. PVC withstands weathering and abrasion, and provides effective moisture resistance, making it acceptable for military-specification cable.

# Fiberoptic Cable Manufacturing

Standard commercial fiberoptic cable contains 1, 2, 6, 12, or 18 fibers, but the number is determined by the application. Two choices can be made in specifying fiberoptic cable: *jacket compound* and *strength member.* The most commonly used jacket compounds for protecting the fiber against impact and abrasion are polyvinyl chloride (PVC), polyethylene, and polyurethane, generally in that order. Strength members are used inside the cable to support the optical fiber and relieve strain on it. Fiberglass rod, Kevlar yarn, and steel wire are commonly used strength members.

# Connectors for Electronics

*Connectors* are mechanical devices for joining two conductors to form a seamless, low-resistance conductive path. Available in many different sizes and shapes, connectors generally have two parts: one terminates a wire or cable and the other is attached to a receptacle, or power outlet. The male part is called the *plug* and the female part is called the *receptacle* or *jack.* They are used to join conductors within circuits, products, and systems.

Electronic connectors are generally classed as follows:

- Card-edge connectors
- Post and box connectors
- Rectangular multipin connectors
- Mass-termination, insulation-displacement connectors (MTIDCs)
- Coaxial or radio-frequency connectors
- Circular or cylindrical connectors

The increased acceptance of international standards in telecommunications protocols and bus structures has led to increased acceptance of international standards for connectors. In addition, reference to applicable military specifications and commercial standards, at least for dimensions, ensures connector interchangeability for many different kinds of commercial, industrial, and consumer products.

## CARD-EDGE CONNECTORS

A *card-edge connector* is a one-piece female connector that mates with the edges of a circuit board or card to complete an interconnection. The card edge has plated contacts on one or both sides that mate with contacts within the edge connector when it is inserted. The connector contacts are spring loaded and oppose each other across the insertion slot. Tension provided by the contacts clamps the edge connector securely to the edge of the board. Where contact pressure will be insufficient to retain the card edge because of shock or vibration displacement, the connector is fitted with clamping screws which fasten it securely to the card edge. These connectors terminate cables or interconnect motherboards with daughterboards.

## POST-AND-BOX CONNECTORS

*Post-and-box connectors* are two-piece multipin connectors made to minimize the possibility of accidental separation when the assembly is subjected to shock and vibration. The multiple pins and sockets provide strong interconnections. These connectors are manufactured to conform to European as well as U.S. commercial and military specifications.

## RECTANGULAR MULTIPIN CONNECTORS

A *rectangular multipin connector* consists of a *plug* terminating a cable and a *receptacle* mounted on a panel or chassis. A popular style, the *D-type connector,* as shown in Fig. 29-4, is recognizable by its trapezoidal shell. The connector pin layout shown is that of a 25-pin RS-232C interface. These connectors are widely used for interconnecting computers to printers, monitors, and other peripherals. There are many variations of these connectors, also called *rack-and-panel connectors.* The D-shaped shell assures that the plug and receptacle will align correctly when mated. Versions of these connectors have shields to block EMI and RFI, and some have clamping screws to assure a secure connection.

## MASS-TERMINATION INSULATION-DISPLACEMENT CONNECTORS (MTIDCs)

A *mass-termination, insulation-displacement connector* (MTIDC) is a one-piece unit similar in appearance to a card-edge connector. It is designed to accept flat or ribbon cable. An enlarged view of one of as many as 36 fork-shaped contacts that are mounted inside the MTIDC connector cavity like teeth is shown in Fig. 29-5a. The contacts have enough freedom of movement to permit them to self-align over individual insulated wires when external pressure is applied. The contacts bite down on the mating conductors of the flat cable.

The internal knife edges of the contacts shear the insulation and embed themselves in the stranded copper wires. An end view in Fig. 29-5b shows how the contacts force the copper strands of round wire into alignment while piercing their outer surfaces to form a gas-tight seal which keeps out contaminants. All terminations are made simultaneously when the

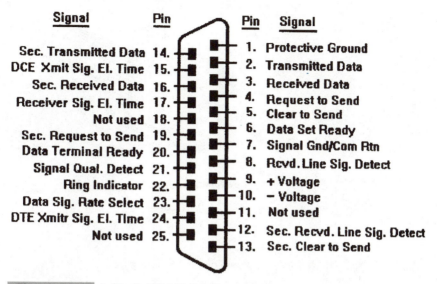

| Signal | Pin | | Pin | Signal |
|---|---|---|---|---|
| Sec. Transmitted Data | 14. | | 1. | Protective Ground |
| DCE Xmit Sig. El. Time | 15. | | 2. | Transmitted Data |
| Sec. Received Data | 16. | | 3. | Received Data |
| Receiver Sig. El. Time | 17. | | 4. | Request to Send |
| Not used | 18. | | 5. | Clear to Send |
| Sec. Request to Send | 19. | | 6. | Data Set Ready |
| Data Terminal Ready | 20. | | 7. | Signal Gnd/Com Rtn |
| Signal Qual. Detect | 21. | | 8. | Rcvd. Line Sig. Detect |
| Ring Indicator | 22. | | 9. | + Voltage |
| Data Sig. Rate Select | 23. | | 10. | − Voltage |
| DTE Xmitr Sig. El. Time | 24. | | 11. | Not used |
| Not used | 25. | | 12. | Sec. Recvd. Line Sig. Detect |
| | | | 13. | Sec. Clear to Send |

**Figure 29-4** A 25-pin RS-232 D-type computer connector.

contacts are forced down together by a press, causing them to displace the insulation and form the seals. The press can be manually or automatically operated.

## COAXIAL OR RADIO-FREQUENCY CONNECTORS

A *coaxial connector* is a cylindrical or circular connector for terminating coaxial and semi-rigid cable. The plug and jack of a military-style coaxial connector is shown in Fig. 29-6. This connector pair has a bayonet-style (pin and curved slot) locking mechanism. The

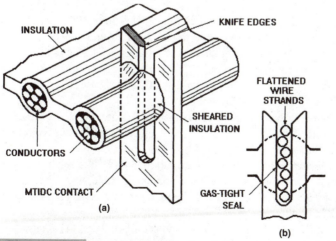

**Figure 29-5** Mass-termination insulation-displacement connector (MTIDC): (*a*) enlarged individual contact, and (*b*) contact end view.

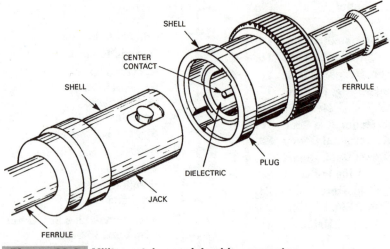

**Figure 29-6**    Military-style coaxial cable connector.

cylindrical shells protect the interconnection against abrasion and crushing forces as well as moisture, salt spray, dust, airborne contaminants, and emitted or received RF energy. The plugs include provisions for terminating the coaxial inner conductor of the cable while also clamping the braid or outer conductor and dielectric core to the connector.

Coaxial connectors are grouped in four product categories:

1. Standard (C, N, twinax, triax, and UHF)
2. Miniature (BNC and TNC)
3. Subminiature (SMA, SMB, and SMC)
4. Precision (APC-3.5 and APC-7). The 3.5 and 7 are mating cable diameters in millimeters.

UHF connectors are designed to operate at frequencies up to 300 MHz. They have peak voltage ratings of 500 V and are available in two sizes, *N* and *C*. The *N connectors* are medium-size, weatherproof units with threaded couplings, useful up to about 1 GHz. They are impedance matched to either 50-ohm or 70-ohm cables. *C connectors* are available in two styles: *standard,* with peak voltage ratings of 1500 V, useful to 10 Hz; and *high-voltage,* with ratings to 4000 V, useful to 2 GHz.

*BNC coaxial connectors* are miniature, bayonet-locked units, designed to operate at frequencies up to 11 GHz. *TNC connectors,* with the same electrical specifications, are weatherproof units with threaded couplings.

*SMA coaxial connectors* are semiprecision, subminiature units for frequencies of up to 18 GHz on semirigid cable and up to 12.4 GHz on flexible cable. Slightly smaller *SMB connectors* can be used across the DC to 4-GHz frequency range. *SMC connectors* can operate over the wider DC to 6-GHz range.

There are four different kinds of coaxial connector products: (1) *plugs,* (2) *jacks,* (3) *receptacles,* and (4) *adapters. Plugs* are the male connectors and *jacks* are the female connectors, used primarily to terminate unsupported cable. *Receptacles* are attached to panels or chassis. *Adapters* match connector elements with different couplings. There are also isolated ground panel and printed-circuit board mounting styles.

*Bulkhead connectors* act as transitions between cables terminated on opposite sides of a metal case or bulkhead in aircraft or ships. These connectors permit watertight or pressurized integrity to be maintained at the bulkhead.

The dimensions, materials selection, manufacturing methods, and test procedures for military-specification coaxial connectors are dictated by MIL-C-39012, a triservice coordinated specification. Another applicable specification is MIL-C-22557.

The key dimensions of coaxial cable connectors are determined by the diameters of the conductors and dielectric cores of the coaxial cable to be terminated, the transmission frequencies to be supported, and the properties of the dielectric. The electrical characteristics of RF cable also influence the design and application of coaxial connectors. These include resistance and inductance of the conductors and the capacitance and leakage between them.

To avoid RF energy losses, the coaxial cable must be terminated in its characteristic impedance. This condition exists when the voltage is the same at all points on the line and there are no voltage standing waves. To meet these requirements, the dimensions of coaxial connectors are highly standardized.

The shells or bodies of standard and miniature coaxial connectors are typically made of nickel- or silver-plated brass. However, the bodies, coupling nuts, and other metal parts of subminiature SMA connectors are made of nonmagnetic stainless steel. Female contacts are made from gold-plated beryllium copper.

Series SMB and SMC connectors have bodies, coupling nuts, and other metal parts made of half-hard brass, while female contacts are made of beryllium copper. Series SMC connectors are mated with threaded couplings, while series SMB connectors are spring mated with snap-fit couplings for quick-connect disconnect applications. Insulators for coaxial connectors are typically made from Teflon tetrafluoroethylene (TFE) or Teflon-glass combinations. Some low-cost commercial coaxial connectors are die cast from zinc alloy. Nevertheless, all the machined or cast surfaces must be free of rough surfaces or burrs that would cause RF losses and make coupling and decoupling difficult.

Coaxial connectors must be disassembled to be attached to the coaxial cable in a sequence of steps. There are at least 10 different methods for attaching the connectors to the cable. For example, they can be clamped, crimped, or soldered on flexible cable and soldered or clamped on semirigid cable.

Low-cost coaxial connectors are now widely used to terminate cables carrying lower-frequency audio and video signals in stereo and test equipment as well as digital data in computers and communications systems. The inner conductor carries the signal, and the outer conductor is connected to a common ground. The external braid forms a continuous shield to protect the signal conductor from externally produced electromagnetic interference (EMI), while simultaneously preventing the radiation of high-frequency data transmission signals.

## CIRCULAR MULTICONDUCTOR CONNECTORS

A *circular multconductor connector,* also called a *cylindrical connector,* is used to terminate round multiconductor cables. These connectors can have from 4 to 128 contacts, meaning that they can accommodate up to 128 individual wires and coaxial or fiberoptic cables in various combinations. The *plug* is the connector part attached to the free or moving end of the cable, and the *receptacle* is the part attached to a bulkhead, case, or console. A cutaway view of a mated plug and receptacle of a military-style circular multiconductor connector is shown in Fig. 29-7.

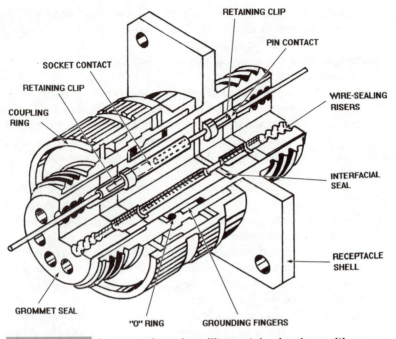

**Figure 29-7** Cutaway view of a military-style circular multiconductor connector.

An important feature of all circular connectors is the rapidity with which they can be aligned and coupled or uncoupled without the aid of tools. The metal shells of the receptacles and plugs provide protection against abrasion and external forces which would crush rectangular connectors. The shells also protect the mating contacts from airborne contaminants, dust, moisture, and salt spray, while also shielding the enclosed conductors against unwanted external EMI or RFI. In addition, the metal shells extend the cable shielding to prevent the conductors from acting as antennas and radiating internal RF energy.

Circular connectors are widely used in military and commercial aircraft for interconnecting modular avionics cases. They are also widely used on navy ships and submarines, military combat vehicles, and equipment shelters. Commercial versions are used in radio and television broadcast equipment, industrial robots, and process control systems.

The integral locking mechanisms, either complete or partially threaded shells (breech lock) or bayonet style (pin and curved slot), provide a self-supporting interconnection. Neither internal nor external screws are required for clamping mating parts. Circular connectors can also withstand the high shock and vibration forces encountered by aircraft, ships, and combat vehicles. A knurled outer ring is twisted to unlock the mating shells. In addition, built-in keys assure rapid and correct alignment or polarization of the mating contacts. Bulkhead connectors serve as transitions between two cables located on the opposite sides of a metal case or bulkhead in aircraft or ships. These permit watertight or pressurized integrity to be maintained at the bulkhead.

Commercial versions of military circular connector families are available. They do not require the same costly Qualified Parts List (QPL) testing and traceability documentation as the military-specification connectors, and thinner plating might be acceptable to commercial customers. As a result, their unit prices are lower.

## MILITARY CIRCULAR MULTICONDUCTOR CONNECTORS

*Circular multiconductor pin connectors* for military applications are made in three size categories: (1) *standard,* (2) *miniature,* and (3) *subminiature.* The internal construction and testing and the test requirements for all military circular connectors are similar. Standard connector sizes have been specified for shipboard and ground-based electronic systems, while subminiature connectors are favored for avionics because of their lighter weight and smaller size. However, the miniaturization of many shipboard, vehicular, and ground-based electronic systems has led to the specification of subminiature connectors for those systems.

Five military connector families have been procured in the largest volumes: (1) MIL-C-38999, (2) MIL-C-26482, (3) MIL-C-5015, (4) IL-C-83723, and (5) MIL-C-22992. The four versions of MIL-C-38999 are Series I, II, III, and IV. Each has mechanical differences indicating modifications by the originating military service working with one of the many connector manufacturers.

Subminiature circular multiconductor connectors are specified for military fixed-wing aircraft and helicopters as well as for comparable functions in large commercial aircraft. Examples are the interconnection of radio, radar, and airborne navigational equipment. But in military aircraft they also interconnect weapons fire control, electronics countermeasures, missile guidance, and other strictly military electronic systems. They are in use in all kinds of missiles, combat ships and submarines, tanks and other combat vehicles, and mobile equipment shelters.

Quick-release or breakaway cylindrical connectors were designed to permit the rapid disengagement of electronic and power cables from missiles as they are launched. Braided metal lanyards are fastened to the outer shell. As the missile blasts off, it applies tension to the lanyard, which pulls out threaded coupling segments for instant release.

## PINS AND SOCKETS FOR MILITARY CIRCULAR CONNECTORS

*Circular connector pins* for military circular connectors are machined from brass or nickel-silver, and *sockets* are formed from nonferrous materials such as beryllium-copper, phosphor-bronze or nickel-silver. Sockets include flexible inner leaf-spring contact surfaces to grip the pins with sufficient force to obtain high electrical conductivity, even after many engagements and disengagements. The location of pins and sockets in a mated connector is shown in Fig. 29-7. The pins are in the receptacle, where they are easier to protect because the receptacle is fastened to a case or cabinet, and the sockets are in the plug.

Gold is the preferred plating for mating contacts because it permits nondestructive sliding contact and resists corrosion, oxidation, and other contamination which blocks low-level "dry" signals. A minimum gold plating thickness of 50 μin (1.3 μm) is required for military connectors, but 15 to 30 μin (0.4 to 0.8 μm) is acceptable for many commercial avionics and industrial applications.

Pins and sockets are placed in aluminum or stainless-steel shells by inserting them in multihole spacers which establish contact distribution and spacing. The pins must have some freedom of movement to prevent damage to either pins or sockets from misalignment when the parts are coupled. Permissible center-to-center spacing of pins and sockets is determined by the voltage, current, and frequency of the signals to be transmitted.

Military-style connectors have removable contacts that are crimped to the individual wires of the cable. The wires with crimped, "poke-home" contacts are inserted into the shells of the connector with a special hand tool that compresses the springs, locking the contacts into position within the shell. This permits the pins and sockets to be removed easily for inspection, field circuit changes, or repairs. Individual wire terminations can be removed and replaced without disturbing adjacent conductors.

Special coaxial cables with pins and sockets and fiberoptic cables have been adapted to multipin circular connectors. The fiberoptic contacts fit the space allowed by these connectors and both coaxial and fiberoptic cables can be intermixed with lower-frequency contacts.

Hermetically sealed connectors are made with contacts rigidly fixed within a glass or ceramic spacer. Because the contacts are not removable, wires must be attached individually by welding or soldering. Care must be taken in the use of these connectors to prevent cracking the hermetic seals and destroying the hermetic integrity.

# Sockets for Components

The term *socket* has several different meanings in electronics technology. It can refer to the female part of a connector that accepts a plug or a circuit-board component for mounting electronic and electromechanical devices. Figure 29-8 shows a socket for mounting a DIP on a circuit board. It is suitable for mounting an IC, relay, or other component. Standard commercial sockets are available for mounting the following devices:

- Dual-in-line package (DIP) cases
- Single-in-line package (SIP) cases
- Leadless and leaded chip carriers (LCCs)
- Pin-grid arrays (PGAs)
- Electron tubes such as cathode-ray and microwave power tubes
- Miniature electromechanical and solid-state relays

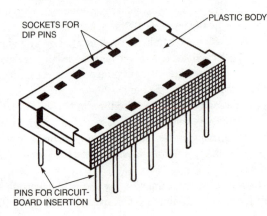

SOCKETS FOR
DIP PINS

PLASTIC BODY

PINS FOR CIRCUIT-
BOARD INSERTION

**Figure 29-8** Socket for a
**14-pin DIP.**

Sockets are generally specified for the installation of complex or costly integrated circuits such as microprocessors or microcontrollers that could be damaged during any of the reflow-solder processes described in Sec. 28. Sockets are also specified where the component is subject to replacement during the service life of the product or system. Field-programmable (PROM) and other memories such as EPROMs must be removed from their host circuits, the PROMs for replacement and the EPROMs for erasure and reprogramming. Sockets are also used for mounting sensitive DIP relays.

*Zero-insertion-force* (ZIF) *sockets* have lever-operated contacts that make the insertion and removal of large multipin integrated circuits easier while avoiding pin damage. The device to be socketed is easily inserted in the open socket, and when the pins are seated, the lever is moved to clamp the device securely in position. However, the drawback of ZIF sockets is their larger size and higher cost.

# Phone and Phono Jacks and Plugs

Specialized jacks and plugs are made to meet many different interconnection requirements. It is important to distinguish between the terms *phone* and *phono* when discussing plugs and jacks. *Phone* is a shorter term for *telephone,* meaning that the plugs and jacks were initially developed for telephone equipment, and *phono* is a shorter form of *phonograph,* meaning that the plugs and jacks were originally developed for interconnecting audio cable.

# Fiberoptic Connectors

Connections must be carefully made between the fiber ends of fiberoptic cables to minimize transmission losses. Special fiberoptic connectors have been designed for splicing fiberoptic cable, and some of these are based on coaxial cable connector designs. No universally accepted standards exist for fiberoptic cable connectors, so many different proprietary commercial products have been introduced. The large telephone service providers and cable companies build their own to fit their installations.

A fiberoptic connector must be able to perform the following functions:

■ Align mating fibers for efficient transfer of optical power.
■ Couple fibers to optical devices.
■ Protect the fiber from the environment and during handling and installation.
■ Terminate the cable strength member that relieves the fiber of tension.
■ Provide for cable strain relief.

The primary problem in connecting mating fibers is axial alignment. Optoelectronic devices such as IREDs, injection laser diodes, and photodiodes are being packaged in connector receptacles to facilitate their connection to the cable.

Many different methods exist for making fiber-to-fiber connections that are intended to assure the axial alignment of the fibers. These include clamping both fiber ends in vee-grooves or compressing them with multiple rods. Efforts have been made to develop simple yet accurate methods for splicing optical fibers in the field to minimize losses. It is obviously impractical to remove lengths of fiberoptic cable and return them to a repair shop.

The optical power loss of a connector-to-connector interface is usually between 0.5 and 2 dB, depending on both the style of the connector and the quality of the workmanship.

# COMPONENT AND CIRCUIT PROTECTION

# Overview

The replacement of receiving tubes by transistors and integrated circuits created new requirements for electronic circuit protection. Tube circuits could operate reliably in elevated temperature environments while simultaneously emitting considerable amounts of heat without much concern for their failure, but all semiconductor devices are vulnerable to excess heat. Moreover, tube circuits operated from high-voltage sources, typically the AC line, and they were tolerant of variations in power supply output, while semiconductor devices can easily be destroyed by overvoltages. In addition, electrostatic discharge (ESD) was not viewed as a threat during the tube-circuit era.

These facts focused more attention on circuit protection beyond the installation of simple fuses and perhaps fans. Semiconductor device vulnerability created new industries for the manufacture of protective devices, tools, and workstation products. And the vulnerability of solid-state circuits became a more critical issue as the component density on circuit boards increased and integrated circuits become more complex and expensive. Some MOS

ICs were found to be particularly susceptible to damage or destruction by ESD and overvoltage.

The circuit breakers and fuses installed in the AC service panels of homes, offices, and factories respond too slowly and imprecisely to guard sensitive devices or circuits against overcurrent, and they are incapable of screening out transient voltage spikes. This means that circuit and system designers have had to bring circuit protection down to the circuit level or at least in close proximity to the circuits. Consequently, electronics manufacturers now include a variety of protective devices such as fuses, metal-oxide varistors (MOVs), circuit protectors, and other devices within their products. In addition, all computer manufacturers recommend that surge protectors be placed between the AC outlet and all powered computer components. The local protection of stereo systems is also recommended.

The threats to ICs and circuit boards during manufacture, inspection, and shipment have spawned many different ESD-protective items, from grounded wrist straps and tools to ionizers and conductive flooring, containers, and clothing. ESD-protection programs have been initiated to train employees in the proper handling of vulnerable semiconductor devices and populated circuit boards.

# Circuit Protectors

A *circuit protector* is a miniature circuit breaker for the protection of electronic circuitry against the damaging effects of overcurrent. It can be mounted within the case or enclosure of a vulnerable electronic product. These miniature circuit breakers, with lower electrical ratings than primary AC-line circuit breakers, normally are placed in series with office, household, or factory primary circuit breakers. They can respond faster to lower overload currents than the AC-line service breakers, so they provide better protection.

Circuit protectors embedded in electronics products are primarily intended to respond to electrical faults within a cabinet or enclosure, such as a faulty component or short circuit. These current-limiting protectors are essentially scaled-down versions of larger circuit breakers and are classed as *instantaneous trip, time delay,* and *thermal circuit protectors.*

## INSTANTANEOUS MAGNETIC CIRCUIT PROTECTORS

*Instantaneous magnetic circuit protectors* are designed so that their contacts open in the presence of the increased magnetic field caused by an overcurrent. They are sensitive to current rather than temperature. The protective element within a magnetic circuit protector is a solenoid with a clapper-type armature linked by collapsible couplings to electrical contacts. When an overcurrent occurs (principally due to a short circuit or component failure), the solenoid is actuated, opening the contacts. The increased magnetic field attracts a hinged armature similar to those in electromagnetic relays.

However, unlike the relay, the protector armature does not directly open the electrical contacts. It collapses a train of mechanical linkage within the protector case, permitting fast-acting springs to snap open the contacts to avoid destructive arcing or sticking. Response time can be as rapid as 2 ms. When the overcurrent condition is cleared, the linkage can be reset manually with a toggle or lever. These protective devices are also made as dual-purpose components combining the functions of a toggle switch and a protective device. The contacts in most protectors will remain open if the fault condition persists, even if the switch is manually held in the "on" position.

## TIME-DELAY MAGNETIC-HYDRAULIC CIRCUIT PROTECTORS

A *time-delay magnetic-hydraulic circuit protector* is an instantaneous magnetic circuit protector that also includes a hydraulic time-delay dashpot, as shown in Fig. 30-1. The current flowing through the windings of the internal coil creates a flux which attracts an armature to the cylinder cap to snap open the protector's contacts. The protector's coil is wound around a hollow dashpot cylinder that houses a piston.

The dashpot slows the response of the protector by allowing the contacts to remain closed during nondestructive overcurrents to minimize nuisance tripping. However, the time delay of the dashpot can be manually set to meet specific circuit-protection requirements. For example, the time delay can be initiated only when the current is between 100 and 125 percent above the rated value. This current is enough to attract the piston at the bottom of the cylinder and pull it upward. As the piston moves up, it adds to the permeability of the magnetic circuit, causing an increase in flux density, thus increasing the attraction of the armature.

The time taken for the piston to travel to the top of the cylinder is determined, in part, by the viscosity of a silicone fluid in the sealed cylinder. When the cylinder reaches the top, the armature will be tripped. If the overcurrent is removed before the piston reaches the top of the cylinder, the breaker will not trip, and a return spring will push the piston back to the bottom of the cylinder. But if the current surge instantly exceeds 6 times the rating, the time-delay dashpot is overridden and the contacts will open instantaneously.

For all practical purposes, a magnetic-hydraulic circuit protector is unaffected by changes in ambient temperature, although changes in viscosity of the dashpot fluid could have a slight effect on delay time.

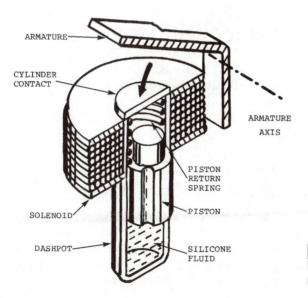

ARMATURE

CYLINDER CONTACT

ARMATURE AXIS

PISTON RETURN SPRING

SOLENOID

PISTON

DASHPOT

SILICONE FLUID

**Figure 30-1** Magnetic-hydraulic time-delay mechanism for a circuit protector.

## THERMAL CIRCUIT PROTECTORS

A *thermal circuit protector,* like a fuse, is a heat-sensitive device. A bimetallic blade sensor opens as a result of heat generated by excess load current flowing through it. This action releases a latch, causing the electrical contact to snap open. The curvature of the bimetallic element depends on the difference in coefficient of expansion between the two different metals in the element. The heat produced is directly related to the product of the square of the current and the time duration of that current $I^2t$, assuming constant resistance.

Because of their slow response, thermal circuit protectors are usually installed to protect wiring from overheating and the deterioration of insulation that could result and are not recommended for protecting sensitive semiconductor devices and circuits. Nevertheless, some thermal circuit protectors, like all magnetic-hydraulic types, can function as power switches, thus saving the expense of a separate switch.

There are two kinds of thermal circuit protectors: *standard* and *positive pressure.* The bimetallic element of the *standard* unit bends upward, causing the pressure between the contact surfaces to decrease, until the blade finally snaps open and separates the contacts. With this design, the heat rise due to overcurrent can be excessive before the blade opens because of decreasing contact pressure. This could lead to early failure or even contact welding. The *positive-pressure* blade is designed so that contact pressure increases with overcurrent heat. When the critical point is reached, the blade snaps open instantly, and the arc is extinguished. Thermal circuit protectors, like magnetic-hydraulic protectors, offer circuit resistance that helps to limit circuit-fault current.

Some thermal protectors, like magnetic-hydraulic protectors, can be manually reset. But the protector cannot be reset until the bimetal element cools down enough to return to its room-temperature position. The thermal protector has a slower response time than the magnetic protector, and there is no provision for a time delay. Thus it is more prone to false triggering. Moreover, its established set point can be altered by changes in ambient temperature.

# Overvoltage Protection Devices

There are three general classes of passive devices designed for protecting semiconductors against overvoltages caused by natural and humanmade causes:

1. Transient voltage suppressor (TVS) diodes
2. Metal-oxide varistors (MOVs)
3. Surge voltage protector (SVP) tubes

## TRANSIENT VOLTAGE SUPPRESSORS (TVSs)

A *transient voltage suppressor* (TVS) is a zener avalanche PN-junction diode optimized to protect circuits by reverse-bias voltage clamping. The TVS diode breaks down and becomes a short circuit when the applied voltage exceeds the device's rated avalanche level. When the voltage (reverse bias) falls below breakdown level, current is restored to its normal level. Silicon TVSs have better surge-handling capabilities than conventional zener diodes, low series-resistance values, and response times measurable in picoseconds.

A standard zener diode is not designed for circuit protection, but a TVS can perform the dual functions of voltage regulation and protective clamping. A TVS can protect a DC circuit, but two TVSs must be positioned back to back to protect an AC circuit. However, dual TVSs are available in a single package. Surface-mount TVSs are also available. The three most important characteristics of the TVS are the following:

**1.** Pulse power (peak pulse power multiplied by the clamping voltage)
**2.** Standoff voltage
**3.** Maximum clamping voltage

*Standoff voltages* range from 5 to 170 V, and *clamping voltages* range from 7 to 210 V. TVSs can perform the following circuit functions:

■ Protect ICs from ESD, power supply reversals, or during power supply switch-on.
■ Protect output transistors, as well as ICs, from transients caused by inductive-load switching.
■ Replace a crowbar network on the output side of a voltage regulator.
■ Prevent inductive-load transients from being transmitted to the output.
■ Protect MOSFET devices from transients conducted in the power supply line.
■ Absorb high energy levels on signal and data lines.

## METAL-OXIDE VARISTORS (MOVs)

A *metal-oxide varistor* (MOV) is a variable resistor that can protect electronic circuits against AC voltage transients because it acts like back-to-back diodes. A MOV is a nonlinear resistor whose resistance value changes as a function of the applied voltage. The body of the MOV is molded from zinc-oxide grains and furnace fired to form a monolithic block. As a result of its symmetrical bilateral characteristics, a MOV can clamp AC voltage during both positive and negative swings. When the applied voltage exceeds the MOV's rating, its resistance drops sharply and it becomes a short circuit. When the voltage transient is bypassed, the MOV will be restored because its body is able to absorb the energy from the transient without destroying the device. MOVs are packaged as conformally coated radial-leaded disks or as epoxy-encapsulated blocks with terminal posts.

## SURGE VOLTAGE PROTECTORS (SVPs)

A *surge voltage protector* (SVP) is a gas tube that can act as a circuit-protective device because it can withstand higher surge voltages than a TVS or a MOV. The SVP provides a low-resistance path for successive voltage transients when its voltage ratings are exceeded. Gas within the metal or ceramic tube ionizes during an overvoltage, causing the SVP to change from a nonconducting to a conducting state. The arc that is formed shorts out the SVP, grounding all high currents as well. After the transient has passed, the gas deionizes and the SVP is reset. It has a slower response time than either a TVS or a MOV.

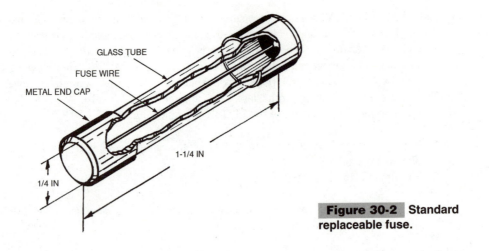

GLASS TUBE

FUSE WIRE

METAL END CAP

1-1/4 IN

1/4 IN

**Figure 30-2** Standard replaceable fuse.

# Fuses

A *fuse,* as shown in the cutaway Fig. 30-2, is a one-time-only expendable protective device that contains an axial resistive wire or ribbon with a low melting temperature. When exposed to an overcurrent, the element melts and opens the circuit. Fuses protect electrical and electronic circuits when they are in series between the load and the power line.

Standard American fuses have ¼-in (6 mm) diameters and are 1¼ in (32 mm) long. They have voltage ratings of 125- and 250-VAC and current ratings of ¹⁄₁₅ to 15 A. European standard fuses have 0.2-in (5-mm) diameters and are 0.8 in (20 mm) long.

The response times for all fuses can be measured in milliseconds. Commonly used fuses are *fast acting* and *slow blow* or *time delay.* Some fuses that conserve circuit-board space are made as leaded components like resistors that are soldered on a circuit board. If they have blown, they must be unsoldered to be replaced.

A new class of resettable fuses has been introduced for direct connection in series with the circuit to be disabled. Identified as *polymer PTC resettable fuses,* they automatically reset after the fault is removed and the power is shut off.

# Surge-Protection Assemblies

A *surge protector* or *surge-protection assembly* is a protective package that is inserted in series with the AC line and the line cord of equipment to be protected from voltage transients and current surges. A *surge strip* is a factory-made product that contains up to seven protected AC outlets and has an attached AC-line cord to plug into the AC outlet. The strips include small, resettable circuit protectors to protect such products as electronic telephones, fax machines, and stand-alone modems. Typical models are rated for 330 to 400 V, and they can dissipate up to 210 J. Some have audible alarms and lamp indicators. *Surge suppressors* are similar to surge strips but generally offer higher energy-dissipation ratings to 1950 J for the protection of computers.

# Electrostatic-Discharge (ESD) Protection

*Electrostatic discharge* (ESD) is the rapid grounding of electrical charge that has built up on nonconductive surfaces which results in the generation of high voltages. The charge can build up on such surfaces as floors, tables, decks, or bench tops, and on paper and plastic. When a conductor such as a human hand or a wire touches a surface with an accumulated charge, the charge is quickly drained and can result in a visible electrical arc. Although low in current, ESD can generate voltages that are high enough to disable or destroy susceptible semiconductor devices and circuits.

ESD can be experienced by scuffing one's shoes on a carpet in a room with low humidity and then touching a doorknob. A mild but painful electrical shock will be delivered as a spark, perhaps a half-inch long, jumps from the finger to the knob.

To combat the nuisance effect of ESD, particularly during the winter months when room heating reduces the humidity below 50 percent, many protective measures have been developed and preventative programs have been devised. Low-cost instruments can determine the seriousness of the ESD threat.

Among the ESD-protective products readily available are wrist and foot grounding straps; electrically conductive plastic bags, boxes, and tote trays; conductive floor mats; and even spray cans containing solutions that will dissipate ESD on nonconductive surfaces. Moreover, instruments have been developed to determine if the tools being used at the workstations and the workers themselves are properly grounded. Other instruments can provide calibrated electrical discharges that simulate actual ESD characteristics for product testing. Also, air humidifiers and ionizing fans condition the air in controlled workspaces.

A comprehensive ESD-protection plan is recommended for factories and warehouses where ESD-sensitive devices and circuits are being handled, inspected, packaged, or installed. Instructions are available for setting up an ESD-controlled workstation in an environmentally controlled room, as shown in Fig. 30-3.

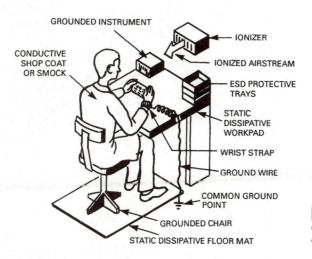

GROUNDED INSTRUMENT

IONIZER

CONDUCTIVE SHOP COAT OR SMOCK

IONIZED AIRSTREAM

ESD PROTECTIVE TRAYS

STATIC DISSIPATIVE WORKPAD

WRIST STRAP

GROUND WIRE

COMMON GROUND POINT

GROUNDED CHAIR

STATIC DISSIPATIVE FLOOR MAT

**Figure 30-3** Electrostatic-discharge- (ESD-) protected workstation.

The important requirements in a coordinated ESD-protection program include the following:

- Transporting all sensitive devices or circuit boards in containers, tote trays, or bags made from conductive materials so that any static charge is dissipated.
- Performing all work such as packing, unpacking, inspecting, and assembly on grounded conductive work surfaces.
- Requiring that all persons handling sensitive devices wear grounded, conductive wrist straps and, where warranted, conductive shop coats or smocks.
- Maintaining the relative humidity in the workplace at 50 percent or higher.
- Circulating air from an ionizing fan over the work area to neutralize ions in the air. This will prevent the buildup of static electricity on walls, cabinets, and other surfaces within or adjacent to the work area.

# BIBLIOGRAPHY

Beards, P. H.: *Analog and Digital Electronics,* 2d ed., Prentice Hall, Englewood Cliffs, N.J., 1986.

Blake, L. V.: *Antennas,* John Wiley & Sons, New York, 1966.

Christiansen, D., et al.: *Electronics Engineer's Handbook,* 4th ed., McGraw Hill, New York, 1997.

Chryssis, G.: *High-Frequency Switching Power Supplies,* McGraw-Hill, New York, 1984.

Clifford, M.: *New Handbook of Electronic Data,* Prentice Hall, Englewood Cliffs, N.J., 1988.

————: *Master Handbook of Electronic Tables and Formulas,* 5th ed., McGraw-Hill, New York, 1992.

Connelly, J. A.: *Analog Integrated Circuits,* John Wiley & Sons, New York, 1975.

Coombs, C. F., Jr.: *Printed Circuits Handbook,* 3d ed., McGraw-Hill, New York, 1988.

————: *Electronic Instrument Handbook,* 2d ed. McGraw-Hill, New York, 1994.

Harper, C. A., and R. M. Sampson: *Electronic Materials and Processes Handbook,* McGraw-Hill, New York, 1994.

Johnson, R. C.: *Antenna Engineering Handbook,* 3d ed., McGraw-Hill, New York, 1993.

Jones, L. D., and A. F. Chin, *Electronic Instruments and Measurements,* 2d ed., Prentice Hall, Englewood Cliffs, N.J., 1991.

Kaufman, M., and A. H. Seidman: *Electronics Sourcebook for Technicians and Engineers,* McGraw-Hill, New York, 1988.

Lee, W. C. Y.: *Mobile Cellular Telecommunication, Analog and Digital Systems,* McGraw-Hill, New York, 1995.

Liao, S. Y.: *Microwave Electron-Tube Devices,* Prentice Hall, Englewood Cliffs, N.J., 1986.

Linden, D.: *Handbook of Batteries,* 2d ed., McGraw-Hill, New York, 1995.

Logsdon, T.: *Mobile Communications Satellites,* McGraw-Hill, New York, 1995.

Ng, K. K.: *Complete Guide to Semiconductor Devices,* McGraw Hill, New York, 1995.

Prince, B., and G. Due-Gunderson: *Semiconductor Memories,* John Wiley & Sons, New York, 1983.

Richaria, M.: *Satellite Communications Systems: Design Principles,* McGraw-Hill, New York, 1995.

Ristic, L.: *Sensor Technology and Devices,* Artech House, Boston, 1994.

Rizzi, P. A.: *Microwave Engineering Passive Circuits,* Prentice Hall, Englewood Cliffs, N.J., 1988.

Sclater, N.: *Electrostatic Discharge Protection for Electronics,* McGraw-Hill, New York, 1990.

———: *Wire & Cable for Electronics: A User's Handbook,* McGraw-Hill, New York, 1991.

——— and J. Markus: *McGraw-Hill Electronics Dictionary,* 6th ed., McGraw-Hill, New York, 1997.

Sheingold, D., et al.: *Analog-Digital Conversion Handbook,* 3d ed., Prentice Hall, Englewood Cliffs, N.J., 1986.

Van Zant, P.: *Microchip Fabrication,* 3d ed., McGraw-Hill, New York, 1997.

Veley, V. F. C.: *The Benchtop Electronics Reference Manual,* 3d ed., McGraw-Hill, New York, 1992.

# INDEX

## About the Author

Neil Sclater began his career as a radar-tube engineer for military and aerospace systems at Raytheon Company and Microwave Associates before switching to writing and editing as a profession. He started as a regional editor for *Electronic Design* and then became the electronics editor for *Product Engineering* magazine.

Later he became an independent consultant in technical marketing communications, and remained active in that field for more than 25 years. During that time he served a list of clients that included manufacturers of circuit boards, power supplies, batteries, light-emitting diodes, microwave communications links, and computer workstations. His work for them included the preparation of illustrated articles for publication, marketing studies, new product literature, and news releases. He was also retained by several technical public relations agencies for special writing assignments on behalf of their clients.

Over the years he has also been a contributing editor to such publications as *Electronic Engineering Times, Control Engineering,* and *Electronic Buyers' News,* and he has authored or coauthored nine books on electronics and electromechanical subjects.

In addition to this handbook, McGraw-Hill has published six of his books on electronics: *The Encyclopedia of Electronics* (second edition, coauthor), *Gallium Arsenide IC Technology, Electrostatic Discharge Protection for Electronics, Wire and Cable for Electronics,* and the fifth and sixth editions of the *McGraw-Hill Electronics Dictionary.*

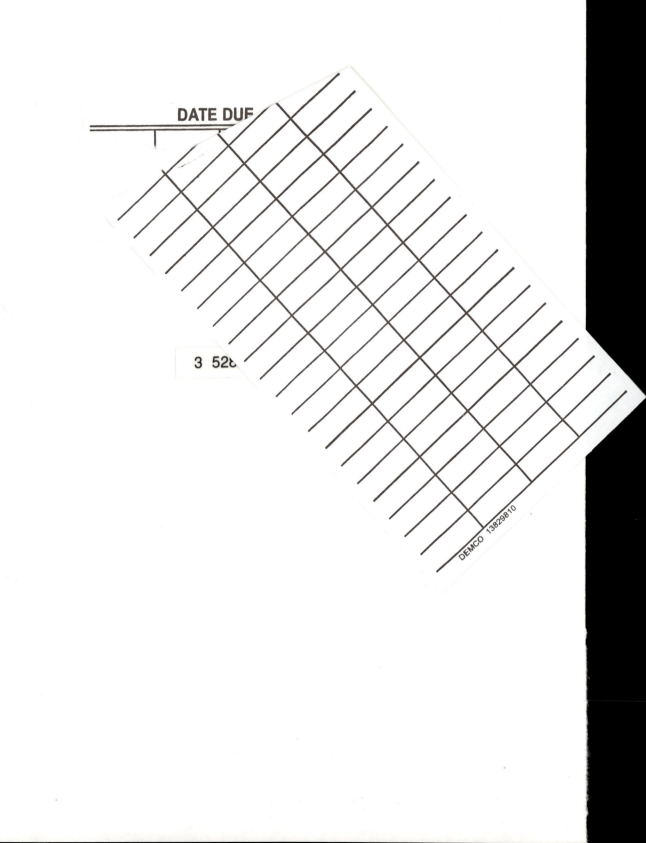